Novum Organum and Other Writings
By Lord Francis Bacon
Edited by Anthony Uyl

Devoted Publishing
Woodstock, Ontario, 2016

Novum Organum and Other Writings
By Lord Francis Bacon
Edited by Anthony Uyl

. What kind of philosophies do you have? Let us know!

Contact us at: devotedpub@hotmail.com
Visit us on Facebook: Devoted Publishing
Get more products via our gaming division website: www.solacegames.ca

Published in Woodstock, Ontario, Canada 2016.

For bulk educational rates, please contact us at the email address above.

ISBN: 978-1-988297-55-2

Table of Contents

Novum Organum
OR TRUE SUGGESTIONS FOR THE INTERPRETATION OF NATURE

Original Edits By Joseph Devey, M.A.

Originally Published By:
New York; P. F. Collier & Son; MCMII; 22; SCIENCE

PREFACE

They who have presumed to dogmatize on nature, as on some well investigated subject, either from self-conceit or arrogance, and in the professorial style, have inflicted the greatest injury on philosophy and learning. For they have tended to stifle and interrupt inquiry exactly in proportion as they have prevailed in bringing others to their opinion: and their own activity has not counterbalanced the mischief they have occasioned by corrupting and destroying that of others. They again who have entered upon a contrary course, and asserted that nothing whatever can be known, whether they have fallen into this opinion from their hatred of the ancient sophists, or from the hesitation of their minds, or from an exuberance of learning, have certainly adduced reasons for it which are by no means contemptible. They have not, however, derived their opinion from true sources, and, hurried on by their zeal and some affectation, have certainly exceeded due moderation. But the more ancient Greeks (whose writings have perished), held a more prudent mean, between the arrogance of dogmatism, and the despair of scepticism; and though too frequently intermingling complaints and indignation at the difficulty of inquiry, and the obscurity of things, and champing, as it were, the bit, have still persisted in pressing their point, and pursuing their intercourse with nature; thinking, as it seems, that the better method was not to dispute upon the very point of the possibility of anything being known, but to put it to the test of experience. Yet they themselves, by only employing the power of the understanding, have not adopted a fixed rule, but have laid their whole stress upon intense meditation, and a continual exercise and perpetual agitation of the mind.

Our method, though difficult in its operation, is easily explained. It consists in determining the degrees of certainty, while we, as it were, restore the senses to their former rank, but generally reject that operation of the mind which follows close upon the senses, and open and establish a new and certain course for the mind from the first actual perceptions of the senses themselves. This, no doubt, was the view taken by those who have assigned so much to logic; showing clearly thereby that they sought some support for the mind, and suspected its natural and spontaneous mode of action. But this is now employed too late as a remedy, when all is clearly lost, and after the mind, by the daily habit and intercourse of life, has come prepossessed with corrupted doctrines, and filled with the vainest idols. The art of logic therefore being (as we have mentioned), too late a precaution,[1] and in no way remedying the matter, has tended more to confirm errors, than to disclose truth. Our only remaining hope and salvation is to begin the whole labor of the mind again; not leaving it to itself, but directing it perpetually from the very first, and attaining our end as it were by mechanical aid. If men, for instance, had attempted mechanical labors with their hands alone, and without the power and aid of instruments, as they have not hesitated to carry on the labors of their understanding with the unaided efforts of their mind, they would have been able to move and overcome but little, though they had exerted their utmost and united powers. And just to pause awhile on this comparison, and look into it as a mirror; let us ask, if any obelisk of a remarkable size were perchance required to be moved, for the purpose of gracing a triumph or any similar pageant, and men were to attempt it with their bare hands, would not any sober spectator avow it to be an act of the greatest madness? And if they should increase the number of workmen, and imagine that they could thus succeed, would he not think so still more? But if they chose to make a selection, and to remove the weak, and only employ the strong and vigorous, thinking by this means, at any rate, to achieve their object, would he not say that they were more fondly deranged? Nay, if not content with this, they were to determine on consulting the athletic art, and were to give orders for all to appear with their hands, arms, and muscles regularly oiled and prepared, would he not exclaim that they were taking pains to rave by method and design? Yet men are hurried on with the same senseless energy and useless combination in intellectual matters, as long as they expect great results either from the number and agreement, or the excellence and acuteness of their wits; or even strengthen their minds with logic, which may be considered as an athletic preparation, but yet do not desist (if we rightly consider the matter) from applying their own understandings merely with all this zeal and effort. While nothing is more clear, than that in every great work executed by the hand of man without machines or implements, it is impossible for the strength of individuals to be increased, or for that of the multitude to combine.

Having premised so much, we lay down two points on which we would admonish mankind, lest they should fail to see or to observe them. The first of these is, that it is our good fortune (as we consider it), for the sake of extinguishing and removing contradiction and irritation of mind, to leave the honor and reverence due to the ancients untouched and undiminished, so that we can perform our intended work, and yet enjoy the benefit of our respectful moderation. For if we should profess to offer something better than the ancients, and yet should pursue the same course as they have done, we could never, by any artifice, contrive to avoid the imputation of having engaged in a contest or rivalry as to our respective wits, excellences, or talents; which, though neither inadmissible nor new (for why should we not blame and point out anything that is imperfectly discovered or laid down by them, of our own right, a right common to all?), yet however just and allowable, would perhaps be scarcely an equal match, on account of the disproportion of our strength. But since our present plan leads up to open an entirely different course to the understanding, and one unattempted and unknown to them, the case is altered. There is an end to party zeal, and we only take upon ourselves the character of a guide, which requires a moderate share of authority and good fortune, rather than talents and excellence. The first admonition relates to persons, the next to things.

We make no attempt to disturb the system of philosophy that now prevails, or any other which may or will exist, either more correct or more complete. For we deny not that the received system of philosophy, and others of a similar nature, encourage discussion, embellish harangues, are employed, and are of service in the duties of the professor, and the affairs of civil life. Nay, we openly express and declare that the philosophy we offer will not be very useful in such respects. It is not obvious, nor to be understood in a cursory view, nor does it flatter the mind in its preconceived notions, nor will it descend to the level of the generality of mankind unless by its advantages and effects.

Let there exist then (and may it be of advantage to both), two sources, and two distributions of learning, and in like manner two tribes, and as it were kindred families of contemplators or philosophers, without any hostility or alienation between them; but rather allied and united by mutual assistance. Let there be in short one method of cultivating the sciences, and another of discovering them. And as for those who prefer and more readily receive the former, on account of their haste or from motives arising from their ordinary life, or because they are unable from weakness of mind to comprehend and embrace the other (which must necessarily be the case with by far the greater number), let us wish that they may prosper as they desire in their undertaking, and attain what they pursue. But if any individual desire, and is anxious not merely to adhere to, and make use of present discoveries, but to penetrate still further, and not to overcome his adversaries in disputes, but nature by labor, not in short to give elegant and specious opinions, but to know to a certainty and demonstration, let him, as a true son of science (if such be his wish), join with us; that when he has left the antechambers of nature trodden by the multitude, an entrance may at last be discovered to her inner apartments. And in order to be better understood, and to render our meaning more familiar by assigning determinate names, we have accustomed ourselves to call the one method the anticipation of the mind, and the other the interpretation of nature.

We have still one request left. We have at least reflected and taken pains in order to render our propositions not only true, but of easy and familiar access to men's minds, however wonderfully prepossessed and limited. Yet it is but just that we should obtain this favor from mankind (especially in so great a restoration of learning and the sciences), that whosoever may be desirous of forming any determination upon an opinion of this our work either from his own perceptions, or the crowd of authorities, or the forms of demonstrations, he will not expect to be able to do so in a cursory manner, and while attending to other matters; but in order to have a thorough knowledge of the subject, will himself by degrees attempt the course which we describe and maintain; will be accustomed to the subtilty of things which is manifested by experience; and will correct the depraved and deeply rooted habits of his mind by a seasonable, and, as it were, just hesitation: and then, finally (if he will), use his judgment when he has begun to be master of himself.

Footnote:

1. Because it was idle to draw a logical conclusion from false principles, error being propagated as much by false premises, which logic does not pretend to examine, as by illegitimate inference. Hence, as Bacon says further on, men being easily led to confound legitimate inference with truth, were confirmed in their errors by the very subtilty of their genius.--Ed.

APHORISMS - BOOK I - ON THE INTERPRETATION OF NATURE AND THE EMPIRE OF MAN

I. Man, as the minister and interpreter of nature, does and understands as much as his observations on the order of nature, either with regard to things or the mind, permit him, and neither knows nor is capable of more.

II. The unassisted hand and the understanding left to itself possess but little power. Effects are produced by the means of instruments and helps, which the understanding requires no less than the hand; and as instruments either promote or regulate the motion of the hand, so those that are applied to the mind prompt or protect the understanding.

III. Knowledge and human power are synonymous, since the ignorance of the cause frustrates the effect; for nature is only subdued by submission, and that which in contemplative philosophy corresponds with the cause in practical science becomes the rule.

IV. Man while operating can only apply or withdraw natural bodies; nature internally performs the rest.

V. Those who become practically versed in nature are, the mechanic, the mathematician, the physician, the alchemist, and the magician,[2] but all (as matters now stand) with faint efforts and meagre success.

VI. It would be madness and inconsistency to suppose that things which have never yet been performed can be performed without employing some hitherto untried means.

VII. The creations of the mind and hand appear very numerous, if we judge by books and manufactures; but all that variety consists of an excessive refinement, and of deductions from a few well known matters--not of a number of axioms.[3]

VIII. Even the effects already discovered are due to chance and experiment rather than to the sciences; for our present sciences are nothing more than peculiar arrangements of matters already discovered, and not methods for discovery or plans for new operations.

IX. The sole cause and root of almost every defect in the sciences is this, that while we falsely admire and extol the powers of the human mind, we do not search for its real helps.

X. The subtilty of nature is far beyond that of sense or of the understanding: so that the specious meditations, speculations, and theories of mankind are but a kind of insanity, only there is no one to stand by and observe it.

XI. As the present sciences are useless for the discovery of effects, so the present system of logic[4] is useless for the discovery of the sciences.

XII. The present system of logic rather assists in confirming and rendering inveterate the errors founded on vulgar notions than in searching after truth, and is therefore more hurtful than useful.

XIII. The syllogism is not applied to the principles of the sciences, and is of no avail in intermediate axioms,[5] as being very unequal to the subtilty of nature. It forces assent, therefore, and not things.

XIV. The syllogism consists of propositions; propositions of words; words are the signs of notions. If, therefore, the notions (which form the basis of the whole) be confused and carelessly abstracted from things, there is no solidity in the superstructure. Our only hope, then, is in genuine induction.

XV. We have no sound notions either in logic or physics; substance, quality, action, passion, and existence are not clear notions; much less weight, levity, density, tenuity, moisture, dryness, generation, corruption, attraction, repulsion, element, matter, form, and the like. They are all fantastical and ill-defined.

XVI. The notions of less abstract natures, as man, dog, dove, and the immediate perceptions of sense, as heat, cold, white, black, do not deceive us materially, yet even these are sometimes confused by the mutability of matter and the intermixture of things. All the rest which men have hitherto employed are errors, and improperly abstracted and deduced from things.

XVII. There is the same degree of licentiousness and error in forming axioms as in abstracting notions, and that in the first principles, which depend on common induction; still more is this the case in axioms and inferior propositions derived from syllogisms.

XVIII. The present discoveries in science are such as lie immediately beneath the surface of common notions. It is necessary, however, to penetrate the more secret and remote parts of nature, in order to abstract both notions and axioms from things by a more certain and guarded method.

XIX. There are and can exist but two ways of investigating and discovering truth. The one hurries on rapidly from the senses and particulars to the most general axioms, and from them, as principles and their supposed indisputable truth, derives and discovers the intermediate axioms. This is the way now in use. The other constructs its axioms from the senses and particulars, by ascending continually and gradually, till it finally arrives at the most general axioms, which is the true but unattempted way.

XX. The understanding when left to itself proceeds by the same way as that which it would have adopted under the guidance of logic, namely, the first; for the mind is fond of starting off to generalities, that it may avoid labor, and after dwelling a little on a subject is fatigued by experiment. But those evils are augmented by logic, for the sake of the ostentation of dispute.

XXI. The understanding, when left to itself in a man of a steady, patient, and reflecting disposition (especially when unimpeded by received doctrines), makes some attempt in the right way, but with little effect, since the understanding, undirected and unassisted, is unequal to and unfit for the task of vanquishing the obscurity of things.

XXII. Each of these two ways begins from the senses and particulars, and ends in the greatest generalities. But they are immeasurably different; for the one merely touches cursorily the limits of experiment and particulars, while the other runs duly and regularly through them--the one from the very outset lays down some abstract and useless generalities, the other gradually rises to those principles which are really the most common in nature.[6]

XXIII. There is no small difference between the idols of the human mind and the ideas of the Divine mind--that is to say, between certain idle dogmas and the real stamp and impression of created objects, as they are found in nature.

XXIV. Axioms determined upon in argument can never assist in the discovery of new effects; for the subtilty of nature is vastly superior to that of argument. But axioms properly and regularly abstracted from particulars easily point out and define new particulars, and therefore impart activity to the sciences.

XXV. The axioms now in use are derived from a scanty handful, as it were, of experience, and a few particulars of frequent occurrence, whence they are of much the same dimensions or extent as their origin. And if any neglected or unknown instance occurs, the axiom is saved by some frivolous distinction, when it would be more consistent with truth to amend it.

XXVI. We are wont, for the sake of distinction, to call that human reasoning which we apply to nature the anticipation of nature (as being rash and premature), and that which is properly deduced from things the interpretation of nature.

XXVII. Anticipations are sufficiently powerful in producing unanimity, for if men were all to become even uniformly mad, they might agree tolerably well with each other.

XXVIII. Anticipations again, will be assented to much more readily than interpretations, because being deduced from a few instances, and these principally of familiar occurrence, they immediately hit the understanding and satisfy the imagination; while, on the contrary, interpretations, being deduced from various subjects, and these widely dispersed, cannot suddenly strike the understanding, so that in common estimation they must appear difficult and discordant, and almost like the mysteries of faith.

XXIX. In sciences founded on opinions and dogmas, it is right to make use of anticipations and logic if you wish to force assent rather than things.

XXX. If all the capacities of all ages should unite and combine and transmit their labors, no great progress will be made in learning by anticipations, because the radical errors, and those which occur in the first process of the mind, are not cured by the excellence of subsequent means and remedies.

XXXI. It is in vain to expect any great progress in the sciences by the superinducing or ingrafting new matters upon old. An instauration must be made from the very foundations, if we do not wish to revolve forever in a circle, making only some slight and contemptible progress.

XXXII. The ancient authors and all others are left in undisputed possession of their honors; for we enter into no comparison of capacity or talent, but of method, and assume the part of a guide rather than of a critic.

XXXIII. To speak plainly, no correct judgment can be formed either of our method or its discoveries by those anticipations which are now in common use; for it is not to be required of us to submit ourselves to the judgment of the very method we ourselves arraign.

XXXIV. Nor is it an easy matter to deliver and explain our sentiments; for those things which are in themselves new can yet be only understood from some analogy to what is old.

XXXV. Alexander Borgia[7] said of the expedition of the French into Italy that they came with chalk in their hands to mark up their lodgings, and not with weapons to force their passage. Even so do we wish our philosophy to make its way quietly into those minds that are fit for it, and of good capacity; for we have no need of contention where we differ in first principles, and in our very notions, and even in our forms of demonstration.

XXXVI. We have but one simple method of delivering our sentiments, namely, we must bring men to particulars and their regular series and order, and they must for a while renounce their notions, and begin to form an acquaintance with things.

XXXVII. Our method and that of the sceptics[8] agree in some respects at first setting out, but differ most widely, and are completely opposed to each other in their conclusion; for they roundly assert that nothing can be known; we, that but a small part of nature can be known, by the present method; their next step, however, is to destroy the authority of the senses and understanding, while we invent and supply them with assistance.

XXXVIII. The idols and false notions which have already preoccupied the human understanding, and are deeply rooted in it, not only so beset men's minds that they become difficult of access, but even when access is obtained will again meet and trouble us in the instauration of the sciences, unless mankind when forewarned guard themselves with all possible care against them.

XXXIX. Four species of idols beset the human mind,[9] to which (for distinction's sake) we have assigned names, calling the first Idols of the Tribe, the second Idols of the Den, the third Idols of the Market, the fourth Idols of the Theatre.

XL. The formation of notions and axioms on the foundation of true induction is the only fitting remedy by which we can ward off and expel these idols. It is, however, of great service to point them out; for the doctrine of idols bears the same relation to the interpretation of nature as that of the confutation of sophisms does to common logic.[10]

XLI. The idols of the tribe are inherent in human nature and the very tribe or race of man; for man's sense is falsely asserted to be the standard of things; on the contrary, all the perceptions both of the senses and the mind bear reference to man and not to the universe, and the human mind resembles those uneven mirrors which impart their own properties to different objects, from which rays are emitted and distort and disfigure them.[11]

XLII. The idols of the den are those of each individual; for everybody (in addition to the errors common to the race of man) has his own individual den or cavern, which intercepts and corrupts the light of nature, either from his own peculiar and singular disposition, or from his education and intercourse with others, or from his reading, and the authority acquired by those whom he reverences and admires, or from the different impressions produced on the mind, as it happens to be preoccupied and predisposed, or equable and tranquil, and the like; so that the spirit of man (according to its several dispositions), is variable, confused, and as it were actuated by chance; and Heraclitus said well that men search for knowledge in lesser worlds, and not in the greater or common world.

XLIII. There are also idols formed by the reciprocal intercourse and society of man with man, which we call idols of the market, from the commerce and association of men with each other; for men converse by means of language, but words are formed at the will of the generality, and there arises from a bad and unapt formation of words a wonderful obstruction to the mind. Nor can the definitions and explanations with which learned men are wont to guard and protect themselves in some instances afford a complete remedy--words still manifestly force the understanding, throw everything into confusion, and lead mankind into vain and innumerable controversies and fallacies.

XLIV. Lastly, there are idols which have crept into men's minds from the various dogmas of peculiar systems of philosophy, and also from the perverted rules of demonstration, and these we denominate idols of the theatre: for we regard all the systems of philosophy hitherto received or imagined, as so many plays brought out and performed, creating fictitious and theatrical worlds. Nor do we speak only of the present systems, or of the philosophy and sects of the ancients, since numerous other plays of a similar nature can be still composed and made to agree with each other, the causes of the most opposite errors being generally the same. Nor, again, do we allude merely to general systems, but also to many elements and axioms of sciences which have become inveterate by tradition, implicit credence, and neglect. We must, however, discuss each species of idols more fully and distinctly in order to guard the human understanding against them.

XLV. The human understanding, from its peculiar nature, easily supposes a greater degree of order and equality in things than it really finds; and although many things in nature be sui generis and most irregular, will yet invent parallels and conjugates and relatives, where no such thing is. Hence the fiction, that all celestial bodies move in perfect circles, thus rejecting entirely spiral and serpentine lines (except as explanatory terms).[12] Hence also the element of fire is introduced with its peculiar orbit,[13] to keep square with those other three which are objects of our senses. The relative rarity of the elements (as they are called) is arbitrarily made to vary in tenfold progression, with many other dreams of the like nature.[14] Nor is this folly confined to theories, but it is to be met with even in simple notions.

XLVI. The human understanding, when any proposition has been once laid down (either from general admission and belief, or from the pleasure it affords), forces everything else to add fresh support and confirmation; and although most cogent and abundant instances may exist to the contrary, yet either does not observe or despises them, or gets rid of and rejects them by some distinction, with violent and injurious prejudice, rather than sacrifice the authority of its first conclusions. It was well answered by him[15] who was shown in a temple the votive tablets suspended by such as had escaped the peril of shipwreck, and was pressed as to whether he would then recognize the power of the gods, by an inquiry, But where are the portraits of those who have perished in spite of their vows? All superstition is much the same, whether it be that of astrology, dreams, omens, retributive judgment, or the like, in all of which the deluded believers observe events which are fulfilled, but neglect and pass over their failure, though it be much more common. But this evil insinuates itself still more craftily in philosophy and the sciences, in which a settled maxim vitiates and governs every other circumstance, though the latter be much more worthy of confidence. Besides, even in the absence of that eagerness and want of thought (which we have mentioned), it is the peculiar and perpetual error of the human understanding to be more moved and excited by affirmatives than negatives, whereas it ought duly and regularly to be impartial; nay, in establishing any true axiom the negative instance is the most powerful.

XLVII. The human understanding is most excited by that which strikes and enters the mind at once and suddenly, and by which the imagination is immediately filled and inflated. It then begins almost imperceptibly to conceive and suppose that everything is similar to the few objects which have taken possession of the mind, while it is very slow and unfit for the transition to the remote and heterogeneous instances by which axioms are tried as by fire, unless the office be imposed upon it by severe regulations and a powerful authority.

XLVIII. The human understanding is active and cannot halt or rest, but even, though without effect, still presses forward. Thus we cannot conceive of any end or external boundary of the world, and it seems necessarily to occur to us that there must be something beyond. Nor can we imagine how eternity has flowed on down to the present day, since the usually received distinction of an infinity, a parte ante and a parte post,[16] cannot hold good; for it would thence follow that one infinity is greater than another, and also that infinity is wasting away and tending to an end. There is the same difficulty in considering the infinite divisibility of lines, arising from the weakness of our minds, which weakness interferes to still greater disadvantage with the discovery of causes; for although the greatest generalities in nature must be positive, just as they are found, and in fact not causable, yet the human understanding, incapable of resting, seeks for something more intelligible. Thus, however, while aiming at further progress, it falls back to what is actually less advanced, namely, final causes; for they are clearly more allied to man's own nature, than the system of the universe, and from this source they have wonderfully corrupted philosophy. But he would be an unskilful and shallow philosopher who should seek for causes in the greatest generalities, and not be anxious to discover them in subordinate objects.

XLIX. The human understanding resembles not a dry light, but admits a tincture of the will[17] and passions, which generate their own system accordingly; for man always believes more readily that which he prefers. He, therefore, rejects difficulties for want of patience in investigation; sobriety, because it limits his hope; the depths of nature, from superstition; the light of experiment, from arrogance and pride, lest his mind should appear to be occupied with common and varying objects; paradoxes, from a fear of the opinion of the vulgar; in short, his feelings imbue and corrupt his understanding in innumerable and sometimes imperceptible ways.

L. But by far the greatest impediment and aberration of the human understanding proceeds from the dulness, incompetence, and errors of the senses; since whatever strikes the senses preponderates over everything, however superior, which does not immediately strike them. Hence contemplation mostly ceases with sight, and a very scanty, or perhaps no regard is paid to invisible objects. The entire operation, therefore, of spirits inclosed in tangible bodies[18] is concealed, and escapes us. All that more delicate change of formation in the parts of coarser substances (vulgarly called alteration, but in fact a change of position in the smallest particles) is equally unknown; and yet, unless the two matters we have mentioned be explored and brought to light, no great effect can be produced in nature. Again, the very nature of common air, and all bodies of less density (of which there are many) is almost unknown; for the senses are weak and erring, nor can instruments be of great use in extending their sphere or acuteness--all the better interpretations of nature are worked out by instances, and fit and apt experiments, where the senses only judge of the experiment, the experiment of nature and the thing itself.

LI. The human understanding is, by its own nature, prone to abstraction, and supposes that which is fluctuating to be fixed. But it is better to dissect than abstract nature: such was the method employed by the school of Democritus,[19] which made greater progress in penetrating nature than the rest. It is best to consider matter, its conformation, and the changes of that conformation, its own action,[20] and the law of this action or motion; for forms are a mere fiction of the human mind, unless you will call the laws of action by that name.[21]

LII. Such are the idols of the tribe, which arise either from the uniformity of the constitution of man's spirit, or its prejudices, or its limited faculties or restless agitation, or from the interference of the passions, or the incompetence of the senses, or the mode of their impressions.

LIII. The idols of the den derive their origin from the peculiar nature of each individual's mind and body, and also from education, habit, and accident; and although they be various and manifold, yet we will treat of some that require the greatest caution, and exert the greatest power in polluting the understanding.

LIV. Some men become attached to particular sciences and contemplations, either from supposing themselves the authors and inventors of them, or from having bestowed the greatest pains upon such subjects, and thus become most habituated to them.[22] If men of this description apply themselves to philosophy and contemplations of a universal nature, they wrest and corrupt them by their preconceived fancies, of which Aristotle affords us a single instance, who made his natural philosophy completely subservient to his logic, and thus rendered it little more than useless and disputatious. The chemists, again, have formed a fanciful philosophy with the most confined views, from a few experiments of the furnace. Gilbert,[23] too, having employed himself most assiduously in the consideration of the magnet, immediately established a system of philosophy to coincide with his favorite pursuit.

LV. The greatest and, perhaps, radical distinction between different men's dispositions for philosophy and the sciences is this, that some are more vigorous and active in observing the differences of things, others in observing their resemblances; for a steady and acute disposition can fix its thoughts, and dwell upon and adhere to a point, through all the refinements of differences, but those that are sublime and discursive recognize and compare even the most delicate and general resemblances; each of them readily falls into excess, by catching either at nice distinctions or shadows of resemblance.

LVI. Some dispositions evince an unbounded admiration of antiquity, others eagerly embrace novelty, and but few can preserve the just medium, so as neither to tear up what the ancients have correctly laid down, nor to despise the just innovations of the moderns. But this is very prejudicial to the sciences and philosophy, and instead of a correct judgment we have but the factions of the ancients and moderns. Truth is not to be sought in the good fortune of any particular conjuncture of time, which is uncertain, but in the light of nature and experience, which is eternal. Such factions, therefore, are to be abjured, and the understanding must not allow them to hurry it on to assent.

LVII. The contemplation of nature and of bodies in their individual form distracts and weakens the understanding; but the contemplation of nature and of bodies in their general composition and formation stupefies and relaxes it. We have a good instance of this in the school of Leucippus and Democritus compared with others, for they applied themselves so much to particulars as almost to neglect the general structure of things, while the others were so astounded while gazing on the structure that they did not penetrate the simplicity of nature. These two species of contemplation must, therefore, be interchanged, and each employed in its turn, in order to render the understanding at once penetrating and capacious, and to avoid the inconveniences we have mentioned, and the idols that result from them.

LVIII. Let such, therefore, be our precautions in contemplation, that we may ward off and expel the idols of the den, which mostly owe their birth either to some predominant pursuit, or, secondly, to an excess in synthesis and analysis, or, thirdly, to a party zeal in favor of certain ages, or, fourthly, to the extent or narrowness of the subject. In general, he who contemplates nature should suspect whatever particularly takes and fixes his understanding, and should use so much the more caution to preserve it equable and unprejudiced.

LIX. The idols of the market are the most troublesome of all, those namely which have entwined themselves round the understanding from the associations of words and names. For men imagine that their reason governs words, while, in fact, words react upon the understanding; and this has rendered philosophy and the sciences sophistical and inactive. Words are generally formed in a popular sense, and define things by those broad lines which are most obvious to the vulgar mind; but when a more acute understanding or more diligent observation is anxious to vary those lines, and to adapt them more accurately to nature, words oppose it. Hence the great and solemn disputes of learned men often terminate in controversies about words and names, in regard to which it would be better (imitating the caution of mathematicians) to proceed more advisedly in the first instance, and to bring such disputes to a regular issue by definitions. Such definitions, however, cannot remedy the evil in natural and material objects, because they consist themselves of words, and these words produce others;[24] so that we must necessarily have recourse to particular instances, and their regular series and arrangement, as we shall mention when we come to the mode and scheme of determining notions and axioms.

LX. The idols imposed upon the understanding by words are of two kinds. They are either the names of things which have no existence (for as some objects are from inattention left without a name, so names are formed by fanciful imaginations which are without an object), or they are the names of actual objects, but confused, badly defined, and hastily and irregularly abstracted from things. Fortune, the primum mobile, the planetary orbits,[25] the element of fire, and the like fictions, which owe their birth to futile and false theories, are instances of the first kind. And this species of idols is removed with greater facility, because it can be exterminated by the constant refutation or the desuetude of the theories themselves. The others, which are created by vicious and unskilful abstraction, are intricate and deeply rooted. Take some word, for instance, as moist, and let us examine how far the different significations of this word are consistent. It will be found that the word moist is nothing but a confused sign of different actions admitted of no settled and defined uniformity. For it means that which easily diffuses itself over another body; that which is indeterminable and cannot be brought to a consistency; that which yields easily in every direction; that which is easily divided and dispersed; that which is easily united and collected; that which easily flows and is put in motion; that which easily adheres to, and wets another body; that which is easily reduced to a liquid state though previously solid. When, therefore, you come to predicate or impose this name, in one sense flame is moist, in another air is not moist, in another fine powder is moist, in another glass is moist; so that it is quite clear that this notion is hastily abstracted from water only, and common ordinary liquors, without any due verification of it.

There are, however, different degrees of distortion and mistake in words. One of the least faulty classes is that of the names of substances, particularly of the less abstract and more defined species (those then of chalk and mud are good, of earth bad); words signifying actions are more faulty, as to generate, to corrupt, to change; but the most faulty are those denoting qualities (except the immediate objects of sense), as heavy, light, rare, dense. Yet in all of these there must be some notions a little better than others, in proportion as a greater or less number of things come before the senses.

LXI. The idols of the theatre are not innate, nor do they introduce themselves secretly into the understanding, but they are manifestly instilled and cherished by the fictions of theories and depraved rules of demonstration. To attempt, however, or undertake their confutation would not be consistent with our declarations. For since we neither agree in our principles nor our demonstrations, all argument is out of the question. And it is fortunate that the ancients are left in possession of their honors. We detract nothing from them, seeing our whole doctrine relates only to the path to be pursued. The lame (as they say) in the path outstrip the swift who wander from it, and it is clear that the very skill and swiftness of him who runs not in the right direction must increase his aberration.

Our method of discovering the sciences is such as to leave little to the acuteness and strength of wit, and indeed rather to level wit and intellect. For as in the drawing of a straight line, or accurate circle by the hand, much depends on its steadiness and practice, but if a ruler or compass be employed there is little occasion for either; so it is with our method. Although, however, we enter into no individual confutations, yet a little must be said, first, of the sects and general divisions of these species of theories; secondly, something further to show that there are external signs of their weakness; and, lastly, we must consider the causes of so great a misfortune, and so long and general a unanimity in error, that we may thus render the access to truth less difficult, and that the human understanding may the more readily be purified, and brought to dismiss its idols.

LXII. The idols of the theatre, or of theories, are numerous, and may, and perhaps will, be still more so. For unless men's minds had been now occupied for many ages in religious and theological considerations, and civil governments (especially monarchies), had been averse to novelties of that nature even in theory (so that men must apply to them with some risk and injury to their own fortunes, and not only without reward, but subject to contumely and envy), there is no doubt that many other sects of philosophers and theorists would have been introduced, like those which formerly flourished in such diversified abundance among the Greeks. For as many imaginary theories of the heavens can be deduced from the phenomena of the sky, so it is even more easy to found many dogmas upon the phenomena of philosophy--and the plot of this our theatre resembles those of the poetical, where the plots which are invented for the stage are more consistent, elegant, and pleasurable than those taken from real history.

In general, men take for the groundwork of their philosophy either too much from a few topics, or too little from many; in either case their philosophy is founded on too narrow a basis of experiment and natural history, and decides on too scanty grounds. For the theoretic philosopher seizes various common circumstances by experiment, without reducing them to certainty or examining and frequently considering them, and relies for the rest upon meditation and the activity of his wit.

There are other philosophers who have diligently and accurately attended to a few experiments, and have thence presumed to deduce and invent systems of philosophy, forming everything to conformity with them.

A third set, from their faith and religious veneration, introduce theology and traditions; the absurdity of some among them having proceeded so far as to seek and derive the sciences from spirits and genii. There are, therefore, three sources of error and three species of false philosophy; the sophistic, empiric, and superstitious.

LXIII. Aristotle affords the most eminent instance of the first; for he corrupted natural philosophy by logic--thus he formed the world of categories, assigned to the human soul, the noblest of substances, a genus determined by words of secondary operation, treated of density and rarity (by which bodies occupy a greater or lesser space), by the frigid distinctions of action and power, asserted that there was a peculiar and proper motion in all bodies, and that if they shared in any other motion, it was owing to an external moving cause, and imposed innumerable arbitrary distinctions upon the nature of things; being everywhere more anxious as to definitions in teaching and the accuracy of the wording of his propositions, than the internal truth of things. And this is best shown by a comparison of his philosophy with the others of greatest repute among the Greeks. For the similar parts of Anaxagoras, the atoms of Leucippus and Democritus, the heaven and earth of Parmenides, the discord and concord of Empedocles,[26] the resolution of bodies into the common nature of fire, and their condensation according to Heraclitus, exhibit some sprinkling of natural philosophy, the nature of things, and experiment; while Aristotle's physics are mere logical terms, and he remodelled the same subject in his metaphysics under a more imposing title, and more as a realist than a nominalist. Nor is much stress to be laid on his frequent recourse to experiment in his books on animals, his problems, and other treatises; for he had already decided, without having properly consulted experience as the basis of his decisions and axioms, and after having so decided, he drags experiment along as a captive constrained to accommodate herself to his decisions: so that he is even more to be blamed than his modern followers (of the scholastic school) who have deserted her altogether.

LXIV. The empiric school produces dogmas of a more deformed and monstrous nature than the sophistic or theoretic school; not being founded in the light of common notions (which, however poor and superstitious, is yet in a manner universal, and of a general tendency), but in the confined obscurity of a few experiments. Hence this species of philosophy appears probable, and almost certain to those who are daily practiced in such experiments, and have thus corrupted their imagination, but incredible and futile to others. We have a strong instance of this in the alchemists and their dogmas; it would be difficult to find another in this age, unless perhaps in the philosophy of Gilbert.[27] We could not, however, neglect to caution others against this school, because we already foresee and augur, that if men be hereafter induced by our exhortations to apply seriously to experiments (bidding farewell to the sophistic doctrines), there will then be imminent danger from empirics, owing to the premature and forward haste of the understanding, and its jumping or flying to generalities and the principles of things. We ought, therefore, already to meet the evil.

LXV. The corruption of philosophy by the mixing of it up with superstition and theology, is of a much wider extent, and is most injurious to it both as a whole and in parts. For the human understanding is no less exposed to the impressions of fancy, than to those of vulgar notions. The disputatious and sophistic school entraps the understanding, while the fanciful, bombastic, and, as it were, poetical school, rather flatters it.

There is a clear example of this among the Greeks, especially in Pythagoras, where, however, the superstition is coarse and overcharged, but it is more dangerous and refined in Plato and his school. This evil is found also in some branches of other systems of philosophy, where it introduces abstracted forms, final and first causes, omitting frequently the intermediate and the like. Against it we must use the greatest caution; for the apotheosis of error is the greatest evil of all, and when folly is worshipped, it is, as it were, a plague spot upon the understanding. Yet some of the moderns have indulged this folly with such consummate inconsiderateness, that they have endeavored to build a system of natural philosophy on the first chapter of Genesis, the book of Job, and other parts of Scripture; seeking thus the dead among the living.[28] And this folly is the more to be prevented and restrained, because not only fantastical philosophy, but heretical religion spring from the absurd mixture of matters divine and human. It is therefore most wise soberly to render unto faith the things that are faith's.

LXVI. Having spoken of the vicious authority of the systems founded either on vulgar notions, or on a few experiments, or on superstition, we must now consider the faulty subjects for contemplation, especially in natural philosophy. The human understanding is perverted by observing the power of mechanical arts, in which bodies are very materially changed by composition or separation, and is induced to suppose that something similar takes place in the universal nature of things. Hence the fiction of elements, and their co-operation in forming natural bodies.[29] Again, when man reflects upon the entire liberty of nature, he meets with particular species of things, as animals, plants, minerals, and is thence easily led to imagine that there exist in nature certain primary forms which she strives to produce, and that all variation from them arises from some impediment or error which she is exposed to in completing her work, or from the collision or metamorphosis of different species. The first hypothesis has produced the doctrine of elementary properties, the second that of occult properties and specific powers; and both lead to trifling courses of reflection, in which the mind acquiesces, and is thus diverted from more important subjects. But physicians exercise a much more useful labor in the consideration of the secondary qualities of things, and the operations of attraction, repulsion, attenuation, inspissation, dilatation, astringency, separation, maturation, and the like; and would do still more if they would not corrupt these proper observations by the two systems I have alluded to, of elementary qualities and specific powers, by which they either reduce the secondary to first qualities, and their subtile and immeasurable composition, or at any rate neglect to advance by greater and more diligent observation to the third and fourth qualities, thus terminating their contemplation prematurely. Nor are these powers (or the like) to be investigated only among the medicines for the human body, but also in all changes of other natural bodies.

A greater evil arises from the contemplation and investigation rather of the stationary principles of things from which, than of the active by which things themselves are created. For the former only serve for discussion, the latter for practice. Nor is any value to be set on those common differences of motion which are observed in the received system of natural philosophy, as generation, corruption, augmentation, diminution, alteration, and translation. For this is their meaning: if a body, unchanged in other respects, is moved from its place, this is translation; if the place and species be given, but the quantity changed, it is alteration; but if, from such a change, the mass and quantity of the body do not continue the same, this is the motion of augmentation and diminution; if the change be continued so as to vary the species and substance, and transfuse them to others, this is generation and corruption. All this is merely popular,.and by no means penetrates into nature; and these are but the measures and bounds of motion, and not different species of it; they merely suggest how far, and not how or whence. For they exhibit neither the affections of bodies nor the process of their parts, but merely establish a division of that motion, which coarsely exhibits to the senses matter in its varied form. Even when they wish to point out something relative to the causes of motion, and to establish a division of them, they most absurdly introduce natural and violent motion, which is also a popular notion, since every violent motion is also in fact natural, that is to say, the external efficient puts nature in action in a different manner to that which she had previously employed.

But if, neglecting these, any one were, for instance, to observe that there is in bodies a tendency of adhesion, so as not to suffer the unity of nature to be completely separated or broken, and a vacuum[30] to be formed, or that they have a tendency to return to their natural dimensions or tension, so that, if compressed or extended within or beyond it, they immediately strive to recover themselves, and resume their former volume and extent; or that they have a tendency to congregate into masses with similar bodies--the dense, for instance, toward the circumference of the earth, the thin and rare toward that of the heavens. These and the like are true physical genera of motions, but the others are clearly logical and scholastic, as appears plainly from a comparison of the two.

Another considerable evil is, that men in their systems and contemplations bestow their labor upon the investigation and discussion of the principles of things and the extreme limits of nature, although all utility and means of action consist in the intermediate objects. Hence men cease not to abstract nature till they arrive at potential and shapeless matter,[31] and still persist in their dissection, till they arrive at atoms; and yet were all this true, it would be of little use to advance man's estate.

LXVII. The understanding must also be cautioned against the intemperance of systems, so far as regards its giving or withholding its assent; for such intemperance appears to fix and perpetuate idols, so as to leave no means of removing them.

These excesses are of two kinds. The first is seen in those who decide hastily, and render the sciences positive and dictatorial. The other in those who have introduced scepticism, and vague unbounded inquiry. The former subdues, the latter enervates the understanding. The Aristotelian philosophy, after destroying other systems (as the Ottomans[32] do their brethren) by its disputatious confutations, decided upon everything, and Aristotle himself then raises up questions at will, in order to settle them; so that everything should be certain and decided, a method now in use among his successors.

The school of Plato introduced scepticism, first, as it were in joke and irony, from their dislike to Protagoras, Hippias,[33] and others, who were ashamed of appearing not to doubt upon any subject. But the new academy dogmatized in their scepticism, and held it as their tenet. Although this method be more honest than arbitrary decision (for its followers allege that they by no means confound all inquiry, like Pyrrho and his disciples, but hold doctrines which they can follow as probable, though they cannot maintain them to be true), yet when the human mind has once despaired of discovering truth, everything begins to languish. Hence men turn aside into pleasant controversies and discussions, and into a sort of wandering over subjects rather than sustain any rigorous investigation. But as we observed at first, we are not to deny the authority of the human senses and understanding, although weak, but rather to furnish them with assistance.

LXVIII. We have now treated of each kind of idols, and their qualities, all of which must be abjured and renounced with firm and solemn resolution, and the understanding must be completely freed and cleared of them, so that the access to the kingdom of man, which is founded on the sciences, may resemble that to the kingdom of heaven, where no admission is conceded except to children.

LXIX. Vicious demonstrations are the muniments and support of idols, and those which we possess in logic, merely subject and enslave the world to human thoughts, and thoughts to words. But demonstrations are in some manner themselves systems of philosophy and science; for such as they are, and accordingly as they are regularly or improperly established, such will be the resulting systems of philosophy and contemplation. But those which we employ in the whole process leading from the senses and things to axioms and conclusions, are fallacious and incompetent. This process is fourfold, and the errors are in equal number. In the first place the impressions of the senses are erroneous, for they fail and deceive us. We must supply defects by substitutions, and fallacies by their correction.

Secondly, notions are improperly abstracted from the senses, and indeterminate and confused when they ought to be the reverse. Thirdly, the induction that is employed is improper, for it determines the principles of sciences by simple enumeration,[34] without adopting exclusions and resolutions, or just separations of nature. Lastly, the usual method of discovery and proof, by first establishing the most general propositions, then applying and proving the intermediate axioms according to them, is the parent of error and the calamity of every science. But we will treat more fully of that which we now slightly touch upon, when we come to lay down the true way of interpreting nature, after having gone through the above expiatory process and purification of the mind.

LXX. But experience is by far the best demonstration, provided it adhere to the experiment actually made, for if that experiment be transferred to other subjects apparently similar, unless with proper and methodical caution it becomes fallacious. The present method of experiment is blind and stupid; hence men wandering and roaming without any determined course, and consulting mere chance, are hurried about to various points, and advance but little--at one time they are happy, at another their attention is distracted, and they always find that they want something further. Men generally make their experiments carelessly, and as it were in sport, making some little variation in a known experiment, and then if they fail they become disgusted and give up the attempt; nay, if they set to work more seriously, steadily, and assiduously, yet they waste all their time on probing some solitary matter, as Gilbert on the magnet, and the alchemists on gold. But such conduct shows their method to be no less unskilful than mean; for nobody can successfully investigate the nature of any object by considering that object alone; the inquiry must be more generally extended.

Even when men build any science and theory upon experiment, yet they almost always turn with premature and hasty zeal to practice, not merely on account of the advantage and benefit to be derived from it, but in order to seize upon some security in a new undertaking of their not employing the remainder of their labor unprofitably, and by making themselves conspicuous, to acquire a greater name for their pursuit. Hence, like Atalanta, they leave the course to pick up the golden apple, interrupting their speed, and giving up the victory. But in the true course of experiment, and in extending it to new effects, we should imitate the Divine foresight and order; for God on the first day only created light, and assigned a whole day to that work without creating any material substance thereon. In like manner we must first, by every kind of experiment, elicit the discovery of causes and true axioms, and seek for experiments which may afford light rather than profit. Axioms, when rightly investigated and established, prepare us not for a limited but abundant practice, and bring in their train whole troops of effects. But we will treat hereafter of the ways of experience, which are not less beset and interrupted than those of judgment; having spoken at present of common experience only as a bad species of demonstration, the order of our subject now requires some mention of those external signs of the weakness in practice of the received systems of philosophy and contemplation[35] which we referred to above, and of the causes of a circumstance at first sight so wonderful and incredible. For the knowledge of these external signs prepares the way for assent, and the explanation of the causes removes the wonder; and these two circumstances are of material use in extirpating more easily and gently the idols from the understanding.

LXXI. The sciences we possess have been principally derived from the Greeks; for the addition of the Roman, Arabic, or more modern writers, are but few and of small importance, and such as they are, are founded on the basis of Greek invention. But the wisdom of the Greeks was professional and disputatious, and thus most adverse to the investigation of truth. The name, therefore, of sophists, which the contemptuous spirit of those who deemed themselves philosophers, rejected and transferred to the rhetoricians--Gorgias,[36] Protagoras, Hippias, Polus--might well suit the whole tribe, such as Plato, Aristotle, Zeno, Epicurus, Theophrastus, and their successors--Chrysippus, Carneades, and the rest. There was only this difference between them--the former were mercenary vagabonds, travelling about to different states, making a show of their wisdom, and requiring pay; the latter more dignified and noble, in possession of fixed habitations, opening schools, and teaching philosophy gratuitously. Both, however (though differing in other respects), were professorial, and reduced every subject to controversy, establishing and defending certain sects and dogmas of philosophy, so that their doctrines were nearly (what Dionysius not unaptly objected to Plato) the talk of idle old men to ignorant youths. But the more ancient Greeks, as Empedocles, Anaxagoras, Leucippus, Democritus, Parmenides, Heraclitus, Xenophanes, Philolaus, and the rest[37] (for I omit Pythagoras as being superstitious), did not (that we are aware) open schools, but betook themselves to the investigation of truth with greater silence and with more severity and simplicity, that is, with less affectation and ostentation. Hence in our opinion they acted more advisedly, however their works may have been eclipsed in course of time by those lighter productions which better correspond with and please the apprehensions and passions of the vulgar; for time, like a river,[38] bears down to us that which is light and inflated, and sinks that which is heavy and solid. Nor were even these more ancient philosophers free from the national defect, but inclined too much to the ambition and vanity of forming a sect, and captivating public opinion, and we must despair of any inquiry after truth when it condescends to such trifles. Nor must we omit the opinion, or rather prophecy, of an Egyptian priest with regard to the Greeks, that they would forever remain children, without any antiquity of knowledge or knowledge of antiquity; for they certainly have this in common with children, that they are prone to talking, and incapable of generation, their wisdom being loquacious and unproductive of effects. Hence the external signs derived from the origin and birthplace of our present philosophy are not favorable.

LXXII. Nor are those much better which can be deduced from the character of the time and age, than the former from that of the country and nation; for in that age the knowledge both of time and of the world was confined and meagre, which is one of the worst evils for those who rely entirely on experience--they had not a thousand years of history worthy of that name, but mere fables and ancient traditions; they were acquainted with but a small portion of the regions and countries of the world, for they indiscriminately called all nations situated far toward the north Scythians, all those to the west Celts; they knew nothing of Africa but the nearest part of Ethiopia, or of Asia beyond the Ganges, and had not even heard any sure and clear tradition of the regions of the New World. Besides, a vast number of climates and zones, in which innumerable nations live and breathe, were pronounced by them to be uninhabitable; nay, the travels of Democritus, Plato, and Pythagoras, which were not extensive, but rather mere excursions from home, were considered as something vast. But in our times many parts of the New World, and every extremity of the Old, are well known, and the mass of experiments has been infinitely increased; wherefore, if external signs were to be taken from the time of the nativity or procreation (as in astrology), nothing extraordinary could be predicted of these early systems of philosophy.

LXXIII. Of all signs there is none more certain or worthy than that of the fruits produced, for the fruits and effects are the sureties and vouchers, as it were, for the truth of philosophy. Now, from the systems of the Greeks, and their subordinate divisions in particular branches of the sciences during so long a period, scarcely one single experiment can be culled that has a tendency to elevate or assist mankind, and can be fairly set down to the speculations and doctrines of their philosophy. Celsus candidly and wisely confesses as much, when he observes that experiments were first discovered in medicine, and that men afterward built their philosophical systems upon them, and searched for and assigned causes, instead of the inverse method of discovering and deriving experiments from philosophy and the knowledge of causes; it is not, therefore, wonderful that the Egyptians (who bestowed divinity and sacred honors on the authors of new inventions) should have consecrated more images of brutes than of men, for the brutes by their natural instinct made many discoveries, while men derived but few from discussion and the conclusions of reason.

The industry of the alchemists has produced some effect, by chance, however, and casualty, or from varying their experiments (as mechanics also do), and not from any regular art or theory, the theory they have imagined rather tending to disturb than to assist experiment. Those, too, who have occupied themselves with natural magic (as they term it) have made but few discoveries, and those of small import, and bordering on imposture; for which reason, in the same manner as we are cautioned by religion to show our faith by our works, we may very properly apply the principle to philosophy, and judge of it by its works, accounting that to be futile which is unproductive, and still more so if, instead of grapes and olives, it yield but the thistle and thorns of dispute and contention.

LXXIV. Other signs may be selected from the increase and progress of particular systems of philosophy and the sciences; for those which are founded on nature grow and increase, while those which are founded on opinion change and increase not. If, therefore, the theories we have mentioned were not like plants, torn up by the roots, but grew in the womb of nature, and were nourished by her, that which for the last two thousand years has taken place would never have happened, namely, that the sciences still continue in their beaten track, and nearly stationary, without having received any important increase, nay, having, on the contrary, rather bloomed under the hands of their first author, and then faded away. But we see that the case is reversed in the mechanical arts, which are founded on nature and the light of experience, for they (as long as they are popular) seem full of life, and uninterruptedly thrive and grow, being at first rude, then convenient, lastly polished, and perpetually improved.

LXXV. There is yet another sign (if such it may be termed, being rather an evidence, and one of the strongest nature), namely, the actual confession of those very authorities whom men now follow; for even they who decide on things so daringly, yet at times, when they reflect, betake themselves to complaints about the subtilty of nature, the obscurity of things, and the weakness of man's wit. If they would merely do this, they might perhaps deter those who are of a timid disposition from further inquiry, but would excite and stimulate those of a more active and confident turn to further advances. They are not, however, satisfied with confessing so much of themselves, but consider everything which has been either unknown or unattempted by themselves or their teachers, as beyond the limits of possibility, and thus, with most consummate pride and envy, convert the defects of their own discoveries into a calumny on nature and a source of despair to every one else. Hence arose the New Academy, which openly professed scepticism,[39] and consigned mankind to eternal darkness; hence the notion that forms, or the true differences of things (which are in fact the laws of simple action), are beyond man's reach, and cannot possibly be discovered; hence those notions in the active and operative branches, that the heat of the sun and of fire are totally different, so as to prevent men from supposing that they can elicit or form, by means of fire, anything similar to the operations of nature; and again, that composition only is the work of man and mixture of nature, so as to prevent men from expecting the generation or transformation of natural bodies by art. Men will, therefore, easily allow themselves to be persuaded by this sign not to engage their fortunes and labor in speculations, which are not only desperate, but actually devoted to desperation.

LXXVI. Nor should we omit the sign afforded by the great dissension formerly prevalent among philosophers, and the variety of schools, which sufficiently show that the way was not well prepared that leads from the senses to the understanding, since the same groundwork of philosophy (namely, the nature of things), was torn and divided into such widely differing and multifarious errors. And although in these days the dissensions and differences of opinions with regard to first principles and entire systems are nearly extinct,[40] yet there remain innumerable questions and controversies with regard to particular branches of philosophy. So that it is manifest that there is nothing sure or sound either in the systems themselves or in the methods of demonstration.[41]

LXXVII. With regard to the supposition that there is a general unanimity as to the philosophy of Aristotle, because the other systems of the ancients ceased and became obsolete on its promulgation, and nothing better has been since discovered; whence it appears that it is so well determined and founded, as to have united the suffrages of both ages; we will observe--1st. That the notion of other ancient systems having ceased after the publication of the works of Aristotle is false, for the works of the ancient philosophers subsisted long after that event, even to the time of Cicero, and the subsequent ages. But at a later period, when human learning had, as it were, been wrecked in the inundation of barbarians into the Roman empire, then the systems of Aristotle and Plato were preserved in the waves of ages, like planks of a lighter and less solid nature. 2d. The notion of unanimity, on a clear inspection, is found to be fallacious. For true unanimity is that which proceeds from a free judgment, arriving at the same conclusion, after an investigation of the fact. Now, by far the greater number of those who have assented to the philosophy of Aristotle, have bound themselves down to it from prejudice and the authority of others, so that it is rather obsequiousness and concurrence than unanimity. But even if it were real and extensive unanimity, so far from being esteemed a true and solid confirmation, it should even lead to a violent presumption to the contrary. For there is no worse augury in intellectual matters than that derived from unanimity, with the exception of divinity and politics, where suffrages are allowed to decide. For nothing pleases the multitude, unless it strike the imagination or bind down the understanding, as we have observed above, with the shackles of vulgar notions. Hence we may well transfer Phocion's remark from morals to the intellect: "That men should immediately examine what error or fault they have committed, when the multitude concurs with, and applauds them."[42] This then is one of the most unfavorable signs. All the signs, therefore, of the truth and soundness of the received systems of philosophy and the sciences are unpropitious, whether taken from their origin, their fruits, their progress, the confessions of their authors, or from unanimity.

LXXVIII. We now come to the causes of errors,[43] and of such perseverance in them for ages. These are sufficiently numerous and powerful to remove all wonder, that what we now offer should have so long been concealed from, and have escaped the notice of mankind, and to render it more worthy of astonishment, that it should even now have entered any one's mind, or become the subject of his thoughts; and that it should have done so, we consider rather the gift of fortune than of any extraordinary talent, and às the offspring of time rather than wit. But, in the first place, the number of ages is reduced to very narrow limits, on a proper consideration of the matter. For out of twenty-five[44] centuries, with which the memory and learning of man are conversant, scarcely six can be set apart and selected as fertile in science and favorable to its progress. For there are deserts and wastes in times as in countries, and we can only reckon up three revolutions and epochs of philosophy. 1. The Greek. 2. The Roman. 3. Our own, that is the philosophy of the western nations of Europe: and scarcely two centuries can with justice be assigned to each. The intermediate ages of the world were unfortunate both in the quantity and richness of the sciences produced. Nor need we mention the Arabs, or the scholastic philosophy, which, in those ages, ground down the sciences by their numerous treatises, more than they increased their weight. The first cause, then, of such insignificant progress in the sciences, is rightly referred to the small proportion of time which has been favorable thereto.

LXXIX. A second cause offers itself, which is certainly of the greatest importance; namely, that in those very ages in which men's wit and literature flourished considerably, or even moderately, but a small part of their industry was bestowed on natural philosophy, the great mother of the sciences. For every art and science torn from this root may, perhaps, be polished, and put into a serviceable shape, but can admit of little growth. It is well known, that after the Christian religion had been acknowledged, and arrived at maturity, by far the best wits were busied upon theology, where the highest rewards offered themselves, and every species of assistance was abundantly supplied, and the study of which was the principal occupation of the western European nations during the third epoch; the rather because literature flourished about the very time when controversies concerning religion first began to bud forth. 2. In the preceding ages, during the second epoch (that of the Romans), philosophical meditation and labor was chiefly occupied and wasted in moral philosophy (the theology of the heathens): besides, the greatest minds in these times applied themselves to civil affairs, on account of the magnitude of the Roman empire, which required the labor of many. 3. The age during which natural philosophy appeared principally to flourish among the Greeks, was but a short period, since in the more ancient times the seven sages (with the exception of Thales), applied themselves to moral philosophy and politics, and at a later period, after Socrates had brought down philosophy from heaven to earth, moral philosophy became more prevalent, and diverted men's attention from natural. Nay, the very period during which physical inquiries flourished, was corrupted and rendered useless by contradictions, and the ambition of new opinions. Since, therefore, during these three epochs, natural philosophy has been materially neglected or impeded, it is not at all surprising that men should have made but little progress in it, seeing they were attending to an entirely different matter.

LXXX. Add to this that natural philosophy, especially of late, has seldom gained exclusive possession of an individual free from all other pursuits, even among those who have applied themselves to it, unless there may be an example or two of some monk studying in his cell, or some nobleman in his villa.[45] She has rather been made a passage and bridge to other pursuits.

Thus has this great mother of the sciences been degraded most unworthily to the situation of a handmaid, and made to wait upon medicine or mathematical operations, and to wash the immature minds of youth, and imbue them with a first dye, that they may afterward be more ready to receive and retain another. In the meantime, let no one expect any great progress in the sciences (especially their operative part), unless natural philosophy be applied to particular sciences, and particular sciences again referred back to natural philosophy. For want of this, astronomy, optics, music, many mechanical arts, medicine itself, and (what perhaps is more wonderful), moral and political philosophy, and the logical sciences have no depth, but only glide over the surface and variety of things; because these sciences, when they have been once partitioned out and established, are no longer nourished by natural philosophy, which would have imparted fresh vigor and growth to them from the sources and genuine contemplation of motion, rays, sounds, texture, and conformation of bodies, and the affections and capacity of the understanding. But we can little wonder that the sciences grow not when separated from their roots.

LXXXI. There is another powerful and great cause of the little advancement of the sciences, which is this; it is impossible to advance properly in the course when the goal is not properly fixed. But the real and legitimate goal of the sciences is the endowment of human life with new inventions and riches. The great crowd of teachers know nothing of this, but consist of dictatorial hirelings; unless it so happen that some artisan of an acute genius, and ambitious of fame, gives up his time to a new discovery, which is generally attended with a loss of property. The majority, so far from proposing to themselves the augmentation of the mass of arts and sciences, make no other use of an inquiry into the mass already before them, than is afforded by the conversion of it to some use in their lectures, or to gain, or to the acquirement of a name, and the like. But if one out of the multitude be found, who courts science from real zeal, and on his own account, even he will be seen rather to follow contemplation, and the variety of theories, than a severe and strict investigation of truth. Again, if there even be an unusually strict investigator of truth, yet will he propose to himself, as the test of truth, the satisfaction of his mind and understanding, as to the causes of things long since known, and not such a test as to lead to some new earnest of effects, and a new light in axioms. If, therefore, no one have laid down the real end of science, we cannot wonder that there should be error in points subordinate to that end.

LXXXII. But, in like manner, as the end and goal of science is ill defined, so, even were the case otherwise, men have chosen an erroneous and impassable direction. For it is sufficient to astonish any reflecting mind, that nobody should have cared or wished to open and complete a way for the understanding, setting off from the senses, and regular, well-conducted experiment; but that everything has been abandoned either to the mists of tradition, the whirl and confusion of argument, or the waves and mazes of chance, and desultory, ill-combined experiment. Now, let any one but consider soberly and diligently the nature of the path men have been accustomed to pursue in the investigation and discovery of any matter, and he will doubtless first observe the rude and inartificial manner of discovery most familiar to mankind: which is no other than this. When any one prepares himself for discovery, he first inquires and obtains a full account of all that has been said on the subject by others, then adds his own reflections, and stirs up and, as it were, invokes his own spirit, after much mental labor, to disclose its oracles. All which is a method without foundation, and merely turns on opinion.

Another, perhaps, calls in logic to assist him in discovery, which bears only a nominal relation to his purpose. For the discoveries of logic are not discoveries of principles and leading axioms, but only of what appears to accord with them.[46] And when men become curious and importunate, and give trouble, interrupting her about her proofs, and the discovery of principles or first axioms, she puts them off with her usual answer, referring them to faith, and ordering them to swear allegiance to each art in its own department.

There remains but mere experience, which, when it offers itself, is called chance; when it is sought after, experiment.[47] But this kind of experience is nothing but a loose fagot; and mere groping in the dark, as men at night try all means of discovering the right road, while it would be better and more prudent either to wait for day, or procure a light, and then proceed. On the contrary, the real order of experience begins by setting up a light, and then shows the road by it, commencing with a regulated and digested, not a misplaced and vague course of experiment, and thence deducing axioms, and from those axioms new experiments: for not even the Divine Word proceeded to operate on the general mass of things without due order.

Let men, therefore, cease to wonder if the whole course of science be not run, when all have wandered from the path; quitting it entirely, and deserting experience, or involving themselves in its mazes, and wandering about, while a regularly combined system would lead them in a sure track through its wilds to the open day of axioms.

LXXXIII. The evil, however, has been wonderfully increased by an opinion, or inveterate conceit, which is both vainglorious and prejudicial, namely, that the dignity of the human mind is lowered by long and frequent intercourse with experiments and particulars, which are the objects of sense, and confined to matter; especially since such matters generally require labor in investigation, are mean subjects for meditation, harsh in discourse, unproductive in practice, infinite in number, and delicate in their subtilty. Hence we have seen the true path not only deserted, but intercepted and blocked up, experience being rejected with disgust, and not merely neglected or improperly applied.

LXXXIV. Again, the reverence for antiquity,[48] and the authority of men who have been esteemed great in philosophy, and general unanimity, have retarded men from advancing in science, and almost enchanted them. As to unanimity, we have spoken of it above.

The opinion which men cherish of antiquity is altogether idle, and scarcely accords with the term. For the old age and increasing years of the world should in reality be considered as antiquity, and this is rather the character of our own times than of the less advanced age of the world in those of the ancients; for the latter, with respect to ourselves, are ancient and elder, with respect to the world modern and younger. And as we expect a greater knowledge of human affairs, and more mature judgment from an old man than from a youth, on account of his experience, and the variety and number of things he has seen, heard, and meditated upon, so we have reason to expect much greater things of our own age (if it knew but its strength and would essay and exert it) than from antiquity, since the world has grown older, and its stock has been increased and accumulated with an infinite number of experiments and observations.

We must also take into our consideration that many objects in nature fit to throw light upon philosophy have been exposed to our view, and discovered by means of long voyages and travels, in which our times have abounded. It would, indeed, be dishonorable to mankind, if the regions of the material globe, the earth, the sea, and stars, should be so prodigiously developed and illustrated in our age, and yet the boundaries of the intellectual globe should be confined to the narrow discoveries of the ancients.

With regard to authority, it is the greatest weakness to attribute infinite credit to particular authors, and to refuse his own prerogative to time, the author of all authors, and, therefore, of all authority. For truth is rightly named the daughter of time, not of authority. It is not wonderful, therefore, if the bonds of antiquity, authority, and unanimity, have so enchained the power of man, that he is unable (as if bewitched) to become familiar with things themselves.

LXXXV. Nor is it only the admiration of antiquity, authority, and unanimity, that has forced man's industry to rest satisfied with present discoveries, but, also, the admiration of the effects already placed within his power. For whoever passes in review the variety of subjects, and the beautiful apparatus collected and introduced by the mechanical arts for the service of mankind, will certainly be rather inclined to admire our wealth than to perceive our poverty: not considering that the observations of man and operations of nature (which are the souls and first movers of that variety) are few, and not of deep research; the rest must be attributed merely to man's patience, and the delicate and well-regulated motion of the hand or of instruments. To take an instance, the manufacture of clocks is delicate and accurate, and appears to imitate the heavenly bodies in its wheels, and the pulse of animals in its regular oscillation, yet it only depends upon one or two axioms of nature.

Again, if one consider the refinement of the liberal arts, or even that exhibited in the preparation of natural bodies in mechanical arts and the like, as the discovery of the heavenly motions in astronomy, of harmony in music, of the letters of the alphabet[49] (still unadopted by the Chinese) in grammar; or, again, in mechanical operations, the productions of Bacchus and Ceres, that is, the preparation of wine and beer, the making of bread, or even the luxuries of the table, distillation, and the like; if one reflect also, and consider for how long a period of ages (for all the above, except distillation, are ancient) these things have been brought to their present state of perfection, and (as we instanced in clocks) to how few observations and axioms of nature they may be referred, and how easily, and as it were, by obvious chance or contemplation, they might be discovered, one would soon cease to admire and rather pity the human lot on account of its vast want and dearth of things and discoveries for so many ages. Yet even the discoveries we have mentioned were more ancient than philosophy and the intellectual arts; so that (to say the truth) when contemplation and doctrinal science began, the discovery of useful works ceased.

But if any one turn from the manufactories to libraries, and be inclined to admire the immense variety of books offered to our view, let him but examine and diligently inspect the matter and contents of these books, and his astonishment will certainly change its object: for when he finds no end of repetitions, and how much men do and speak the same thing over again, he will pass from admiration of this variety to astonishment at the poverty and scarcity of matter, which has hitherto possessed and filled men's minds.

But if any one should condescend to consider such sciences as are deemed rather curious than sound, and take a full view of the operations of the alchemists or magii, he will perhaps hesitate whether he ought rather to laugh or to weep. For the alchemist cherishes eternal hope, and when his labors succeed not, accuses his own mistakes, deeming, in his self-accusation, that he has not properly understood the words of art or of his authors; upon which he listens to tradition and vague whispers, or imagines there is some slight unsteadiness in the minute details of his practice, and then has recourse to an endless repetition of experiments: and in the meantime, when, in his casual experiments, he falls upon something in appearance new, or of some degree of utility, he consoles himself with such an earnest, and ostentatiously publishes them, keeping up his hope of the final result. Nor can it be denied that the alchemists have made several discoveries, and presented mankind with useful inventions. But we may well apply to them the fable of the old man, who bequeathed to his sons some gold buried in his garden, pretending not to know the exact spot, whereupon they worked diligently in digging the vineyard, and though they found no gold, the vintage was rendered more abundant by their labor.

The followers of natural magic, who explain everything by sympathy and antipathy, have assigned false powers and marvellous operations to things by gratuitous and idle conjectures: and if they have ever produced any effects, they are rather wonderful and novel than of any real benefit or utility.

In superstitious magic (if we say anything at all about it) we must chiefly observe, that there are only some peculiar and definite objects with which the curious and superstitious arts have, in every nation and age, and even under every religion, been able to exercise and amuse themselves. Let us, therefore, pass them over. In the meantime we cannot wonder that the false notion of plenty should have occasioned want.

LXXXVI. The admiration of mankind with regard to the arts and sciences, which is of itself sufficiently simple and almost puerile, has been increased by the craft and artifices of those who have treated the sciences, and delivered them down to posterity. For they propose and produce them to our view so fashioned, and as it were, masked, as to make them pass for perfect and complete. For if you consider their method and divisions, they appear to embrace and comprise everything which can relate to the subject. And although this frame be badly filled up and resemble an empty bladder, yet it presents to the vulgar understanding the form and appearance of a perfect science.

The first and most ancient investigators of truth were wont, on the contrary, with more honesty and success, to throw all the knowledge they wished to gather from contemplation, and to lay up for use, into aphorisms, or short scattered sentences unconnected by any method, and without pretending or professing to comprehend any entire art. But according to the present system, we cannot wonder that men seek nothing beyond that which is handed down to them as perfect, and already extended to its full complement.

LXXXVII. The ancient theories have received additional support and credit from the absurdity and levity of those who have promoted the new, especially in the active and practical part of natural philosophy. For there have been many silly and fantastical fellows who, from credulity or imposture, have loaded mankind with promises, announcing and boasting of the prolongation of life, the retarding of old age, the alleviation of pains, the remedying of natural defects, the deception of the senses, the restraint and excitement of the passions, the illumination and exaltation of the intellectual faculties, the transmutation of substances, the unlimited intensity and multiplication of motion, the impressions and changes of the air, the bringing into our power the management of celestial influences, the divination of future events, the representation of distant objects, the revelation of hidden objects, and the like. One would not be very wrong in observing with regard to such pretenders, that there is as much difference in philosophy, between their absurdity and real science, as there is in history between the exploits of Cæsar or Alexander, and those of Amadis de Gaul and Arthur of Britain. For those illustrious generals are found to have actually performed greater exploits than such fictitious heroes are even pretended to have accomplished, by the means, however, of real action, and not by any fabulous and portentous power. Yet it is not right to suffer our belief in true history to be diminished, because it is sometimes injured and violated by fables. In the meantime we cannot wonder that great prejudice has been excited against any new propositions (especially when coupled with any mention of effects to be produced), by the conduct of impostors who have made a similar attempt; for their extreme absurdity, and the disgust occasioned by it, has even to this day overpowered every spirited attempt of the kind.

LXXXVIII. Want of energy, and the littleness and futility of the tasks that human industry has undertaken, have produced much greater injury to the sciences: and yet (to make it still worse) that very want of energy manifests itself in conjunction with arrogance and disdain.

For, in the first place, one excuse, now from its repetition become familiar, is to be observed in every art, namely, that its promoters convert the weakness of the art itself into a calumny upon nature: and whatever it in their hands fails to effect, they pronounce to be physically impossible. But how can the art ever be condemned while it acts as judge in its own cause? Even the present system of philosophy cherishes in its bosom certain positions or dogmas, which (it will be found on diligent inquiry) are calculated to produce a full conviction that no difficult, commanding, and powerful operation upon nature ought to be anticipated through the means of art; we instanced[50] above the alleged different quality of heat in the sun and fire, and composition and mixture. Upon an accurate observation the whole tendency of such positions is wilfully to circumscribe man's power, and to produce a despair of the means of invention and contrivance, which would not only confound the promises of hope, but cut the very springs and sinews of industry, and throw aside even the chances of experience. The only object of such philosophers is to acquire the reputation of perfection for their own art, and they are anxious to obtain the most silly and abandoned renown, by causing a belief that whatever has not yet been invented and understood can never be so hereafter. But if any one attempt to give himself up to things, and to discover something new; yet he will only propose and destine for his object the investigation and discovery of some one invention, and nothing more; as the nature of the magnet, the tides, the heavenly system, and the like, which appear enveloped in some degree of mystery, and have hitherto been treated with but little success. Now it is the greatest proof of want of skill, to investigate the nature of any object in itself alone; for that same nature, which seems concealed and hidden in some instances, is manifest and almost palpable in others, and excites wonder in the former, while it hardly attracts attention in the latter.[51] Thus the nature of consistency is scarcely observed in wood or stone, but passed over by the term solid without any further inquiry about the repulsion of separation or the solution of continuity. But in water-bubbles the same circumstance appears matter of delicate and ingenious research, for they form themselves into thin pellicles, curiously shaped into hemispheres, so as for an instant to avoid the solution of continuity.

In general those very things which are considered as secret are manifest and common in other objects, but will never be clearly seen if the experiments and contemplation of man be directed to themselves only. Yet it commonly happens, that if, in the mechanical arts, any one bring old discoveries to a finer polish, or more elegant height of ornament, or unite and compound them, or apply them more readily to practice, or exhibit them on a less heavy and voluminous scale, and the like, they will pass off as new.

We cannot, therefore, wonder that no magnificent discoveries, worthy of mankind, have been brought to light, while men are satisfied and delighted with such scanty and puerile tasks, nay, even think that they have pursued or attained some great object in their accomplishment.

LXXXIX. Nor should we neglect to observe that natural philosophy has, in every age, met with a troublesome and difficult opponent: I mean superstition, and a blind and immoderate zeal for religion. For we see that, among the Greeks, those who first disclosed the natural causes of thunder and storms to the yet untrained ears of man were condemned as guilty of impiety toward the gods.[52] Nor did some of the old fathers of Christianity treat those much better who showed by the most positive proofs (such as no one now disputes) that the earth is spherical, and thence asserted that there were antipodes.[53]

Even in the present state of things the condition of discussions on natural philosophy is rendered more difficult and dangerous by the summaries and methods of divines, who, after reducing divinity into such order as they could, and brought it into a scientific form, have proceeded to mingle an undue proportion of the contentious and thorny philosophy of Aristotle with the substance of religion.[54]

The fictions of those who have not feared to deduce and confirm the truth of the Christian religion by the principles and authority of philosophers, tend to the same end, though in a different manner.[55] They celebrate the union of faith and the senses as though it were legitimate, with great pomp and solemnity, and gratify men's pleasing minds with a variety, but in the meantime confound most improperly things divine and human. Moreover, in these mixtures of divinity and philosophy the received doctrines of the latter are alone included, and any novelty, even though it be an improvement, scarcely escapes banishment and extermination.

In short, you may find all access to any species of philosophy, however pure, intercepted by the ignorance of divines. Some in their simplicity are apprehensive that a too deep inquiry into nature may penetrate beyond the proper bounds of decorum, transferring and absurdly applying what is said of sacred mysteries in Holy Writ against those who pry into divine secrets, to the mysteries of nature, which are not forbidden by any prohibition. Others with more cunning imagine and consider, that if secondary causes be unknown, everything may more easily be referred to the Divine hand and wand, a matter, as they think, of the greatest consequence to religion, but which can only really mean that God wishes to be gratified by means of falsehood. Others fear, from past example, lest motion and change in philosophy should terminate in an attack upon religion. Lastly, there are others who appear anxious lest there should be something discovered in the investigation of nature to overthrow, or at least shake, religion, particularly among the unlearned. The last two apprehensions appear to resemble animal instinct, as if men were diffident, in the bottom of their minds and secret meditations, of the strength of religion and the empire of faith over the senses, and therefore feared that some danger awaited them from an inquiry into nature. But any one who properly considers the subject will find natural philosophy to be, after the Word of God, the surest remedy against superstition, and the most approved support of faith. She is, therefore, rightly bestowed upon religion as a most faithful attendant, for the one exhibits the will and the other the power of God. Nor was he wrong who observed, "Ye err, not knowing the Scriptures and the power of God," thus uniting in one bond the revelation of his will and the contemplation of his power. In the meanwhile, it is not wonderful that the progress of natural philosophy has been restrained, since religion, which has so much influence on men's minds, has been led and hurried to oppose her through the ignorance of some and the imprudent zeal of others.

XC. Again, in the habits and regulations of schools, universities, and the like assemblies, destined for the abode of learned men and the improvement of learning, everything is found to be opposed to the progress of the sciences; for the lectures and exercises are so ordered, that anything out of the common track can scarcely enter the thoughts and contemplations of the mind. If, however, one or two have perhaps dared to use their liberty, they can only impose the labor on themselves, without deriving any advantage from the association of others; and if they put up with this, they will find their industry and spirit of no slight disadvantage to them in making their fortune; for the pursuits of men in such situations are, as it were, chained down to the writings of particular authors, and if any one dare to dissent from them he is immediately attacked as a turbulent and revolutionary spirit. Yet how great is the difference between civil matters and the arts, for there is not the same danger from new activity and new light. In civil matters even a change for the better is suspected on account of the commotion it occasions, for civil government is supported by authority, unanimity, fame, and public opinion, and not by demonstration. In the arts and sciences, on the contrary, every department should resound, as in mines, with new works and advances. And this is the rational, though not the actual view of the case, for that administration and government of science we have spoken of is wont too rigorously to repress its growth.

XCI. And even should the odium I have alluded to be avoided, yet it is sufficient to repress the increase of science that such attempts and industry pass unrewarded; for the cultivation of science and its reward belong not to the same individual. The advancement of science is the work of a powerful genius, the prize and reward belong to the vulgar or to princes, who (with a few exceptions) are scarcely moderately well informed. Nay, such progress is not only deprived of the rewards and beneficence of individuals, but even of popular praise; for it is above the reach of the generality, and easily overwhelmed and extinguished by the winds of common opinions. It is not wonderful, therefore, that little success has attended that which has been little honored.

XCII. But by far the greatest obstacle to the advancement of the sciences, and the undertaking of any new attempt or department, is to be found in men's despair and the idea of impossibility; for men of a prudent and exact turn of thought are altogether diffident in matters of this nature, considering the obscurity of nature, the shortness of life, the deception of the senses, and weakness of the judgment. They think, therefore, that in the revolutions of ages and of the world there are certain floods and ebbs of the sciences, and that they grow and flourish at one time, and wither and fall off at another, that when they have attained a certain degree and condition they can proceed no further.

If, therefore, any one believe or promise greater things, they impute it to an uncurbed and immature mind, and imagine that such efforts begin pleasantly, then become laborious, and end in confusion. And since such thoughts easily enter the minds of men of dignity and excellent judgment, we must really take heed lest we should be captivated by our affection for an excellent and most beautiful object, and relax or diminish the severity of our judgment; and we must diligently examine what gleam of hope shines upon us, and in what direction it manifests itself, so that, banishing her lighter dreams, we may discuss and weigh whatever appears of more sound importance. We must consult the prudence of ordinary life, too, which is diffident upon principle, and in all human matters augurs the worst. Let us, then, speak of hope, especially as we are not vain promisers, nor are willing to enforce or insnare men's judgment, but would rather lead them willingly forward. And although we shall employ the most cogent means of enforcing hope when we bring them to particulars, and especially those which are digested and arranged in our Tables of Invention (the subject partly of the second, but principally of the fourth part of the Instauration), which are, indeed, rather the very object of our hopes than hope itself; yet to proceed more leniently we must treat of the preparation of men's minds, of which the manifestation of hope forms no slight part; for without it all that we have said tends rather to produce a gloom than to encourage activity or quicken the industry of experiment, by causing them to have a worse and more contemptuous opinion of things as they are than they now entertain, and to perceive and feel more thoroughly their unfortunate condition. We must, therefore, disclose and prefix our reasons for not thinking the hope of success improbable, as Columbus, before his wonderful voyage over the Atlantic, gave the reasons of his conviction that new lands and continents might be discovered besides those already known; and these reasons, though at first rejected, were yet proved by subsequent experience, and were the causes and beginnings of the greatest events.

XCIII. Let us begin from God, and show that our pursuit from its exceeding goodness clearly proceeds from him, the author of good and father of light. Now, in all divine works the smallest beginnings lead assuredly to some result, and the remark in spiritual matters that "the kingdom of God cometh without observation," is also found to be true in every great work of Divine Providence, so that everything glides quietly on without confusion or noise, and the matter is achieved before men either think or perceive that it is commenced. Nor should we neglect to mention the prophecy of Daniel, of the last days of the world, "Many shall run to and fro, and knowledge shall be increased,"[56] thus plainly hinting and suggesting that fate (which is Providence) would cause the complete circuit of the globe (now accomplished, or at least going forward by means of so many distant voyages), and the increase of learning to happen at the same epoch.

XCIV. We will next give a most potent reason for hope deduced from the errors of the past, and the ways still unattempted; for well was an ill-governed state thus reproved, "That which is worst with regard to the past should appear most consolatory for the future; for if you had done all that your duty commanded, and your affairs proceeded no better, you could not even hope for their improvement; but since their present unhappy situation is not owing to the force of circumstances, but to your own errors, you have reason to hope that by banishing or correcting the latter you can produce a great change for the better in the former." So if men had, during the many years that have elapsed, adhered to the right way of discovering and cultivating the sciences without being able to advance, it would be assuredly bold and presumptuous to imagine it possible to improve; but if they have mistaken the way and wasted their labor on improper objects, it follows that the difficulty does not arise from things themselves, which are not in our power, but from the human understanding, its practice and application, which is susceptible of remedy and correction. Our best plan, therefore, is to expose these errors; for in proportion as they impeded the past, so do they afford reason to hope for the future. And although we have touched upon them above, yet we think it right to give a brief, bare, and simple enumeration of them in this place.

XCV. Those who have treated of the sciences have been either empirics or dogmatical.[57] The former like ants only heap up and use their store, the latter like spiders spin out their own webs. The bee, a mean between both, extracts matter from the flowers of the garden and the field, but works and fashions it by its own efforts. The true labor of philosophy resembles hers, for it neither relies entirely or principally on the powers of the mind, nor yet lays up in the memory the matter afforded by the experiments of natural history and mechanics in its raw state, but changes and works it in the understanding. We have good reason, therefore, to derive hope from a closer and purer alliance of these faculties (the experimental and rational) than has yet been attempted.

XCVI. Natural philosophy is not yet to be found unadulterated, but is impure and corrupted--by logic in the school of Aristotle, by natural theology in that of Plato,[58] by mathematics in the second school of Plato (that of Proclus and others)[59] which ought rather to terminate natural philosophy than to generate or create it. We may, therefore, hope for better results from pure and unmixed natural philosophy.

XCVII. No one has yet been found possessed of sufficient firmness and severity to resolve upon and undertake the task of entirely abolishing common theories and notions, and applying the mind afresh, when thus cleared and levelled, to particular researches; hence our human reasoning is a mere farrago and crude mass made up of a great deal of credulity and accident, and the puerile notions it originally contracted.

But if a man of mature age, unprejudiced senses, and clear mind, would betake himself anew to experience and particulars, we might hope much more from such a one; in which respect we promise ourselves the fortune of Alexander the Great, and let none accuse us of vanity till they have heard the tale, which is intended to check vanity.

For Æschines spoke thus of Alexander and his exploits: "We live not the life of mortals, but are born at such a period that posterity will relate and declare our prodigies"; as if he considered the exploits of Alexander to be miraculous.

But in succeeding ages[60] Livy took a better view of the fact, and has made some such observation as this upon Alexander: "That he did no more than dare to despise insignificance." So in our opinion posterity will judge of us, that we have achieved no great matters, but only set less account upon what is considered important; for the meantime (as we have before observed) our only hope is in the regeneration of the sciences, by regularly raising them on the foundation of experience and building them anew, which I think none can venture to affirm to have been already done or even thought of.

XCVIII. The foundations of experience (our sole resource) have hitherto failed completely or have been very weak; nor has a store and collection of particular facts, capable of informing the mind or in any way satisfactory, been either sought after or amassed. On the contrary, learned, but idle and indolent, men have received some mere reports of experience, traditions as it were of dreams, as establishing or confirming their philosophy, and have not hesitated to allow them the weight of legitimate evidence. So that a system has been pursued in philosophy with regard to experience resembling that of a kingdom or state which would direct its councils and affairs according to the gossip of city and street politicians, instead of the letters and reports of ambassadors and messengers worthy of credit. Nothing is rightly inquired into, or verified, noted, weighed, or measured, in natural history; indefinite and vague observation produces fallacious and uncertain information. If this appear strange, or our complaint somewhat too unjust (because Aristotle himself, so distinguished a man and supported by the wealth of so great a king, has completed an accurate history of animals, to which others with greater diligence but less noise have made considerable additions, and others again have composed copious histories and notices of plants, metals, and fossils), it will arise from a want of sufficiently attending to and comprehending our present observations; for a natural history compiled on its own account, and one collected for the mind's information as a foundation for philosophy, are two different things. They differ in several respects, but principally in this--the former contains only the varieties of natural species without the experiments of mechanical arts; for as in ordinary life every person's disposition, and the concealed feelings of the mind and passions are most drawn out when they are disturbed--so the secrets of nature betray themselves more readily when tormented by art than when left to their own course. We must begin, therefore, to entertain hopes of natural philosophy then only, when we have a better compilation of natural history, its real basis and support.

XCIX. Again, even in the abundance of mechanical experiments, there is a very great scarcity of those which best inform and assist the understanding. For the mechanic, little solicitous about the investigation of truth, neither directs his attention, nor applies his hand to anything that is not of service to his business. But our hope of further progress in the sciences will then only be well founded, when numerous experiments shall be received and collected into natural history, which, though of no use in themselves, assist materially in the discovery of causes and axioms; which experiments we have termed enlightening, to distinguish them from those which are profitable. They possess this wonderful property and nature, that they never deceive or fail you; for being used only to discover the natural cause of some object, whatever be the result, they equally satisfy your aim by deciding the question.

C. We must not only search for, and procure a greater number of experiments, but also introduce a completely different method, order, and progress of continuing and promoting experience. For vague and arbitrary experience is (as we have observed), mere groping in the dark, and rather astonishes than instructs. But when experience shall proceed regularly and uninterruptedly by a determined rule, we may entertain better hopes of the sciences.

CI. But after having collected and prepared an abundance and store of natural history, and of the experience required for the operations of the understanding or philosophy, still the understanding is as incapable of acting on such materials of itself, with the aid of memory alone, as any person would be of retaining and achieving, by memory, the computation of an almanac. Yet meditation has hitherto done more for discovery than writing, and no experiments have been committed to paper. We cannot, however, approve of any mode of discovery without writing, and when that comes into more general use, we may have further hopes.

CII. Besides this, there is such a multitude and host, as it were, of particular objects, and lying so widely dispersed, as to distract and confuse the understanding; and we can, therefore, hope for no advantage from its skirmishing, and quick movements and incursions, unless we put its forces in due order and array, by means of proper and well arranged, and, as it were, living tables of discovery of these matters, which are the subject of investigation, and the mind then apply itself to the ready prepared and digested aid which such tables afford.

CIII. When we have thus properly and regularly placed before the eyes a collection of particulars, we must not immediately proceed to the investigation and discovery of new particulars or effects, or, at least, if we do so, must not rest satisfied therewith. For, though we do not deny that by transferring the experiments from one art to another (when all the experiments of each have been collected and arranged, and have been acquired by the knowledge, and subjected to the judgment of a single individual), many new experiments may be discovered tending to benefit society and mankind, by what we term literate experience; yet comparatively insignificant results are to be expected thence, while the more important are to be derived from the new light of axioms, deduced by certain method and rule from the above particulars, and pointing out and defining new particulars in their turn. Our road is not a long plain, but rises and falls, ascending to axioms, and descending to effects.

CIV. Nor can we suffer the understanding to jump and fly from particulars to remote and most general axioms (such as are termed the principles of arts and things), and thus prove and make out their intermediate axioms according to the supposed unshaken truth of the former. This, however, has always been done to the present time from the natural bent of the understanding, educated too, and accustomed to this very method, by the syllogistic mode of demonstration. But we can then only augur well for the sciences, when the assent shall proceed by a true scale and successive steps, without interruption or breach, from particulars to the lesser axioms, thence to the intermediate (rising one above the other), and lastly, to the most general. For the lowest axioms differ but little from bare experiment;[61] the highest and most general (as they are esteemed at present), are notional, abstract, and of no real weight. The intermediate are true, solid, full of life, and upon them depend the business and fortune of mankind; beyond these are the really general, but not abstract, axioms, which are truly limited by the intermediate.

We must not then add wings, but rather lead and ballast to the understanding, to prevent its jumping or flying, which has not yet been done; but whenever this takes place, we may entertain greater hopes of the sciences. .

CV. In forming axioms, we must invent a different form of induction from that hitherto in use; not only for the proof and discovery of principles (as they are called), but also of minor, intermediate, and, in short, every kind of axioms. The induction which proceeds by simple enumeration is puerile, leads to uncertain conclusions, and is exposed to danger from one contradictory instance, deciding generally from too small a number of facts, and those only the most obvious. But a really useful induction for the discovery and demonstration of the arts and sciences, should separate nature by proper rejections and exclusions, and then conclude for the affirmative, after collecting a sufficient number of negatives. Now this has not been done, nor even attempted, except perhaps by Plato, who certainly uses this form of induction in some measure, to sift definitions and ideas. But much of what has never yet entered the thoughts of man must necessarily be employed, in order to exhibit a good and legitimate mode of induction or demonstration, so as even to render it essential for us to bestow more pains upon it than have hitherto been bestowed on syllogisms. The assistance of induction is to serve us not only in the discovery of axioms, but also in defining our notions. Much indeed is to be hoped from such an induction as has been described.

CVI. In forming our axioms from induction, we must examine and try whether the axiom we derive be only fitted and calculated for the particular instances from which it is deduced, or whether it be more extensive and general. If it be the latter, we must observe, whether it confirm its own extent and generality by giving surety, as it were, in pointing out new particulars, so that we may neither stop at actual discoveries, nor with a careless grasp catch at shadows and abstract forms, instead of substances of a determinate nature: and as soon as we act thus, well authorized hope may with reason be said to beam upon us.

CVII. Here, too, we may again repeat what we have said above, concerning the extending of natural philosophy and reducing particular sciences to that one, so as to prevent any schism or dismembering of the sciences; without which we cannot hope to advance.

CVIII. Such are the observations we would make in order to remove despair and excite hope, by bidding farewell to the errors of past ages, or by their correction. Let us examine whether there be other grounds for hope. And, first, if many useful discoveries have occurred to mankind by chance or opportunity, without investigation or attention on their part, it must necessarily be acknowledged that much more may be brought to light by investigation and attention, if it be regular and orderly, not hasty and interrupted. For although it may now and then happen that one falls by chance upon something that had before escaped considerable efforts and laborious inquiries, yet undoubtedly the reverse is generally the case. We may, therefore, hope for further, better, and more frequent results from man's reason, industry, method, and application, than from chance and mere animal instinct, and the like, which have hitherto been the sources of invention.

CIX. We may also derive some reason for hope from the circumstance of several actual inventions being of such a nature, that scarcely any one could have formed a conjecture about them previously to their discovery, but would rather have ridiculed them as impossible. For men are wont to guess about new subjects from those they are already acquainted with, and the hasty and vitiated fancies they have thence formed: than which there cannot be a more fallacious mode of reasoning, because much of that which is derived from the sources of things does not flow in their usual channel.

If, for instance, before the discovery of cannon, one had described its effects in the following manner: There is a new invention by which walls and the greatest bulwarks can be shaken and overthrown from a considerable distance; men would have begun to contrive various means of multiplying the force of projectiles and machines by means of weights and wheels, and other modes of battering and projecting. But it is improbable that any imagination or fancy would have hit upon a fiery blast, expanding and developing itself so suddenly and violently, because none would have seen an instance at all resembling it, except perhaps in earthquakes or thunder, which they would have immediately rejected as the great operations of nature, not to be imitated by man.

So, if before the discovery of silk thread, any one had observed, that a species of thread had been discovered, fit for dresses and furniture, far surpassing the thread of worsted or flax in fineness, and at the same time in tenacity, beauty, and softness; men would have begun to imagine something about Chinese plants, or the fine hair of some animals, or the feathers or down of birds, but certainly would never have had an idea of its being spun by a small worm, in so copious a manner, and renewed annually. But if any one had ventured to suggest the silkworm, he would have been laughed at as if dreaming of some new manufacture from spiders.

So again, if before the discovery of the compass, any one had said, that an instrument had been invented, by which the quarters and points of the heavens could be exactly taken and distinguished, men would have entered into disquisitions on the refinement of astronomical instruments, and the like, from the excitement of their imaginations; but the thought of anything being discovered, which, not being a celestial body, but a mere mineral or metallic substance, should yet in its motion agree with that of such bodies, would have appeared absolutely incredible. Yet were these facts, and the like (unknown for so many ages) not discovered at last either by philosophy or reasoning, but by chance and opportunity; and (as we have observed), they are of a nature most heterogeneous, and remote from what was hitherto known, so that no previous knowledge could lead to them.

We may, therefore, well hope[62] that many excellent and useful matters are yet treasured up in the bosom of nature, bearing no relation or analogy to our actual discoveries, but out of the common track of our imagination, and still undiscovered, and which will doubtless be brought to light in the course and lapse of years, as the others have been before them; but in the way we now point out, they may rapidly and at once be both represented and anticipated.

CX. There are, moreover, some inventions which render it probable that men may pass and hurry over the most noble discoveries which lie immediately before them. For however the discovery of gunpowder, silk, the compass, sugar, paper, or the like, may appear to depend on peculiar properties of things and nature, printing at least involves no contrivance which is not clear and almost obvious. But from want of observing that although the arrangement of the types of letters required more trouble than writing with the hand, yet these types once arranged serve for innumerable impressions, while manuscript only affords one copy; and again, from want of observing that ink might be thickened so as to stain without running (which was necessary, seeing the letters face upward, and the impression is made from above), this most beautiful invention (which assists so materially the propagation of learning) remained unknown for so many ages.

The human mind is often so awkward and ill-regulated in the career of invention that it is at first diffident, and then despises itself. For it appears at first incredible that any such discovery should be made, and when it has been made, it appears incredible that it should so long have escaped men's research. All which affords good reason for the hope that a vast mass of inventions yet remains, which may be deduced not only from the investigation of new modes of operation, but also from transferring, comparing, and applying these already known, by the method of what we have termed literate experience.

CXI. Nor should we omit another ground of hope. Let men only consider (if they will) their infinite expenditure of talent, time, and fortune, in matters and studies of far inferior importance and value; a small portion of which applied to sound and solid learning would be sufficient to overcome every difficulty. And we have thought right to add this observation, because we candidly own that such a collection of natural and experimental history as we have traced in our own mind, and as is really necessary, is a great and as it were royal work, requiring much labor and expense.

CXII. In the meantime let no one be alarmed at the multitude of particulars, but rather inclined to hope on that very account. For the particular phenomena of the arts and nature are in reality but as a handful, when compared with the fictions of the imagination removed and separated from the evidence of facts. The termination of our method is clear, and I had almost said near at hand; the other admits of no termination, but only of infinite confusion. For men have hitherto dwelt but little, or rather only slightly touched upon experience, while they have wasted much time on theories and the fictions of the imagination. If we had but any one who could actually answer our interrogations of nature, the invention of all causes and sciences would be the labor of but a few years.

CXIII. We think some ground of hope is afforded by our own example, which is not mentioned for the sake of boasting, but as a useful remark. Let those who distrust their own powers observe myself, one who have among my contemporaries been the most engaged in public business, who am not very strong in health (which causes a great loss of time), and am the first explorer of this course, following the guidance of none, nor even communicating my thoughts to a single individual; yet having once firmly entered in the right way, and submitting the powers of my mind to things, I have somewhat advanced (as I make bold to think) the matter I now treat of. Then let others consider what may be hoped from men who enjoy abundant leisure, from united labors, and the succession of ages, after these suggestions on our part, especially in a course which is not confined, like theories, to individuals, but admits of the best distribution and union of labor and effect, particularly in collecting experiments. For men will then only begin to know their own power, when each performs a separate part, instead of undertaking in crowds the same work.

CXIV. Lastly, though a much more faint and uncertain breeze of hope were to spring up from our new continent, yet we consider it necessary to make the experiment, if we would not show a dastard spirit. For the risk attending want of success is not to be compared with that of neglecting the attempt; the former is attended with the loss of a little human labor, the latter with that of an immense benefit. For these and other reasons it appears to us that there is abundant ground to hope, and to induce not only those who are sanguine to make experiment, but even those who are cautious and sober to give their assent.

CXV. Such are the grounds for banishing despair, hitherto one of the most powerful causes of the delay and restraint to which the sciences have been subjected; in treating of which we have at the same time discussed the signs and causes of the errors, idleness, and ignorance that have prevailed; seeing especially that the more refined causes, which are not open to popular judgment and observation, may be referred to our remarks on the idols of the human mind.

Here, too, we should close the demolishing branch of our Instauration, which is comprised in three confutations: 1, the confutation of natural human reason left to itself; 2, the confutation of demonstration; 3, the confutation of theories, or received systems of philosophy and doctrines. Our confutation has followed such a course as was open to it, namely, the exposing of the signs of error, and the producing evidence of the causes of it: for we could adopt no other, differing as we do both in first principles and demonstrations from others.

It is time for us therefore to come to the art itself, and the rule for the interpretation of nature: there is, however, still something which must not be passed over. For the intent of this first book of aphorisms being to prepare the mind for understanding, as well as admitting, what follows, we must now, after having cleansed, polished, and levelled its surface, place it in a good position, and as it were a benevolent aspect toward our propositions; seeing that prejudice in new matters may be produced not only by the strength of preconceived notions, but also by a false anticipation or expectation of the matter proposed. We shall therefore endeavor to induce good and correct opinions of what we offer, although this be only necessary for the moment, and as it were laid out at interest, until the matter itself be well understood.

CXVI. First, then, we must desire men not to suppose that we are ambitious of founding any philosophical sect, like the ancient Greeks, or some moderns, as Telesius, Patricius, and Severinus.[63] For neither is this our intention, nor do we think that peculiar abstract opinions on nature and the principles of things are of much importance to men's fortunes, since it were easy to revive many ancient theories, and to introduce many new ones; as, for instance, many hypotheses with regard to the heavens can be formed, differing in themselves, and yet sufficiently according with the phenomena.

We bestow not our labor on such theoretical and, at the same time, useless topics. On the contrary, our determination is that of trying, whether we can lay a firmer foundation, and extend to a greater distance the boundaries of human power and dignity. And although here and there, upon some particular points, we hold (in our own opinion) more true and certain, and I might even say, more advantageous tenets than those in general repute (which we have collected in the fifth part of our Instauration), yet we offer no universal or complete theory. The time does not yet appear to us to be arrived, and we entertain no hope of our life being prolonged to the completion of the sixth part of the Instauration (which is destined for philosophy discovered by the interpretation of nature), but are content if we proceed quietly and usefully in our intermediate pursuit, scattering, in the meantime, the seeds of less adulterated truth for posterity, and, at least, commence the great work.

CXVII. And, as we pretend not to found a sect, so do we neither offer nor promise particular effects; which may occasion some to object to us, that since we so often speak of effects, and consider everything in its relation to that end, we ought also to give some earnest of producing them. Our course and method, however (as we have often said, and again repeat), is such as not to deduce effects from effects, nor experiments from experiments (as the empirics do), but in our capacity of legitimate interpreters of nature, to deduce causes and axioms from effects and experiments; and new effects and experiments from those causes and axioms.

And although any one of moderate intelligence and ability will observe the indications and sketches of many noble effects in our tables of inventions (which form the fourth part of the Instauration), and also in the examples of particular instances cited in the second part, as well as in our observations on history (which is the subject of the third part); yet we candidly confess that our present natural history, whether compiled from books or our own inquiries, is not sufficiently copious and well ascertained to satisfy, or even assist, a proper interpretation.

If, therefore, there be any one who is more disposed and prepared for mechanical art, and ingenious in discovering effects, than in the mere management of experiment, we allow him to employ his industry in gathering many of the fruits of our history and tables in this way, and applying them to effects, receiving them as interest till he can obtain the principal. For our own part, having a greater object in view, we condemn all hasty and premature rest in such pursuits as we would Atalanta's apple (to use a common allusion of ours); for we are not childishly ambitious of golden fruit, but use all our efforts to make the course of art outstrip nature, and we hasten not to reap moss or the green blade, but wait for a ripe harvest.

CXVIII. There will be some, without doubt, who, on a perusal of our history and tables of invention, will meet with some uncertainty, or perhaps fallacy, in the experiments themselves, and will thence perhaps imagine that our discoveries are built on false foundations and principles. There is, however, really nothing in this, since it must needs happen in beginnings.[64] For it is the same as if in writing or printing one or two letters were wrongly turned or misplaced, which is no great inconvenience to the reader, who can easily by his own eye correct the error; let men in the same way conclude, that many experiments in natural history may be erroneously believed and admitted, which are easily expunged and rejected afterward, by the discovery of causes and axioms. It is, however, true, that if these errors in natural history and experiments become great, frequent, and continued, they cannot be corrected and amended by any dexterity of wit or art. If then, even in our natural history, well examined and compiled with such diligence, strictness, and (I might say) reverential scruples, there be now and then something false and erroneous in the details, what must we say of the common natural history, which is so negligent and careless when compared with ours? or of systems of philosophy and the sciences, based on such loose soil (or rather quicksand)? Let none then be alarmed by such observations.

CXIX. Again, our history and experiments will contain much that is light and common, mean and illiberal, too refined and merely speculative, and, as it were, of no use, and this perhaps may divert and alienate the attention of mankind.

With regard to what is common; let men reflect, that they have hitherto been used to do nothing but refer and adapt the causes of things of rare occurrence to those of things which more frequently happen, without any investigation of the causes of the latter, taking them for granted and admitted.

Hence, they do not inquire into the causes of gravity, the rotation of the heavenly bodies, heat, cold, light, hardness, softness, rarity, density, liquidity, solidity, animation, inanimation, similitude, difference, organic formation, but taking them to be self-evident, manifest, and admitted, they dispute and decide upon other matters of less frequent and familiar occurrence.

But we (who know that no judgment can be formed of that which is rare or remarkable, and much less anything new brought to light, without a previous regular examination and discovery of the causes of that which is common, and the causes again of those causes) are necessarily compelled to admit the most common objects into our history. Besides, we have observed that nothing has been so injurious to philosophy as this circumstance, namely, that familiar and frequent objects do not arrest and detain men's contemplation, but are carelessly admitted, and their causes never inquired after; so that information on unknown subjects is not more often wanted than attention to those which are known.

CXX. With regard to the meanness, or even the filthiness of particulars, for which (as Pliny observes), an apology is requisite, such subjects are no less worthy of admission into natural history than the most magnificent and costly; nor do they at all pollute natural history, for the sun enters alike the palace and the privy, and is not thereby polluted. We neither dedicate nor raise a capitol or pyramid to the pride of man, but rear a holy temple in his mind, on the model of the universe, which model therefore we imitate. For that which is deserving of existence is deserving of knowledge, the image of existence. Now the mean and splendid alike exist. Nay, as the finest odors are sometimes produced from putrid matter (such as musk and civet), so does valuable light and information emanate from mean and sordid instances. But we have already said too much, for such fastidious feelings are childish and effeminate.

CXXI. The next point requires a more accurate consideration, namely, that many parts of our history will appear to the vulgar, or even any mind accustomed to the present state of things, fantastically and uselessly refined. Hence, we have in regard to this matter said from the first, and must again repeat, that we look for experiments that shall afford light rather than profit, imitating the divine creation, which, as we have often observed, only produced light on the first day, and assigned that whole day to its creation, without adding any material work.

If any one, then, imagine such matters to be of no use, he might equally suppose light to be of no use, because it is neither solid nor material. For, in fact, the knowledge of simple natures, when sufficiently investigated and defined, resembles light, which, though of no great use in itself, affords access to the general mysteries of effects, and with a peculiar power comprehends and draws with it whole bands and troops of effects, and the sources of the most valuable axioms. So also the elements of letters have of themselves separately no meaning, and are of no use, yet are they, as it were, the original matter in the composition and preparation of speech. The seeds of substances, whose effect is powerful, are of no use except in their growth, and the scattered rays of light itself avail not unless collected.

But if speculative subtilties give offence, what must we say of the scholastic philosophers who indulged in them to such excess? And those subtilties were wasted on words, or, at least, common notions (which is the same thing), not on things or nature, and alike unproductive of benefit in their origin and their consequences: in no way resembling ours, which are at present useless, but in their consequences of infinite benefit. Let men be assured that all subtile disputes and discursive efforts of the mind are late and preposterous, when they are introduced subsequently to the discovery of axioms, and that their true, or, at any rate, chief opportunity is, when experiment is to be weighed and axioms to be derived from it. They otherwise catch and grasp at nature, but never seize or detain her: and we may well apply to nature that which has been said of opportunity or fortune, that she wears a lock in front, but is bald behind.

In short, we may reply decisively to those who despise any part of natural history as being vulgar, mean, or subtile, and useless in its origin, in the words of a poor woman to a haughty prince,[65] who had rejected her petition as unworthy, and beneath the dignity of his majesty: "Then cease to reign"; for it is quite certain that the empire of nature can neither be obtained nor administered by one who refuses to pay attention to such matters as being poor and too minute.

CXXII. Again, it may be objected to us as being singular and harsh, that we should with one stroke and assault, as it were, banish all authorities and sciences, and that too by our own efforts, without requiring the assistance and support of any of the ancients.

Now we are aware, that had we been ready to act otherwise than sincerely, it was not difficult to refer our present method to remote ages, prior to those of the Greeks (since the sciences in all probability flourished more in their natural state, though silently, than when they were paraded with the fifes and trumpets of the Greeks); or even (in parts, at least) to some of the Greeks themselves, and to derive authority and honor from thence; as men of no family labor to raise and form nobility for themselves in some ancient line, by the help of genealogies. Trusting, however, to the evidence of facts, we reject every kind of fiction and imposture; and think it of no more consequence to our subject, whether future discoveries were known to the ancients, and set or rose according to the vicissitudes of events and lapse of ages, than it would be of importance to mankind to know whether the new world be the island of Atlantis,[66] and known to the ancients, or be now discovered for the first time.

With regard to the universal censure we have bestowed, it is quite clear, to any one who properly considers the matter, that it is both more probable and more modest than any partial one could have been. For if the errors had not been rooted in the primary notions, some well conducted discoveries must have corrected others that were deficient. But since the errors were fundamental, and of such a nature, that men may be said rather to have neglected or passed over things, than to have formed a wrong or false judgment of them, it is little to be wondered at, that they did not obtain what they never aimed at, nor arrive at a goal which they had not determined, nor perform a course which they had neither entered upon nor adhered to.

With regard to our presumption, we allow that if we were to assume a power of drawing a more perfect straight line or circle than any one else, by superior steadiness of hand or acuteness of eye, it would lead to a comparison of talent; but if one merely assert that he can draw a more perfect line or circle with a ruler or compasses, than another can by his unassisted hand or eye, he surely cannot be said to boast of much. Now this applies not only to our first original attempt, but also to those who shall hereafter apply themselves to the pursuit. For our method of discovering the sciences merely levels men's wits, and leaves but little to their superiority, since it achieves everything by the most certain rules and demonstrations. Whence (as we have often observed), our attempt is to be attributed to fortune rather than talent, and is the offspring of time rather than of wit. For a certain sort of chance has no less effect upon our thoughts than on our acts and deeds.

CXXIII. We may, therefore, apply to ourselves the joke of him who said, that water and wine drinkers could not think alike,[67] especially as it hits the matter so well. For others, both ancients and moderns, have in the sciences drank a crude liquor like water, either flowing of itself from the understanding, or drawn up by logic as the wheel draws up the bucket. But we drink and pledge others with a liquor made of many well-ripened grapes, collected and plucked from particular branches, squeezed in the press, and at last clarified and fermented in a vessel. It is not, therefore, wonderful that we should not agree with others.

CXXIV. Another objection will without doubt be made, namely, that we have not ourselves established a correct, or the best goal or aim of the sciences (the very defect we blame in others). For they will say that the contemplation of truth is more dignified and exalted than any utility or extent of effects; but that our dwelling so long and anxiously on experience and matter, and the fluctuating state of particulars, fastens the mind to earth, or rather casts it down into an abyss of confusion and disturbance, and separates and removes it from a much more divine state, the quiet and tranquillity of abstract wisdom. We willingly assent to their reasoning, and are most anxious to effect the very point they hint at and require. For we are founding a real model of the world in the understanding, such as it is found to be, not such as man's reason has distorted. Now this cannot be done without dissecting and anatomizing the world most diligently; but we declare it necessary to destroy completely the vain, little and, as it were, apish imitations of the world, which have been formed in various systems of philosophy by men's fancies. Let men learn (as we have said above) the difference that exists between the idols of the human mind and the ideas of the divine mind. The former are mere arbitrary abstractions; the latter the true marks of the Creator on his creatures, as they are imprinted on, and defined in matter, by true and exquisite touches. Truth, therefore, and utility, are here perfectly identical, and the effects are of more value as pledges of truth than from the benefit they confer on men.

CXXV. Others may object that we are only doing that which has already been done, and that the ancients followed the same course as ourselves. They may imagine, therefore, that, after all this stir and exertion, we shall at last arrive at some of those systems that prevailed among the ancients: for that they, too, when commencing their meditations, laid up a great store of instances and particulars, and digested them under topics and titles in their commonplace books, and so worked out their systems and arts, and then decided upon what they discovered, and related now and then some examples to confirm and throw light upon their doctrine; but thought it superfluous and troublesome to publish their notes, minutes, and commonplaces, and therefore followed the example of builders who remove the scaffolding and ladders when the building is finished. Nor can we indeed believe the case to have been otherwise. But to any one, not entirely forgetful of our previous observations, it will be easy to answer this objection or rather scruple; for we allow that the ancients had a particular form of investigation and discovery, and their writings show it. But it was of such a nature, that they immediately flew from a few instances and particulars (after adding some common notions, and a few generally received opinions most in vogue) to the most general conclusions or the principles of the sciences, and then by their intermediate propositions deduced their inferior conclusions, and tried them by the test of the immovable and settled truth of the first, and so constructed their art. Lastly, if some new particulars and instances were brought forward, which contradicted their dogmas, they either with great subtilty reduced them to one system, by distinctions or explanations of their own rules, or got rid of them clumsily as exceptions, laboring most pertinaciously in the meantime to accommodate the causes of such as were not contradictory to their own principles. Their natural history and their experience were both far from being what they ought to have been, and their flying off to generalities ruined everything.

CXXVI. Another objection will be made against us, that we prohibit decisions and the laying down of certain principles, till we arrive regularly at generalities by the intermediate steps, and thus keep the judgment in suspense and lead to uncertainty. But our object is not uncertainty but fitting certainty, for we derogate not from the senses but assist them, and despise not the understanding but direct it. It is better to know what is necessary, and not to imagine we are fully in possession of it, than to imagine that we are fully in possession of it, and yet in reality to know nothing which we ought.

CXXVII. Again, some may raise this question rather than objection, whether we talk of perfecting natural philosophy alone according to our method, or the other sciences also, such as logic, ethics, politics. We certainly intend to comprehend them all. And as common logic, which regulates matters by syllogisms, is applied not only to natural, but also to every other science, so our inductive method likewise comprehends them all.[68] For we form a history and tables of invention for anger, fear, shame, and the like, and also for examples in civil life, and the mental operations of memory, composition, division, judgment, and the rest, as well as for heat and cold, light, vegetation, and the like. But since our method of interpretation, after preparing and arranging a history, does not content itself with examining the operations and disquisitions of the mind like common logic, but also inspects the nature of things, we so regulate the mind that it may be enabled to apply itself in every respect correctly to that nature. On that account we deliver numerous and various precepts in our doctrine of interpretation, so that they may apply in some measure to the method of discovering the quality and condition of the subject matter of investigation.

CXXVIII. Let none even doubt whether we are anxious to destroy and demolish the philosophy, arts, and sciences, which are now in use. On the contrary, we readily cherish their practice, cultivation, and honor; for we by no means interfere to prevent the prevalent system from encouraging discussion, adorning discourses, or bèing employed serviceable in the chair of the professor or the practice of common life, and being taken, in short, by general consent as current coin. Nay, we plainly declare, that the system we offer will not be very suitable for such purposes, not being easily adapted to vulgar apprehensions, except by effects and works. To show our sincerity in professing our regard and friendly disposition toward the received sciences, we can refer to the evidence of our published writings (especially our books on the Advancement of Learning). We will not, therefore, endeavor to evince it any further by words; but content ourselves with steadily and professedly premising, that no great progress can be made by the present methods in the theory or contemplation of science, and that they cannot be made to produce any very abundant effects.

CXXIX. It remains for us to say a few words on the excellence of our proposed end. If we had done so before, we might have appeared merely to express our wishes, but now that we have excited hope and removed prejudices, it will perhaps have greater weight. Had we performed and completely accomplished the whole, without frequently calling in others to assist in our labors, we should then have refrained from saying any more, lest we should be thought to extol our own deserts. Since, however, the industry of others must be quickened, and their courage roused and inflamed, it is right to recall some points to their memory.

First, then, the introduction of great inventions appears one of the most distinguished of human actions, and the ancients so considered it; for they assigned divine honors to the authors of inventions, but only heroic honors to those who displayed civil merit (such as the founders of cities and empire legislators, the deliverers of their country from lasting misfortunes, the quellers of tyrants, and the like). And if any one rightly compare them, he will find the judgment of antiquity to be correct; for the benefits derived from inventions may extend to mankind in general, but civil benefits to particular spots alone; the latter, moreover, last but for a time, the former forever. Civil reformation seldom is carried on without violence and confusion, while inventions are a blessing and a benefit without injuring or afflicting any.

Inventions are also, as it were, new creations and imitations of divine works, as was expressed by the poet:[69]

"Primum frugiferos fœtus mortalibus ægris   Dididerant quondam præstanti nomine Athenæ   Et recreaverunt vitam legesque rogarunt."

And it is worthy of remark in Solomon, that while he flourished in the possession of his empire, in wealth, in the magnificence of his works, in his court, his household, his fleet, the splendor of his name, and the most unbounded admiration of mankind, he still placed his glory in none of these, but declared[70] that it is the glory of God to conceal a thing, but the glory of a king to search it out.

Again, let any one but consider the immense difference between men's lives in the most polished countries of Europe, and in any wild and barbarous region of the new Indies, he will think it so great, that man may be said to be a god unto man, not only on account of mutual aid and benefits, but from their comparative states--the result of the arts, and not of the soil or climate.

Again, we should notice the force, effect, and consequences of inventions, which are nowhere more conspicuous than in those three which were unknown to the ancients; namely, printing, gunpowder, and the compass. For these three have changed the appearance and state of the whole world: first in literature, then in warfare, and lastly in navigation; and innumerable changes have been thence derived, so that no empire, sect, or star, appears to have exercised a greater power and influence on human affairs than these mechanical discoveries.

It will, perhaps, be as well to distinguish three species and degrees of ambition. First, that of men who are anxious to enlarge their own power in their country, which is a vulgar and degenerate kind; next, that of men who strive to enlarge the power and empire of their country over mankind, which is more dignified but not less covetous; but if one were to endeavor to renew and enlarge the power and empire of mankind in general over the universe, such ambition (if it may be so termed) is both more sound and more noble than the other two. Now the empire of man over things is founded on the arts and sciences alone, for nature is only to be commanded by obeying her.

Besides this, if the benefit of any particular invention has had such an effect as to induce men to consider him greater than a man, who has thus obliged the whole race, how much more exalted will that discovery be, which leads to the easy discovery of everything else! Yet (to speak the truth) in the same manner as we are very thankful for light which enables us to enter on our way, to practice arts, to read, to distinguish each other, and yet sight is more excellent and beautiful than the various uses of light; so is the contemplation of things as they are, free from superstition or imposture, error or confusion, much more dignified in itself than all the advantage to be derived from discoveries.

Lastly, let none be alarmed at the objection of the arts and sciences becoming depraved to malevolent or luxurious purposes and the like, for the same can be said of every worldly good; talent, courage, strength, beauty, riches, light itself, and the rest. Only let mankind regain their rights over nature, assigned to them by the gift of God, and obtain that power, whose exercise will be governed by right reason and true religion.

CXXX. But it is time for us to lay down the art of interpreting nature, to which we attribute no absolute necessity (as if nothing could be done without it) nor perfection, although we think that our precepts are most useful and correct. For we are of opinion, that if men had at their command a proper history of nature and experience, and would apply themselves steadily to it, and could bind themselves to two things: 1, to lay aside received opinions and notions; 2, to restrain themselves, till the proper season, from generalization, they might, by the proper and genuine exertion of their minds, fall into our way of interpretation without the aid of any art. For interpretation is the true and natural act of the mind, when all obstacles are removed: certainly, however, everything will be more ready and better fixed by our precepts.

Yet do we not affirm that no addition can be made to them; on the contrary, considering the mind in its connection with things, and not merely relatively to its own powers, we ought to be persuaded that the art of invention can be made to grow with the inventions themselves.

Footnotes:

2. Bacon uses the term in its ancient sense, and means one who, knowing the occult properties of bodies, is able to startle the ignorant by drawing out of them wonderful and unforeseen changes. See the 85th aphorism of this book, and the 5th cap. book iii. of the De Augmentis Scientiarum, where he

speaks more clearly.--Ed.

3. By this term axiomata, Bacon here speaks of general principles, or universal laws. In the 19th aphorism he employs the term to express any proposition collected from facts by induction, and thus fitted to become the starting-point of deductive reasoning. In the last and more rigorous sense of the term, Bacon held they arose from experience. See Whewell's "Philosophy of the Inductive Sciences," vol. i. p. 74; and Mill's "Logic," vol. i. p. 311; and the June "Quarterly," 1841, for the modern phase of the discussion.--Ed.

4. Bacon here attributes to the Aristotelian logic the erroneous consequences which sprung out of its abuse. The demonstrative forms it exhibits, whether verbally or mathematically expressed, are necessary to the support, verification, and extension of induction, and when the propositions they embrace are founded on an accurate and close observation of facts, the conclusions to which they lead, even in moral science, may be regarded as certain as the facts wrested out of nature by direct experiment. In physics such forms are absolutely required to generalize the results of experience, and to connect intermediate axioms with laws still more general, as is sufficiently attested by the fact, that no science since Bacon's day has ceased to be experimental by the mere method of induction, and that all become exact only so far as they rise above experience, and connect their isolated phenomena with general laws by the principles of deductive reasoning. So far, then, are these forms from being useless, that they are absolutely essential to the advancement of the sciences, and in no case can be looked on as detrimental, except when obtruded in the place of direct experiment, or employed as a means of deducing conclusions about nature from imaginary hypotheses and abstract conceptions. This had been unfortunately the practice of the Greeks. From the rapid development geometry received in their hands, they imagined the same method would lead to results equally brilliant in natural science, and snatching up some abstract principle, which they carefully removed from the test of experiment, imagined they could reason out from it all the laws and external appearances of the universe. The scholastics were impelled along the same path, not only by precedent, but by profession. Theology was the only science which received from them a consistent development, and the à priori grounds on which it rested prevented them from employing any other method in the pursuit of natural phenomena. Thus, forms of demonstration, in themselves accurate, and of momentous value in their proper sphere, became confounded with fable, and led men into the idea they were exploring truth when they were only accurately deducing error from error. One principle ever so slightly deflected, like a false quantity in an equation, could be sufficient to infect the whole series of conclusions of which it was the base; and though the philosopher might subsequently deduce a thousand consecutive inferences with the utmost accuracy or precision, he would only succeed in drawing out very methodically nine hundred and ninety-nine errors.--Ed.

5. It would appear from this and the two preceding aphorisms, that Bacon fell into the error of denying the utility of the syllogism in the very part of inductive science where it is essentially required. Logic, like mathematics, is purely a formal process, and must, as the scaffolding to the building, be employed to arrange facts in the structure of a science, and not to form any portion of its groundwork, or to supply the materials of which the system is to be composed. The word syllogism, like most other psychological terms, has no fixed or original signification, but is sometimes employed, as it was by the Greeks, to denote general reasoning, and at others to point out the formal method of deducing a particular inference from two or more general propositions. Bacon does not confine the term within the boundaries of express definition, but leaves us to infer that he took it in the latter sense, from his custom of associating the term with the wranglings of the schools. The scholastics, it is true, abused the deductive syllogism, by employing it in its naked, skeleton-like form, and confounding it with the whole breadth of logical theory; but their errors are not to be visited on Aristotle, who never dreamed of playing with formal syllogisms, and, least of all, mistook the descending for the ascending series of inference. In our mind we are of accord with the Stagyrite, who propounds, as far as we can interpret him, two modes of investigating truth--the one by which we ascend from particular and singular facts to general laws and axioms, and the other by which we descend from universal propositions to the individual cases which they virtually include. Logic, therefore, must equally vindicate the formal purity of the synthetic illation by which it ascends to the whole, as the analytic process by which it descends to the parts. The deductive and inductive syllogism are of equal significance in building up any body of truth, and whoever restricts logic to either process, mistakes one-half of its province for the whole; and if he acts upon his error, will paralyze his methods, and strike the noblest part of science with sterility.--Ed.

6. The Latin is, ad ea quæ revera sunt naturæ notiora. This expression, naturæ notiora, naturæ notior, is so frequently employed by Bacon, that we may conclude it to point to some distinguishing feature in the Baconian physics. It properly refers to the most evident principles and laws of nature, and springs from that system which regards the material universe as endowed with intelligence, and acting according to rules either fashioned or clearly understood by itself.--Ed.

7. This Borgia was Alexander VI., and the expedition alluded to that in which Charles VIII.

overran the Italian peninsula in five months. Bacon uses the same illustration in concluding his survey of natural philosophy, in the second book of the "De Augmentis."--Ed.

8. Ratio eorum qui acatalepsiam tenuerunt. Bacon alludes to the members of the later academy, who held the ἀκατάληψια, or the impossibility of comprehending anything. His translator, however, makes him refer to the sceptics, who neither dogmatized about the known or the unknown, but simply held, that as all knowledge was relative, πρὸς πάντα τι, man could never arrive at absolute truth, and therefore could not with certainty affirm or deny anything.--Ed.

9. It is argued by Hallam, with some appearance of truth, that idols is not the correct translation of εἴδωλα, from which the original idola is manifestly derived; but that Bacon used it in the literal sense attached to it by the Greeks, as a species of illusion, or false appearance, and not as a species of divinity before which the mind bows down. If Hallam be right, Bacon is saved from the odium of an analogy which his foreign commentators are not far wrong in denouncing as barbarous; but this service is rendered at the expense of the men who have attached an opposite meaning to the word, among whom are Brown, Playfair and Dugald Stewart.--Ed.

10. We cannot see how these idols have less to do with sophistical paralogisms than with natural philosophy. The process of scientific induction involves only the first elements of reasoning, and presents such a clear and tangible surface, as to allow no lurking-place for prejudice; while questions of politics and morals, to which the deductive method, or common logic, as Bacon calls it, is peculiarly applicable, are ever liable to be swayed or perverted by the prejudices he enumerates. After mathematics, physical science is the least amenable to the illusions of feeling; each portion having been already tested by experiment and observation, is fitted into its place in the system, with all the rigor of the geometrical method; affection or prejudice cannot, as in matters of taste, history or religion, select fragmentary pieces, and form a system of their own. The whole must be admitted, or the structure of authoritative reason razed to the ground. It is needless to say that the idols enumerated present only another interpretation of the substance of logical fallacies.--Ed.

11. The propensity to this illusion may be viewed in the spirit of system, or hasty generalization, which is still one of the chief obstacles in the path of modern science.--Ed.

12. Though Kepler had, when Bacon wrote this, already demonstrated his three great laws concerning the elliptical path of the planets, neither Bacon nor Descartes seems to have known or assented to his discoveries. Our author deemed the startling astronomical announcements of his time to be mere theoretic solutions of the phenomena of the heavens, not so perfect as those advanced by antiquity, but still deserving a praise for the ingenuity displayed in their contrivance. Bacon believed a hundred such systems might exist, and though true in their explanation of phenomena, yet might all more or less differ, according to the preconceived notions which their framers brought to the survey of the heavens. He even thought he might put in his claim to the notice of posterity for his astronomical ingenuity, and, as Ptolemy had labored by means of epicycles and eccentrics, and Kepler with ellipses, to explain the laws of planetary motion, Bacon thought the mystery would unfold itself quite as philosophically through spiral labyrinths and serpentine lines. What the details of his system were, we are left to conjecture, and that from a very meagre but naïve account of one of his inventions which he has left in his Miscellany MSS.--Ed.

13. Hinc elementum ignis cum orbe suo introductum est. Bacon saw in fire the mere result of a certain combination of action, and was consequently led to deny its elementary character. The ancient physicists attributed an orbit to each of the four elements, into which they resolved the universe, and supposed their spheres to involve each other. The orbit of the earth was in the centre, that of fire at the circumference. For Bacon's inquisition into the nature of heat, and its complete failure, see the commencement of the second book of the Novum Organum.--Ed.

14. Robert Fludd is the theorist alluded to, who had supposed the gravity of the earth to be ten times heavier than water, that of water ten times heavier than air, and that of air ten times heavier than fire.--Ed.

15. Diagoras. The same allusion occurs in the second part of the Advancement of Learning, where Bacon treats of the idols of the mind.

16. A scholastic term, to signify the two eternities of past and future duration, that stretch out on both sides of the narrow isthmus (time) occupied by man. It must be remembered that Bacon lived before the doctrine of limits gave rise to the higher calculus, and therefore could have no conception of different denominations of infinities: on the other hand he would have thought the man insane who should have talked to him about lines infinitely great, inclosing angles infinitely little; that a right line, which is a right line so long as it is finite, by changing infinitely little its direction, becomes an infinite curve, and that a curve may become infinitely less than another curve; that there are infinite squares and infinite cubes, and infinites of infinites, all greater than one another, and the last but one of which is nothing in comparison with the last. Yet half a century sufficed from Bacon's time, to make this nomenclature, which would have appeared to him the excess of frenzy, not only reasonable but necessary, to grasp the higher demonstrations of physical science.--Ed.

17. Spinoza, in his letter to Oldenberg (Op. Posth. p. 398), considers this aphorism based on a wrong conception of the origin of error, and, believing it to be fundamental, was led to reject Bacon's method altogether. Spinoza refused to acknowledge in man any such thing as a will, and resolved all his volitions into particular acts, which he considered to be as fatally determined by a chain of physical causes as any effects in nature.--Ed.

18. Operatio spirituum in corporibus tangibilibus. Bacon distinguished with the schools the gross and tangible parts of bodies, from such as were volatile and intangible. These, in conformity with the scholastic language, he terms spirits, and frequently returns to their operations in the 2d book.--Ed.

19. Democritus, of Abdera, a disciple of Leucippus, born B.C. 470, died 360; all his works are destroyed. He is said to be the author of the doctrine of atoms: he denied the immortality of the soul, and first taught that the milky way was occasioned by a confused light from a multitude of stars. He may be considered as the parent of experimental philosophy, in the prosecution of which he was so ardent as to declare that he would prefer the discovery of one of the causes of natural phenomena, to the possession of the diadem of Persia. Democritus imposed on the blind credulity of his contemporaries, and, like Roger Bacon, astonished them by his inventions.--Ed.

20. The Latin is actus purus, another scholastic expression to denote the action of the substance, which composes the essence of the body apart from its accidental qualities. For an exposition of the various kinds of motions he contemplates, the reader may refer to the 48th aphorism of the 2d book.--Ed.

21. The scholastics after Aristotle distinguished in a subject three modes of beings: viz., the power or faculty, the act, and the habitude, or in other words that which is able to exist, what exists actually, and what continues to exist. Bacon means that is necessary to fix our attention not on that which can or ought to be, but on that which actually is; not on the right, but on the fact.--Ed.

22. The inference to be drawn from this is to suspect that kind of evidence which is most consonant to our inclinations, and not to admit any notion as real except we can base it firmly upon that kind of demonstration which is peculiar to the subject, not to our impression. Sometimes the mode of proof may be consonant to our inclinations, and to the subject at the same time, as in the case of Pythagoras, when he applied his beloved numbers to the solution of astronomical phenomena; or in that of Descartes, when he reasoned geometrically concerning the nature of the soul. Such examples cannot be censured with justice, inasmuch as the methods pursued were adapted to the end of the inquiry. The remark in the text can only apply to those philosophers who attempt to build up a moral or theological system by the instruments of induction alone, or who rush, with the geometrical axiom, and the à priori syllogism, to the investigation of nature. The means in such cases are totally inadequate to the object in view.--Ed.

23. Gilbert lived toward the close of the sixteenth century, and was court physician to both Elizabeth and James. In his work alluded to in the text he continually asserts the advantages of the experimental over the à priori method in physical inquiry, and succeeded when his censor failed in giving a practical example of the utility of his precepts. His "De Magnete" contains all the fundamental parts of the science, and these so perfectly treated, that we have nothing to add to them at the present day.

Gilbert adopted the Copernican system, and even spoke of the contrary theory as utterly absurd, grounding his argument on the vast velocities which such a supposition requires us to ascribe to the heavenly bodies.--Ed.

24. The Latin text adds "without end"; but Bacon is scarcely right in supposing that the descent from complex ideas and propositions to those of simple nature, involve the analyst in a series of continuous and interminable definitions. For in the gradual and analytical scale, there is a bar beyond which we cannot go, as there is a summit bounded by the limited variations of our conceptions. Logical definitions, to fulfil their conditions, or indeed to be of any avail, must be given in simpler terms than the object which is sought to be defined; now this, in the case of primordial notions and objects of sense, is impossible; therefore we are obliged to rest satisfied with the mere names of our perceptions.--Ed.

25. The ancients supposed the planets to describe an exact circle round the south. As observations increased and facts were disclosed, which were irreconcilable with this supposition, the earth was removed from the centre to some other point in the circle, and the planets were supposed to revolve in a smaller circle (epicycle) round an imaginary point, which in its turn described a circle of which the earth was the centre. In proportion as observation elicited fresh facts, contradictory to these representations, other epicycles and eccentrics were added, involving additional confusion. Though Kepler had swept away all these complicated theories in the preceding century, by the demonstration of his three laws, which established the elliptical course of the planets, Bacon regarded him and Copernicus in the same light as Ptolemy and Xenophanes.--Ed.

26. Empedocles, of Agrigentum, flourished 444 B.C. He was the disciple of Telanges the Pythagorean, and warmly adopted the doctrine of transmigration. He resolved the universe into the four

ordinary elements, the principles of whose composition were life and happiness, or concord and amity, but whose decomposition brought forth death and evil, or discord and hatred. Heraclitus held matter to be indifferent to any peculiar form, but as it became rarer or more dense, it took the appearance of fire, air, earth and water. Fire, however, he believed to be the elementary principle out of which the others were evolved. This was also the belief of Lucretius. See book i. 783, etc.

27. It is thus the Vulcanists and Neptunians have framed their opposite theories in geology. Phrenology is a modern instance of hasty generalization.--Ed.

28. In Scripture everything which concerns the passing interests of the body is called dead; the only living knowledge having regard to the eternal interest of the soul.--Ed.

29. In mechanics and the general sciences, causes compound their effects, or in other words, it is generally possible to deduce à priori the consequence of introducing complex agencies into any experiment, by allowing for the effect of each of the simple causes which enter into their composition. In chemistry and physiology a contrary law holds; the causes which they embody generally uniting to form distinct substances, and to introduce unforeseen laws and combinations. The deductive method here is consequently inapplicable, and we are forced back upon experiment.

Bacon in the text is hardly consistent with himself, as he admits in the second book the doctrine, to which modern discovery points, of the reciprocal transmutation of the elements. What seemed poetic fiction in the theories of Pythagoras and Seneca, assumes the appearance of scientific fact in the hands of Baron Caynard.--Ed.

30. Galileo had recently adopted the notion that nature abhorred a vacuum for an axiomatic principle, and it was not till Torricelli, his disciple, had given practical proof of the utility of Bacon's method, by the discovery of the barometer (1643) that this error, as also that expressed below, and believed by Bacon, concerning the homœopathic tendencies of bodies, was destroyed.--Ed.

31. Donec ad materiam potentialem et informem ventum fuerit. Nearly all the ancient philosophers admitted the existence of a certain primitive and shapeless matter as the substratum of things which the creative power had reduced to fixed proportions, and resolved into specific substances. The expression potential matter refers to that substance forming the basis of the Peripatetic system, which virtually contained all the forms that it was in the power of the efficient cause to draw out of it.--Ed.

32. An allusion to the humanity of the Sultans, who, in their earlier histories are represented as signalizing their accession to the throne by the destruction of their family, to remove the danger of rivalry and the terrors of civil war.--Ed.

33. The text is "in odium veterum sophistarum, Protagoræ, Hippiæ, et reliquorum." Those were called sophists, who, ostentationis aut questus causa philosophabantur. (Acad. Prior. ii. 72.) They had corrupted and degraded philosophy before Socrates. Protagoras of Abdera (Ἄβδηρα), the most celebrated, taught that man is the measure of all things, by which he meant not only that all which can be known is known only as it related to our faculties, but also that apart from our faculties nothing can be known. The sceptics equally held that knowledge was probable only as it related to our faculties, but they stopped there, and did not, like the sophist, dogmatize about the unknown. The works of Protagoras were condemned for their impiety, and publicly burned by the ædiles of Athens, who appear to have discharged the office of common hangmen to the literary blasphemers of their day.--Ed.

34. Bacon is hardly correct in implying that the enumerationem per simplicem was the only light in which the ancients looked upon induction, as they appear to have regarded it as only one, and that the least important, of its species. Aristotle expressly considers induction in a perfect or dialectic sense, and in an imperfect or rhetorical sense. Thus if a genus (=G=), contains four species (=A=, =B=, =C=, =D=), the syllogism would lead us to infer, that what is true of =G=, is true of any one of the four. But perfect induction would reason, that what we can prove of =A=, =B=, =C=, =D=, separately, we may properly state as true of =G=, the whole genus. This is evidently a formal argument as demonstrative as the syllogism. In necessary matters, however, legitimate induction may claim a wider province, and infer of the whole genus what is only apparent in a part of the species. Such are those inductive inferences which concern the laws of nature, the immutability of forms, by which Bacon strove to erect his new system of philosophy. The Stagyrite, however, looked upon enumerationem per simplicem, without any regard to the nature of the matter, or to the completeness of the species, with as much reprehensive caution as Bacon, and guarded his readers against it as the source of innumerable errors.--Ed.

35. See Ax. lxi. toward the end. This subject extends to Ax. lxxviii.

36. Gorgias of Leontium went to Athens in 424 B.C. He and Polus were disciples of Empedocles, whom we have already noticed (Aphorism 63), where he sustained the three famous propositions, that nothing exists, that nothing can be known, and that it is out of the power of man to transmit or communicate intelligence. He is reckoned one of the earliest writers on the art of rhetoric, and for that reason, Plato called his elegant dialogue on that subject after his name.

37. Chrysippus, a stoic philosopher of Soli in Cilicia, Campestris, born in 280, died in the 143d Olympiad, 208 B.C. He was equally distinguished for natural abilities and industry, seldom suffering a

day to elapse without writing 500 lines. He wrote several hundred volumes, of which three hundred were on logical subjects; but in all, borrowed largely from others. He was very fond of the sorites in argument, which is hence called by Persius the heap of Chrysippus. He was called the Column of the Portico, a name given to the Stoical School from Zeno, its founder, who had given his lessons under the portico.

Carneades, born about 215, died in 130. He attached himself to Chrysippus, and sustained with éclat the scepticism of the academy. The Athenians sent him with Critolaus and Diogenes as ambassador to Rome, where he attracted the attention of his new auditory by the subtilty of his reasoning, and the fluency and vehemence of his language. Before Galba and Cato the Censor, he harangued with great variety of thought and copiousness of diction in praise of justice. The next day, to establish his doctrine of the uncertainty of human knowledge, he undertook to refute all his arguments. He maintained with the New Academy, that the senses, the imagination, and the understanding frequently deceive us, and therefore cannot be infallible judges of truth, but that from the impressions produced on the mind by means of the senses, we infer appearances of truth or probabilities. Nevertheless, with respect to the conduct of life, Carneades held that probable opinions are a sufficient guide.

Xenophanes, a Greek philosopher, of Colophon, born in 556, the founder of the Eleatic school, which owes its fame principally to Parmenides. Wild in his opinions about astronomy, he supposed that the stars were extinguished every morning, and rekindled at night; that eclipses were occasioned by the temporary extinction of the sun, and that there were several suns for the convenience of the different climates of the earth. Yet this man held the chair of philosophy at Athens for seventy years.

Philolaus, a Pythagorean philosopher of Crotona, B.C. 374. He first supported the diurnal motion of the earth round its axis, and its annual motion round the sun. Cicero (Acad. iv. 39) has ascribed this opinion to the Syracusan philosopher Nicetas, and likewise to Plato. From this passage, it is most probable that Copernicus got the idea of the system he afterward established. Bacon, in the Advancement of Human Learning, charges Gilbert with restoring the doctrines of Philolaus, because he ventured to support the Copernican theory.--Ed.

38. Bacon is equally conspicuous for the use and abuse of analogical illustrations. The levity, as Stuart Mill very properly observes, by which substances float on a stream, and the levity which is synonymous with worthlessness, have nothing beside the name in common; and to show how little value there is in the figure, we need only change the word into buoyancy, to turn the semblance of Bacon's argument against himself.--Ed.

39. We have before observed, that the New Academy did not profess skepticism, but the ἀκατάληψια, or incomprehensibility of the absolute essences of things. Even modern physicists are not wanting, to assert with this school that the utmost knowledge we can obtain is relative, and necessarily short of absolute certainty. It is not without an appearance of truth that these philosophers maintain that our ideas and perceptions do not express the nature of the things which they represent, but only the effects of the peculiar organs by which they are conveyed to the understanding, so that were these organs changed, we should have different conceptions of their nature. That constitution of air which is dark to man is luminous to bats and owls.

40. Owing to the universal prevalence of Aristotelism.

41. It must be remembered, that when Bacon wrote, algebra was in its infancy, and the doctrine of units and infinitesimals undiscovered.

42. Because the vulgar make up the overwhelming majority in such decisions, and generally allow their judgments to be swayed by passion or prejudice.

43. See end of Axiom lxi. The subject extends to Axiom xc.

44. If we adopt the statement of Herodotus, who places the Homeric era 400 years back from his time, Homer lived about 900 years before Christ. On adding this number to the sixteen centuries of the Christian era which had elapsed up to Bacon's time, we get the twenty-five centuries he mentions. The Homeric epoch is the furthest point in antiquity from which Bacon could reckon with any degree of certainty. Hesiod, if he were not contemporary, immediately preceded him.

The epoch of Greek philosophy may be included between Thales and Plato, that is, from the 35th to the 88th Olympiad; that of the Roman, between Terence and Pliny. The modern revolution, in which Bacon is one of the central figures, took its rise from the time of Dante and Petrarch, who lived at the commencement of the fourteenth century; and to which, on account of the invention of printing, and the universal spread of literature, which has rendered a second destruction of learning impossible, it is difficult to foresee any other end than the extinction of the race of man.--Ed.

45. The allusion is evidently to Roger Bacon and Réné Descartes.--Ed.

46. From the abuse of the scholastics, who mistook the à priori method, the deductive syllogism, for the entire province of logic.--Ed.

47. See Aphorism xcv.

48. The incongruity to which Bacon alludes appears to spring from confounding two things, which

are not only distinct, but affect human knowledge in inverse proportion, viz., the experience which terminates with life, with that experience which one century transmits to another.--Ed.

49. The Chinese characters resemble, in many respects, the hieroglyphics of the Egyptians, being adapted to represent ideas, not sounds.

50. See Axiom 75.

51. The methods by which Newton carried the rule and compass to the boundaries of creation is a sufficient comment on the sagacity of the text. The same cause which globulizes a bubble, has rounded the earth, and the same law which draws a stone to its surface, keeps the moon in her orbit. It was by calculating and ascertaining these principles upon substances entirely at his disposal that this great philosopher was enabled to give us a key to unlock the mysteries of the universe.--Ed.

52. See the "Clouds" of Aristophanes, where Socrates is represented as chasing Jupiter out of the sky, by resolving thunderstorms into aërial concussions and whirlwinds.--Ed.

53. Robespierre is the latest victim of this bigotry. In his younger days he attempted to introduce Franklin's lightning conductor into France, but was persecuted by those whose lives he sought to protect, as one audaciously striving to avert the designs of Providence.--Ed.

54. We can hardly agree with the text. The scholastics, in building up a system of divinity, certainly had recourse to the deductive syllogism, because the inductive was totally inapplicable, except as a verificatory process. With regard to the technical form in which they marshalled their arguments, which is what our author aims at in his censure, they owed nothing at all to Aristotle, the conducting a dispute in naked syllogistic fashion having originated entirely with themselves.--Ed.

55. Bacon cannot be supposed to allude to those divines who have attempted to show that the progress of physical science is confirmatory of revelation, but only to such as have built up a system of faith out of their own refinements on nature and revelation, as Patricius and Emanuel Swedenborg.--Ed.

56. Daniel xii. 4.

57. Bacon, in this Aphorism, appears to have entertained a fair idea of the use of the inductive and deductive methods in scientific inquiry, though his want of geometrical knowledge must have hindered him from accurately determining the precise functions of each, as it certainly led him in other parts of the Organon (V. Aph. 82), to undervalue the deductive, and, as he calls it, the dogmatic method, and to rely too much upon empiricism.--Ed.

58. The reader may consult the note of the 23d Aphorism for the fault which Bacon censures, and, if he wish to pursue the subject further, may read Plato's Timæus, where that philosopher explains his system in detail. Bacon, however, is hardly consistent in one part of his censure, for he also talks about the spirit and appetites of inanimate substances, and that so frequently, as to preclude the supposition that he is employing metaphor.--Ed.

59. Proclus flourished about the beginning of the fifth century, and was the successor of Plotinus, Porphyry and Iamblicus, who, in the two preceding centuries, had revived the doctrines of Plato, and assailed the Christian religion. The allusion in the text must be assigned to Iamblicus, who, in the fourth century, had republished the Pythagorean theology of numbers, and endeavored to construct the world out of arithmetic, thinking everything could be solved by the aid of proportions and geometry. Bacon must not be understood in the text to censure the use but the abuse of mathematics and physical investigations, as in the "De Augmentis" (lib. iv. c. 6), he enumerates the multiplicity of demonstration scientific facts admit of, from this source.--Ed.

60. See Livy, lib. ix. c. 17, where, in a digression on the probable effect of a contest between Rome and Alexander the Great, he says: "Non cum Dario rem esse dixisset: quem mulierum ac spadonum agmen trahentêm inter purpuram atque aurum, oneratum fortunæ apparatibus, prædam veriùs quam hostem, nihil aliud quam ausus vana contemnere, incruentus devicit."

61. The lowest axioms are such as spring from simple experience--such as in chemistry, that animal substances yield no fixed salt by calcination; in music, that concords intermixed with discords make harmony, etc. Intermediate axioms advance a step further, being the result of reflection, which, applied to our experimental knowledge, deduces laws from them, such as in optics of the first degree of generality, that the angle of incidence is equal to the angle of reflection; and in mechanics, Kepler's three laws of motion, while his general law, that all bodies attract each other with forces proportional to their masses, and inversely as the squares of their distances, may be taken as one of the highest axioms. Yet so far is this principle from being only notional or abstract, it has presented us with a key which fits into the intricate wards of the heavens, and has laid bare to our gaze the principal mechanism of the universe. But natural philosophy in Bacon's day had not advanced beyond intermediate axioms, and the term notional or abstract is applied to those general axioms then current, not founded on the solid principles of inductive inquiry, but based upon à priori reasoning and airy metaphysics.--Ed.

62. This hope has been abundantly realized in the discovery of gravity and the decomposition of light, mainly by the inductive method. To a better philosophy we may also attribute the discovery of electricity, galvanism and their mutual connection with each other, and magnetism, the inventions of the air-pump, steam-engine and the chronometer.

63. As Bacon very frequently cites these authors, a slight notice of their labors may not be unacceptable to the reader. Bernardinus Telesius, born at Cosenza, in 1508, combated the Aristotelian system in a work entitled "De Rerum Natura juxta propria principia," i.e., according to principles of his own. The proem of the work announces his design was to show that "the construction of the world, the magnitude and nature of the bodies contained in it, are not to be investigated by reasoning, which was done by the ancients, but are to be apprehended by the senses, and collected from the things themselves." He had, however, no sooner laid down this principle than he departed from it in practice, and pursued the deductive method he so much condemned in his predecessors. His first step was an assumption of principles as arbitrary as any of the empirical notions of antiquity; at the outset of his book he very quietly takes it for granted that heat is the principle of motion, cold of immobility, matter being assumed as the corporeal substratum, in which these incorporeal and active agents carry on their operations. Out of these abstract and ill-defined conceptions Telesius builds up a system quite as complete, symmetrical, and imaginative as any of the structures of antiquity.

Francis Patricius, born at Cherso, in Dalmatia, about 1529, was another physicist who rose up against Aristotle, and announced the dawn of a new philosophy. In 1593 appeared his "Nova de Universis Philosophia." He lays down a string of axioms, in which scholastic notions, physical discoveries, and theological dogmas, are strangely commingled, and erects upon them a system which represents all the grotesque features of theological empiricism.

Severinus, born in Jutland, in 1529, published an attack on Aristotle's natural history, but adopted fantasies which the Stagyrite ridiculed in his own day. He was a follower of Paracelsus, a Swiss enthusiast of the fifteenth century, who ignored the ancient doctrine of the four elements for salt, sulphur and mercury, and allied chemistry and medicine with mysticism.--Ed.

64. Bacon's apology is sound, and completely answers these German and French critics, who have refused him a niche in the philosophical pantheon. One German commentator, too modest to reveal his name, accuses Bacon of ignorance of the calculus, though, in his day, Wallis had not yet stumbled upon the laws of continuous fractions; while Count de Maistre, in a coarse attack upon his genius, expresses his astonishment at finding Bacon unacquainted with discoveries which were not heard of till a century after his death.--Ed.

65. Philip of Macedon.

66. See Plato's Timæus.

67. The saying of Philocrates when he differed from Demosthenes.--Ed.

68. The old error of placing the deductive syllogism in antagonism to the inductive, as if they were not both parts of one system or refused to cohere together. So far from there being any radical opposition between them, it would not be difficult to show that Bacon's method was syllogistic in his sense of the term. For the suppressed premise of every Baconian enthymeme, viz., the acknowledged uniformity of the laws of nature as stated in the axiom, whatever has once occurred will occur again, must be assumed as the basis of every conclusion which he draws before we can admit its legitimacy. The opposition, therefore, of Bacon's method could not be directed against the old logic, for it assumed and exemplified its principles, but rather to the abusive application which the ancients made of this science, on turning its powers to the development of abstract principles which they imagined to be pregnant with the solution of the latent mysteries of the universe. Bacon justly overthrew these ideal notions, and accepted of no principle as a basis which was not guaranteed by actual experiment and observation; and so far he laid the foundations of a sound philosophy by turning the inductive logic to its proper account in the interpretation of nature.

69. This is the opening of the Sixth Book of Lucretius. Bacon probably quoted from memory; the lines are--

"Primæ frugiferos fœtus mortalibus ægris   Dididerunt quondam præclaro nomine Athenæ   Et recreaverunt," etc.

The teeming corn, that feeble mortals crave,   First, and long since, renowned Athens gave,
And cheered their life--then taught to frame their laws.

70. Prov. xxv. 2.

APHORISMS – BOOK II – ON THE INTERPRETATION OF NATURE, OR THE REIGN OF MAN

I. To generate and superinduce a new nature or new natures, upon a given body, is the labor and aim of human power: while to discover the form or true difference of a given nature, or the nature[71] to which such nature is owing, or source from which it emanates (for these terms approach nearest to an explanation of our meaning), is the labor and discovery of human knowledge; and subordinate to these primary labors are two others of a secondary nature and inferior stamp. Under the first must be ranked the transformation of concrete bodies from one to another, which is possible within certain limits; under the second, the discovery, in every species of generation and motion, of the latent and uninterrupted process from the manifest efficient and manifest subject matter up to the given form: and a like discovery of the latent conformation of bodies which are at rest instead of being in motion.

II. The unhappy state of man's actual knowledge is manifested even by the common assertions of the vulgar. It is rightly laid down that true knowledge is that which is deduced from causes. The division of four causes also is not amiss: matter, form, the efficient, and end or final cause.[72] Of these, however, the latter is so far from being beneficial, that it even corrupts the sciences, except in the intercourse of man with man. The discovery of form is considered desperate. As for the efficient cause and matter (according to the present system of inquiry and the received opinions concerning them, by which they are placed remote from, and without any latent process toward form), they are but desultory and superficial, and of scarcely any avail to real and active knowledge. Nor are we unmindful of our having pointed out and corrected above the error of the human mind, in assigning the first qualities of essence to forms.[73] For although nothing exists in nature except individual bodies,[74] exhibiting clear individual effects according to particular laws, yet in each branch of learning, that very law, its investigation, discovery, and development, are the foundation both of theory and practice. This law, therefore, and its parallel in each science, is what we understand by the term form,[75] adopting that word because it has grown into common use, and is of familiar occurrence.

III. He who has learned the cause of a particular nature (such as whiteness or heat), in particular subjects only, has acquired but an imperfect knowledge: as he who can induce a certain effect upon particular substances only, among those which are susceptible of it, has acquired but an imperfect power. But he who has only learned the efficient and material cause (which causes are variable and mere vehicles conveying form to particular substances) may perhaps arrive at some new discoveries in matters of a similar nature, and prepared for the purpose, but does not stir the limits of things which are much more deeply rooted; while he who is acquainted with forms, comprehends the unity of nature in substances apparently most distinct from each other. He can disclose and bring forward, therefore (though it has never yet been done), things which neither the vicissitudes of nature, nor the industry of experiment, nor chance itself, would ever have brought about, and which would forever have escaped man's thoughts; from the discovery of forms, therefore, results genuine theory and free practice.

IV. Although there is a most intimate connection, and almost an identity between the ways of human power and human knowledge, yet, on account of the pernicious and inveterate habit of dwelling upon abstractions, it is by far the safest method to commence and build up the sciences from those foundations which bear a relation to the practical division, and to let them mark out and limit the theoretical. We must consider, therefore, what precepts, or what direction or guide, a person would most desire, in order to generate and superinduce any nature upon a given body: and this not in abstruse, but in the plainest language.

For instance, if a person should wish to superinduce the yellow color of gold upon silver, or an additional weight (observing always the laws of matter) or transparency on an opaque stone, or tenacity in glass, or vegetation on a substance which is not vegetable, we must (I say) consider what species of precept or guide this person would prefer. And, first, he will doubtless be anxious to be shown some method that will neither fail in effect, nor deceive him in the trial of it; secondly, he will be anxious that the prescribed method should not restrict him and tie him down to peculiar means, and certain particular methods of acting; for he will, perhaps, be at loss, and without the power or opportunity of collecting and procuring such means. Now if there be other means and methods (besides those prescribed) of creating such a nature, they will perhaps be of such a kind as are in his power, yet by the confined limits of the precept he will be deprived of reaping any advantage from them; thirdly, he will be anxious to be shown something not so difficult as the required effect itself, but approaching more nearly to practice.

We will lay this down, therefore, as the genuine and perfect rule of practice, that it should be certain, free and preparatory, or having relation to practice. And this is the same thing as the discovery of a true form; for the form of any nature is such, that when it is assigned the particular nature infallibly follows. It is, therefore, always present when that nature is present, and universally attests such presence, and is inherent in the whole of it. The same form is of such a character, that if it be removed the particular nature infallibly vanishes. It is, therefore, absent, whenever that nature is absent, and perpetually testifies such absence, and exists in no other nature. Lastly, the true form is such, that it deduces the particular nature from some source of essence existing in many subjects, and more known (as they term it) to nature, than the form itself. Such, then, is our determination and rule with regard to a genuine and perfect theoretical axiom, that a nature be found convertible with a given nature, and yet such as to limit the more known nature, in the manner of a real genus. But these two rules, the practical and theoretical, are in fact the same, and that which is most useful in practice is most correct in theory.

V. But the rule or axiom for the transformation of bodies is of two kinds. The first regards the body as an aggregate or combination of simple natures. Thus, in gold are united the following circumstances: it is yellow, heavy, of a certain weight, malleable and ductile to a certain extent; it is not volatile, loses part of its substance by fire, melts in a particular manner, is separated and dissolved by particular methods, and so of the other natures observable in gold. An axiom, therefore, of this kind deduces the subject from the forms of simple natures; for he who has acquired the forms and methods of superinducing yellowness, weight, ductility, stability, deliquescence, solution, and the like, and their degrees and modes, will consider and contrive how to unite them in any body, so as to transform[76] it into gold. And this method of operating belongs to primary action; for it is the same thing to produce one or many simple natures, except that man is more confined and restricted in his operations, if many be required, on account of the difficulty of uniting many natures together. It must, however, be observed, that this method of operating (which considers natures as simple though in a concrete body) sets out from what is constant, eternal, and universal in nature, and opens such broad paths to human power, as the thoughts of man can in the present state of things scarcely comprehend or figure to itself.

The second kind of axiom (which depends on the discovery of the latent process) does not proceed by simple natures, but by concrete bodies, as they are found in nature and in its usual course. For instance, suppose the inquiry to be, from what beginnings, in what manner, and by what process gold or any metal or stone is generated from the original menstruum, or its elements, up to the perfect mineral: or, in like manner, by what process plants are generated, from the first concretion of juices in the earth, or from seeds, up to the perfect plant, with the whole successive motion, and varied and uninterrupted efforts of nature; and the same inquiry be made as to a regularly deduced system of the generation of animals from coition to birth, and so on of other bodies.

Nor is this species of inquiry confined to the mere generation of bodies, but it is applicable to other changes and labors of nature. For instance, where an inquiry is made into the whole series and continued operation of the nutritive process, from the first reception of the food to its complete assimilation to the recipient;[77] or into the voluntary motion of animals, from the first impression of the imagination, and the continuous effects of the spirits, up to the bending and motion of the joints; or into the free motion of the tongue and lips, and other accessories which give utterance to articulate sounds. For all these investigations relate to concrete or associated natures artificially brought together, and take into consideration certain particular and special habits of nature, and not those fundamental and general laws which constitute forms. It must, however, be plainly owned, that this method appears more prompt and easy, and of greater promise than the primary one.

In like manner the operative branch, which answers to this contemplative branch, extends and advances its operation from that which is usually observed in nature, to other subjects immediately connected with it, or not very remote from such immediate connection. But the higher and radical operations upon nature depend entirely on the primary axioms. Besides, even where man has not the means of acting, but only of acquiring knowledge, as in astronomy (for man cannot act upon, change, or transform the heavenly bodies), the investigation of facts or truth, as well as the knowledge of causes and coincidences, must be referred to those primary and universal axioms that regard simple natures; such as the nature of spontaneous rotation, attraction, or the magnetic force, and many others which are more common than the heavenly bodies themselves. For let no one hope to determine the question whether the earth or heaven revolve in the diurnal motion, unless he have first comprehended the nature of spontaneous rotation.

VI. But the latent process of which we speak, is far from being obvious to men's minds, beset as they now are. For we mean not the measures, symptoms, or degrees of any process which can be exhibited in the bodies themselves, but simply a continued process, which, for the most part, escapes the observation of the senses.

For instance, in all generations and transformations of bodies, we must inquire, what is in the act of being lost and escaping, what remains, what is being added, what is being diluted, what is being contracted, what is being united, what is being separated, what is continuous, what is broken off, what is urging forward, what impedes, what predominates, what is subservient, and many other circumstances.

Nor are these inquiries again to be made in the mere generation and transformation of bodies only, but in all other alterations and fluctuations we must in like manner inquire; what precedes, what succeeds, what is quick, what is slow, what produces and what governs motion, and the like. All which matters are unknown and unattempted by the sciences, in their present heavy and inactive state. For, since every natural act is brought about by the smallest efforts,[78] or at least such as are too small to strike our senses, let no one hope that he will be able to direct or change nature unless he have properly comprehended and observed these efforts.

VII. In like manner, the investigation and discovery of the latent conformation in bodies is no less new, than the discovery of the latent process and form. For we as yet are doubtless only admitted to the antechamber of nature, and do not prepare an entrance into her presence-room. But nobody can endue a given body with a new nature, or transform it successfully and appropriately into a new body, without possessing a complete knowledge of the body so to be changed or transformed. For he will run into vain, or, at least, into difficult and perverse methods, ill adapted to the nature of the body upon which he operates. A clear path, therefore, toward this object also must be thrown open, and well supported.

Labor is well and usefully bestowed upon the anatomy of organized bodies, such as those of men and animals, which appears to be a subtile matter, and a useful examination of nature. The species of anatomy, however, is that of first sight, open to the senses, and takes place only in organized bodies. It is obvious, and of ready access, when compared with the real anatomy of latent conformation in bodies which are considered similar, particularly in specific objects and their parts; as those of iron, stone, and the similar parts of plants and animals, as the root, the leaf, the flower, the flesh, the blood, and bones, etc. Yet human industry has not completely neglected this species of anatomy; for we have an instance of it in the separation of similar bodies by distillation, and other solutions, which shows the dissimilarity of the compound by the union of the homogeneous parts. These methods are useful, and of importance to our inquiry, although attended generally with fallacy: for many natures are assigned and attributed to the separate bodies, as if they had previously existed in the compound, which, in reality, are recently bestowed and superinduced by fire and heat, and the other modes of separation. Besides, it is, after all, but a small part of the labor of discovering the real conformation in the compound, which is so subtile and nice, that it is rather confused and lost by the operation of the fire, than discovered and brought to light.

A separation and solution of bodies, therefore, is to be effected, not by fire indeed, but rather by reasoning and true induction, with the assistance of experiment, and by a comparison with other bodies, and a reduction to those simple natures and their forms which meet, and are combined in the compound; and we must assuredly pass from Vulcan to Minerva, if we wish to bring to light the real texture and conformation of bodies, upon which every occult and (as it is sometimes called) specific property and virtue of things depends, and whence also every rule of powerful change and transformation is deduced.

For instance, we must examine what spirit is in every body,[79] what tangible essence; whether that spirit is copious and exuberant, or meagre and scarce, fine or coarse, aëriform or igniform, active or sluggish, weak or robust, progressive or retrograde, abrupt or continuous, agreeing with external and surrounding objects, or differing from them, etc. In like manner must we treat tangible essence (which admits of as many distinctions as the spirit), and its hairs, fibres, and varied texture. Again, the situation of the spirit in the corporeal mass, its pores, passages, veins, and cells, and the rudiments or first essays of the organic body, are subject to the same examination. In these, however, as in our former inquiries, and therefore in the whole investigation of latent conformation, the only genuine and clear light which completely dispels all darkness and subtile difficulties, is admitted by means of the primary axioms.

VIII. This method will not bring us to atoms,[80] which takes for granted the vacuum, and immutability of matter (neither of which hypotheses is correct), but to the real particles such as we discover them to be. Nor is there any ground for alarm at this refinement as if it were inexplicable, for, on the contrary, the more inquiry is directed to simple natures, the more will everything be placed in a plain and perspicuous light, since we transfer our attention from the complicated to the simple, from the incommensurable to the commensurable, from surds to rational quantities, from the indefinite and vague to the definite and certain; as when we arrive at the elements of letters, and the simple tones of concords. The investigation of nature is best conducted when mathematics are applied to physics. Again, let none be alarmed at vast numbers and fractions, for in calculation it is as easy to set down or to reflect upon a thousand as a unit, or the thousandth part of an integer as an integer itself.

IX.[81] From the two kinds of axioms above specified, arise the two divisions of philosophy and the sciences, and we will use the commonly adopted terms which approach the nearest to our meaning, in our own sense. Let the investigation of forms, which (in reasoning at least, and after their own laws), are eternal and immutable, constitute metaphysics,[82] and let the investigation of the efficient cause of matter, latent process, and latent conformation (which all relate merely to the ordinary course of nature, and not to her fundamental and eternal laws), constitute physics. Parallel to these, let there be two practical divisions; to physics that of mechanics, and to metaphysics that of magic, in the purest sense of the term, as applied to its ample means, and its command over nature.

X. The object of our philosophy being thus laid down, we proceed to precepts, in the most clear and regular order. The signs for the interpretation of nature comprehend two divisions; the first regards the eliciting or creating of axioms from experiment, the second the deducing or deriving of new experiments from axioms. The first admits of three subdivisions into ministrations. 1. To the senses. 2. To the memory. 3. To the mind or reason.

For we must first prepare as a foundation for the whole, a complete and accurate natural and experimental history. We must not imagine or invent, but discover the acts and properties of nature.

But natural and experimental history is so varied and diffuse, that it confounds and distracts the understanding unless it be fixed and exhibited in due order. We must, therefore, form tables and co-ordinations of instances, upon such a plan, and in such order that the understanding may be enabled to act upon them.

Even when this is done, the understanding, left to itself and to its own operation, is incompetent and unfit to construct its axioms without direction and support. Our third ministration, therefore, must be true and legitimate induction, the very key of interpretation. We must begin, however, at the end, and go back again to the others.

XI. The investigation of forms proceeds thus: a nature being given, we must first present to the understanding all the known instances which agree in the same nature, although the subject matter be considerably diversified. And this collection must be made as a mere history, and without any premature reflection, or too great degree of refinement. For instance; take the investigation of the form of heat.

Instances agreeing in the Form of Heat

1. The rays of the sun, particularly in summer, and at noon.
2. The same reflected and condensed, as between mountains, or along walls, and particularly in burning mirrors.
3. Ignited meteors.
4. Burning lightning.
5. Eruptions of flames from the cavities of mountains, etc.
6. Flame of every kind.
7. Ignited solids.
8. Natural warm baths.
9. Warm or heated liquids.
10. Warm vapors and smoke; and the air itself, which admits a most powerful and violent heat if confined, as in reverberating furnaces.
11. Damp hot weather, arising from the constitution of the air, without any reference to the time of the year.
12. Confined and subterraneous air in some caverns, particularly in winter.
13. All shaggy substances, as wool, the skins of animals, and the plumage of birds, contain some heat.
14. All bodies, both solid and liquid, dense and rare (as the air itself), placed near fire for any time.
15. Sparks arising from the violent percussion of flint and steel.
16. All bodies rubbed violently, as stone, wood, cloth, etc., so that rudders, and axles of wheels, sometimes catch fire, and the West Indians obtain fire by attrition.
17. Green and moist vegetable matter confined and rubbed together, as roses, peas in baskets; so hay, if it be damp when stacked, often catches fire.
18. Quicklime sprinkled with water.
19. Iron, when first dissolved by acids in a glass, and without any application to fire; the same of tin, but not so intensely.
20. Animals, particularly internally; although the heat is not perceivable by the touch in insects, on account of their small size.
21. Horse dung, and the like excrement from other animals, when fresh.
22. Strong oil of sulphur and of vitriol exhibit the operation of heat in burning linen.
23. As does the oil of marjoram, and like substances, in burning the bony substance of the teeth.
24. Strong and well rectified spirits of wine exhibit the same effects; so that white of eggs when thrown into it grows hard and white, almost in the same manner as when boiled, and bread becomes burned and brown as if toasted.
25. Aromatic substances and warm plants, as the dracunculus [arum], old nasturtium, etc., which, though they be not warm to the touch (whether whole or pulverized), yet are discovered by the tongue and palate to be warm and almost burning when slightly masticated.

26. Strong vinegar and all acids, on any part of the body not clothed with the epidermis, as the eye, tongue, or any wounded part, or where the skin is removed, excite a pain differing but little from that produced by heat.

27. Even a severe and intense cold produces a sensation of burning.[83]

"Nec Boreæ penetrabile frigus adurit."

28. Other instances.

We are wont to call this a table of existence and presence.

XII. We must next present to the understanding instances which do not admit of the given nature, for form (as we have observed) ought no less to be absent where the given nature is absent, than to be present where it is present. If, however, we were to examine every instance, our labor would be infinite.

Negatives, therefore, must be classed under the affirmatives, and the want of the given nature must be inquired into more particularly in objects which have a very close connection with those others in which it is present and manifest. And this we are wont to term a table of deviation or of absence in proximity.

Proximate Instances wanting the Nature of Heat

The rays of the moon, stars, and comets, are not found to be warm to the touch, nay, the severest cold has been observed to take place at the full of the moon. Yet the larger fixed stars are supposed to increase and render more intense the heat of the sun, as he approaches them, when the sun is in the sign of the Lion, for instance, and in the dog-days.[84]

The rays of the sun in what is called the middle region of the air give no heat, to account for which the commonly assigned reason is satisfactory; namely, that that region is neither sufficiently near to the body of the sun whence the rays emanate, nor to the earth whence they are reflected. And the fact is manifested by snow being perpetual on the tops of mountains, unless extremely lofty. But it is observed, on the other hand, by some, that at the Peak of Teneriffe, and also among the Andes of Peru, the tops of the mountains are free from snow, which only lies in the lower part as you ascend. Besides, the air on the summit of these mountains is found to be by no means cold, but only thin and sharp; so much so, that in the Andes it pricks and hurts the eyes from its extreme sharpness, and even excites the orifice of the stomach and produces vomiting. The ancients also observed, that the rarity of the air on the summit of Olympus was such, that those who ascended it were obliged to carry sponges moistened with vinegar and water, and to apply them now and then to their nostrils, as the air was not dense enough for their respiration; on the summit of which mountain it is also related, there reigned so great a serenity and calm, free from rain, snow, or wind, that the letters traced upon the ashes of the sacrifices on the altar of Jupiter, by the fingers of those who had offered them, would remain undisturbed till the next year. Those even, who at this day go to the top of the Peak of Teneriffe, walk by night and not in the daytime, and are advised and pressed by their guides, as soon as the sun rises, to make haste in their descent, on account of the danger (apparently arising from the rarity of the atmosphere), lest their breathing should be relaxed and suffocated.[85]

The reflection of the solar rays in the polar regions is found to be weak and inefficient in producing heat, so that the Dutch, who wintered in Nova Zembla, and expected that their vessel would be freed about the beginning of July from the obstruction of the mass of ice which had blocked it up, were disappointed and obliged to embark in their boat. Hence the direct rays of the sun appear to have but little power even on the plain, and when reflected, unless they are multiplied and condensed, which takes place when the sun tends more to the perpendicular; for, then, the incidence of the rays occurs at more acute angles, so that the reflected rays are nearer to each other, while, on the contrary, when the sun is in a very oblique position, the angles of incidence are very obtuse, and the reflected rays at a greater distance. In the meantime it must be observed, that there may be many operations of the solar rays, relating, too, to the nature of heat, which are not proportioned to our touch, so that, with regard to us, they do not tend to produce warmth, but, with regard to some other bodies, have their due effect in producing it.

Let the following experiment be made. Take a lens the reverse of a burning-glass, and place it between the hand and the solar rays, and observe whether it diminish the heat of the sun as a burning-glass increases it. For it is clear, with regard to the visual rays, that in proportion as the lens is made of unequal thickness in the middle and at its sides, the images appear either more diffused or contracted. It should be seen, therefore, if the same be true with regard to heat.

Let the experiment be well tried, whether the lunar rays can be received and collected by the strongest and best burning-glasses, so as to produce even the least degree of heat.[86] But if that degree be, perhaps, so subtile and weak, as not to be perceived or ascertained by the touch, we must have recourse to those glasses which indicate the warm or cold state of the atmosphere, and let the lunar rays fall through the burning-glass on the top of this thermometer, and then notice if the water be depressed by the heat.[87]

Let the burning-glass be tried on warm objects which emit no luminous rays, as heated but not ignited iron or stone, or hot water, or the like; and observe whether the heat become increased and condensed, as happens with the solar rays.

Let it be tried on common flame.

The effect of comets (if we can reckon them among meteors[88]) in augmenting the heat of the season is not found to be constant or clear, although droughts have generally been observed to follow them. However, luminous lines, and pillars, and openings, and the like, appear more often in winter than in summer, and especially with the most intense cold but joined with drought. Lightning, and coruscations, and thunder, however, rarely happen in winter; and generally at the time of the greatest heats. The appearances we term falling stars are generally supposed to consist of some shining and inflamed viscous substance, rather than of violently hot matter; but let this be further investigated.

Some coruscations emit light without burning, but are never accompanied by thunder.

Eructations and eruptions of flame are to be found in cold climates as well as in hot, as in Iceland and Greenland; just as the trees of cold countries are sometimes inflammable and more pitchy and resinous than in warm, as the fir, pine, and the like. But the position and nature of the soil, where such eruptions are wont to happen, is not yet sufficiently investigated to enable us to subjoin a negative instance to the affirmative.

All flame is constantly more or less warm, and this instance is not altogether negative; yet it is said that the ignis fatuus (as it is called), and which sometimes is driven against walls, has but little heat; perhaps it resembles that of spirits of wine, which is mild and gentle. That flame, however, appears yet milder, which in some well authenticated and serious histories is said to have appeared round the head and hair of boys and virgins, and instead of burning their hair, merely to have played about it. And it is most certain that a sort of flash, without any evident heat, has sometimes been seen about a horse when sweating at night, or in damp weather. It is also a well known fact,[89] and it was almost considered as a miracle, that a few years since a girl's apron sparkled when a little shaken or rubbed, which was, perhaps, occasioned by the alum or salts with which the apron was imbued, and which, after having been stuck together and incrusted rather strongly, were broken by the friction. It is well known that all sugar, whether candied or plain, if it be hard, will sparkle when broken or scraped in the dark. In like manner sea and salt water is sometimes found to shine at night when struck violently by the oar. The foam of the sea when agitated by tempests also sparkles at night, and the Spaniards call this appearance the sea's lungs. It has not been sufficiently ascertained what degree of heat attends the flame which the ancient sailors called Castor and Pollux, and the moderns call St. Ermus' fire.

Every ignited body that is red-hot is always warm, although without flame, nor is any negative instance subjoined to this affirmative. Rotten wood, however, approaches nearly to it, for it shines at night, and yet is not found to be warm; and the putrefying scales of fish which shine in the same manner are not warm to the touch, nor the body of the glowworm, or of the fly called Lucciola.[90]

The situation and nature of the soil of natural warm baths has not been sufficiently investigated, and therefore a negative instance is not subjoined.

To the instances of warm liquids we may subjoin the negative one of the peculiar nature of liquids in general; for no tangible liquid is known that is at once warm in its nature and constantly continues warm; but their heat is only superinduced as an adventitious nature for a limited time, so that those which are extremely warm in their power and effect, as spirits of wine, chemical aromatic oils, the oils of vitriol and sulphur, and the like, and which speedily burn, are yet cold at first to the touch, and the water of natural baths, poured into any vessel and separated from its source, cools down like water heated by the fire. It is, however, true that oily substances are rather less cold to the touch than those that are aqueous, oil for instance than water, silk than linen; but this belongs to the table of degrees of cold.

In like manner we may subjoin a negative instance to that of warm vapor, derived from the nature of vapor itself, as far as we are acquainted with it. For exhalations from oily substances, though easily inflammable, are yet never warm unless recently exhaled from some warm substance.

The same may be said of the instance of air; for we never perceive that air is warm unless confined or pressed, or manifestly heated by the sun, by fire, or some other warm body.

A negative instance is exhibited in weather by its coldness with an east or north wind, beyond what the season would lead us to expect, just as the contrary takes place with the south or west winds. An inclination to rain (especially in winter) attends warm weather, and to frost cold weather.

A negative instance as to air confined in caverns may be observed in summer. Indeed, we should make a more diligent inquiry into the nature of confined air. For in the first place the qualities of air in its own nature with regard to heat and cold may reasonably be the subject of doubt; for air evidently derives its heat from the effects of celestial bodies, and possibly its cold from the exhalation of the earth, and in the mid region of air (as it is termed) from cold vapors and snow, so that no judgment can be formed of the nature of air by that which is out of doors and exposed, but a more correct one might be derived from confined air. It is necessary, however, that the air should be inclosed in a vessel of such materials as would not imbue it with heat or cold of themselves, nor easily admit the influence of the external atmosphere. The experiment should be made, therefore, with an earthen jar, covered with folds of leather to protect it from the external air, and the air should be kept three or four days in this vessel well closed. On opening the jar, the degree of heat may be ascertained either by the hand or a graduated glass tube.

There is a similar doubt as to whether the warmth of wool, skins, feathers, and the like, is derived from a slight inherent heat, since they are animal excretions, or from their being of a certain fat and oily nature that accords with heat, or merely from the confinement and separation of air which we spoke of in the preceding paragraph;[91] for all air appears to possess a certain degree of warmth when separated from the external atmosphere. Let an experiment be made, therefore, with fibrous substances of linen, and not of wool, feathers, or silk, which are animal excretions. For it is to be observed that all powders (where air is manifestly inclosed) are less cold than the substances when whole, just as we imagine froth (which contains air) to be less cold than the liquid itself.

We have here no exactly negative instance, for we are not acquainted with any body tangible or spirituous which does not admit of heat when exposed to the fire. There is, however, this difference, that some admit it more rapidly, as air, oil, and water, others more slowly, as stone and metals.[92] This, however, belongs to the table of degrees.

No negative is here subjoined, except the remark that sparks are not kindled by flint and steel, or any other hard substance, unless some small particles of the stone or metal are struck off, and that the air never forms them by friction, as is commonly supposed; besides, the sparks from the weight of the ignited substance have a tendency to descend rather than to rise, and when extinguished become a sort of dark ash.

We are of opinion that here again there is no negative; for we are not acquainted with any tangible body which does not become decidedly warm by friction, so that the ancients feigned that the gods had no other means or power of creating heat than the friction of air, by rapid and violent rotation. On this point, however, further inquiry must be made, whether bodies projected by machines (as balls from cannon) do not derive some degree of heat from meeting the air, which renders them somewhat warm when they fall. The air in motion rather cools than heats, as in the winds, the bellows, or breath when the mouth is contracted. The motion, however, in such instances is not sufficiently rapid to excite heat, and is applied to a body of air, and not to its component parts, so that it is not surprising that heat should not be generated.

We must make a more diligent inquiry into this instance; for herbs and green and moist vegetables appear to possess a latent heat, so small, however, as not to be perceived by the touch in single specimens, but when they are united and confined, so that their spirit cannot exhale into the air, and they rather warm each other, their heat is at once manifested, and even flame occasionally in suitable substances.

Here, too, we must make a more diligent inquiry; for quicklime, when sprinkled with water, appears to conceive heat, either from its being collected into one point (as we observed of herbs when confined), or from the irritation and exasperation of the fiery spirit by water, which occasions a conflict and struggle. The true reason will more readily be shown if oil be used instead of water, for oil will equally tend to collect the confined spirit, but not to irritate. The experiment may be made more general, both by using the ashes and calcined products of different bodies and by pouring different liquids upon them.

A negative instance may be subjoined of other metals which are more soft and soluble; for leaf gold dissolved by aqua regia, or lead by aqua fortis, are not warm to the touch while dissolving, no more is quicksilver (as far as I remember), but silver excites a slight heat, and so does copper, and tin yet more plainly, and most of all iron and steel, which excite not only a powerful heat, but a violent bubbling. The heat, therefore, appears to be occasioned by the struggle which takes place when these strong dissolvents penetrate, dig into, and tear asunder the parts of those substances, while the substances themselves resist. When, however, the substances yield more easily, scarcely any heat is excited.

There is no negative instance with regard to the heat of animals, except in insects (as has been observed), owing to their small size; for in fishes, as compared with land animals, a lower degree rather than a deprivation of heat is observable. In plants and vegetables, both as to their exudations and pith when freshly exposed, there is no sensible degree of heat. But in animals there is a great difference in the degree, both in particular parts (for the heat varies near the heart, the brain, and the extremities) and in the circumstances in which they are placed, such as violent exercise and fevers.

Here, again, there is scarcely a negative instance. I might add that the excrements of animals, even when they are no longer fresh, possess evidently some effective heat, as is shown by their enriching the soil.

Such liquids (whether oily or watery) as are intensely acrid exhibit the effects of heat, by the separation and burning of bodies after some little action upon them, yet they are not at first warm to the touch, but they act according to their affinity and the pores of the substances to which they are applied; for aqua regia dissolves gold but not silver--on the contrary, aqua fortis dissolves silver but not gold;

neither of them dissolves glass, and so of the rest.

Let spirits of wine be tried on wood, or butter, wax, or pitch, to see if this will melt them at all by their heat; for the twenty-fourth instance shows that they possess properties resembling those of heat in causing incrustation. Let an experiment also be made with a graduated glass or calendar,[93] concave at the top, by pouring well-rectified spirits of wine into the cavity, and covering it up in order that they may the better retain their heat, then observe whether their heat make the water descend.

Spices and acrid herbs are sensibly warm to the palate, and still more so when taken internally; one should see, therefore, on what other substances they exhibit the effects of heat. Now, sailors tell us that when large quantities of spices are suddenly opened, after having been shut up for some time, there is some danger of fever and inflammation to those who stir them or take them out. An experiment might, therefore, be made whether such spices and herbs, when produced, will, like smoke, dry fish and meat hung up over them.

There is an acrid effect and a degree of penetration in cold liquids, such as vinegar and oil of vitriol, as well as in warm, such as oil of marjoram and the like; they have, therefore, an equal effect in causing animated substances to smart, and separating and consuming inanimate parts. There is not any negative instance as to this, nor does there exist any animal pain unaccompanied by the sensation of heat.

There are many effects common to cold and heat, however different in their process; for snowballs appear to burn boys' hands after a little time, and cold no less than fire preserves bodies from putrefaction--besides both heat and cold contract bodies. But it is better to refer these instances and the like to the investigation of cold.

XIII. In the third place we must exhibit to the understanding the instances in which that nature, which is the object of our inquiries, is present in a greater or less degree, either by comparing its increase and decrease in the same object, or its degree in different objects; for since the form of a thing is its very essence, and the thing only differs from its form as the apparent from the actual object, or the exterior from the interior, or that which is considered with relation to man from that which is considered with relation to the universe; it necessarily follows that no nature can be considered a real form which does not uniformly diminish and increase with the given nature. We are wont to call this our Table of Degrees, or Comparative Instances.

Table of the Degrees or Comparative Instances of Heat

We will first speak of those bodies which exhibit no degree of heat sensible to the touch, but appear rather to possess a potential heat, or disposition and preparation for it. We will then go on to others, which are actually warm to the touch, and observe the strength and degree of it.

1. There is no known solid or tangible body which is by its own nature originally warm; for neither stone, metal, sulphur, fossils, wood, water, nor dead animal carcasses are found warm. The warm springs in baths appear to be heated accidentally, by flame, subterraneous fire (such as is thrown up by Etna and many other mountains), or by the contact of certain bodies, as heat is exhibited in the dissolution of iron and tin. The degree of heat, therefore, in inanimate objects is not sensible to our touch; but they differ in their degrees of cold, for wood and metal are not equally cold.[94] This, however, belongs to the Table of Degrees of Cold.

2. But with regard to potential heat and predisposition to flame, we find many inanimate substances wonderfully adapted to it, as sulphur, naphtha, and saltpetre.

3. Bodies which have previously acquired heat, as horse dung from the animal, or lime, and perhaps ashes and soot from fire, retain some latent portion of it. Hence distillations and separations of substances are effected by burying them in horse dung, and heat is excited in lime by sprinkling it with water (as has been before observed).

4. In the vegetable world we know of no plant, nor part of any plant (as the exudations or pith) that is warm to man's touch. Yet (as we have before observed) green weeds grow warm when confined, and some vegetables are warm and others cold to our internal touch, i.e., the palate and stomach, or even after a while to our external skin (as is shown in plasters and ointments).

5. We know of nothing in the various parts of animals, when dead or detached from the rest, that is warm to the touch; for horse dung itself does not retain its heat, unless it be confined and buried. All dung, however, appears to possess a potential heat, as in manuring fields; so also dead bodies are endued with this latent and potential heat to such a degree, that in cemeteries where people are interred daily the earth acquires a secret heat, which consumes any recently deposited body much sooner than pure earth; and they tell you that the people of the East are acquainted with a fine soft cloth, made of the down of birds, which can melt butter wrapped gently up in it by its own warmth.

6. Manures, such as every kind of dung, chalk, sea-sand, salt and the like, have some disposition toward heat.

7. All putrefaction exhibits some slight degree of heat, though not enough to be perceptible by the touch; for neither the substances which by putrefaction are converted into animalculæ,[95] as flesh and cheese, nor rotten wood which shines in the dark, are warm to the touch. The heat, however, of putrid

substances displays itself occasionally in a disgusting and strong scent.

8. The first degree of heat, therefore, in substances which are warm to the human touch appears to be that of animals, and this admits of a great variety of degrees, for the lowest (as in insects) is scarcely perceptible, the highest scarcely equals that of the sun's rays in warm climates and weather, and is not so acute as to be insufferable to the hand. It is said, however, of Constantius, and some others of a very dry constitution and habit of body, that when attacked with violent fevers, they became so warm as to appear almost to burn the hand applied to them.

9. Animals become more warm by motion and exercise, wine and feasting, venery, burning fevers, and grief.

10. In the paroxysm of intermittent fevers the patients are at first seized with cold and shivering, but soon afterward become more heated than at first--in burning and pestilential fevers they are hot from the beginning.

11. Let further inquiry be made into the comparative heat of different animals, as fishes, quadrupeds, serpents, birds, and also of the different species, as the lion, the kite, or man; for, according to the vulgar opinion, fishes are the least warm internally, and birds the most, particularly doves, hawks, and ostriches.

12. Let further inquiry be made as to the comparative heat in different parts and limbs of the same animal; for milk, blood, seed, and eggs are moderately warm, and less hot than the outward flesh of the animal when in motion or agitated. The degree of heat of the brain, stomach, heart, and the rest, has not yet been equally well investigated.

13. All animals are externally cold in winter and cold weather, but are thought to be internally warmer.

14. The heat of the heavenly bodies, even in the warmest climates and seasons, never reaches such a pitch as to light or burn the driest wood or straw, or even tinder without the aid of burning-glasses. It can, however, raise vapor from moist substances.

15. Astronomers tell us that some stars are hotter than others. Mars is considered the warmest after the Sun, then Jupiter, then Venus. The Moon and, above all, Saturn, are considered to be cold. Among the fixed stars Sirius is thought the warmest, then Cor Leonis or Regulus, then the lesser Dog-star.

16. The sun gives out more heat as it approaches toward the perpendicular or zenith, which may be supposed to be the case with the other planets, according to their degree of heat; for instance, that Jupiter gives out more heat when situated beneath Cancer or Leo than when he is beneath Capricorn and Aquarius.

17. It is to be supposed that the sun and other planets give more heat in perigee, from their approximation to the earth, than when in apogee. But if in any country the sun should be both in its perigee and nearer to the perpendicular at the same time, it must necessarily give out more heat than in a country where it is also in perigee, but situated more obliquely; so that the comparative altitude of the planets should be observed, and their approach to or declination from the perpendicular in different countries.

18. The sun and other planets are thought also to give out more heat in proportion as they are nearer to the larger fixed stars, as when the sun is in Leo he is nearer Cor Leonis, Cauda Leonis, Spica Virginis, Sirius, and the lesser Dog-star, than when he is in Cancer, where, however, he approaches nearer to the perpendicular. It is probable, also, that the quarters of the heavens produce a greater heat (though not perceptibly), in proportion as they are adorned with a greater number of stars, particularly those of the first magnitude.

19. On the whole, the heat of the heavenly bodies is augmented in three ways: 1. The approach to the perpendicular; 2. Proximity or their perigee; 3. The conjunction or union of stars.

20. There is a very considerable difference between the degree of heat in animals, and even in the rays of the heavenly bodies (as they reach us), and the heat of the most gentle flame, and even of all ignited substances, nay, liquids, or the air itself when unusually heated by fire. For the flame of spirit of wine, though diffused and uncollected, is yet able to set straw, linen, or paper on fire, which animal heat, or that of the sun, will never accomplish without a burning-glass.

21. There are, however, many degrees of strength and weakness in flame and ignited bodies: but no diligent inquiry has been made in this respect, and we must, therefore, pass it hastily over. Of all flames, that of spirits of wine appears to be the most gentle, except perhaps the ignis fatuus, or the flashes from the perspiration of animals. After this we should be inclined to place the flame of light and porous vegetables, such as straw, reeds, and dried leaves; from which the flame of hair or feathers differs but little. Then, perhaps, comes the flame of wood, particularly that which contains but little rosin or pitch; that of small wood, however (such as is usually tied up in fagots), is milder than that of the trunks or roots of trees. This can be easily tried in iron furnaces, where a fire of fagots or branches of trees is of little service. Next follows the flame of oil, tallow, wax, and the like oily and fat substances, which are not very violent. But a most powerful heat is found in pitch and rosin, and a still greater in sulphur, camphor, naphtha, saltpetre, and salts (after they have discharged their crude matter), and in

their compounds; as in gunpowder, Greek fire (vulgarly called wild fire), and its varieties, which possess such a stubborn heat as scarcely to be extinguished by water.

22. We consider that the flame which results from some imperfect metals is very strong and active; but on all these points further inquiry should be made.

23. The flame of vivid lightning appears to exceed all the above, so as sometimes to have melted even wrought iron into drops, which the other flames cannot accomplish.

24. In ignited bodies there are different degrees of heat, concerning which, also, a diligent inquiry has not been made. We consider the faintest heat to be that of tinder, touchwood, and dry rope match, such as is used for discharging cannon. Next follows that of ignited charcoal or cinders, and even bricks, and the like; but the most violent is that of ignited metals, as iron, copper, and the like. Further inquiry, however, must be made into this also.

25. Some ignited bodies are found to be much warmer than some flames; for instance, red hot iron is much warmer, and burns more than the flame of spirits of wine.

26. Some bodies even not ignited, but only heated by the fire, as boiling water, and the air confined in reverberatories, surpass in heat many flames and ignited substances.

27. Motion increases heat,[96] as is shown in the bellows and the blowpipe; for the harder metals are not dissolved or melted by steady quiet fire, without the aid of the blowpipe.

28. Let an experiment be made with burning-glasses; in which respect I have observed, that if a glass be placed at the distance of ten inches, for instance, from the combustible object, it does not kindle or burn it so readily, as if the glass be placed at the distance of five inches (for instance), and be then gradually and slowly withdrawn to the distance of ten inches. The cone and focus of the rays, however, are the same, but the mere motion increases the effect of the heat.

29. Conflagrations, which take place with a high wind, are thought to make greater way against than with the wind, because when the wind slackens, the flame recoils more rapidly than it advances when the wind is favorable.

30. Flame does not burst out or arise unless it have some hollow space to move and exert itself in, except in the exploding flame of gunpowder, and the like, where the compression and confinement of the flame increase its fury.

31. The anvil becomes so hot by the hammer, that if it were a thin plate it might probably grow red, like ignited iron by repeated strokes. Let the experiment be tried.

32. But in ignited bodies that are porous, so as to leave room for the fire to move itself, if its motion be prevented by strong compression, the fire is immediately extinguished; thus it is with tinder, or the burning snuff of a candle or lamp, or even hot charcoal or cinders; for when they are squeezed by snuffers, or the foot, and the like, the effect of the fire instantly ceases.

33. The approach toward a hot body increases heat in proportion to the approximation; a similar effect to that of light, for the nearer any object is placed toward the light, the more visible it becomes.

34. The[97] union of different heats increases heat, unless the substances be mixed; for a large and small fire in the same spot tend mutually to increase each other's heat, but lukewarm water poured into boiling water cools it.

35. The continued neighborhood of a warm body increases heat. For the heat, which perpetually passes and emanates from it, being mixed with that which preceded it, multiplies the whole. A fire, for instance, does not warm a room in half an hour as much as the same fire would in an hour. This does not apply to light, for a lamp or candle placed in a spot gives no more light by remaining there, than it did at first.

36. The irritation of surrounding cold increases heat, as may be seen in fires during a sharp frost. We think that this is owing not merely to the confinement and compression of the heat (which forms a sort of union), but also by the exasperation of it, as when the air or a stick are violently compressed or bent, they recoil, not only to the point they first occupied, but still further back. Let an accurate experiment, therefore, be made with a stick, or something of the kind, put into the flame, in order to see whether it be not sooner burned at the sides than in the middle of it.[98]

37. There are many degrees in the susceptibility of heat. And, first, it must be observed how much a low gentle heat changes and partially warms even the bodies least susceptible of it. For even the heat of the hand imparts a little warmth to a ball of lead or other metal held a short time in it; so easily is heat transmitted and excited, without any apparent change in the body.

38. Of all bodies that we are acquainted with, air admits and loses heat the most readily, which is admirably seen in weather-glasses, whose construction is as follows: Take a glass with a hollow belly, and a thin and long neck; turn it upside down, and place it with its mouth downward into another glass vessel containing water; the end of the tube touching the bottom of the vessel, and the tube itself leaning a little on the edge, so as to be fixed upright. In order to do this more readily, let a little wax be applied to the edge, not, however, so as to block up the orifice, lest, by preventing the air from escaping, the motion, which we shall presently speak of, and which is very gentle and delicate, should be impeded.

Before the first glass be inserted in the other, its upper part (the belly) should be warmed at the fire. Then upon placing it as we have described, the air (which was dilated by the heat), after a sufficient time has been allowed for it to lose the additional temperature, will restore and contract itself to the same dimensions as that of the external or common atmosphere at the moment of immersion, and the water will be attracted upward in the tube to a proportionate extent. A long narrow slip of paper should be attached to the tube, divided into as many degrees as you please. You will then perceive, as the weather grows warmer or colder, that the air contracts itself into a narrower space in cold weather and dilates in the warm, which will be exhibited by the rising of the water as the air contracts itself, and its depression as the air dilates. The sensibility of the air with regard to heat or cold is so delicate and exquisite, that it far exceeds the human touch, so that a ray of sunshine, the heat of the breath, and much more, that of the hand placed on the top of the tube, immediately causes an evident depression of the water. We think, however, that the spirit of animals possesses a much more delicate susceptibility of heat and cold, only that it is impeded and blunted by the grossness of their bodies.

39. After air, we consider those bodies to be most sensible of heat, which have been recently changed and contracted by cold, as snow and ice; for they begin to be dissolved and melt with the first mild weather. Next, perhaps, follows quicksilver; then greasy substances, as oil, butter, and the like; then wood; then water; lastly, stones and metals, which do not easily grow hot, particularly toward their centre.[99] When heated, however, they retain their temperature for a very long time; so that a brick or stone, or hot iron, plunged in a basin of cold water, and kept there for a quarter of an hour or thereabout, retains such a heat as not to admit of being touched.

40. The less massive the body is, the more readily it grows warm at the approach of a heated body, which shows that heat with us is somewhat averse to a tangible mass.[100]

41. Heat with regard to the human senses and touch is various and relative, so that lukewarm water appears hot if the hand be cold, and cold if the hand be hot.

XIV. Any one may readily see how poor we are in history, since in the above tables, besides occasionally inserting traditions and report instead of approved history and authentic instances (always, however, adding some note if their credit or authority be doubtful), we are often forced to subjoin, "Let the experiment be tried--Let further inquiry be made."

XV. We are wont to term the office and use of these three tables the presenting a review of instances to the understanding; and when this has been done, induction itself is to be brought into action. For on an individual review of all the instances a nature is to be found, such as always to be present and absent with the given nature, to increase and decrease with it, and, as we have said, to form a more common limit of the nature. If the mind attempt this affirmatively from the first (which it always will when left to itself), there will spring up phantoms, mere theories and ill-defined notions, with axioms requiring daily correction. These will, doubtless, be better or worse, according to the power and strength of the understanding which creates them. But it is only for God (the bestower and creator of forms), and perhaps for angels and intelligences, at once to recognize forms affirmatively at the first glance of contemplation: man, at lest, is unable to do so, and is only allowed to proceed first by negatives, and then to conclude with affirmatives, after every species of exclusion.

XVI. We must, therefore, effect a complete solution and separation of nature; not by fire, but by the mind, that divine fire. The first work of legitimate induction, in the discovery of forms, is rejection, or the exclusive instances of individual natures, which are not found in some one instance where the given nature is present, or are found in any one instance where it is absent, or are found to increase in any one instance where the given nature decreases, or the reverse. After an exclusion correctly effected, an affirmative form will remain as the residuum, solid, true, and well defined, while all volatile opinions go off in smoke. This is readily said; but we must arrive at it by a circuitous route. We shall perhaps, however, omit nothing that can facilitate our progress.

XVII. The first and almost perpetual precaution and warning which we consider necessary is this; that none should suppose from the great part assigned by us to forms, that we mean such forms as the meditations and thoughts of men have hitherto been accustomed to. In the first place, we do not at present mean the concrete forms, which (as we have observed) are in the common course of things compounded of simple natures, as those of a lion, an eagle, a rose, gold, or the like. The moment for discussing these will arrive when we come to treat of the latent process and latent conformation, and the discovery of them as they exist in what are called substances, or concrete natures.

Nor again, would we be thought to mean (even when treating of simple natures) any abstract forms or ideas, either undefined or badly defined in matter. For when we speak of forms, we mean nothing else than those laws and regulations of simple action which arrange and constitute any simple nature, such as heat, light, weight, in every species of matter, and in a susceptible subject. The form of heat or form of light, therefore, means no more than the law of heat or the law of light. Nor do we ever abstract or withdraw ourselves from things, and the operative branch of philosophy. When, therefore, we say (for instance) in our investigation of the form of heat, Reject rarity, or, Rarity is not of the form of heat, it is the same as if we were to say, Man can superinduce heat on a dense body, or the reverse, Man can abstract or ward off heat from a rare body.

But if our forms appear to any one to be somewhat abstracted, from their mingling and uniting heterogeneous objects (the heat, for instance, of the heavenly bodies appears to be very different from that of fire; the fixed red of the rose and the like, from that which is apparent in the rainbow, or the radiation of opal or the diamond;[101] death by drowning, from that by burning, the sword, apoplexy, or consumption; and yet they all agree in the common natures of heat, redness, and death), let him be assured that his understanding is inthralled by habit, by general appearances and hypotheses. For it is most certain that, however heterogeneous and distinct, they agree in the form or law which regulates heat, redness, or death; and that human power cannot be emancipated and freed from the common course of nature, and expanded and exalted to new efficients and new modes of operation, except by the revelation and invention of forms of this nature. But after this[102] union of nature, which is the principal point, we will afterward, in its proper place, treat of the divisions and ramifications of nature, whether ordinary or internal and more real.

XVIII. We must now offer an example of the exclusion or rejection of natures found by the tables of review, not to be of the form of heat; first premising that not only each table is sufficient for the rejection of any nature, but even each single instance contained in them. For it is clear from what has been said that every contradictory instance destroys a hypothesis as to the form. Still, however, for the sake of clearness, and in order to show more plainly the use of the tables, we redouble or repeat the exclusive.

An Example of the Exclusive Table, or of the Rejection of Natures from the Form of Heat

1. On account of the sun's rays, reject elementary (or terrestrial) nature.

2. On account of common fire, and particularly subterranean fires (which are the most remote and secluded from the rays of the heavenly bodies), reject celestial nature.

3. On account of the heat acquired by every description of substances (as minerals, vegetables, the external parts of animals, water, oil, air, etc.) by mere approximation to the fire or any warm body, reject all variety and delicate texture of bodies.

4. On account of iron and ignited metals, which warm other bodies, and yet neither lose their weight nor substance, reject the imparting or mixing of the substance of the heating body.

5. On account of boiling water and air, and also those metals and other solid bodies which are heated, but not to ignition, or red heat, reject flame or light.

6. On account of the rays of the moon and other heavenly bodies (except the sun), again reject flame or light.

7. On account of the comparison between red-hot iron and the flame of spirits of wine (for the iron is more hot and less bright, while the flame of spirits of wine is more bright and less hot), again reject flame and light.

8. On account of gold and other ignited metals, which are of the greatest specific density, reject rarity.

9. On account of air, which is generally found to be cold and yet continues rare, reject rarity.

10. On account of ignited iron,[103] which does not swell in bulk, but retains the same apparent dimension, reject the absolute expansive motion of the whole.

11. On account of the expansion of the air in thermometers and the like, which is absolutely moved and expanded to the eye, and yet acquires no manifest increase of heat, again reject absolute or expansive motion of the whole.

12. On account of the ready application of heat to all substances without any destruction or remarkable alteration of them, reject destructive nature or the violent communication of any new nature.

13. On account of the agreement and conformity of the effects produced by cold and heat, reject both expansive and contracting motion as regards the whole.

14. On account of the heat excited by friction, reject principal nature, by which we mean that which exists positively, and is not caused by a preceding nature.

There are other natures to be rejected; but we are merely offering examples, and not perfect tables.

None of the above natures are of the form of heat; and man is freed from them all in his operation upon heat.

XIX. In the exclusive table are laid the foundations of true induction, which is not, however, completed until the affirmative be attained. Nor is the exclusive table perfect, nor can it be so at first. For it is clearly a rejection of simple natures; but if we have not as yet good and just notions of simple natures, how can the exclusive table be made correct? Some of the above, as the notion of elementary and celestial nature, and rarity, are vague and ill defined. We, therefore, who are neither ignorant nor forgetful of the great work which we attempt, in rendering the human understanding adequate to things and nature, by no means rest satisfied with what we have hitherto enforced, but push the matter further, and contrive and prepare more powerful aid for the use of the understanding, which we will next subjoin. And, indeed, in the interpretation of nature the mind is to be so prepared and formed, as to rest itself on proper degrees of certainty, and yet to remember (especially at first) that what is present depends much upon what remains behind.

XX. Since, however, truth emerges more readily from error than confusion, we consider it useful to leave the understanding at liberty to exert itself and attempt the interpretation of nature in the affirmative, after having constructed and weighed the three tables of preparation, such as we have laid them down, both from the instances there collected, and others occurring elsewhere. Which attempt we are wont to call the liberty of the understanding, or the commencement of interpretation, or the first vintage.

The First Vintage of the Form of Heat

It must be observed that the form of anything is inherent (as appears clearly from our premises) in each individual instance in which the thing itself is inherent, or it would not be a form. No contradictory instance, therefore, can be alleged. The form, however, is found to be much more conspicuous and evident in some instances than in others; in those (for example) where its nature is less restrained and embarrassed, and reduced to rule by other natures. Such instances we are wont to term coruscations, or conspicuous instances. We must proceed, then, to the first vintage of the form of heat.

From the instances taken collectively, as well as singly, the nature whose limit is heat appears to be motion. This is chiefly exhibited in flame, which is in constant motion, and in warm or boiling liquids, which are likewise in constant motion. It is also shown in the excitement or increase of heat by motion, as by bellows and draughts: for which see Inst. 29, Tab. 3, and by other species of motion, as in Inst. 28 and 31, Tab. 3. It is also shown by the extinction of fire and heat upon any strong pressure, which restrains and puts a stop to motion; for which see Inst. 30 and 32, Tab. 3. It is further shown by this circumstance, namely, that every substance is destroyed, or at least materially changed, by strong and powerful fire and heat: whence it is clear that tumult and confusion are occasioned by heat, together with a violent motion in the internal parts of bodies; and this gradually tends to their dissolution.

What we have said with regard to motion must be thus understood, when taken as the genus of heat: it must not be thought that heat generates motion, or motion heat (though in some respects this be true), but that the very essence of heat, or the substantial self[104] of heat, is motion and nothing else, limited, however, by certain differences which we will presently add, after giving some cautions for avoiding ambiguity.

Sensible heat is relative, and regards man, not universe; and is rightly held to be merely the effect of heat on animal spirit. It is even variable in itself, since the same body (in different states of sensation) excites the feeling of heat and of cold; this is shown by Inst. 41, Tab. 3.

Nor should we confound the communication of heat or its transitive nature, by which a body grows warm at the approach of a heated body, with the form of heat; for heat is one thing and heating another. Heat can be excited by friction without any previous heating body, and, therefore, heating is excluded from the form of heat. Even when heat is excited by the approach of a hot body, this depends not on the form of heat, but on another more profound and common nature; namely, that of assimilation and multiplication, about which a separate inquiry must be made.

The notion of fire is vulgar, and of no assistance; it is merely compounded of the conjunction of heat and light in any body, as in ordinary flame and red-hot substances.

Laying aside all ambiguity, therefore, we must lastly consider the true differences which limit motion and render it the form of heat.

I. The first difference is, that heat is an expansive motion, by which the body strives to dilate itself, and to occupy a greater space than before. This difference is principally seen in flame, where the smoke or thick vapor is clearly dilated and bursts into flame.

It is also shown in all boiling liquids, which swell, rise, and boil up to the sight, and the process of expansion is urged forward till they are converted into a much more extended and dilated body than the liquid itself, such as steam, smoke, or air.

It is also shown in wood and combustibles where exudation sometimes takes place, and evaporation always.

It is also shown in the melting of metals, which, being very compact, do not easily swell and dilate, but yet their spirit, when dilated and desirous of further expansion, forces and urges its thicker parts into dissolution, and if the heat be pushed still further, reduces a considerable part of them into a volatile state.

It is also shown in iron or stones, which though not melted or dissolved, are however softened. The same circumstance takes place in sticks of wood, which become flexible when a little heated in warm ashes.

It is most readily observed in air, which instantly and manifestly expands with a small degree of heat, as in Inst. 38, Tab. 3.

It is also shown in the contrary nature of cold; for cold contracts and narrows every substance;[105] so that in intense frosts nails fall out of the wall and brass cracks, and heated glass exposed suddenly to the cold cracks and breaks. So the air, by a slight degree of cold, contracts itself, as in Inst. 38, Tab. 3. More will be said of this in the inquiry into cold.

Nor is it to be wondered at if cold and heat exhibit many common effects (for which see Inst. 32,

Tab. 2), since two differences, of which we shall presently speak, belong to each nature: although in the present difference the effects be diametrically opposed to each other. For heat occasions an expansive and dilating motion, but cold a contracting and condensing motion.

II. The second difference is a modification of the preceding, namely, that heat is an expansive motion, tending toward the exterior, but at the same time bearing the body upward. For there is no doubt that there be many compound motions, as an arrow or dart, for instance, has both a rotatory and progressive motion. In the same way the motion of heat is both expansive and tending upward.

This difference is shown by putting the tongs or poker into the fire. If placed perpendicularly with the hand above, they soon burn it, but much less speedily if the hand hold them sloping or from below.

It is also conspicuous in distillations per descensum, which men are wont to employ with delicate flowers, whose scent easily evaporates. Their industry has devised placing the fire above instead of below, that it may scorch less; for not only flame but all heat has an upward tendency.

Let an experiment be made on the contrary nature of cold, whether its contraction be downward, as the expansion of heat is upward. Take, therefore, two iron rods or two glass tubes, alike in other respects, and warm them a little, and place a sponge, dipped in cold water, or some snow, below the one and above the other. We are of opinion that the extremities will grow cold in that rod first where it is placed beneath, as the contrary takes place with regard to heat.

III. The third difference is this; that heat is not a uniform expansive motion of the whole, but of the small particles of the body; and this motion being at the same time restrained, repulsed, and reflected, becomes alternating, perpetually hurrying, striving, struggling, and irritated by the repercussion, which is the source of the violence of flame and heat.

But this difference is chiefly shown in flame and boiling liquids, which always hurry, swell, and subside again in detached parts.

It is also shown in bodies of such hard texture as not to swell or dilate in bulk, such as red-hot iron, in which the heat is most violent.

It is also shown by the fires burning most briskly in the coldest weather.

It is also shown by this, that when the air is dilated in the thermometer uniformly and equably, without any impediment or repulsion, the heat is not perceptible. In confined draughts also, although they break out very violently, no remarkable heat is perceived, because the motion affects the whole, without any alternating motion in the particles; for which reason try whether flame do not burn more at the sides than in its centre.

It is also shown in this, that all burning proceeds by the minute pores of bodies--undermining, penetrating, piercing, and pricking them as if with an infinite number of needle-points. Hence all strong acids (if adapted to the body on which they act) exhibit the effects of fire, from their corroding and pungent nature.

The difference of which we now speak is common also to the nature of cold, in which the contracting motion is restrained by the resistance of expansion, as in heat the expansive motion is restrained by the resistance of contraction.

Whether, therefore, the particles of matter penetrate inward or outward, the reasoning is the same, though the power be very different, because we have nothing on earth which is intensely cold.

IV. The fourth difference is a modification of the preceding, namely, that this stimulating or penetrating motion should be rapid and never sluggish, and should take place not in the very minutest particles, but rather in those of some tolerable dimensions.

It is shown by comparing the effects of fire with those of time. Time dries, consumes, undermines, and reduces to ashes as well as fire, and perhaps to a much finer degree; but as its motion is very slow, and attacks very minute particles, no heat is perceived.

It is also shown in a comparison of the dissolution of iron and gold; for gold is dissolved without the excitement of any heat, but iron with a vehement excitement of it, although most in the same time, because in the former the penetration of the separating acid is mild, and gently insinuates itself, and the particles of gold yield easily, but the penetration of iron is violent, and attended with some struggle, and its particles are more obstinate.

It is partially shown, also, in some gangrenes and mortifications of flesh, which do not excite great heat or pain, from the gentle nature of the putrefaction.

Let this suffice for a first vintage, or the commencement of the interpretation of the form of heat by the liberty of the understanding.

From this first vintage the form or true definition of heat (considered relatively to the universe and not to the sense) is briefly thus--Heat is an expansive motion restrained, and striving to exert itself in the smaller particles.[106] The expansion is modified by its tendency to rise, though expanding toward the exterior; and the effort is modified by its not being sluggish, but active and somewhat violent.

With regard to the operative definition, the matter is the same. If you are able to excite a dilating or expansive motion in any natural body, and so to repress that motion and force it on itself as not to allow the expansion to proceed equally, but only to be partially exerted and partially repressed, you will beyond all doubt produce heat, without any consideration as to whether the body be of earth (or elementary, as they term it), or imbued with celestial influence, luminous or opaque, rare or dense, locally expanded or contained within the bounds of its first dimensions, verging to dissolution or remaining fixed, animal, vegetable, or mineral, water, or oil, or air, or any other substance whatever susceptible of such motion. Sensible heat is the same, but considered relatively to the senses. Let us now proceed to further helps.

XXI. After our tables of first review, our rejection or exclusive table, and the first vintage derived from them, we must advance to the remaining helps of the understanding with regard to the interpretation of nature, and a true and perfect induction, in offering which we will take the examples of cold and heat where tables are necessary, but where fewer instances are required we will go through a variety of others, so as neither to confound investigation nor to narrow our doctrine.

In the first place, therefore, we will treat of prerogative instances;[107] 2. Of the supports of induction; 3. Of the correction of induction; 4. Of varying the investigation according to the nature of the subject; 5. Of the prerogative natures with respect to investigation, or of what should be the first or last objects of our research; 6. Of the limits of investigation, or a synopsis of all natures that exist in the universe; 7. Of the application to practical purposes, or of what relates to man; 8. Of the preparations for investigation; 9. And lastly, of the ascending and descending scale of axioms.[108]

XXII. Among the prerogative instances we will first mention solitary instances. Solitary instances are those which exhibit the required nature in subjects that have nothing in common with any other subject than the nature in question, or which do not exhibit the required nature in subjects resembling others in every respect except that of the nature in question; for these instances manifestly remove prolixity, and accelerate and confirm exclusion, so that a few of them are of as much avail as many.

For instance, let the inquiry be the nature of color. Prisms, crystalline gems, which yield colors not only internally but on the wall, dews, etc., are solitary instances; for they have nothing in common with the fixed colors in flowers and colored gems, metals, woods, etc., except the color itself. Hence we easily deduce that color is nothing but a modification of the image of the incident and absorbed light, occasioned in the former case by the different degrees of incidence, in the latter by the various textures and forms of bodies.[109] These are solitary instances as regards similitude.

Again, in the same inquiry the distinct veins of white and black in marble, and the variegated colors of flowers of the same species, are solitary instances; for the black and white of marble, and the spots of white and purple in the flowers of the stock, agree in every respect but that of color. Thence we easily deduce that color has not much to do with the intrinsic natures of any body, but depends only on the coarser and as it were mechanical arrangement of the parts. These are solitary instances as regards difference. We call them both solitary or wild, to borrow a word from the astronomers.

XXIII. In the second rank of prerogative instances we will consider migrating instances. In these the required nature passes toward generation, having no previous existence, or toward corruption, having first existed. In each of these divisions, therefore, the instances are always twofold, or rather it is one instance, first in motion or on its passage, and then brought to the opposite conclusion. These instances not only hasten and confirm exclusion, but also reduce affirmation, or the form itself, to a narrow compass; for the form must be something conferred by this migration, or, on the contrary, removed and destroyed by it; and although all exclusion advances affirmation, yet this takes place more directly in the same than in different subjects; but if the form (as it is quite clear from what has been advanced) exhibit itself in one subject, it leads to all. The more simple the migration is, the more valuable is the instance. These migrating instances are, moreover, very useful in practice, for since they manifest the form, coupled with that which causes or destroys it, they point out the right practice in some subjects, and thence there is an easy transition to those with which they are most allied. There is, however, a degree of danger which demands caution, namely, lest they should refer the form too much to its efficient cause, and imbue, or at least tinge, the understanding with a false notion of the form from the appearance of such cause, which is never more than a vehicle or conveyance of the form. This may easily be remedied by a proper application of exclusion.

Let us then give an example of a migrating instance. Let whiteness be the required nature. An instance which passes toward generation is glass in its entire and in its powdered state, or water in its natural state, and when agitated to froth; for glass when entire, and water in its natural state, are transparent and not white, but powdered glass and the froth of water are white and not transparent. We must inquire, therefore, what has happened to the glass or water in the course of this migration; for it is manifest that the form of whiteness is conveyed and introduced by the bruising of the glass and the agitation of the water; but nothing is found to have been introduced but a diminishing of the parts of the glass and water and the insertion of air. Yet this is no slight progress toward discovering the form of whiteness, namely, that two bodies, in themselves more or less transparent (as air and water, or air and glass), when brought into contact in minute portions, exhibit whiteness from the unequal refraction of the rays of light.

But here we must also give an example of the danger and caution of which we spoke; for instance, it will readily occur to an understanding perverted by efficients, that air is always necessary for producing the form of whiteness, or that whiteness is only generated by transparent bodies, which suppositions are both false, and proved to be so by many exclusions; nay, it will rather appear (without any particular regard to air or the like), that all bodies which are even in such of their parts as affect the sight exhibit transparency, those which are uneven and of simple texture whiteness, those which are uneven and of compound but regular texture all the other colors except black, but those which are uneven and of a compound irregular and confused texture exhibit blackness. An example has been given, therefore, of an instance migrating toward generation in the required nature of whiteness. An instance migrating toward corruption in the same nature is that of dissolving froth or snow, for they lose their whiteness and assume the transparency of water in its pure state without air.

Nor should we by any means omit to state, that under migrating instances we must comprehend not only those which pass toward generation and destruction, but also those which pass toward increase or decrease, for they, too, assist in the discovery of the form, as is clear from our definition of a form and the Table of Degrees. Hence paper, which is white when dry, is less white when moistened (from the exclusion of air and admission of water), and tends more to transparency. The reason is the same as in the above instances.[110]

XXIV. In the third rank of prerogative instances we will class conspicuous instances, of which we spoke in our first vintage of the form of heat, and which we are also wont to call coruscations, or free and predominant instances. They are such as show the required nature in its bare substantial shape, and at its height or greatest degree of power, emancipated and free from all impediments, or at least overcoming, suppressing, and restraining them by the strength of its qualities; for since every body is susceptible of many united forms of natures in the concrete, the consequence is that they mutually deaden, depress, break, and confine each other, and the individual forms are obscured. But there are some subjects in which the required nature exists in its full vigor rather than in others, either from the absence of any impediment, or the predominance of its quality. Such instances are eminently conspicuous. But even in these care must be taken, and the hastiness of the understanding checked, for whatever makes a show of the form, and forces it forward, is to be suspected, and recourse must be had to severe and diligent exclusion.

For example, let heat be the required nature. The thermometer is a conspicuous instance of the expansive motion, which (as has been observed) constitutes the chief part of the form of heat; for although flame clearly exhibits expansion, yet from its being extinguished every moment, it does not exhibit the progress of expansion. Boiling water again, from its rapid conversion into vapor, does not so well exhibit the expansion of water in its own shape, while red-hot iron and the like are so far from showing this progress, that, on the contrary, the expansion itself is scarcely evident to the senses, on account of its spirit being repressed and weakened by the compact and coarse particles which subdue and restrain it. But the thermometer strikingly exhibits the expansion of the air as being evident and progressive, durable and not transitory.[111]

Take another example. Let the required nature be weight. Quicksilver is a conspicuous instance of weight; for it is far heavier than any other substance except gold, which is not much heavier, and it is a better instance than gold for the purpose of indicating the form of weight; for gold is solid and consistent, which qualities must be referred to density, but quicksilver is liquid and teeming with spirit, yet much heavier than the diamond and other substances considered to be most solid; whence it is shown that the form of gravity or weight predominates only in the quantity of matter, and not in the close fitting of it.[112]

XXV. In the fourth rank of prerogative instances we will class clandestine instances, which we are also wont to call twilight instances; they are as it were opposed to the conspicuous instances, for they show the required nature in its lowest state of efficacy, and as it were its cradle and first rudiments, making an effort and a sort of first attempt, but concealed and subdued by a contrary nature. Such instances are, however, of great importance in discovering forms, for as the conspicuous tend easily to differences, so do the clandestine best lead to genera, that is, to those common natures of which the required natures are only the limits.

As an example, let consistency, or that which confines itself, be the required nature, the opposite of which is a liquid or flowing state. The clandestine instances are such as exhibit some weak and low degree of consistency in fluids, as a water bubble, which is a sort of consistent and bounded pellicle formed out of the substance of the water. So eaves' droppings, if there be enough water to follow them, draw themselves out into a thin thread, not to break the continuity of the water, but if there be not enough to follow, the water forms itself into a round drop, which is the best form to prevent a breach of continuity; and at the moment the thread ceases, and the water begins to fall in drops, the thread of water recoils upward to avoid such a breach. Nay, in metals, which when melted are liquid but more tenacious, the melted drops often recoil and are suspended. There is something similar in the instance of the child's looking-glass, which little boys will sometimes form of spittle between rushes, and where the same pellicle of water is observable; and still more in that other amusement of children, when they take some water rendered a little more tenacious by soap, and inflate it with a pipe, forming the water into a sort of castle of bubbles, which assumes such consistency, by the interposition of the air, as to admit of being thrown some little distance without bursting. The best example is that of froth and snow, which assume such consistency as almost to admit of being cut, although composed of air and water, both liquids. All these circumstances clearly show that the terms liquid and consistent are merely vulgar notions adapted to the sense, and that in reality all bodies have a tendency to avoid a breach of continuity, faint and weak in bodies composed of homogeneous parts (as is the case with liquids), but more vivid and powerful in those composed of heterogeneous parts, because the approach of heterogeneous matter binds bodies together, while the insinuation of homogeneous matter loosens and relaxes them.

Again, to take another example, let the required nature be attraction or the cohesion of bodies. The most remarkable conspicuous instance with regard to its form is the magnet. The contrary nature to attraction is non-attraction, though in a similar substance. Thus iron does not attract iron, lead lead, wood wood, nor water water. But the clandestine instance is that of the magnet armed with iron, or rather that of iron in the magnet so armed. For its nature is such that the magnet when armed does not attract iron more powerfully at any given distance than when unarmed; but if the iron be brought in contact with the armed magnet, the latter will sustain a much greater weight than the simple magnet, from the resemblance of substance in the two portions of iron, a quality altogether clandestine and hidden in the iron until the magnet was introduced. It is manifest, therefore, that the form of cohesion is something which is vivid and robust in the magnet, and hidden and weak in the iron. It is to be observed, also, that small wooden arrows without an iron point, when discharged from large mortars, penetrate further into wooden substances (such as the ribs of ships or the like), than the same arrows pointed with iron,[113] owing to the similarity of substance, though this quality was previously latent in the wood. Again, although in the mass air does not appear to attract air, nor water water, yet when one bubble is brought near another, they are both more readily dissolved, from the tendency to contact of the water with the water, and the air with the air.[114] These clandestine instances (which are, as has been observed, of the most important service) are principally to be observed in small portions of bodies, for the larger masses observe more universal and general forms, as will be mentioned in its proper place.[115]

XXVI. In the fifth rank of prerogative instances we will class constitutive instances, which we are wont also to call collective instances. They constitute a species or lesser form, as it were, of the required nature. For since the real forms (which are always convertible with the given nature) lie at some depth, and are not easily discovered, the necessity of the case and the infirmity of the human understanding require that the particular forms, which collect certain groups of instances (but by no means all) into some common notion, should not be neglected, but most diligently observed. For whatever unites nature, even imperfectly, opens the way to the discovery of the form. The instances, therefore, which are serviceable in this respect are of no mean power, but endowed with some degree of prerogative.

Here, nevertheless, great care must be taken that, after the discovery of several of these particular forms, and the establishing of certain partitions or divisions of the required nature derived from them, the human understanding do not at once rest satisfied, without preparing for the investigation of the great or leading form, and taking it for granted that nature is compound and divided from its very root, despise and reject any further union as a point of superfluous refinement, and tending to mere abstraction.

For instance, let the required nature be memory, or that which excites and assists memory. The constitutive instances are order or distribution, which manifestly assists memory: topics or commonplaces in artificial memory, which may be either places in their literal sense, as a gate, a corner, a window, and the like, or familiar persons and marks, or anything else (provided it be arranged in a determinate order), as animals, plants, and words, letters, characters, historical persons, and the like, of which, however, some are more convenient than others. All these commonplaces materially assist memory, and raise it far above its natural strength. Verse, too, is recollected and learned more easily than prose. From this group of three instances--order, the commonplaces of artificial memory, and verses--is constituted one species of aid for the memory,[116] which may be well termed a separation from infinity. For when a man strives to recollect or recall anything to memory, without a preconceived notion or perception of the object of his search, he inquires about, and labors, and turns from point to point, as if involved in infinity. But if he have any preconceived notion, this infinity is separated off, and the range of his memory is brought within closer limits. In the three instances given above, the preconceived notion is clear and determined. In the first, it must be something that agrees with order; in the second, an image which has some relation or agreement with the fixed commonplaces; in the third, words which fall into a verse: and thus infinity is divided off. Other instances will offer another species, namely, that whatever brings the intellect into contact with something that strikes the sense (the principal point of artificial memory), assists the memory. Others again offer another species, namely, whatever excites an impression by any powerful passion, as fear, shame, wonder, delight, assists the memory. Other instances will afford another species: thus those impressions remain most fixed in the memory which are taken from the mind when clear and least occupied by preceding or succeeding notions, such as the things we learn in childhood, or imagine before sleep, and the first time of any circumstance happening. Other instances afford the following species: namely, that a multitude of circumstances or handles assist the memory, such as writing in paragraphs, reading aloud, or recitation. Lastly, other instances afford still another species: thus the things we anticipate, and which rouse our attention, are more easily remembered than transient events; as if you read any work twenty times over, you will not learn it by heart so readily as if you were to read it but ten times, trying each time to repeat it, and when your memory fails you looking into the book. There are, therefore, six lesser forms, as it were, of things which assist the memory: namely--1, the separation of infinity; 2, the connection of the mind with the senses; 3, the impression in strong passion; 4, the impression on the mind when pure; 5, the multitude of handles; 6, anticipation.

Again, for example's sake, let the required nature be taste or the power of tasting. The following instances are constitutive: 1. Those who do not smell, but are deprived by nature of that sense, do not perceive or distinguish rancid or putrid food by their taste, nor garlic from roses, and the like. 2. Again, those whose nostrils are obstructed by accident (such as a cold) do not distinguish any putrid or rancid matter from anything sprinkled with rose-water. 3. If those who suffer from a cold blow their noses violently at the very moment in which they have anything fetid or perfumed in their mouth, or on their palate, they instantly have a clear perception of the fetor or perfume. These instances afford and constitute this species or division of taste, namely, that it is in part nothing else than an internal smelling, passing and descending through the upper passages of the nostrils to the mouth and palate. But, on the other hand, those whose power of smelling is deficient or obstructed, perceive what is salt, sweet, pungent, acid, rough, and bitter, and the like, as well as any one else: so that the taste is clearly something compounded of the internal smelling, and an exquisite species of touch which we will not here discuss.

Again, as another example, let the required nature be the communication of quality, without intermixture of substance. The instance of light will afford or constitute one species of communication, heat and the magnet another. For the communication of light is momentary and immediately arrested upon the removal of the original light. But heat, and the magnetic force, when once transmitted to or excited in another body, remain fixed for a considerable time after the removal of the source.

In fine, the prerogative of constitutive instances is considerable, for they materially assist the definitions (especially in detail) and the divisions or partitions of natures, concerning which Plato has well said, "He who can properly define and divide is to be considered a god."[117]

XXVII. In the sixth rank of prerogative instances we will place similar or proportionate instances, which we are also wont to call physical parallels, or resemblances. They are such as exhibit the resemblances and connection of things, not in minor forms (as the constitutive do), but at once in the concrete. They are, therefore, as it were, the first and lowest steps toward the union of nature; nor do they immediately establish any axiom, but merely indicate and observe a certain relation of bodies to each other. But although they be not of much assistance in discovering forms, yet they are of great advantage in disclosing the frame of parts of the universe, upon whose members they practice a species of anatomy, and thence occasionally lead us gently on to sublime and noble axioms, especially such as relate to the construction of the world, rather than to simple natures and forms.

As an example, take the following similar instances: a mirror and the eye; the formation of the ear, and places which return an echo. From such similarity, besides observing the resemblance (which is useful for many purposes), it is easy to collect and form this axiom. That the organs of the senses, and bodies which produce reflections to the senses, are of a similar nature. Again, the understanding once informed of this, rises easily to a higher and nobler axiom; namely, that the only distinction between sensitive and inanimate bodies, in those points in which they agree and sympathize, is this: in the former, animal spirit is added to the arrangement of the body, in the latter it is wanting. So that there might be as many senses in animals as there are points of agreement with inanimate bodies, if the animated body were perforated, so as to allow the spirit to have access to the limb properly disposed for action, as a fit organ. And, on the other hand, there are, without doubt, as many motions in an inanimate as there are senses in the animated body, though the animal spirit be absent. There must, however, be many more motions in inanimate bodies than senses in the animated, from the small number of organs of sense. A very plain example of this is afforded by pains. For, as animals are liable to many kinds and various descriptions of pains (such as those of burning, of intense cold, of pricking, squeezing, stretching, and the like), so is it most certain, that the same circumstances, as far as motion is concerned, happen to inanimate bodies, such as wood or stone when burned, frozen, pricked, cut, bent, bruised, and the like; although there be no sensation, owing to the absence of animal spirit.

Again, wonderful as it may appear, the roots and branches of trees are similar instances. For every vegetable swells and throws out its constituent parts toward the circumference, both upward and downward. And there is no difference between the roots and branches, except that the root is buried in the earth, and the branches are exposed to the air and sun. For if one take a young and vigorous shoot, and bend it down to a small portion of loose earth, although it be not fixed to the ground, yet will it immediately produce a root, and not a branch. And, vice versâ, if earth be placed above, and so forced down with a stone or any hard substance, as to confine the plant and prevent its branching upward, it will throw out branches into the air downward.

The gums of trees, and most rock gems, are similar instances; for both of them are exudations and filtered juices, derived in the former instance from trees, in the latter from stones; the brightness and clearness of both arising from a delicate and accurate filtering. For nearly the same reason, the hair of animals is less beautiful and vivid in its color than the plumage of most birds, because the juices are less delicately filtered through the skin than through the quills.

The scrotum of males and matrix of females are also similar instances; so that the noble formation which constitutes the difference of the sexes appears to differ only as to the one being internal and the other external; a greater degree of heat causing the genitals to protrude in the male, while the heat of the female being too weak to effect this, they are retained internally.

The fins of fishes and the feet of quadrupeds, or the feet and wings of birds, are similar instances; to which Aristotle adds the four folds in the motion of serpents;[118] so that in the formation of the universe, the motion of animals appears to be chiefly effected by four joints or bendings.

The teeth of land animals, and the beaks of birds, are similar instances, whence it is clear, that in all perfect animals there is a determination of some hard substance toward the mouth.

Again, the resemblance and conformity of man to an inverted plant is not absurd. For the head is the root of the nerves and animal faculties, and the seminal parts are the lowest, not including the extremities of the legs and arms. But in the plant, the root (which resembles the head) is regularly placed in the lowest, and the seeds in the highest part.[119]

Lastly, we must particularly recommend and suggest, that man's present industry in the investigation and compilation of natural history be entirely changed, and directed to the reverse of the present system. For it has hitherto been active and curious in noting the variety of things, and explaining the accurate differences of animals, vegetables, and minerals, most of which are the mere sport of nature, rather than of any real utility as concerns the sciences. Pursuits of this nature are certainly agreeable, and sometimes of practical advantage, but contribute little or nothing to the thorough investigation of nature. Our labor must therefore be directed toward inquiring into and observing resemblances and analogies, both in the whole and its parts, for they unite nature, and lay the foundation of the sciences.

Here, however, a severe and rigorous caution must be observed, that we only consider as similar and proportionate instances, those which (as we first observed) point out physical resemblances; that is, real and substantial resemblances, deeply founded in nature, and not casual and superficial, much less superstitious or curious; such as those which are constantly put forward by the writers on natural magic (the most idle of men, and who are scarcely fit to be named in connection with such serious matters as we now treat of), who, with much vanity and folly, describe, and sometimes too, invent, unmeaning resemblances and sympathies.

But leaving such to themselves, similar instances are not to be neglected, in the greater portions of the world's conformation; such as Africa and the Peruvian continent, which reaches to the Straits of Magellan; both of which possess a similar isthmus and similar capes, a circumstance not to be attributed to mere accident.

Again, the New and Old World are both of them broad and expanded toward the north, and narrow and pointed toward the south.

Again, we have verÿ remarkable similar instances in the intense cold, toward the middle regions (as it is termed) of the air, and the violent fires which are often found to burst from subterraneous spots, the similarity consisting in both being ends and extremes; the extreme of the nature of cold, for instance, is toward the boundary of heaven, and that of the nature of heat toward the centre of the earth, by a similar species of opposition or rejection of the contrary nature.

Lastly, in the axioms of the sciences, there is a similarity of instances worthy of observation. Thus the rhetorical trope which is called surprise, is similar to that of music termed the declining of a cadence. Again--the mathematical postulate, that things which are equal to the same are equal to one another, is similar to the form of the syllogism in logic, which unites things agreeing in the middle term.[120] Lastly, a certain degree of sagacity in collecting and searching for physical points of similarity, is very useful in many respects.[121]

XXVIII. In the seventh rank of prerogative instances, we will place singular instances, which we are also wont to call irregular or heteroclite (to borrow a term from the grammarians). They are such as exhibit bodies in the concrete, of an apparently extravagant and separate nature, agreeing but little with other things of the same species. For, while the similar instances resemble each other, those we now speak of are only like themselves. Their use is much the same with that of clandestine instances: they bring out and unite nature, and discover genera or common natures, which must afterward be limited by real differences. Nor should we desist from inquiry, until the properties and qualities of those things, which may be deemed miracles, as it were, of nature, be reduced to, and comprehended in, some form or certain law; so that all irregularity or singularity may be found to depend on some common form; and the miracle only consists in accurate differences, degree, and rare coincidence, not in the species itself. Man's meditation proceeds no further at present, than just to consider things of this kind as the secrets and vast efforts of nature, without an assignable cause, and, as it were, exceptions to general rules.

As examples of singular instances, we have the sun and moon among the heavenly bodies; the magnet among minerals; quicksilver among metals; the elephant among quadrupeds; the venereal sensation among the different kinds of touch; the scent of sporting dogs among those of smell. The letter S, too, is considered by the grammarians as sui generis, from its easily uniting with double or triple consonants, which no other letter will. These instances are of great value, because they excite and keep alive inquiry, and correct an understanding depraved by habit and the common course of things.

XXIX. In the eighth rank of prerogative instances, we will place deviating instances, such as the errors of nature, or strange and monstrous objects, in which nature deviates and turns from her ordinary course. For the errors of nature differ from singular instances, inasmuch as the latter are the miracles of species, the former of individuals. Their use is much the same, for they rectify the understanding in opposition to habit, and reveal common forms. For with regard to these, also, we must not desist from inquiry, till we discern the cause of the deviation. The cause does not, however, in such cases rise to a regular form, but only to the latent process toward such a form. For he who is acquainted with the paths of nature, will more readily observe her deviations; and, vice versâ, he who has learned her deviations will be able more accurately to describe her paths.

They differ again from singular instances, by being much more apt for practice and the operative branch. For it would be very difficult to generate new species, but less so to vary known species, and thus produce many rare and unusual results.[122] The passage from the miracles of nature to those of art is easy; for if nature be once seized in her variations, and the cause be manifest, it will be easy to lead her by art to such deviation as she was at first led to by chance; and not only to that but others, since deviations on the one side lead and open the way to others in every direction. Of this we do not require any examples, since they are so abundant. For a compilation, or particular natural history, must be made of all monsters and prodigious births of nature; of everything, in short, which is new, rare and unusual in nature. This should be done with a rigorous selection, so as to be worthy of credit. Those are most to be suspected which depend upon superstition, as the prodigies of Livy, and those perhaps, but little less, which are found in the works of writers on natural magic, or even alchemy, and the like; for such men, as it were, are the very suitors and lovers of fables; but our instances should be derived from some grave and credible history, and faithful narration.

XXX. In the ninth rank of prerogative instances, we will place bordering instances, which we are also wont to term participants. They are such as exhibit those species of bodies which appear to be composed of two species, or to be the rudiments between the one and the other. They may well be classed with the singular or heteroclite instances; for in the whole system of things, they are rare and extraordinary. Yet from their dignity, they must be treated of and classed separately, for they point out admirably the order and constitution of things, and suggest the causes of the number and quality of the more common species in the universe, leading the understanding from that which is, to that which is possible.

We have examples of them in moss, which is something between putrescence and a plant;[123] in some comets, which hold a place between stars and ignited meteors; in flying fishes, between fishes and birds; and in bats, between birds and quadrupeds.[124] Again,

Simia quam similis turpissima bestia nobis.

We have also biformed fœtus, mingled species and the like.

XXXI. In the tenth rank of prerogative instances, we will place the instances of power, or the fasces (to borrow a term from the insignia of empire), which we are also wont to call the wit or hands of man. These are such works as are most noble and perfect, and, as it were, the masterpieces in every art. For since our principal object is to make nature subservient to the state and wants of man, it becomes us well to note and enumerate the works, which have long since been in the power of man, especially those which are most polished and perfect: because the passage from these to new and hitherto undiscovered works, is more easy and feasible. For if any one, after an attentive contemplation of such works as are extant, be willing to push forward in his design with alacrity and vigor, he will undoubtedly either advance them, or turn them to something within their immediate reach, or even apply and transfer them to some more noble purpose.

Nor is this all: for as the understanding is elevated and raised by rare and unusual works of nature, to investigate and discover the forms which include them also, so is the same effect frequently produced by the excellent and wonderful works of art; and even to a greater degree, because the mode of effecting and constructing the miracles of art is generally plain, while that of effecting the miracles of nature is more obscure. Great care, however, must be taken, that they do not depress the understanding, and fix it, as it were, to earth.

For there is some danger, lest the understanding should be astonished and chained down, and as it were bewitched, by such works of art, as appear to be the very summit and pinnacle of human industry, so as not to become familiar with them, but rather to suppose that nothing of the kind can be accomplished, unless the same means be employed, with perhaps a little more diligence, and more accurate preparation.

Now, on the contrary, it may be stated as a fact, that the ways and means hitherto discovered and observed, of effecting any matter or work, are for the most part of little value, and that all really efficient power depends, and is really to be deduced from the sources of forms, none of which have yet been discovered.

Thus (as we have before observed), had any one meditated on ballistic machines, and battering rams, as they were used by the ancients, whatever application he might have exerted, and though he might have consumed a whole life in the pursuit, yet would he never have hit upon the invention of flaming engines, acting by means of gunpowder; nor would any person, who had made woollen manufactories and cotton the subject of his observation and reflection, have ever discovered thereby the nature of the silkworm or of silk.

Hence all the most noble discoveries have (if you observe) come to light, not by any gradual improvement and extension of the arts, but merely by chance; while nothing imitates or anticipates chance (which is wont to act at intervals of ages) but the invention of forms.

There is no necessity for adducing any particular examples of these instances, since they are abundant. The plan to be pursued is this: all the mechanical, and even the liberal arts (as far as they are practical), should be visited and thoroughly examined, and thence there should be formed a compilation or particular history of the great masterpieces, or most finished works in each, as well as of the mode of carrying them into effect.

Nor do we confine the diligence to be used in such a compilation to the leading works and secrets only of every art, and such as excite wonder; for wonder is engendered by rarity, since that which is rare, although it be compounded of ordinary natures, always begets wonder.

On the contrary, that which is really wonderful, from some specific difference distinguishing it from other species, is carelessly observed, if it be but familiar. Yet the singular instances of art should be observed no less than those of nature, which we have before spoken of: and as in the latter we have classed the sun, the moon, the magnet, and the like, all of them most familiar to us, but yet in their nature singular, so should we proceed with the singular instances of art.

For example: paper, a very common substance, is a singular instance of art; for if you consider the subject attentively, you will find that artificial substances are either woven by straight and transverse lines, as silk, woollen, or linen cloth, and the like; or coagulated from concrete juices, such as brick, earthenware, glass, enamel, porcelain and the like, which admit of a polish if they be compact, but if not, become hard without being polished; all which latter substances are brittle, and not adherent or tenacious. On the contrary, paper is a tenacious substance, which can be cut and torn, so as to resemble and almost rival the skin of any animal, or the leaf of vegetables, and the like works of nature; being neither brittle like glass, nor woven like cloth, but having fibres and not distinct threads, just as natural substances, so that scarcely anything similar can be found among artificial substances, and it is absolutely singular. And in artificial works we should certainly prefer those which approach the nearest to an imitation of nature, or, on the other hand, powerfully govern and change her course.

Again, in these instances which we term the wit and hands of man, charms and conjuring should not be altogether despised, for although mere amusements, and of little use, yet they may afford considerable information.

Lastly, superstition and magic (in its common acceptation) are not to be entirely omitted; for although they be overwhelmed by a mass of lies and fables, yet some investigation should be made, to see if there be really any latent natural operation in them; as in fascination, and the fortifying of the imagination, the sympathy of distant objects, the transmission of impressions from spirit to spirit no less than from body to body, and the like.

XXXII. From the foregoing remarks, it is clear that the last five species of instances (the similar, singular, deviating and bordering instances, and those of power) should not be reserved for the investigation of any given nature, as the preceding and many of the succeeding instances must, but a collection of them should be made at once, in the style of a particular history, so that they may arrange the matter which enters the understanding, and correct its depraved habit, for it is necessarily imbued, corrupted, perverted and distorted by daily and habitual impressions.

They are to be used, therefore, as a preparative, for the purpose of rectifying and purifying the understanding; for whatever withdraws it from habit, levels and planes down its surface for the reception of the dry and pure light of true notions.

These instances, moreover, level and prepare the way for the operative branch, as we will mention in its proper place when speaking of the practical deductions.

XXXIII. In the eleventh rank of prerogative instances we will place accompanying and hostile instances. These are such as exhibit any body or concrete, where the required nature is constantly found, as an inseparable companion, or, on the contrary, where the required nature is constantly avoided, and excluded from attendance, as an enemy. From these instances may be formed certain and universal propositions, either affirmative or negative; the subject of which will be the concrete body, and the predicate the required nature. For particular propositions are by no means fixed, when the required nature is found to fluctuate and change in the concrete, either approaching and acquired, or receding and laid aside. Hence particular propositions have no great prerogative, except in the case of migration, of which we have spoken above. Yet such particular propositions are of great use, when compared with the universal, as will be mentioned in its proper place. Nor do we require absolute affirmation or negation, even in universal propositions, for if the exceptions be singular or rare, it is sufficient for our purpose.

The use of accompanying instances is to narrow the affirmative of form; for as it is narrowed by the migrating instances, where the form must necessarily be something communicated or destroyed by the act of migration, so it is narrowed by accompanying instances, where the form must necessarily be something which enters into the concretion of the body, or, on the contrary, is repugnant to it; and one who is well acquainted with the constitution or formation of the body, will not be far from bringing to light the form of the required nature.

For example: let the required nature be heat. Flame is an accompanying instance; for in water, air, stone, metal, and many other substances, heat is variable, and can approach or retire; but all flame is hot, so that heat always accompanies the concretion of flame. We have no hostile instance of heat; for the senses are unacquainted with the interior of the earth, and there is no concretion of any known body which is not susceptible of heat.

Again, let solidity be the required nature. Air is a hostile instance; for metals may be liquid or solid, so may glass; even water may become solid by congelation, but air cannot become solid or lose its fluidity.

With regard to these instances of fixed propositions, there are two points to be observed, which are of importance. First, that if there be no universal affirmative or negative, it be carefully noted as not existing. Thus, in heat, we have observed that there exists no universal negative, in such substances, at least, as have come to our knowledge. Again, if the required nature be eternity or incorruptibility, we have no universal affirmative within our sphere, for these qualities cannot be predicated of any bodies below the heavens, or above the interior of the earth. Secondly, to our general propositions as to any concrete, whether affirmative or negative, we should subjoin the concretes which appear to approach nearest to the non-existing substances; such as the most gentle or least-burning flames in heat, or gold in incorruptibility, since it approaches nearest to it. For they all serve to show the limit of existence and non-existence, and circumscribe forms, so that they cannot wander beyond the conditions of matter.

XXXIV. In the twelfth rank of prerogative instances, we will class those subjunctive instances, of which we spoke in the last aphorism, and which we are also wont to call instances of extremity or limits; for they are not only serviceable when subjoined to fixed propositions, but also of themselves and from their own nature. They indicate with sufficient precision the real divisions of nature, and measures of things, and the "how far" nature effects or allows of anything, and her passage thence to something else. Such are gold in weight, iron in hardness, the whale in the size of animals, the dog in smell, the flame of gunpowder in rapid expansion, and others of a like nature. Nor are we to pass over the extremes in defect, as well as in abundance, as spirits of wine in weight, the touchstone in softness, the worms upon the skin in the size of animals, and the like.

XXXV. In the thirteenth rank of prerogative instances we will place those of alliance or union. They are such as mingle and unite natures held to be heterogeneous, and observed and marked as such in received classifications.

These instances show that the operation and effect, which is considered peculiar to some one of such heterogeneous natures, may also be attributed to another nature styled heterogeneous, so as to prove that the difference of the natures is not real nor essential, but a mere modification of a common nature. They are very serviceable, therefore, in elevating and carrying on the mind, from differences to genera, and in removing those phantoms and images of things, which meet it in disguise in concrete substances.

For example: let the required nature be heat. The classification of heat into three kinds, that of the celestial bodies, that of animals, and that of fire, appears to be settled and admitted; and these kinds of heat, especially one of them compared with the other two, are supposed to be different, and clearly heterogeneous in their essence and species, or specific nature, since the heat of the heavenly bodies and of animals generates and cherishes, while that of fire corrupts and destroys. We have an instance of alliance, then, in a very common experiment, that of a vine branch admitted into a building where there is a constant fire, by which the grapes ripen a whole month sooner than in the air; so that fruit upon the tree can be ripened by fire, although this appear the peculiar effect of the sun. From this beginning, therefore, the understanding rejects all essential difference, and easily ascends to the investigation of the real differences between the heat of the sun and that of fire, by which their operation is rendered dissimilar, although they partake of a common nature.

These differences will be found to be four in number. 1. The heat of the sun is much milder and gentler in degree than that of fire. 2. It is much more moist in quality, especially as it is transmitted to us through the air. 3. Which is the chief point, it is very unequal, advancing and increased at one time, retiring and diminished at another, which mainly contributes to the generation of bodies. For Aristotle rightly asserted, that the principal cause of generation and corruption on the surface of the earth was the oblique path of the sun in the zodiac, whence its heat becomes very unequal, partly from the alternation of night and day, partly from the succession of summer and winter. Yet must he immediately corrupt and pervert his discovery, by dictating to nature according to his habit, and dogmatically assigning the cause of generation to the approach of the sun, and that of corruption to its retreat; while, in fact, each circumstance indifferently and not respectively contributes both to generation and corruption; for unequal heat tends to generate and corrupt, as equable heat does to preserve. 4. The fourth difference between the heat of the sun and fire is of great consequence; namely, that the sun, gradually, and for a length of time, insinuates its effects, while those of fire (urged by the impatience of man) are brought to a termination in a shorter space of time. But if any one were to pay attention to the tempering of fire, and reducing it to a more moderate and gentle degree (which may be done in various ways), and then were to sprinkle and mix a degree of humidity with it; and, above all, were to imitate the sun in its inequality; and, lastly, were patiently to suffer some delay (not such, however, as is proportioned to the effects of the sun, but more than men usually admit of in those of fire), he would soon banish the notion of any difference, and would attempt, or equal, or perhaps sometimes surpass the effect of the sun, by the heat of fire. A like instance of alliance is that of reviving butterflies, benumbed and nearly dead from cold, by the gentle warmth of fire; so that fire is no less able to revive animals than to ripen vegetables. We may also mention the celebrated invention of Fracastorius, of applying a pan considerably heated to the head in desperate cases of apoplexy, which clearly expands the animal spirits, when compressed and almost extinguished by the humors and obstructions of the brain, and excites them to action, as the fire would operate on water or air, and in the result produces life. Eggs are sometimes hatched by the heat of fire, an exact imitation of animal heat; and there are many instances of the like nature, so that no one can doubt that the heat of fire, in many cases, can be modified till it resemble that of the heavenly bodies and of animals.

Again, let the required natures be motion and rest. There appears to be a settled classification, grounded on the deepest philosophy, that natural bodies either revolve, move in a straight line, or stand still and rest. For there is either motion without limit, or continuance within a certain limit, or a translation toward a certain limit. The eternal motion of revolution appears peculiar to the heavenly bodies, rest to this our globe, and the other bodies (heavy and light, as they are termed, that is to say, placed out of their natural position) are borne in a straight line to masses or aggregates which resemble them, the light toward the heaven, the heavy toward the earth; and all this is very fine language.

But we have an instance of alliance in low comets, which revolve, though far below the heavens; and the fiction of Aristotle, of the comet being fixed to, or necessarily following some star, has been long since exploded; not only because it is improbable in itself, but from the evident fact of the discursive and irregular motion of comets through various parts of the heavens.[125]

Another instance of alliance is that of the motion of air, which appears to revolve from east to west within the tropics, where the circles of revolution are the greatest.

The flow and ebb of the sea would perhaps be another instance, if the water were once found to have a motion of revolution, though slow and hardly perceptible, from east to west, subject, however, to a reaction twice a day. If this be so, it is clear that the motion of revolution is not confined to the celestial bodies, but is shared, also, by air and water.

Again--the supposed peculiar disposition of light bodies to rise is rather shaken; and here we may find an instance of alliance in a water bubble. For if air be placed under water, it rises rapidly toward the surface by that striking motion (as Democritus terms it) with which the descending water strikes the air and raises it, not by any struggle or effort of the air itself; and when it has reached the surface of the water, it is prevented from ascending any further, by the slight resistance it meets with in the water, which does not allow an immediate separation of its parts, so that the tendency of the air to rise must be very slight.

Again, let the required nature be weight. It is certainly a received classification, that dense and solid bodies are borne toward the centre of the earth, and rare and light bodies to the circumference of the heavens, as their appropriate places. As far as relates to places (though these things have much weight in the schools), the notion of there being any determinate place is absurd and puerile. Philosophers trifle, therefore, when they tell you, that if the earth were perforated, heavy bodies would stop on their arrival at the centre. This centre would indeed be an efficacious nothing, or mathematical point, could it affect bodies or be sought by them, for a body is not acted upon except by a body.[126] In fact, this tendency to ascend and descend is either in the conformation of the moving body, or in its harmony and sympathy with another body. But if any dense and solid body be found, which does not, however, tend toward the earth, the classification is at an end. Now, if we allow of Gilbert's opinion, that the magnetic power of the earth, in attracting heavy bodies, is not extended beyond the limit of its peculiar virtue (which operates always at a fixed distance and no further),[127] and this be proved by some instance, such an instance will be one of alliance in our present subject. The nearest approach to it is that of waterspouts, frequently seen by persons navigating the Atlantic toward either of the Indies. For the force and mass of the water suddenly effused by waterspouts, appears to be so considerable, that the water must have been collected previously, and have remained fixed where it was formed, until it was afterward forced down by some violent cause, rather than made to fall by the natural motion of gravity: so that it may be conjectured that a dense and compact mass, at a great distance from the earth, may be suspended as the earth itself is, and would not fall, unless forced down. We do not, however, affirm this as certain. In the meanwhile, both in this respect and many others, it will readily be seen how deficient we are in natural history, since we are forced to have recourse to suppositions for examples, instead of ascertained instances.

Again, let the required nature be the discursive power of the mind. The classification of human reason and animal instinct appears to be perfectly correct. Yet there are some instances of the actions of brutes which seem to show that they, too, can syllogize. Thus it is related, that a crow, which had nearly perished from thirst in a great drought, saw some water in the hollow trunk of a tree, but as it was too narrow for him to get into it, he continued to throw in pebbles, which made the water rise till he could drink; and it afterward became a proverb.

Again, let the required nature be vision. The classification appears real and certain, which considers light as that which is originally visible, and confers the power of seeing; and color, as being secondarily visible, and not capable of being seen without light, so as to appear a mere image or modification of light. Yet there are instances of alliance in each respect; as in snow when in great quantities, and in the flame of sulphur; the one being a color originally and in itself light, the other a light verging toward color.[128]

XXXVI. In the fourteenth rank of prerogative instances, we will place the instances of the cross, borrowing our metaphor from the crosses erected where two roads meet, to point out the different directions. We are wont also to call them decisive and judicial instances, and in some cases instances of the oracle and of command. Their nature is as follows: When in investigating any nature the understanding is, as it were, balanced, and uncertain to which of two or more natures the cause of the required nature should be assigned, on account of the frequent and usual concurrence of several natures, the instances of the cross show that the union of one nature with the required nature is firm and indissoluble, while that of the other is unsteady and separable; by which means the question is decided, and the first is received as the cause, while the other is dismissed and rejected. Such instances, therefore, afford great light, and are of great weight, so that the course of interpretation sometimes terminates, and is completed in them. Sometimes, however, they are found among the instances already observed, but they are generally new, being expressly and purposely sought for and applied, and brought to light only by attentive and active diligence.

For example: let the required nature be the flow and ebb of the sea, which is repeated twice a day, at intervals of six hours between each advance and retreat, with some little difference, agreeing with the motion of the moon. We have here the following crossways:

This motion must be occasioned either by the advancing and the retiring of the sea, like water shaken in a basin, which leaves one side while it washes the other; or by the rising of the sea from the bottom, and its again subsiding, like boiling water. But a doubt arises, to which of these causes we should assign the flow and ebb. If the first assertion be admitted, it follows, that when there is a flood on one side, there must at the same time be an ebb on another, and the question therefore is reduced to this. Now Acosta, and some others, after a diligent inquiry, have observed that the flood tide takes place on the coast of Florida, and the opposite coasts of Spain and Africa, at the same time, as does also the ebb; and that there is not, on the contrary, a flood tide at Florida when there is an ebb on the coasts of Spain and Africa. Yet if one consider the subject attentively, this does not prove the necessity of a rising motion, nor refute the notion of a progressive motion. For the motion may be progressive, and yet inundate the opposite shores of a channel at the same time; as if the waters be forced and driven together from some other quarter, for instance, which takes place in rivers, for they flow and ebb toward each bank at the same time, yet their motion is clearly progressive, being that of the waters from the sea entering their mouths. So it may happen, that the waters coming in a vast body from the eastern Indian Ocean are driven together, and forced into the channel of the Atlantic, and therefore inundate both coasts at once. We must inquire, therefore, if there be any other channel by which the waters can at the same time sink and ebb; and the Southern Ocean at once suggests itself, which is not less than the Atlantic, but rather broader and more extensive than is requisite for this effect.

We at length arrive, then, at an instance of the cross, which is this. If it be positively discovered, that when the flood sets in toward the opposite coasts of Florida and Spain in the Atlantic, there is at the same time a flood tide on the coasts of Peru and the back part of China, in the Southern Ocean, then assuredly, from this decisive instance, we must reject the assertion, that the flood and ebb of the sea, about which we inquire, takes place by progressive motion; for no other sea or place is left where there can be an ebb. But this may most easily be learned, by inquiring of the inhabitants of Panama and Lima (where the two oceans are separated by a narrow isthmus), whether the flood and ebb takes place on the opposite sides of the isthmus at the same time, or the reverse. This decision or rejection appears certain, if it be granted that the earth is fixed; but if the earth revolves, it may perhaps happen, that from the unequal revolution (as regards velocity) of the earth and the waters of the sea, there may be a violent forcing of the waters into a mass, forming the flood, and a subsequent relaxation of them (when they can no longer bear the accumulation), forming the ebb. A separate inquiry must be made into this. Even with this hypothesis, however, it remains equally true, that there must be an ebb somewhere, at the same time that there is a flood in another quarter.

Again, let the required nature be the latter of the two motions we have supposed; namely, that of a rising and subsiding motion, if it should happen that upon diligent examination the progressive motion be rejected. We have, then, three ways before us, with regard to this nature. The motion, by which the waters raise themselves, and again fall back, in the floods and ebbs, without the addition of any other water rolled toward them, must take place in one of the three following ways: Either the supply of water emanates from the interior of the earth, and returns back again; or there is really no greater quantity of water, but the same water (without any augmentation of its quantity) is extended or rarefied, so as to occupy a greater space and dimension, and again contracts itself; or there is neither an additional supply nor any extension, but the same waters (with regard to quantity, density, or rarity) raise themselves and fall from sympathy, by some magnetic power attracting and calling them up, as it were, from above. Let us then (passing over the first two motions) reduce the investigation to the last, and inquire if there be any such elevation of the water by sympathy or a magnetic force; and it is evident, in the first place, that the whole mass of water being placed in the trench or cavity of the sea, cannot be raised at once, because there would not be enough to cover the bottom, so that if there be any tendency of this kind in the water to raise itself, yet it would be interrupted and checked by the cohesion of things, or (as the common expression is) that there may be no vacuum. The water, therefore, must rise on one side, and for that reason be diminished and ebb on another. But it will again necessarily follow that the magnetic power not being able to operate on the whole, operates most intensely on the centre, so as to raise the waters there, which, when thus raised successively, desert and abandon the sides.[129]

We at length arrive, then, at an instance of the cross, which is this: if it be found that during the ebb the surface of the waters at sea is more curved and round, from the waters rising in the middle, and sinking at the sides or coast, and if, during a flood, it be more even and level, from the waters returning to their former position, then assuredly, by this decisive instance, the raising of them by a magnetic force can be admitted; if otherwise, it must be entirely rejected. It is not difficult to make the experiment (by sounding in straits), whether the sea be deeper toward the middle in ebbs, than in floods. But it must be observed, if this be the case, that (contrary to common opinion) the waters rise in ebbs, and only return to their former position in floods, so as to bathe and inundate the coast.

Again, let the required nature be the spontaneous motion of revolution, and particularly, whether the diurnal motion, by which the sun and stars appear to us to rise and set, be a real motion of revolution in the heavenly bodies, or only apparent in them, and real in the earth. There may be an instance of the cross of the following nature. If there be discovered any motion in the ocean from east to west, though very languid and weak, and if the same motion be discovered rather more swift in the air (particularly within the tropics, where it is more perceptible from the circles being greater). If it be discovered also in the low comets, and be already quick and powerful in them; if it be found also in the planets, but so tempered and regulated as to be slower in those nearest the earth, and quicker in those at the greatest distance, being quickest of all in the heavens, then the diurnal motion should certainly be considered as real in the heavens, and that of the earth must be rejected; for it will be evident that the motion from east to west is part of the system of the world and universal; since it is most rapid in the height of the heavens, and gradually grows weaker, till it stops and is extinguished in rest at the earth.

Again, let the required nature be that other motion of revolution, so celebrated among astronomers, which is contrary to the diurnal, namely, from west to east--and which the ancient astronomers assign to the planets, and even to the starry sphere, but Copernicus and his followers to the earth also--and let it be examined whether any such motion be found in nature, or it be rather a fiction and hypothesis for abridging and facilitating calculation, and for promoting that fine notion of effecting the heavenly motions by perfect circles; for there is nothing which proves such a motion in heavenly objects to be true and real, either in a planet's not returning in its diurnal motion to the same point of the starry sphere, or in the pole of the zodiac being different from that of the world, which two circumstances have occasioned this notion. For the first phenomenon is well accounted for by the spheres overtaking or falling behind each other, and the second by spiral lines; so that the inaccuracy of the return and declination to the tropics may be rather modifications of the one diurnal motion than contrary motions, or about different poles. And it is most certain, if we consider ourselves for a moment as part of the vulgar (setting aside the fictions of astronomers and the school, who are wont undeservedly to attack the senses in many respects, and to affect obscurity), that the apparent motion is such as we have said, a model of which we have sometimes caused to be represented by wires in a sort of a machine.

We may take the following instances of the cross upon this subject. If it be found in any history worthy of credit, that there has existed any comet, high or low, which has not revolved in manifest harmony (however irregularly) with the diurnal motion, then we may decide so far as to allow such a motion to be possible in nature. But if nothing of the sort be found, it must be suspected, and recourse must be had to other instances of the cross.

Again, let the required nature be weight or gravity. Heavy and ponderous bodies must, either of their own nature, tend toward the centre of the earth by their peculiar formation, or must be attracted and hurried by the corporeal mass of the earth itself, as being an assemblage of similar bodies, and be drawn to it by sympathy. But if the latter be the cause, it follows that the nearer bodies approach to the earth, the more powerfully and rapidly they must be borne toward it, and the further they are distant, the more faintly and slowly (as is the case in magnetic attractions), and that this must happen within a given distance; so that if they be separated at such a distance from the earth that the power of the earth cannot act upon them, they will remain suspended like the earth, and not fall at all.[130]

The following instance of the cross may be adopted. Take a clock moved by leaden weights,[131] and another by a spring, and let them be set well together, so that one be neither quicker nor slower than the other; then let the clock moved by weights be placed on the top of a very high church, and the other be kept below, and let it be well observed, if the former move slower than it did, from the diminished power of the weights. Let the same experiment be made at the bottom of mines worked to a considerable depth, in order to see whether the clock move more quickly from the increased power of the weights. But if this power be found to diminish at a height, and to increase in subterraneous places, the attraction of the corporeal mass of the earth may be taken as the cause of weight.

Again, let the required nature be the polarity of the steel needle when touched with the magnet. We have these two ways with regard to this nature--Either the touch of the magnet must communicate polarity to the steel toward the north and south, or else it may only excite and prepare it, while the actual motion is occasioned by the presence of the earth, which Gilbert considers to be the case, and endeavors to prove with so much labor. The particulars he has inquired into with such ingenious zeal amount to this--1. An iron bolt placed for a long time toward the north and south acquires polarity from this habit, without the touch of the magnet, as if the earth itself operating but weakly from its distance (for the surface or outer crust of the earth does not, in his opinion, possess the magnetic power), yet, by long continued motion, could supply the place of the magnet, excite the iron, and convert and change it when excited. 2. Iron, at a red or white heat, when quenched in a direction parallel to the north and south, also acquires polarity without the touch of the magnet, as if the parts of iron being put in motion by ignition, and afterward recovering themselves, were, at the moment of being quenched, more susceptible and sensitive of the power emanating from the earth, than at other times, and therefore as it were excited. But these points, though well observed, do not completely prove his assertion.

An instance of the cross on this point might be as follows: Let a small magnetic globe be taken, and its poles marked, and placed toward the east and west, not toward the north and south, and let it continue thus. Then let an untouched needle be placed over it, and suffered to remain so for six or seven days. Now, the needle (for this is not disputed), while it remains over the magnet, will leave the poles of the world and turn to those of the magnet, and therefore, as long as it remains in the above position, will turn to the east and west. But if the needle, when removed from the magnet and placed upon a pivot, be found immediately to turn to the north and south, or even by degrees to return thither, then the presence of the earth must be considered as the cause, but if it remains turned as at first, toward the east and west, or lose its polarity, then that cause must be suspected, and further inquiry made.

Again, let the required nature be the corporeal substance of the moon, whether it be rare, fiery, and aërial (as most of the ancient philosophers have thought), or solid and dense (as Gilbert and many of the moderns, with some of the ancients, hold).[132] The reasons for this latter opinion are grounded chiefly upon this, that the moon reflects the sun's rays, and that light does not appear capable of being reflected except by solids. The instances of the cross will therefore (if any) be such as to exhibit reflection by a rare body, such as flame, if it be but sufficiently dense. Now, certainly, one of the reasons of twilight is the reflection[133] of the rays of the sun by the upper part of the atmosphere. We see the sun's rays also reflected on fine evenings by streaks of moist clouds, with a splendor not less, but perhaps more bright and glorious than that reflected from the body of the moon, and yet it is not clear that those clouds have formed into a dense body of water. We see, also, that the dark air behind the windows at night reflects the light of a candle in the same manner as a dense body would do.[134] The experiment should also be made of causing the sun's rays to fall through a hole upon some dark and bluish flame. The unconfined rays of the sun, when falling on faint flames, do certainly appear to deaden them, and render them more like white smoke than flames. These are the only instances which occur at present of the nature of those of the cross, and better perhaps can be found. But it must always be observed that reflection is not to be expected from flame, unless it be of some depth, for otherwise it becomes nearly transparent. This at least may be considered certain, that light is always either received and transmitted or reflected by an even surface.

Again, let the required nature be the motion of projectiles (such as darts, arrows, and balls) through the air. The school, in its usual manner, treats this very carelessly, considering it enough to distinguish it by the name of violent motion, from that which they term natural, and as far as regards the first percussion or impulse, satisfies itself by its axiom, that two bodies cannot exist in one place, or there would be a penetration of dimensions. With regard to this nature we have these two crossways-- The motion must arise either from the air carrying the projected body, and collecting behind it, like a stream behind boats, or the wind behind straws; or from the parts of the body itself not supporting the impression, but pushing themselves forward in succession to ease it. Fracastorius, and nearly all those who have entered into any refined inquiry upon the subject, adopt the first. Nor can it be doubted that the air has some effect, yet the other motion is without doubt real, as is clear from a vast number of experiments. Among others we may take this instance of the cross, namely, that a thin plate or wire of iron rather stiff, or even a reed or pen split in two, when drawn up and bent between the finger and thumb, will leap forward; for it is clear that this cannot be attributed to the air's being collected behind the body, because the source of motion is in the centre of the plate or pen, and not in its extremities.

Again, let the required nature be the rapid and powerful motion of the explosion of gunpowder, by which such vast masses are upheaved, and such weights discharged as we observe in large mines and mortars, there are two crossways before us with regard to this nature. This motion is excited either by the mere effort of the body expanding itself when inflamed, or by the assisting effort of the crude spirit, which escapes rapidly from fire, and bursts violently from the surrounding flame as from a prison. The school, however, and common opinion only consider the first effort; for men think that they are great philosophers when they assert that flame, from the form of the element, is endowed with a kind of necessity of occupying a greater space than the same body had occupied when in the form of powder, and that thence proceeds the motion in question. In the meantime they do not observe, that although this may be true, on the supposition of flame being generated, yet the generation may be impeded by a weight of sufficient force to compress and suffocate it, so that no such necessity exists as they assert. They are right, indeed, in imagining that the expansion and the consequent emission or removal of the opposing body, is necessary if flame be once generated, but such a necessity is avoided if the solid opposing mass suppress the flame before it be generated; and we in fact see that flame, especially at the moment of its generation, is mild and gentle, and requires a hollow space where it can play and try its force. The great violence of the effect, therefore, cannot be attributed to this cause; but the truth is, that the generation of these exploding flames and fiery blasts arises from the conflict of two bodies of a decidedly opposite nature--the one very inflammable, as is the sulphur, the other having an antipathy to flame, namely, the crude spirit of the nitre; so that an extraordinary conflict takes place while the sulphur is becoming inflamed as far as it can (for the third body, the willow charcoal, merely incorporates and conveniently unites the two others), and the spirit of nitre is escaping, as far also as it can, and at the same time expanding itself (for air, and all crude substances, and water are expanded by heat), fanning thus, in every direction, the flame of the sulphur by its escape and violence, just as if by invisible bellows.

Two kinds of instances of the cross might here be used--the one of very inflammable substances, such as sulphur and camphor, naphtha and the like, and their compounds, which take fire more readily and easily than gunpowder if left to themselves (and this shows that the effort to catch fire does not of itself produce such a prodigious effect); the other of substances which avoid and repel flame, such as all salts; for we see that when they are cast into the fire, the aqueous spirit escapes with a crackling noise before flame is produced, which also happens in a less degree in stiff leaves, from the escape of the aqueous part before the oily part has caught fire. This is more particularly observed in quicksilver, which is not improperly called mineral water, and which, without any inflammation, nearly equals the force of gunpowder by simple explosion and expansion, and is said, when mixed with gunpowder, to increase its force.

Again, let the required nature be the transitory nature of flame and its momentaneous extinction; for to us the nature of flame does not appear to be fixed or settled, but to be generated from moment to moment, and to be every instant extinguished; it being clear that those flames which continue and last, do not owe their continuance to the same mass of flame, but to a continued succession of new flame regularly generated, and that the same identical flame does not continue. This is easily shown by removing the food or source of the flame, when it at once goes out. We have the two following crossways with regard to this nature:

This momentary nature either arises from the cessation of the cause which first produced it, as in light, sounds, and violent motions, as they are termed, or flame may be capable, by its own nature, of duration, but is subjected to some violence from the contrary natures which surround it, and is destroyed.

We may therefore adopt the following instance of the cross. We see to what a height the flames rise in great conflagrations; for as the base of the flame becomes more extensive, its vertex is more lofty. It appears, then, that the commencement of the extinction takes place at the sides, where the flame is compressed by the air, and is ill at ease; but the centre of the flame, which is untouched by the air and surrounded by flame, continues the same, and is not extinguished until compressed by degrees by the air attacking it from the sides. All flame, therefore, is pyramidal, having its base near the source, and its vertex pointed from its being resisted by the air, and not supplied from the source. On the contrary, the smoke, which is narrow at the base, expands in its ascent, and resembles an inverted pyramid, because the air admits the smoke, but compresses the flame; for let no one dream that the lighted flame is air, since they are clearly heterogeneous.

The instance of the cross will be more accurate, if the experiment can be made by flames of different colors. Take, therefore, a small metal sconce, and place a lighted taper in it, then put it in a basin, and pour a small quantity of spirits of wine round the sconce, so as not to reach its edge, and light the spirit. Now the flame of the spirit will be blue, and that of the taper yellow; observe, therefore, whether the latter (which can easily be distinguished from the former by its color, for flames do not mix immediately, as liquids do) continue pyramidal, or tend more to a globular figure, since there is nothing to destroy or compress it. If the latter result be observed, it must be considered as settled, that flame continues positively the same, while inclosed within another flame, and not exposed to the resisting force of the air.

Let this suffice for the instances of the cross. We have dwelt the longer upon them in order gradually to teach and accustom mankind to judge of nature by these instances, and enlightening experiments, and not by probable reasons.[135]

XXXVII. We will treat of the instances of divorce as the fifteenth of our prerogative instances. They indicate the separation of natures of the most common occurrence. They differ, however, from those subjoined to the accompanying instances; for the instances of divorce point out the separation of a particular nature from some concrete substance with which it is usually found in conjunction, while the hostile instances point out the total separation of one nature from another. They differ, also, from the instances of the cross, because they decide nothing, but only inform us that the one nature is capable of being separated from the other. They are of use in exposing false forms, and dissipating hasty theories derived from obvious facts; so that they add ballast and weight, as it were, to the understanding.

For instance, let the acquired natures be those four which Telesius terms associates, and of the same family, namely, heat, light, rarity, and mobility, or promptitude to motion; yet many instances of divorce can be discovered between them. Air is rare and easily moved, but neither hot nor light; the moon is light but not hot; boiling water is warm but not light; the motion of the needle in the compass is swift and active, and yet its substance is cold, dense, and opaque; and there are many similar examples.

Again, let the required natures be corporeal nature and natural action. The latter appears incapable of subsisting without some body, yet may we, perhaps, even here find an instance of divorce, as in the magnetic motion, which draws the iron to the magnet, and heavy bodies to the globe of the earth; to which we may add other actions which operate at a distance. For such action takes place in time, by distinct moments, not in an instant; and in space, by regular degrees and distances. There is, therefore, some one moment of time and some interval of space, in which the power or action is suspended between the two bodies creating the motion. Our consideration, then, is reduced to this, whether the bodies which are the extremes of motion prepare or alter the intermediate bodies, so that the power advances from one extreme to the other by succession and actual contact, and in the meantime exists in some intermediate body; or whether there exists in reality nothing but the bodies, the power, and the space? In the case of the rays of light, sounds, and heat, and some other objects which operate at a distance, it is indeed probable that the intermediate bodies are prepared and altered, the more so because a qualified medium is required for their operation. But the magnetic or attractive power admits of an indifferent medium, and it is not impeded in any. But if that power or action is independent of the intermediate body, it follows that it is a natural power or action existing in a certain time and space without any body, since it exists neither in the extreme nor in the intermediate bodies. Hence the magnetic action may be taken as an instance of divorce of corporeal nature and natural action; to which we may add, as a corollary and an advantage not to be neglected, that it may be taken as a proof of essence and substance being separate and incorporeal, even by those who philosophize according to the senses. For if natural power and action emanating from a body can exist at any time and place entirely without any body, it is nearly a proof that it can also emanate originally from an incorporeal substance; for a corporeal nature appears to be no less necessary for supporting and conveying, than for exciting or generating natural action.

XXXVIII. Next follow five classes of instances which we are wont to call by the general term of instances of the lamp, or of immediate information. They are such as assist the senses; for since every interpretation of nature sets out from the senses, and leads, by a regular fixed and well-established road, from the perceptions of the senses to those of the understanding (which are true notions and axioms), it necessarily follows, that in proportion as the representatives or ministerings of the senses are more abundant and accurate, everything else must be more easy and successful.

The first of these five sets of instances of the lamp, strengthen, enlarge, and correct the immediate operations of the senses; the second reduce to the sphere of the senses such matters as are beyond it; the third indicate the continued process or series of such things and motions, as for the most part are only observed in their termination, or in periods; the fourth supply the absolute wants of the senses; the fifth excite their attention and observation, and at the same time limit the subtilty of things. We will now proceed to speak of them singly.

XXXIX. In the sixteenth rank, then, of prerogative instances, we will place the instances of the door or gate, by which name we designate such as assist the immediate action of the senses. It is obvious, that sight holds the first rank among the senses, with regard to information, for which reason we must seek principally helps for that sense. These helps appear to be threefold, either to enable it to perceive objects not naturally seen, or to see them from a greater distance, or to see them more accurately and distinctly.

We have an example of the first (not to speak of spectacles and the like, which only correct and remove the infirmity of a deficient sight, and therefore give no further information) in the lately invented microscopes, which exhibit the latent and invisible minutiæ of substances, and their hidden formation and motion, by wonderfully increasing their apparent magnitude. By their assistance we behold with astonishment the accurate form and outline of a flea, moss, and animalculæ, as well as their previously invisible color and motion. It is said, also, that an apparently straight line, drawn with a pen or pencil, is discovered by such a microscope to be very uneven and curved, because neither the motion of the hand, when assisted by a ruler, nor the impression of ink or color, are really regular, although the irregularities are so minute as not to be perceptible without the assistance of the microscope. Men have (as is usual in new and wonderful discoveries) added a superstitious remark, that the microscope sheds a lustre on the works of nature, and dishonor on those of art, which only means that the tissue of nature is much more delicate than that of art. For the microscope is only of use for minute objects, and Democritus, perhaps, if he had seen it, would have exulted in the thought of a means being discovered for seeing his atom, which he affirmed to be entirely invisible. But the inadequacy of these microscopes, for the observation of any but the most minute bodies, and even of those if parts of a larger body, destroys their utility; for if the invention could be extended to greater bodies, or the minute parts of greater bodies, so that a piece of cloth would appear like a net, and the latent minutiæ and irregularities of gems, liquids, urine, blood, wounds, and many other things could be rendered visible, the greatest advantage would, without doubt, be derived.

We have an instance of the second kind in the telescope, discovered by the wonderful exertions of Galileo; by the assistance of which a nearer intercourse may be opened (as by boats or vessels) between ourselves and the heavenly objects. For by its aid we are assured that the Milky Way is but a knot or constellation of small stars, clearly defined and separate, which the ancients only conjectured to be the case; whence it appears to be capable of demonstration, that the spaces of the planetary orbits (as they are termed) are not quite destitute of other stars, but that the heaven begins to glitter with stars before we arrive at the starry sphere, although they may be too small to be visible without the telescope. By the telescope, also, we can behold the revolutions of smaller stars round Jupiter, whence it may be conjectured that there are several centres of motion among the stars. By its assistance, also, the irregularity of light and shade on the moon's surface is more clearly observed and determined, so as to allow of a sort of selenography.[136] By the telescope we see the spots in the sun, and other similar phenomena; all of which are most noble discoveries, as far as credit can be safely given to demonstrations of this nature, which are on this account very suspicious, namely, that experiment stops at these few, and nothing further has yet been discovered by the same method, among objects equally worthy of consideration.

We have instances of the third kind in measuring-rods, astrolabes, and the like, which do not enlarge, but correct and guide the sight. If there be other instances which assist the other senses in their immediate and individual action, yet if they add nothing further to their information they are not apposite to our present purpose, and we have therefore said nothing of them.

XL. In the seventeenth rank of prerogative instances we will place citing instances (to borrow a term from the tribunals), because they cite those things to appear, which have not yet appeared. We are wont also to call them invoking instances, and their property is that of reducing to the sphere of the senses objects which do not immediately fall within it.

Objects escape the senses either from their distance, or the intervention of other bodies, or because they are not calculated to make an impression upon the senses, or because they are not in sufficient quantity to strike the senses, or because there is not sufficient time for their acting upon the senses, or because the impression is too violent, or because the senses are previously filled and possessed by the object, so as to leave no room for any new motion. These remarks apply principally to sight, and next to touch, which two senses act extensively in giving information, and that too upon general objects, while the remaining three inform us only, as it were, by their immediate action, and as to specific objects.

There can be no reduction to the sphere of the senses in the first case, unless in the place of the object, which cannot be perceived on account of the distance, there be added or substituted some other object, which can excite and strike the sense from a greater distance, as in the communication of intelligence by fires, bells, and the like.

In the second case we effect this reduction by rendering those things which are concealed by the interposition of other bodies, and which cannot easily be laid open, evident to the senses by means of that which lies at the surface, or proceeds from the interior; thus the state of the body is judged of by the pulse, urine, etc.

The third and fourth cases apply to many subjects, and the reduction to the sphere of the senses must be obtained from every quarter in the investigation of things. There are many examples. It is obvious that air, and spirit, and the like, whose whole substance is extremely rare and delicate, can neither be seen nor touched--a reduction, therefore, to the senses becomes necessary in every investigation relating to such bodies.

Let the required nature, therefore, be the action and motion of the spirit inclosed in tangible bodies; for every tangible body with which we are acquainted contains an invisible and intangible spirit, over which it is drawn, and which it seems to clothe. This spirit being emitted from a tangible substance, leaves the body contracted and dry; when retained, it softens and melts it; when neither wholly emitted nor retained, it models it, endows it with limbs, assimilates, manifests, organizes it, and the like. All these points are reduced to the sphere of the senses by manifest effects.

For in every tangible and inanimate body the inclosed spirit at first increases, and as it were feeds on the tangible parts which are most open and prepared for it; and when it has digested and modified them, and turned them into spirit, it escapes with them. This formation and increase of spirit is rendered sensible by the diminution of weight; for in every desiccation something is lost in quantity, not only of the spirit previously existing in the body, but of the body itself, which was previously tangible, and has been recently changed, for the spirit itself has no weight. The departure or emission of spirit is rendered sensible in the rust of metals, and other putrefactions of a like nature, which stop before they arrive at the rudiments of life, which belong to the third species of process.[137] In compact bodies the spirit does not find pores and passages for its escape, and is therefore obliged to force out, and drive before it, the tangible parts also, which consequently protrude, whence arises rust and the like. The contraction of the tangible parts, occasioned by the emission of part of the spirit (whence arises desiccation), is rendered sensible by the increased hardness of the substance, and still more by the fissures, contractions, shrivelling, and folds of the bodies thus produced. For the parts of wood split and contract, skins become shrivelled, and not only that, but, if the spirit be emitted suddenly by the heat of the fire, become so hastily contracted as to twist and roll themselves up.

On the contrary, when the spirit is retained, and yet expanded and excited by heat or the like (which happens in solid and tenacious bodies), then the bodies are softened, as in hot iron; or flow, as in metals; or melt, as in gums, wax, and the like. The contrary effects of heat, therefore (hardening some substances and melting others), are easily reconciled,[138] because the spirit is emitted in the former, and agitated and retained in the latter; the latter action is that of heat and the spirit, the former that of the tangible parts themselves, after the spirit's emission.

But when the spirit is neither entirely retained nor emitted, but only strives and exercises itself, within its limits, and meets with tangible parts, which obey and readily follow it wherever it leads them, then follows the formation of an organic body, and of limbs, and the other vital actions of vegetables and animals. These are rendered sensible chiefly by diligent observation of the first beginnings, and rudiments or effects of life in animalculæ sprung from putrefaction, as in the eggs of ants, worms, mosses, frogs after rain, etc. Both a mild heat and a pliant substance, however, are necessary for the production of life, in order that the spirit may neither hastily escape, nor be restrained by the obstinacy of the parts, so as not to be able to bend and model them like wax.

Again, the difference of spirit which is important and of effect in many points (as unconnected spirit, branching spirit, branching and cellular spirit, the first of which is that of all inanimate substances, the second of vegetables, and the third of animals), is placed, as it were, before the eyes by many reducing instances.

Again, it is clear that the more refined tissue and conformation of things (though forming the whole body of visible or tangible objects) are neither visible nor tangible. Our information, therefore, must here also be derived from reduction to the sphere of the senses. But the most radical and primary difference of formation depends on the abundance or scarcity of matter within the same space or dimensions. For the other formations which regard the dissimilarity of the parts contained in the same body, and their collocation and position, are secondary in comparison with the former.

Let the required nature then be the expansion or coherence of matter in different bodies, or the quantity of matter relative to the dimensions of each. For there is nothing in nature more true than the twofold proposition--that nothing proceeds from nothing and that nothing is reduced to nothing, but that the quantum, or sum total of matter, is constant, and is neither increased nor diminished. Nor is it less true, that out of this given quantity of matter, there is a greater or less quantity, contained within the same space or dimensions according to the difference of bodies; as, for instance, water contains more than air. So that if any one were to assert that a given content of water can be changed into an equal content of air, it is the same as if he were to assert that something can be reduced into nothing. On the contrary, if any one were to assert that a given content of air can be changed into an equal content of water, it is the same as if he were to assert that something can proceed from nothing. From this abundance or scarcity of matter are properly derived the notions of density and rarity, which are taken in various and promiscuous senses.

This third assertion may be considered as being also sufficiently certain; namely, that the greater or less quantity of matter in this or that body, may, by comparison, be reduced to calculation, and exact, or nearly exact, proportion. Thus, if one should say that there is such an accumulation of matter in a given quantity of gold, that it would require twenty-one times the quantity in dimension of spirits of wine, to make up the same quantity of matter, it would not be far from the truth.

The accumulation of matter, however, and its relative quantity, are rendered sensible by weight; for weight is proportionate to the quantity of matter, as regards the parts of a tangible substance, but spirit and its quantity of matter are not to be computed by weight, which spirit rather diminishes than augments.

We have made a tolerably accurate table of weight, in which we have selected the weights and size of all the metals, the principal minerals, stones, liquids, oils, and many other natural and artificial bodies: a very useful proceeding both as regards theory and practice, and which is capable of revealing many unexpected results. Nor is this of little consequence, that it serves to demonstrate that the whole range of the variety of tangible bodies with which we are acquainted (we mean tolerably close, and not spongy, hollow bodies, which are for a considerable part filled with air), does not exceed the ratio of one to twenty-one. So limited is nature, or at least that part of it to which we are most habituated.

We have also thought it deserving our industry, to try if we could arrive at the ratio of intangible or pneumatic bodies to tangible bodies, which we attempted by the following contrivance. We took a vial capable of containing about an ounce, using a small vessel in order to effect the subsequent evaporation with less heat. We filled this vial, almost to the neck, with spirits of wine, selecting it as the tangible body which, by our table, was the rarest, and contained a less quantity of matter in a given space than all other tangible bodies which are compact and not hollow. Then we noted exactly the weight of the liquid and vial. We next took a bladder, containing about two pints, and squeezed all the air out of it, as completely as possible, and until the sides of the bladder met. We first, however, rubbed the bladder gently with oil, so as to make it air-tight, by closing its pores with the oil. We tied the bladder tightly round the mouth of the vial, which we had inserted in it, and with a piece of waxed thread to make it fit better and more tightly, and then placed the vial on some hot coals in a brazier. The vapor or steam of the spirit, dilated and become aëriform by the heat, gradually swelled out the bladder, and stretched it in every direction like a sail. As soon as that was accomplished, we removed the vial from the fire and placed it on a carpet, that it might not be cracked by the cold; we also pricked the bladder immediately, that the steam might not return to a liquid state by the cessation of heat, and confound the proportions. We then removed the bladder, and again took the weight of the spirit which remained; and so calculated the quantity which had been converted into vapor, or an aëriform shape, and then examined how much space had been occupied by the body in its form of spirits of wine in the vial, and how much, on the other hand, had been occupied by it in its aëriform shape in the bladder, and subtracted the results; from which it was clear that the body, thus converted and changed, acquired an expansion of one hundred times beyond its former bulk.

Again, let the required nature be heat or cold, of such a degree as not to be sensible from its weakness. They are rendered sensible by the thermometer, as we described it above;[139] for the cold and heat are not actually perceived by the touch, but heat expands and cold contracts the air. Nor, again, is that expansion or contraction of the air in itself visible, but the air when expanded depresses the water, and when contracted raises it, which is the first reduction to sight.

Again, let the required nature be the mixture of bodies; namely, how much aqueous, oleaginous or spirituous, ashy or salt parts they contain; or, as a particular example, how much butter, cheese, and whey there is in milk, and the like. These things are rendered sensible by artificial and skilful separations in tangible substances; and the nature of the spirit in them, though not immediately perceptible, is nevertheless discovered by the various motions and efforts of bodies. And, indeed, in this branch men have labored hard in distillations and artificial separations, but with little more success than in their other experiments now in use; their methods being mere guesses and blind attempts, and more industrious than intelligent; and what is worst of all, without any imitation or rivalry of nature, but rather by violent heats and too energetic agents, to the destruction of any delicate conformation, in which principally consist the hidden virtues and sympathies. Nor do men in these separations ever attend to or observe what we have before pointed out; namely, that in attacking bodies by fire, or other methods, many qualities are superinduced by the fire itself, and the other bodies used to effect the separation, which were not originally in the compound. Hence arise most extraordinary fallacies; for the mass of vapor which is emitted from water by fire, for instance, did not exist as vapor or air in the water, but is chiefly created by the expansion of the water by the heat of the fire.

So, in general, all delicate experiments on natural or artificial bodies, by which the genuine are distinguished from the adulterated, and the better from the more common, should be referred to this division; for they bring that which is not the object of the senses within their sphere. They are therefore to be everywhere diligently sought after.

With regard to the fifth cause of objects escaping our senses, it is clear that the action of the sense takes place by motion, and this motion is time. If, therefore, the motion of any body be either so slow or so swift as not to be proportioned to the necessary momentum which operates on the senses, the object is not perceived at all; as in the motion of the hour hand, and that, again, of a musket-ball. The motion which is imperceptible by the senses from its slowness, is readily and usually rendered sensible by the accumulation of motion; that which is imperceptible from its velocity, has not as yet been well measured; it is necessary, however, that this should be done in some cases, with a view to a proper investigation of nature.

The sixth case, where the sense is impeded by the power of the object, admits of a reduction to the sensible sphere, either by removing the object to a greater distance, or by deadening its effects by the interposition of a medium, which may weaken and not destroy the object; or by the admission of its reflection where the direct impression is too strong, as that of the sun in a basin of water.

The seventh case, where the senses are so overcharged with the object as to leave no further room, scarcely occurs except in the smell or taste, and is not of much consequence as regards our present subject. Let what we have said, therefore, suffice with regard to the reduction to the sensible sphere of objects not naturally within its compass.

Sometimes, however, this reduction is not extended to the senses of man, but to those of some other animal, whose senses, in some points, exceed those of man; as (with regard to some scents) to that of the dog, and with regard to light existing imperceptibly in the air, when not illuminated from any extraneous source, to the sense of the cat, the owl, and other animals which see by night. For Telesius has well observed, that there appears to be an original portion of light even in the air itself,[140] although but slight and meagre, and of no use for the most part to the eyes of men, and those of the generality of animals; because those animals to whose senses this light is proportioned can see by night, which does not, in all probability, proceed from their seeing either without light or by any internal light.

Here, too, we would observe, that we at present discuss only the wants of the senses, and their remedies; for their deceptions must be referred to the inquiries appropriated to the senses, and sensible objects; except that important deception, which makes them define objects in their relation to man, and not in their relation to the universe, and which is only corrected by universal reasoning and philosophy.[141]

XLI. In the eighteenth rank of prerogative instances we will class the instances of the road, which we are also wont to call itinerant and jointed instances. They are such as indicate the gradually continued motions of nature. This species of instances escapes rather our observation than our senses; for men are wonderfully indolent upon this subject, consulting nature in a desultory manner, and at periodic intervals, when bodies have been regularly finished and completed, and not during her work. But if any one were desirous of examining and contemplating the talents and industry of an artificer, he would not merely wish to see the rude materials of his art, and then his work when finished, but rather to be present while he is at labor, and proceeding with his work. Something of the same kind should be done with regard to nature. For instance, if any one investigate the vegetation of plants, he should observe from the first sowing of any seed (which can easily be done, by pulling up every day seeds which have been two, three, or four days in the ground, and examining them diligently), how and when the seed begins to swell and break, and be filled, as it were, with spirit; then how it begins to burst the bark and push out fibres, raising itself a little at the same time, unless the ground be very stiff; then how it pushes out these fibres, some downward for roots, others upward for the stem, sometimes also creeping laterally, if it find the earth open and more yielding on one side, and the like. The same should be done in observing the hatching of eggs, where we may easily see the process of animation and organization, and what parts are formed of the yolk, and what of the white of the egg, and the like. The same may be said of the inquiry into the formation of animals from putrefaction; for it would not be so humane to inquire into perfect and terrestrial animals, by cutting the fœtus from the womb; but opportunities may perhaps be offered of abortions, animals killed in hunting, and the like. Nature, therefore, must, as it were, be watched, as being more easily observed by night than by day: for contemplations of this kind may be considered as carried on by night, from the minuteness and perpetual burning of our watch-light.

The same must be attempted with inanimate objects, which we have ourselves done by inquiring into the opening of liquids by fire. For the mode in which water expands is different from that observed in wine, vinegar, or verjuice, and very different, again, from that observed in milk and oil, and the like; and this was easily seen by boiling them with slow heat, in a glass vessel, through which the whole may be clearly perceived. But we merely mention this, intending to treat of it more at large and more closely when we come to the discovery of the latent process; for it should always be remembered that we do not here treat of things themselves, but merely propose examples.[142]

XLII. In the nineteenth rank of prerogative instances we will class supplementary or substitutive instances, which we are also wont to call instances of refuge. They are such as supply information, where the senses are entirely deficient, and we therefore have recourse to them when appropriate instances cannot be obtained. This substitution is twofold, either by approximation or by analogy. For instance, there is no known medium which entirely prevents the effect of the magnet in attracting iron--neither gold, nor silver, nor stone, nor glass, wood, water, oil, cloth, or fibrous bodies, air, flame, or the like. Yet by accurate experiment, a medium may perhaps be found which would deaden its effect, more than another comparatively and in degree; as, for instance, the magnet would not perhaps attract iron through the same thickness of gold as of air, or the same quantity of ignited as of cold silver, and so on; for we have not ourselves made the experiment, but it will suffice as an example. Again, there is no known body which is not susceptible of heat, when brought near the fire; yet air becomes warm much sooner than stone. These are examples of substitution by approximation.

Substitution by analogy is useful, but less sure, and therefore to be adopted with some judgment. It serves to reduce that which is not the object of the senses to their sphere, not by the perceptible operations of the imperceptible body, but by the consideration of some similar perceptible body. For instance, let the subject for inquiry be the mixture of spirits, which are invisible bodies. There appears to be some relation between bodies and their sources or support. Now, the source of flame seems to be oil and fat; that of air, water, and watery substances; for flame increases over the exhalation of oil, and air over that of water. One must therefore consider the mixture of oil and water, which is manifest to the senses, since that of air and flame in general escapes the senses. But oil and water mix very imperfectly by composition or stirring, while they are exactly and nicely mixed in herbs, blood, and the parts of animals. Something similar, therefore, may take place in the mixture of flame and air in spirituous substances, not bearing mixture very well by simple collision, while they appear, however, to be well mixed in the spirits of plants and animals.

Again, if the inquiry do not relate to perfect mixtures of spirits, but merely to their composition, as whether they easily incorporate with each other, or there be rather (as an example) certain winds and exhalations, or other spiritual bodies, which do not mix with common air, but only adhere to and float in it in globules and drops, and are rather broken and pounded by the air, than received into, and incorporated with it; this cannot be perceived in common air, and other aëriform substances, on account of the rarity of the bodies, but an image, as it were, of this process may be conceived in such liquids as quicksilver, oil, water, and even air, when broken and dissipated it ascends in small portions through water, and also in the thicker kinds of smoke; lastly, in dust, raised and remaining in the air, in all of which there is no incorporation: and the above representation in this respect is not a bad one, if it be first diligently investigated, whether there can be such a difference of nature between spirituous substances, as between liquids, for then these images might conveniently be substituted by analogy.

And although we have observed of these supplementary instances, that information is to be derived from them, when appropriate instances are wanting, by way of refuge, yet we would have it understood, that they are also of great use, when the appropriate instances are at hand, in order to confirm the information afforded by them; of which we will speak more at length, when our subject leads us, in due course, to the support of induction.

XLIII. In the twentieth rank of prerogative instances we will place lancing instances, which we are also wont (but for a different reason) to call twitching instances. We adopt the latter name, because they twitch the understanding, and the former because they pierce nature, whence we style them occasionally the instances of Democritus.[143] They are such as warn the understanding of the admirable and exquisite subtilty of nature, so that it becomes roused and awakened to attention, observation, and proper inquiry; as, for instance, that a little drop of ink should be drawn out into so many letters; that silver merely gilt on its surface should be stretched to such a length of gilt wire; that a little worm, such as you may find on the skin, should possess both a spirit and a varied conformation of its parts; that a little saffron should imbue a whole tub of water with its color; that a little musk or aroma should imbue a much greater extent of air with its perfume; that a cloud of smoke should be raised by a little incense; that such accurate differences of sounds as articulate words should be conveyed in all directions through the air, and even penetrate the pores of wood and water (though they become much weakened), that they should be, moreover, reflected, and that with such distinctness and velocity; that light and color should for such an extent and so rapidly pass through solid bodies, such as glass and water, with so great and so exquisite a variety of images, and should be refracted and reflected; that the magnet should attract through every description of body, even the most compact; but (what is still more wonderful) that in all these cases the action of one should not impede that of another in a common medium, such as air; and that there should be borne through the air, at the same time, so many images of visible objects, so many impulses of articulation, so many different perfumes, as of the violet, rose, etc., besides cold and heat, and magnetic attractions; all of them, I say, at once, without any impediment from each other, as if each had its paths and peculiar passage set apart for it, without infringing against or meeting each other.

To these lancing instances, however, we are wont, not without some advantage, to add those which we call the limits of such instances. Thus, in the cases we have pointed out, one action does not disturb or impede another of a different nature, yet those of a similar nature subdue and extinguish each other; as the light of the sun does that of the candle, the sound of a cannon that of the voice, a strong perfume a more delicate one, a powerful heat a more gentle one, a plate of iron between the magnet and other iron the effect of the magnet. But the proper place for mentioning these will be also among the supports of induction.

XLIV. We have now spoken of the instances which assist the senses, and which are principally of service as regards information; for information begins from the senses. But our whole labor terminates in practice, and as the former is the beginning, so is the latter the end of our subject. The following instances, therefore, will be those which are chiefly useful in practice. They are comprehended in two classes, and are seven in number. We call them all by the general name of practical instances. Now there are two defects in practice, and as many divisions of important instances. Practice is either deceptive or too laborious. It is generally deceptive (especially after a diligent examination of natures), on account of the power and actions of bodies being ill defined and determined. Now the powers and actions of bodies are defined and determined either by space or by time, or by the quantity at a given period, or by the predominance of energy; and if these four circumstances be not well and diligently considered, the sciences may indeed be beautiful in theory, but are of no effect in practice. We call the four instances referred to this class, mathematical instances and instances of measure.

Practice is laborious either from the multitude of instruments, or the bulk of matter and substances requisite for any given work. Those instances, therefore, are valuable, which either direct practice to that which is of most consequence to mankind, or lessen the number of instruments or of matter to be worked upon. We assign to the three instances relating to this class, the common name of propitious or benevolent instances. We will now separately discuss these seven instances, and conclude with them that part of our work which relates to the prerogative or illustrious instances.

XLV. In the twenty-first rank of prerogative instances we will place the instances of the rod or rule, which we are also wont to call the instances of completion or non ultrà. For the powers and motions of bodies do not act and take effect through indefinite and accidental, but through limited and certain spaces; and it is of great importance to practice that these should be understood and noted in every nature which is investigated, not only to prevent deception, but to render practice more extensive and efficient. For it is sometimes possible to extend these powers, and bring the distance, as it were, nearer, as in the example of telescopes.

Many powers act and take effect only by actual touch, as in the percussion of bodies, where the one does not remove the other, unless the impelling touch the impelled body. External applications in medicine, as ointment and plasters, do not exercise their efficacy except when in contact with the body. Lastly, the objects of touch and taste only strike those senses when in contact with their organs.

Other powers act at a distance, though it be very small, of which but few have as yet been noted, although there be more than men suspect; this happens (to take everyday instances) when amber or jet attracts straws, bubbles dissolve bubbles, some purgative medicines draw humors from above, and the like. The magnetic power by which iron and the magnet, or two magnets, are attracted together, acts within a definite and narrow sphere, but if there be any magnetic power emanating from the earth a little below its surface, and affecting the needle in its polarity, it must act at a great distance.

Again, if there be any magnetic force which acts by sympathy between the globe of the earth and heavy bodies, or between that of the moon and the waters of the sea (as seems most probable from the particular floods and ebbs which occur twice in the month), or between the starry sphere and the planets, by which they are summoned and raised to their apogees, these must all operate at very great distances.[144]

Again, some conflagrations and the kindling of flames take place at very considerable distances with particular substances, as they report of the naphtha of Babylon. Heat, too, insinuates itself at wide distances, as does also cold, so that the masses of ice which are broken off and float upon the Northern Ocean, and are borne through the Atlantic to the coast of Canada, become perceptible by the inhabitants, and strike them with cold from a distance. Perfumes also (though here there appears to be always some corporeal emission) act at remarkable distances, as is experienced by persons sailing by the coast of Florida, or parts of Spain, where there are whole woods of lemons, oranges, and other odoriferous plants, or rosemary and marjoram bushes, and the like. Lastly, the rays of light and the impressions of sound act at extensive distances.

Yet all these powers, whether acting at a small or great distance, certainly act within definite distances, which are well ascertained by nature, so that there is a limit depending either on the mass or quantity of the bodies, the vigor or faintness of the powers, or the favorable or impeding nature of the medium, all of which should be taken into account and observed. We must also note the boundaries of violent motions, such as missiles, projectiles, wheels and the like, since they are also manifestly confined to certain limits.

Some motions and virtues are to be found of a directly contrary nature to these, which act in contact but not at a distance; namely, such as operate at a distance and not in contact, and again act with less force at a less distance, and the reverse. Sight, for instance, is not easily effective in contact, but requires a medium and distance; although I remember having heard from a person deserving of credit, that in being cured of a cataract (which was done by putting a small silver needle within the first coat of the eye, to remove the thin pellicle of the cataract, and force it into a corner of the eye), he had distinctly seen the needle moving across the pupil. Still, though this may be true, it is clear that large bodies cannot be seen well or distinctly, unless at the vertex of a cone, where the rays from the object meet at some distance from the eye. In old persons the eye sees better if the object be moved a little further, and not nearer. Again, it is certain that in projectiles the impact is not so violent at too short a distance as a little afterward.[145] Such are the observations to be made on the measure of motions as regards distance.

There is another measure of motion in space which must not be passed over, not relating to progressive but spherical motion--that is, the expansion of bodies into a greater, or their contraction into a lesser sphere. For in our measure of this motion we must inquire what degree of compression or extension bodies easily and readily admit of, according to their nature, and at what point they begin to resist it, so as at last to bear it no further--as when an inflated bladder is compressed, it allows a certain compression of the air, but if this be increased, the air does not suffer it, and the bladder is burst.

We have proved this by a more delicate experiment. We took a metal bell, of a light and thin sort, such as is used for salt-cellars, and immersed it in a basin of water, so as to carry the air contained in its interior down with it to the bottom of the basin. We had first, however, placed a small globe at the bottom of the basin, over which we placed the bell. The result was, that if the globe were small compared with the interior of the bell, the air would contract itself, and be compressed without being forced out, but if it were too large for the air readily to yield to it, the latter became impatient of the pressure, raised the bell partly up, and ascended in bubbles.

To prove, also, the extension (as well as the compression) which air admits of, we adopted the following method:--We took a glass egg, with a small hole at one end; we drew out the air by violent suction at this hole, and then closed the hole with the finger, immersed the egg in water, and then removed the finger. The air being constrained by the effort made in suction, and dilated beyond its natural state, and therefore striving to recover and contract itself (so that if the egg had not been immersed in water, it would have drawn in the air with a hissing sound), now drew in a sufficient quantity of water to allow the air to recover its former dimensions.[146]

It is well ascertained that rare bodies (such as air) admit of considerable contraction, as has been before observed; but tangible bodies (such as water) admit of it much less readily, and to a less extent. We investigated the latter point by the following experiment:

We had a leaden globe made, capable of containing about two pints, wine measure, and of tolerable thickness, so as to support considerable pressure. We poured water into it through an aperture, which we afterward closed with melted lead, as soon as the globe was filled with water, so that the whole became perfectly solid. We next flattened the two opposite sides with a heavy hammer, which necessarily caused the water to occupy a less space, since the sphere is the solid of greatest content; and when hammering failed from the resistance of the water, we made use of a mill or press, till at last the water, refusing to submit to a greater pressure, exuded like a fine dew through the solid lead. We then computed the extent to which the original space had been reduced, and concluded that water admitted such a degree of compression when constrained by great violence.

The more solid, dry or compact bodies, such as stones, wood and metals, admit of much less, and indeed scarcely any perceptible compression or expansion, but escape by breaking, slipping forward, or other efforts; as appears in bending wood, or steel for watch-springs, in projectiles, hammering and many other motions, all of which, together with their degrees, are to be observed and examined in the investigation of nature, either to a certainty, or by estimation, or comparison, as opportunity permits.

XLVI. In the twenty-second rank of prerogative instances we will place the instances of the course, which we are also wont to call water instances, borrowing our expression from the water hour-glasses employed by the ancients instead of those with sand. They are such as measure nature by the moments of time, as the last instances do by the degrees of space. For all motion or natural action takes place in time, more or less rapidly, but still in determined moments well ascertained by nature. Even those actions which appear to take effect suddenly, and in the twinkling of an eye (as we express it), are found to admit of greater or less rapidity.

In the first place, then, we see that the return of the heavenly bodies to the same place takes place in regular times, as does the flood and ebb of the sea. The descent of heavy bodies toward the earth, and the ascent of light bodies toward the heavenly sphere, take place in definite times,[147] according to the nature of the body, and of the medium through which it moves. The sailing of ships, the motions of animals, the transmission of projectiles, all take place in times the sums of which can be computed. With regard to heat, we see that boys in winter bathe their hands in the flame without being burned; and conjurers, by quick and regular movements, overturn vessels filled with wine or water, and replace them without spilling the liquid, with several similar instances. The compression, expansion and eruption of several bodies, take place more or less rapidly, according to the nature of the body and its motion, but still in definite moments.

In the explosion of several cannon at once (which are sometimes heard at the distance of thirty miles), the sound of those nearest to the spot is heard before that of the most distant. Even in sight (whose action is most rapid), it is clear that a definite time is necessary for its exertion, which is proved by certain objects being invisible from the velocity of their motion, such as a musket-ball; for the flight of the ball is too swift to allow an impression of its figure to be conveyed to the sight.

This last instance, and others of a like nature, have sometimes excited in us a most marvellous doubt, no less than whether the image of the sky and stars is perceived as at the actual moment of its existence, or rather a little after, and whether there is not (with regard to the visible appearance of the heavenly bodies) a true and apparent time, as well as a true and apparent place, which is observed by astronomers in parallaxes. It appeared so incredible to us, that the images or radiations of heavenly bodies could suddenly be conveyed through such immense spaces to the sight, and it seemed that they ought rather to be transmitted in a definite time.[148] That doubt, however (as far as regards any great difference between the true and apparent time), was subsequently completely set at rest, when we considered the infinite loss and diminution of size as regards the real and apparent magnitude of a star, occasioned by its distance, and at the same time observed at how great a distance (at least sixty miles) bodies which are merely white can be suddenly seen by us. For there is no doubt, that the light of the heavenly bodies not only far surpasses the vivid appearance of white, but even the light of any flame (with which we are acquainted) in the vigor of its radiation. The immense velocity of the bodies themselves, which is perceived in their diurnal motion, and has so astonished thinking men, that they have been more ready to believe in the motion of the earth, renders the motion of radiation from them (marvellous as it is in its rapidity) more worthy of belief. That which has weighed most with us, however, is, that if there were any considerable interval of time between the reality and the appearance, the images would often be interrupted and confused by clouds formed in the meantime, and similar disturbances of the medium. Let this suffice with regard to the simple measures of time.

It is not merely the absolute, but still more the relative measure of motions and actions which must be inquired into, for this latter is of great use and application. We perceive that the flame of firearms is seen sooner than the sound is heard, although the ball must have struck the air before the flame, which was behind it, could escape: the reason of which is, that light moves with greater velocity than sound. We perceive, also, that visible images are received by the sight with greater rapidity than they are dismissed, and for this reason, a violin string touched with the finger appears double or triple, because the new image is received before the former one is dismissed. Hence, also, rings when spinning appear globular, and a lighted torch, borne rapidly along at night, appears to have a tail. Upon the principle of the inequality of motion, also, Galileo attempted an explanation of the flood and ebb of the sea, supposing the earth to move rapidly, and the water slowly, by which means the water, after accumulating, would at intervals fall back, as is shown in a vessel of water made to move rapidly. He has, however, imagined this on data which cannot be granted (namely, the earth's motion), and besides, does not satisfactorily account for the tide taking place every six hours.

An example of our present point (the relative measure of motion), and, at the same time, of its remarkable use of which we have spoken, is conspicuous in mines filled with gunpowder, where immense weights of earth, buildings, and the like, are overthrown and prostrated by a small quantity of powder; the reason of which is decidedly this, that the motion of the expansion of the gunpowder is much more rapid than that of gravity,[149] which would resist it, so that the former has terminated before the latter has commenced. Hence, also, in missiles, a strong blow will not carry them so far as a sharp and rapid one. Nor could a small portion of animal spirit in animals, especially in such vast bodies as those of the whale and elephant, have ever bent or directed such a mass of body, were it not owing to the velocity of the former, and the slowness of the latter in resisting its motion.

In short, this point is one of the principal foundations of the magic experiments (of which we shall presently speak), where a small mass of matter overcomes and regulates a much larger, if there but be an anticipation of motion, by the velocity of one before the other is prepared to act.

Finally, the point of the first and last should be observed in all natural actions. Thus, in an infusion of rhubarb the purgative property is first extracted, and then the astringent; we have experienced something of the same kind in steeping violets in vinegar, which first extracts the sweet and delicate odor of the flower, and then the more earthy part, which disturbs the perfume; so that if the violets be steeped a whole day, a much fainter perfume is extracted than if they were steeped for a quarter of an hour only, and then taken out; and since the odoriferous spirit in the violet is not abundant, let other and fresh violets be steeped in the vinegar every quarter of an hour, as many as six times, when the infusion becomes so strengthened, that although the violets have not altogether remained there for more than one hour and a half, there remains a most pleasing perfume, not inferior to the flower itself, for a whole year. It must be observed, however, that the perfume does not acquire its full strength till about a month after the infusion. In the distillation of aromatic plants macerated in spirits of wine, it is well known that an aqueous and useless phlegm rises first, then water containing more of the spirit, and, lastly, water containing more of the aroma; and many observations of the like kind, well worthy of notice, are to be made in distillations. But let these suffice as examples.[150]

XLVII. In the twenty-third rank of prerogative instances we will place instances of quantity, which we are also wont to call the doses of nature (borrowing a word from medicine). They are such as measure the powers by the quantity of bodies, and point out the effect of the quantity in the degree of power. And in the first place, some powers only subsist in the universal quantity, or such as bears a relation to the confirmation and fabric of the universe. Thus the earth is fixed, its parts fall. The waters in the sea flow and ebb, but not in the rivers, except by the admission of the sea. Then, again, almost all particular powers act according to the greater or less quantity of the body. Large masses of water are not easily rendered foul, small are. New wine and beer become ripe and drinkable in small skins much more readily than in large casks. If a herb be placed in a considerable quantity of liquid, infusion takes place rather than impregnation; if in less, the reverse. A bath, therefore, and a light sprinkling, produce different effects on the human body. Light dew, again, never falls, but is dissipated and incorporated with the air; thus we see that in breathing on gems, the slight quantity of moisture, like a small cloud in the air, is immediately dissolved. Again, a piece of the same magnet does not attract so much iron as the whole magnet did. There are some powers where the smallness of the quantity is of more avail; as in boring, a sharp point pierces more readily than a blunt one; the diamond, when pointed, makes an impression on glass, and the like.

Here, too, we must not rest contented with a vague result, but inquire into the exact proportion of quantity requisite for a particular exertion of power; for one would be apt to suppose that the power bears an exact proportion to the quantity; that if a leaden bullet of one ounce, for instance, would fall in a given time, one of two ounces ought to fall twice as rapidly, which is most erroneous. Nor does the same ratio prevail in every kind of power, their difference being considerable. The measure, therefore, must be determined by experiment, and not by probability or conjecture.

Lastly, we must in all our investigations of nature observe what quantity, or dose, of the body is requisite for a given effect, and must at the same time be guarded against estimating it at too much or too little.

XLVIII. In the twenty-fourth rank of prerogative instances we will place wrestling instances, which we are also wont to call instances of predominance. They are such as point out the predominance and submission of powers compared with each other, and which of them is the more energetic and superior, or more weak and inferior. For the motions and effects of bodies are compounded, decomposed, and combined, no less than the bodies themselves. We will exhibit, therefore, the principal kinds of motions or active powers, in order that their comparative strength, and thence a demonstration and definition of the instances in question, may be rendered more clear.

Let the first motion be that of the resistance of matter, which exists in every particle, and completely prevents its annihilation; so that no conflagration, weight, pressure, violence, or length of time can reduce even the smallest portion of matter to nothing, or prevent it from being something, and occupying some space, and delivering itself (whatever straits it be put to), by changing its form or place, or, if that be impossible, remaining as it is; nor can it ever happen that it should either be nothing or nowhere. This motion is designated by the schools (which generally name and define everything by its effects and inconveniences rather than by its inherent cause) by the axiom, that two bodies cannot exist in the same place, or they call it a motion to prevent the penetration of dimensions. It is useless to give examples of this motion, since it exists in every body.

Let the second motion be that which we term the motion of connection, by which bodies do not allow themselves to be separated at any point from the contact of another body, delighting, as it were, in the mutual connection and contact. This is called by the schools a motion to prevent a vacuum. It takes place when water is drawn up by suction or a syringe, the flesh by cupping, or when the water remains without escaping from perforated jars, unless the mouth be opened to admit the air, and innumerable instances of a like nature.

Let the third be that which we term the motion of liberty, by which bodies strive to deliver themselves from any unnatural pressure or tension, and to restore themselves to the dimensions suited to their mass; and of which, also, there are innumerable examples. Thus, we have examples of their escaping from pressure, in the water in swimming, in the air in flying, in the water again in rowing, and in the air in the undulation of the winds, and in springs of watches. An exact instance of the motion of compressed air is seen in children's popguns, which they make by scooping out elder-branches or some such matter, and forcing in a piece of some pulpy root or the like, at each end; then they force the root or other pellet with a ramrod to the opposite end, from which the lower pellet is emitted and projected with a report, and that before it is touched by the other piece of root or pellet, or by the ramrod. We have examples of their escape from tension, in the motion of the air that remains in glass eggs after suction, in strings, leather, and cloth, which recoil after tension, unless it be long continued. The schools define this by the term of motion from the form of the element; injudiciously enough, since this motion is to be found not only in air, water, or fire, but in every species of solid, as wood, iron, lead, cloth, parchment, etc., each of which has its own proper size, and is with difficulty stretched to any other. Since, however, this motion of liberty is the most obvious of all, and to be seen in an infinite number of cases, it will be as well to distinguish it correctly and clearly; for some most carelessly confound this with the two others of resistance and connection; namely, the freedom from pressure with the former, and that from tension with the latter, as if bodies when compressed yielded or expanded to prevent a penetration of dimensions, and when stretched rebounded and contracted themselves to prevent a vacuum. But if the air, when compressed, could be brought to the density of water, or wood to that of stone, there would be no need of any penetration of dimensions, and yet the compression would be much greater than they actually admit of. So if water could be expanded till it became as rare as air, or stone as rare as wood, there would be no need of a vacuum, and yet the expansion would be much greater than they actually admit of.

We do not, therefore, arrive at a penetration of dimensions or a vacuum before the extremes of condensation and rarefaction, while the motion we speak of stops and exerts itself much within them, and is nothing more than a desire of bodies to preserve their specific density (or, if it be preferred, their form), and not to desert them suddenly, but only to change by degrees, and of their own accord. It is, however, much more necessary to intimate to mankind (because many other points depend upon this), that the violent motion which we call mechanical, and Democritus (who, in explaining his primary motions, is to be ranked even below the middling class of philosophers) termed the motion of a blow, is nothing else than this motion of liberty, namely, a tendency to relaxation from compression. For in all simple impulsion or flight through the air, the body is not displaced or moved in space, until its parts are placed in an unnatural state, and compressed by the impelling force. When that takes place, the different parts urging the other in succession, the whole is moved, and that with a rotatory as well as progressive motion, in order that the parts may, by this means also, set themselves at liberty, or more readily submit. Let this suffice for the motion in question.

Let the fourth be that which we term the motion of matter, and which is opposed to the last; for in the motion of liberty, bodies abhor, reject, and avoid, a new size or volume, or any new expansion or contraction (for these different terms have the same meaning), and strive, with all their power, to rebound and resume their former density; on the contrary, in the motion of matter, they are anxious to acquire a new volume or dimension, and attempt it willingly and rapidly, and occasionally by a most vigorous effort, as in the example of gunpowder. The most powerful, or at least most frequent, though not the only instruments of this motion, are heat and cold. For instance, the air, if expanded by tension (as by suction in the glass egg), struggles anxiously to restore itself; but if heat be applied, it strives, on the contrary, to dilate itself, and longs for a larger volume, regularly passing and migrating into it, as into a new form (as it is termed); nor after a certain degree of expansion is it anxious to return, unless it be invited to do so by the application of cold, which is not indeed a return, but a fresh change. So also water, when confined by compression, resists, and wishes to become as it was before, namely, more expanded; but if there happen an intense and continued cold, it changes itself readily, and of its own accord, into the condensed state of ice; and if the cold be long continued, without any intervening warmth (as in grottoes and deep caves), it is changed into crystal or similar matter, and never resumes its form.

Let the fifth be that which we term the motion of continuity. We do not understand by this simple and primary continuity with any other body (for that is the motion of connection), but the continuity of a particular body in itself; for it is most certain that all bodies abhor a solution of continuity, some more and some less, but all partially. In hard bodies (such as steel and glass) the resistance to an interruption of continuity is most powerful and efficacious, while although in liquids it appears to be faint and languid, yet it is not altogether null, but exists in the lowest degree, and shows itself in many experiments, such as bubbles, the round form of drops, the thin threads which drip from roofs, the cohesion of glutinous substances, and the like. It is most conspicuous, however, if an attempt be made to push this separation to still smaller particles. Thus, in mortars, the pestle produces no effect after a certain degree of contusion, water does not penetrate small fissures, and the air itself, notwithstanding its subtilty, does not penetrate the pores of solid vessels at once, but only by long-continued insinuation.

Let the sixth be that which we term the motion of acquisition, or the motion of need.[151] It is that by which bodies placed among others of a heterogeneous and, as it were, hostile nature, if they meet with the means or opportunity of avoiding them, and uniting themselves with others of a more analogous nature, even when these latter are not closely allied to them, immediately seize and, as it were, select them, and appear to consider it as something acquired (whence we derive the name), and to have need of these latter bodies. For instance, gold, or any other metal in leaf, does not like the neighborhood of air; if, therefore, they meet with any tangible and thick substance (such as the finger, paper, or the like), they immediately adhere to it, and are not easily torn from it. Paper, too, and cloth, and the like, do not agree with the air, which is inherent and mixed in their pores. They readily, therefore, imbibe water or other liquids, and get rid of the air. Sugar, or a sponge, dipped in water or wine, and though part of it be out of the water or wine, and at some height above it, will yet gradually absorb them.[152]

Hence an excellent rule is derived for the opening and dissolution of bodies; for (not to mention corrosive and strong waters, which force their way) if a body can be found which is more adapted, suited, and friendly to a given solid, than that with which it is by some necessity united, the given solid immediately opens and dissolves itself to receive the former, and excludes or removes the latter.[153] Nor is the effect or power of this motion confined to contact, for the electric energy (of which Gilbert and others after him have told so many fables) is only the energy excited in a body by gentle friction, and which does not endure the air, but prefers some tangible substance if there be any at hand.

Let the seventh be that which we term the motion of greater congregation, by which bodies are borne toward masses of a similar nature, for instance, heavy bodies toward the earth, light to the sphere of heaven. The schools termed this natural motion, by a superficial consideration of it, because produced by no external visible agent, which made them consider it innate in the substances; or perhaps because it does not cease, which is little to be wondered at, since heaven and earth are always present, while the causes and sources of many other motions are sometimes absent and sometimes present. They therefore called this perpetual and proper, because it is never interrupted, but instantly takes place when the others are interrupted, and they called the others adscititious. The former, however, is in reality weak and slow, since it yields, and is inferior to the others as long as they act, unless the mass of the body be great; and although this motion have so filled men's minds, as almost to have obscured all others, yet they know but little about it, and commit many errors in its estimate.

Let the eighth be that which we term the motion of lesser congregation, by which the homogeneous parts in any body separate themselves from the heterogeneous and unite together, and whole bodies of a similar substance coalesce and tend toward each other, and are sometimes congregated, attracted, and meet, from some distance; thus in milk the cream rises after a certain time, and in wine the dregs and tartar sink; which effects are not to be attributed to gravity and levity only, so as to account for the rising of some parts and the sinking of others, but much more to the desire of the homogeneous bodies to meet and unite. This motion differs from that of need in two points: 1st, because the latter is the stimulus of a malignant and contrary nature, while in this of which we treat (if there be no impediment or restraint), the parts are united by their affinity, although there be no foreign nature to create a struggle; 2dly, because the union is closer and more select. For in the other motion, bodies which have no great affinity unite, if they can but avoid the hostile body, while in this, substances which are connected by a decided kindred resemblance come together and are molded into one. It is a motion existing in all compound bodies, and would be readily seen in each, if it were not confined and checked by the other affections and necessities of bodies which disturb the union.

This motion is usually confined in the three following manners: by the torpor of the bodies; by the power of the predominating body; by external motion. With regard to the first, it is certain that there is more or less sluggishness in tangible bodies, and an abhorrence of locomotion; so that unless excited they prefer remaining contented with their actual state, to placing themselves in a better position. There are three means of breaking through this sluggishness--heat; the active power of a similar body; vivid and powerful motion. With regard to the first, heat is, on this account, defined as that which separates heterogeneous, and draws together homogeneous substances; a definition of the Peripatetics which is justly ridiculed by Gilbert, who says it is as if one were to define man to be that which sows wheat and plants vineyards; being only a definition deduced from effects, and those but partial. But it is still more to be blamed, because those effects, such as they are, are not a peculiar property of heat, but a mere accident (for cold, as we shall afterward show, does the same), arising from the desire of the homogeneous parts to unite; the heat then assists them in breaking through that sluggishness which before restrained their desire. With regard to the assistance derived from the power of a similar body, it is most conspicuous in the magnet when armed with steel, for it excites in the steel a power of adhering to steel, as a homogeneous substance, the power of the magnet breaking through the sluggishness of the steel. With regard to the assistance of motion, it is seen in wooden arrows or points, which penetrate more deeply into wood than if they were tipped with iron, from the similarity of the substance, the swiftness of the motion breaking through the sluggishness of the wood; of which two last experiments we have spoken above in the aphorism on clandestine instances.[154]

The confinement of the motion of lesser congregation, which arises from the power of the predominant body, is shown in the decomposition of blood and urine by cold. For as long as these substances are filled with the active spirit, which regulates and restrains each of their component parts, as the predominant ruler of the whole, the several different parts do not collect themselves separately on account of the check; but as soon as that spirit has evaporated, or has been choked by the cold, then the decomposed parts unite, according to their natural desire. Hence it happens, that all bodies which contain a sharp spirit (as salts and the like), last without decomposition, owing to the permanent and durable power of the predominating and imperious spirit.

The confinement of the motion of lesser congregation, which arises from external motion, is very evident in that agitation of bodies which preserves them from putrefaction. For all putrefaction depends on the congregation of the homogeneous parts, whence, by degrees, there ensues a corruption of the first form (as it is called), and the generation of another. For the decomposition of the original form, which is itself the union of the homogeneous parts, precedes the putrefaction, which prepares the way for the generation of another. This decomposition, if not interrupted, is simple; but if there be various obstacles, putrefactions ensue, which are the rudiments of a new generation. But if (to come to our present point) a frequent agitation be excited by external motion, the motion toward union (which is delicate and gentle, and requires to be free from all external influence) is disturbed, and ceases; which we perceive to be the case in innumerable instances. Thus, the daily agitation or flowing of water prevents putrefaction; winds prevent the air from being pestilent; corn turned about and shaken in granaries continues clean: in short, everything which is externally agitated will with difficulty rot internally.

We must not omit that union of the parts of bodies which is the principal cause of induration and desiccation. When the spirit or moisture, which has evaporated into spirit, has escaped from a porous body (such as wood, bone, parchment, and the like), the thicker parts are drawn together, and united with a greater effort, and induration or desiccation is the consequence; and this we attribute not so much to the motion of connection (in order to prevent a vacuum), as to this motion of friendship and union.

Union from a distance is rare, and yet is to be met with in more instances than are generally observed. We perceive it when one bubble dissolves another, when medicines attract humors from a similarity of substance, when one string moves another in unison with it on different instruments, and the like. We are of opinion that this motion is very prevalent also in animal spirits, but are quite ignorant of the fact. It is, however, conspicuous in the magnet, and magnetized iron. While speaking of the motions of the magnet, we must plainly distinguish them, for there are four distinct powers or effects of the magnet which should not be confounded, although the wonder and astonishment of mankind has classed them together. 1. The attraction of the magnet to the magnet, or of iron to the magnet, or of magnetized iron to iron. 2. Its polarity toward the north and south, and its variation. 3. Its penetration through gold, glass, stone, and all other substances. 4. The communication of power from the mineral to iron, and from iron to iron, without any communication of the substances. Here, however, we only speak of the first. There is also a singular motion of attraction between quicksilver and gold, so that the gold attracts quicksilver even when made use of in ointment; and those who work surrounded by the vapors of quicksilver, are wont to hold a piece of gold in their mouths, to collect the exhalations, which would otherwise attack their heads and bones, and this piece soon grows white.[155] Let this suffice for the motion of lesser congregation.

Let the ninth be the magnetic motion, which, although of the nature of that last mentioned, yet, when operating at great distances, and on great masses, deserves a separate inquiry, especially if it neither begin in contact, as most motions of congregation do, nor end by bringing the substances into contact, as all do, but only raise them, and make them swell without any further effect. For if the moon raise the waters, or cause moist substances to swell, or if the starry sphere attract the planets toward their apogees, or the sun confine the planets Mercury and Venus to within a certain distance of his mass;[156] these motions do not appear capable of being classed under either of those of congregation, but to be, as it were, intermediately and imperfectly congregative, and thus to form a distinct species.

Let the tenth motion be that of avoidance, or that which is opposed to the motion of lesser congregation, by which bodies, with a kind of antipathy, avoid and disperse, and separate themselves from, or refuse to unite themselves with others of a hostile nature. For although this may sometimes appear to be an accidental motion, necessarily attendant upon that of the lesser congregation, because the homogeneous parts cannot unite, unless the heterogeneous be first removed and excluded, yet it is still to be classed separately,[157] and considered as a distinct species, because, in many cases, the desire of avoidance appears to be more marked than that of union.

It is very conspicuous in the excrements of animals, nor less, perhaps, in objects odious to particular senses, especially the smell and taste; for a fetid smell is rejected by the nose, so as to produce a sympathetic motion of expulsion at the mouth of the stomach; a bitter and rough taste is rejected by the palate or throat, so as to produce a sympathetic concussion and shivering of the head. This motion is visible also in other cases. Thus it is observed in some kinds of antiperistasis, as in the middle region of the air, the cold of which appears to be occasioned by the rejection of cold from the regions of the heavenly bodies; and also in the heat and combustion observed in subterranean spots, which appear to be owing to the rejection of heat from the centre of the earth. For heat and cold, when in small quantities, mutually destroy each other, while in larger quantities, like armies equally matched, they remove and eject each other in open conflict. It is said, also that cinnamon and other perfumes retain their odor longer when placed near privies and foul places, because they will not unite and mix with stinks. It is well known that quicksilver, which would otherwise reunite into a complete mass, is prevented from so doing by man's spittle, pork lard, turpentine and the like, from the little affinity of its parts with those substances, so that when surrounded by them it draws itself back, and its avoidance of these intervening obstacles is greater than its desire of reuniting itself to its homogeneous parts; which is what they term the mortification of quicksilver. Again, the difference in weight of oil and water is not the only reason for their refusing to mix, but it is also owing to the little affinity of the two; for spirits of wine, which are lighter than oil, mix very well with water. A very remarkable instance of the motion in question is seen in nitre, and crude bodies of a like nature, which abhor flame, as may be observed in gunpowder, quicksilver and gold. The avoidance of one pole of the magnet by iron is not (as Gilbert has well observed), strictly speaking, an avoidance, but a conformity, or attraction to a more convenient situation.

Let the eleventh motion be that of assimilation, or self-multiplication, or simple generation, by which latter term we do not mean the simple generation of integral bodies, such as plants or animals, but of homogeneous bodies. By this motion homogeneous bodies convert those which are allied to them, or at least well disposed and prepared, into their own substance and nature. Thus flame multiplies itself over vapors and oily substances and generates fresh flame; the air over water and watery substances multiplies itself and generates fresh air; the vegetable and animal spirit, over the thin particles of a watery or oleaginous spirit contained in its food, multiplies itself and generates fresh spirit; the solid parts of plants and animals, as the leaf, flower, the flesh, bone and the like, each of them assimilate some part of the juices contained in their food, and generate a successive and daily substance. For let none rave with Paracelsus, who (blinded by his distillations) would have it, that nutrition takes place by mere separation, and that the eye, nose, brain and liver lie concealed in bread and meat, the root, leaf and flower, in the juice of the earth; asserting that just as the artist brings out a leaf, flower, eye, nose, hand, foot and the like, from a rude mass of stone or wood by the separation and rejection of what is superfluous; so the great artist within us brings out our several limbs and parts by separation and rejection. But to leave such trifling, it is most certain that all the parts of vegetables and animals, as well the homogeneous as organic, first of all attract those juices contained in their food, which are nearly common, or at least not very different, and then assimilate and convert them into their own nature. Nor does this assimilation, or simple generation, take place in animated bodies only; for the inanimate also participate in the same property (as we have observed of flame and air), and that languid spirit, which is contained in every tangible animated substance, is perpetually working upon the coarser parts, and converting them into spirit, which afterward is exhaled, whence ensues a diminution of weight, and a desiccation of which we have spoken elsewhere.[158]

Nor should we, in speaking of assimilation, neglect to mention the accretion which is usually distinguished from aliment, and which is observed when mud grows into a mass between stones, and is converted into a stony substance, and the scaly substance round the teeth is converted into one no less hard than the teeth themselves; for we are of opinion that there exists in all bodies a desire of assimilation, as well as of uniting with homogeneous masses. Each of these powers, however, is confined, although in different manners, and should be diligently investigated, because they are connected with the revival of old age. Lastly, it is worthy of observation, that in the nine preceding motions, bodies appear to aim at the mere preservation of their nature, while in this they attempt its propagation.

Let the twelfth motion be that of excitement, which appears to be a species of the last, and is sometimes mentioned by us under that name. It is, like that, a diffusive, communicative, transitive and multiplying motion; and they agree remarkably in their effect, although they differ in their mode of action, and in their subject matter. The former proceeds imperiously and with authority; it orders and compels the assimilated to be converted and changed into the assimilating body. The latter proceeds by art, insinuation and stealth, inviting and disposing the excited toward the nature of the exciting body. The former both multiplies and transforms bodies and substances; thus a greater quantity of flame, air, spirit and flesh is formed; but in the latter, the powers only are multiplied and changed, and heat, the magnetic power, and putrefaction, in the above instances, are increased. Heat does not diffuse itself when heating other bodies by any communication of the original heat, but only by exciting the parts of the heated body to that motion which is the form of heat, and of which we spoke in the first vintage of the nature of heat. Heat, therefore, is excited much less rapidly and readily in stone or metal than in air, on account of the inaptitude and sluggishness of those bodies in acquiring that motion, so that it is probable, that there may be some substances, toward the centre of the earth, quite incapable of being heated, on account of their density, which may deprive them of the spirit by which the motion of excitement is usually commenced. Thus also the magnet creates in the iron a new disposition of its parts, and a conformable motion, without losing any of its virtue. So the leaven of bread, yeast, rennet and some poisons, excite and invite successive and continued motion in dough, beer, cheese or the human body; not so much from the power of the exciting, as the predisposition and yielding of the excited body.

Let the thirteenth motion be that of impression, which is also a species of motion of assimilation, and the most subtile of diffusive motions. We have thought it right, however, to consider it as a distinct species, on account of its remarkable difference from the last two; for the simple motion of assimilation transforms the bodies themselves, so that if you remove the first agent, you diminish not the effect of those which succeed; thus, neither the first lighting of flame, nor the first conversion into air, are of any importance to the flame or air next generated. So, also, the motion of excitement still continues for a considerable time after the removal of the first agent, as in a heated body on the removal of the original heat, in the excited iron on the removal of the magnet, and in the dough on the removal of the leaven. But the motion of impression, although diffusive and transitive, appears, nevertheless, to depend on the first agent, so that upon the removal of the latter the former immediately fails and perishes; for which reason also it takes effect in a moment, or at least a very short space of time. We are wont to call the two former motions the motions of the generation of Jupiter, because when born they continue to exist; and the latter, the motion of the generation of Saturn, because it is immediately devoured and absorbed. It may be seen in three instances: 1, in the rays of light; 2, in the percussions of sounds; 3, in magnetic attractions as regards communication. For, on the removal of light, colors and all its other images disappear, as on the cessation of the first percussion and the vibration of the body, sound soon fails, and although sounds are agitated by the wind, like waves, yet it is to be observed, that the same sound does not last during the whole time of the reverberation. Thus, when a bell is struck, the sound appears to be continued for a considerable time, and one might easily be led into the mistake of supposing it to float and remain in the air during the whole time, which is most erroneous.[159] For the reverberation is not one identical sound, but the repetition of sounds, which is made manifest by stopping and confining the sonorous body; thus, if a bell be stopped and held tightly, so as to be immovable, the sound fails, and there is no further reverberation, and if a musical string be touched after the first vibration, either with the finger (as in the harp), or a quill (as in the harpsichord), the sound immediately ceases. If the magnet be removed the iron falls. The moon, however, cannot be removed from the sea, nor the earth from a heavy falling body, and we can, therefore, make no experiment upon them; but the case is the same.

Let the fourteenth motion be that configuration or position, by which bodies appear to desire a peculiar situation, collocation, and configuration with others, rather than union or separation. This is a very abstruse notion, and has not been well investigated; and, in some instances, appears to occur almost without any cause, although we be mistaken in supposing this to be really the case. For if it be asked, why the heavens revolve from east to west, rather than from west to east, or why they turn on poles situate near the Bears, rather than round Orion or any other part of the heaven, such a question appears to be unreasonable, since these phenomena should be received as determinate and the objects of our experience. There are, indeed, some ultimate and self-existing phenomena in nature, but those which we have just mentioned are not to be referred to that class: for we attribute them to a certain harmony and consent of the universe, which has not yet been properly observed. But if the motion of the earth from west to east be allowed, the same question may be put, for it must also revolve round certain poles, and why should they be placed where they are, rather than elsewhere? The polarity and variation of the needle come under our present head. There is also observed in both natural and artificial bodies, especially solids rather than fluids, a particular collocation and position of parts, resembling hairs or fibres, which should be diligently investigated, since, without a discovery of them, bodies cannot be conveniently controlled or wrought upon. The eddies observable in liquids by which, when compressed, they successively raise different parts of their mass before they can escape, so as to equalize the pressure, is more correctly assigned to the motion of liberty.

Let the fifteenth motion be that of transmission or of passage, by which the powers of bodies are more or less impeded or advanced by the medium, according to the nature of the bodies and their effective powers, and also according to that of the medium. For one medium is adapted to light, another to sound, another to heat and cold, another to magnetic action, and so on with regard to the other actions.

Let the sixteenth be that which we term the royal or political motion, by which the predominant and governing parts of any body check, subdue, reduce, and regulate the others, and force them to unite, separate, stand still, move, or assume a certain position, not from any inclination of their own, but according to a certain order, and as best suits the convenience of the governing part, so that there is a sort of dominion and civil government exercised by the ruling part over its subjects. The motion is very conspicuous in the spirits of animals, where, as long as it is in force, it tempers all the motions of the other parts. It is found in a less degree in other bodies, as we have observed in blood and urine, which are not decomposed until the spirit, which mixed and retained their parts, has been emitted or extinguished. Nor is this motion peculiar to spirits only, although in most bodies the spirit predominates, owing to its rapid motion and penetration; for the grosser parts predominate in denser bodies, which are not filled with a quick and active spirit (such as exists in quicksilver or vitriol), so that unless this check or yoke be thrown off by some contrivance, there is no hope of any transformation of such bodies. And let not any one suppose that we have forgotten our subject, because we speak of predominance in this classification of motions, which is made entirely with the view of assisting the investigation of wrestling instances, or instances of predominance. For we do not now treat of the general predominance of motions or powers, but of that of parts in whole bodies, which constitutes the particular species here considered.

Let the seventeenth motion be the spontaneous motion of revolution, by which bodies having a tendency to move, and placed in a favorable situation, enjoy their peculiar nature, pursuing themselves and nothing else, and seeking, as it were, to embrace themselves. For bodies seem either to move without any limit, or to tend toward a limit, arrived at which they either revolve according to their peculiar nature, or rest. Those which are favorably situated, and have a tendency to motion, move in a circle with an eternal and unlimited motion; those which are favorably situated and abhor motion, rest. Those which are not favorably situated move in a straight line (as their shortest path), in order to unite with others of a congenial nature. This motion of revolution admits of nine differences: 1, with regard to the centre about which the bodies move; 2, the poles round which they move; 3, the circumference or orbit relatively to its distance from the centre; 4, the velocity, or greater or less speed with which they revolve; 5, the direction of the motion as from east to west, or the reverse; 6, the deviation from a perfect circle, by spiral lines at a greater or less distance from the centre; 7, the deviation from the circle, by spiral lines at a greater or less distance from the poles; 8, the greater or less distance of these spirals from each other; 9, and lastly, the variation of the poles if they be movable; which, however, only affects revolution when circular. The motion in question is, according to common and long-received opinion, considered to be that of the heavenly bodies. There exists, however, with regard to this, a considerable dispute between some of the ancients as well as moderns, who have attributed a motion of revolution to the earth. A much more reasonable controversy, perhaps, exists (if it be not a matter beyond dispute), whether the motion in question (on the hypothesis of the earth's being fixed) is confined to the heavens, or rather descends and is communicated to the air and water. The rotation of missiles, as in darts, musket-balls, and the like, we refer entirely to the motion of liberty.

Let the eighteenth motion be that of trepidation,[160] to which (in the sense assigned to it by astronomers) we do not give much credit; but in our serious and general search after the tendencies of natural bodies, this motion occurs, and appears worthy of forming a distinct species. It is the motion of an (as it were) eternal captivity; when bodies, for instance, being placed not altogether according to their nature, and yet not exactly ill, constantly tremble, and are restless, not contented with their position, and yet not daring to advance. Such is the motion of the heart and pulse of animals, and it must necessarily occur in all bodies which are situated in a mean state, between conveniences and inconveniences; so that being removed from their proper position, they strive to escape, are repulsed, and again continue to make the attempt.

Let the nineteenth and last motion be one which can scarcely be termed a motion, and yet is one; and which we may call the motion of repose, or of abhorrence of motion. It is by this motion that the earth stands by its own weight, while its extremes move toward the middle, not to an imaginary centre, but in order to unite. It is owing to the same tendency, that all bodies of considerable density abhor motion, and their only tendency is not to move, which nature they preserve, although excited and urged in a variety of ways to motion. But if they be compelled to move, yet do they always appear anxious to recover their former state, and to cease from motion, in which respect they certainly appear active, and attempt it with sufficient swiftness and rapidity, as if fatigued, and impatient of delay. We can only have a partial representation of this tendency, because with us every tangible substance is not only not condensed to the utmost, but even some spirit is added, owing to the action and concocting influence of the heavenly bodies.

We have now, therefore, exhibited the species, or simple elements of the motions, tendencies, and active powers, which are most universal in nature; and no small portion of natural science has been thus sketched out. We do not, however, deny that other instances can perhaps be added, and our divisions changed according to some more natural order of things, and also reduced to a less number; in which respect we do not allude to any abstract classification, as if one were to say, that bodies desire the preservation, exaltation, propagation, or fruition of their nature; or, that motion tends to the preservation and benefit either of the universe (as in the case of those of resistance and connection), or of extensive wholes, as in the case of those of the greater congregation, revolution, and abhorrence of motion, or of particular forms, as in the case of the others. For although such remarks be just, yet, unless they terminate in matter and construction, according to true definitions, they are speculative, and of little use. In the meantime, our classification will suffice, and be of much use in the consideration of the predominance of powers, and examining the wrestling instances which constitute our present subject.

For of the motions here laid down, some are quite invincible, some more powerful than others, which they confine, check, and modify; others extend to a greater distance, others are more immediate and swift, others strengthen, increase, and accelerate the rest.

The motion of resistance is most adamantine and invincible. We are yet in doubt whether such be the nature of that of connection; for we cannot with certainty determine whether there be a vacuum, either extensive or intermixed with matter. Of one thing, however, we are satisfied, that the reason assigned by Leucippus and Democritus for the introduction of a vacuum (namely, that the same bodies could not otherwise comprehend, and fill greater and less spaces) is false. For there is clearly a folding of matter, by which it wraps and unwraps itself in space within certain limits, without the intervention of a vacuum. Nor is there two thousand times more of vacuum in air than in gold, as there should be on this hypothesis; a fact demonstrated by the very powerful energies of fluids (which would otherwise float like fine dust in vacuo), and many other proofs. The other motions direct, and are directed by each other, according to their strength, quantity, excitement, emission, or the assistance or impediments they meet with.

For instance; some armed magnets hold and support iron of sixty times their own weight; so far does the motion of lesser congregation predominate over that of the greater; but if the weight be increased, it yields. A lever of a certain strength will raise a given weight, and so far the motion of liberty predominates over that of the greater congregation, but if the weight be greater, the former motion yields. A piece of leather stretched to a certain point does not break, and so far the motion of continuity predominates over that of tension, but if the tension be greater, the leather breaks, and the motion of continuity yields. A certain quantity of water flows through a chink, and so far the motion of greater congregation predominates over that of continuity, but if the chink be smaller it yields. If a musket be charged with ball and powdered sulphur alone, and fire be applied, the ball is not discharged, in which case the motion of greater congregation overcomes that of matter; but when gunpowder is used, the motion of matter in the sulphur predominates, being assisted by that motion, and the motion of avoidance in the nitre; and so of the rest. For wrestling instances (which show the predominance of powers, and in what manner and proportion they predominate and yield) must be searched for with active and industrious diligence.

The methods and nature of this yielding must also be diligently examined, as for instance, whether the motions completely cease, or exert themselves, but are constrained. For in the bodies with which we are acquainted, there is no real but an apparent rest, either in the whole or in parts. This apparent rest is occasioned either by equilibrium, or the absolute predominance of motions. By equilibrium, as in the scales of the balance, which rest if the weights be equal. By predominance, as in perforated jars, in which the water rests, and is prevented from falling by the predominance of the motion of connection. It is, however, to be observed (as we have said before), how far the yielding motions exert themselves. For if a man be held stretched out on the ground against his will, with arms and legs bound down, or otherwise confined, and yet strive with all his power to get up, the struggle is not the less, although ineffectual. The real state of the case (namely, whether the yielding motion be, as it were, annihilated by the predominance, or there be rather a continued, although an invisible effort) will, perhaps, appear in the concurrence of motions, although it escape our notice in their conflict. For instance: let an experiment be made with muskets; whether a musket-ball, at its utmost range in a straight line, or (as it is commonly called) point-blank, strike with less force when projected upward, where the motion of the blow is simple, than when projected downward, where the motion of gravity concurs with the blow.

The rules of such instances of predominance as occur should be collected: such as the following; the more general the desired advantage is, the stronger will be the motion; the motion of connection, for instance, which relates to the intercourse of the parts of the universe, is more powerful than that of gravity, which relates to the intercourse of dense bodies only. Again, the desire of a private good does not in general prevail against that of a public one, except where the quantities are small. Would that such were the case in civil matters!

XLIX. In the twenty-fifth rank of prerogative instances we will place suggesting instances; such as suggest, or point out, that which is advantageous to mankind; for bare power and knowledge in themselves exalt rather than enrich human nature. We must, therefore, select from the general store such things as are most useful to mankind. We shall have a better opportunity of discussing these when we treat of the application to practice; besides, in the work of interpretation, we leave room, on every subject, for the human or optative chart; for it is a part of science to make judicious inquiries and wishes.

L. In the twenty-sixth rank of prerogative instances we will place the generally useful instances. They are such as relate to various points, and frequently occur, sparing by that means considerable labor and new trials. The proper place for treating of instruments and contrivances, will be that in which we speak of the application to practice, and the methods of experiment. All that has hitherto been ascertained, and made use of, will be described in the particular history of each art. At present, we will subjoin a few general examples of the instances in question.

Man acts, then, upon natural bodies (besides merely bringing them together or removing them) by seven principal methods: 1, by the exclusion of all that impedes and disturbs; 2, by compression, extension, agitation, and the like; 3, by heat and cold; 4, by detention in a suitable place; 5, by checking or directing motion; 6, by peculiar harmonies; 7, by a seasonable and proper alternation, series, and succession of all these, or, at least, of some of them.

1. With regard to the first--common air, which is always at hand, and forces its admission, as also the rays of the heavenly bodies, create much disturbance. Whatever, therefore, tends to exclude them may well be considered as generally useful. The substance and thickness of vessels in which bodies are placed when prepared for operations may be referred to this head. So also may the accurate methods of closing vessels by consolidation, or the lutum sapientiæ, as the chemists call it. The exclusion of air by means of liquids at the extremity is also very useful, as when they pour oil on wine, or the juices of herbs, which by spreading itself upon the top like a cover, preserves them uninjured from the air. Powders, also, are serviceable, for although they contain air mixed up in them, yet they ward off the power of the mass of circumambient air, which is seen in the preservation of grapes and other fruits in sand or flour. Wax, honey, pitch, and other resinous bodies, are well used in order to make the exclusion more perfect, and to remove the air and celestial influence. We have sometimes made an experiment by placing a vessel or other bodies in quicksilver, the most dense of all substances capable of being poured round others. Grottoes and subterraneous caves are of great use in keeping off the effects of the sun, and the predatory action of air, and in the north of Germany are used for granaries. The depositing of bodies at the bottom of water may be also mentioned here; and I remember having heard of some bottles of wine being let down into a deep well in order to cool them, but left there by chance, carelessness, and forgetfulness for several years, and then taken out; by which means the wine not only escaped becoming flat or dead, but was much more excellent in flavor, arising (as it appears) from a more complete mixture of its parts. But if the case require that bodies should be sunk to the bottom of water, as in rivers or the sea, and yet should not touch the water, nor be inclosed in sealed vessels, but surrounded only by air, it would be right to use that vessel which has been sometimes employed under water above ships that have sunk, in order to enable the divers to remain below and breathe occasionally by turns. It was of the following nature: A hollow tub of metal was formed, and sunk so as to have its bottom parallel with the surface of the water; it thus carried down with it to the bottom of the sea all the air contained in the tub. It stood upon three feet (like a tripod), being of rather less height than a man, so that, when the diver was in want of breath, he could put his head into the hollow of the tub, breathe, and then continue his work. We hear that some sort of boat or vessel has now been invented, capable of carrying men some distance under water. Any bodies, however, can easily be suspended under some such vessel as we have mentioned, which has occasioned our remarks upon the experiment.

Another advantage of the careful and hermetical closing of bodies is this--not only the admission of external air is prevented (of which we have treated), but the spirit of bodies also is prevented from making its escape, which is an internal operation. For any one operating on natural bodies must be certain as to their quantity, and that nothing has evaporated or escaped, since profound alterations take place in bodies, when art prevents the loss or escape of any portion, while nature prevents their annihilation. With regard to this circumstance, a false idea has prevailed (which if true would make us despair of preserving quantity without diminution), namely, that the spirit of bodies, and air when rarefied by a great degree of heat, cannot be so kept in by being inclosed in any vessel as not to escape by the small pores. Men are led into this idea by the common experiments of a cup inverted over water, with a candle or piece of lighted paper in it, by which the water is drawn up, and of those cups which, when heated, draw up the flesh. For they think that in each experiment the rarefied air escapes, and that its quantity is therefore diminished, by which means the water or flesh rises by the motion of connection. This is, however, most incorrect. For the air is not diminished in quantity, but contracted in dimensions,[161] nor does this motion of the rising of the water begin till the flame is extinguished, or the air cooled, so that physicians place cold sponges, moistened with water, on the cups, in order to increase their attraction. There is, therefore, no reason why men should fear much from the ready escape of air: for although it be true that the most solid bodies have their pores, yet neither air, nor spirit, readily suffers itself to be rarefied to such an extreme degree; just as water will not escape by a small chink.

2. With regard to the second of the seven above-mentioned methods, we must especially observe, that compression and similar violence have a most powerful effect either in producing locomotion, and other motions of the same nature, as may be observed in engines and projectiles, or in destroying the organic body, and those qualities, which consist entirely in motion (for all life, and every description of flame and ignition are destroyed by compression, which also injures and deranges every machine); or in destroying those qualities which consist in position and a coarse difference of parts, as in colors; for the color of a flower when whole, differs from that it presents when bruised, and the same may be observed of whole and powdered amber; or in tastes, for the taste of a pear before it is ripe, and of the same pear when bruised and softened, is different, since it becomes perceptibly more sweet. But such violence is of little avail in the more noble transformations and changes of homogeneous bodies, for they do not, by such means, acquire any constantly and permanently new state, but one that is transitory, and always struggling to return to its former habit and freedom. It would not, however, be useless to make some

more diligent experiments with regard to this; whether, for instance, the condensation of a perfectly homogeneous body (such as air, water, oil, and the like) or their rarefaction, when effected by violence, can become permanent, fixed, and, as it were, so changed, as to become a nature. This might at first be tried by simple perseverance, and then by means of helps and harmonies. It might readily have been attempted (if we had but thought of it), when we condensed water (as was mentioned above), by hammering and compression, until it burst out. For we ought to have left the flattened globe untouched for some days, and then to have drawn off the water, in order to try whether it would have immediately occupied the same dimensions as it did before the condensation. If it had not done so, either immediately, or soon afterward, the condensation would have appeared to have been rendered constant; if not, it would have appeared that a restitution took place, and that the condensation had been transitory. Something of the same kind might have been tried with the glass eggs; the egg should have been sealed up suddenly and firmly, after a complete exhaustion of the air, and should have been allowed to remain so for some days, and it might then have been tried whether, on opening the aperture, the air would be drawn in with a hissing noise, or whether as much water would be drawn into it when immersed, as would have been drawn into it at first, if it had not continued sealed. For it is probable (or, at least, worth making the experiment) that this might have happened, or might happen, because perseverance has a similar effect upon bodies which are a little less homogeneous. A stick bent together for some time does not rebound, which is not owing to any loss of quantity in the wood during the time, for the same would occur (after a larger time) in a plate of steel, which does not evaporate. If the experiment of simple perseverance should fail, the matter should not be given up, but other means should be employed. For it would be no small advantage, if bodies could be endued with fixed and constant natures by violence. Air could then be converted into water by condensation, with other similar effects; for man is more the master of violent motions than of any other means.

3. The third of our seven methods is referred to that great practical engine of nature, as well as of art, cold and heat. Here, man's power limps, as it were, with one leg. For we possess the heat of fire, which is infinitely more powerful and intense than that of the sun (as it reaches us), and that of animals. But we want cold,[162] except such as we can obtain in winter, in caverns, or by surrounding objects with snow and ice, which, perhaps, may be compared in degree with the noontide heat of the sun in tropical countries, increased by the reflection of mountains and walls. For this degree of heat and cold can be borne for a short period only by animals, yet it is nothing compared with the heat of a burning furnace, or the corresponding degree of cold.[163] Everything with us has a tendency to become rarefied, dry and wasted, and nothing to become condensed or soft, except by mixtures, and, as it were, spurious methods. Instances of cold, therefore, should be searched for most diligently, such as may be found by exposing bodies upon buildings in a hard frost, in subterraneous caverns, by surrounding bodies with snow and ice in deep places excavated for that purpose, by letting bodies down into wells, by burying bodies in quicksilver and metals, by immersing them in streams which petrify wood, by burying them in the earth (which the Chinese are reported to do with their china, masses of which, made for that purpose, are said to remain in the ground for forty or fifty years, and to be transmitted to their heirs as a sort of artificial mine) and the like. The condensations which take place in nature, by means of cold, should also be investigated, that by learning their causes, they may be introduced into the arts; such as are observed in the exudation of marble and stones, in the dew upon the panes of glass in a room toward morning after a frosty night, in the formation and the gathering of vapors under the earth into water, whence spring fountains and the like.

Besides the substances which are cold to the touch, there are others which have also the effect of cold, and condense; they appear, however, to act only upon the bodies of animals, and scarcely any further. Of these we have many instances, in medicines and plasters. Some condense the flesh and tangible parts, such as astringent and inspissating medicines, others the spirits, such as soporifics. There are two modes of condensing the spirits, by soporifics or provocatives to sleep; the one by calming the motion, the other by expelling the spirit. The violet, dried roses, lettuces, and other benign or mild remedies, by their friendly and gently cooling vapors, invite the spirits to unite, and restrain their violent and perturbed motion. Rose-water, for instance, applied to the nostrils in fainting fits, causes the resolved and relaxed spirits to recover themselves, and, as it were, cherishes them. But opiates, and the like, banish the spirits by their malignant and hostile quality. If they be applied, therefore, externally, the spirits immediately quit the part and no longer readily flow into it; but if they be taken internally, their vapor, mounting to the head, expels, in all directions, the spirits contained in the ventricles of the brain, and since these spirits retreat, but cannot escape, they consequently meet and are condensed, and are sometimes completely extinguished and suffocated; although the same opiates, when taken in moderation, by a secondary accident (the condensation which succeeds their union), strengthen the spirits, render them more robust, and check their useless and inflammatory motion, by which means they contribute not a little to the cure of diseases, and the prolongation of life.

The preparations of bodies, also, for the reception of cold should not be omitted, such as that water a little warmed is more easily frozen than that which is quite cold, and the like.

Moreover, since nature supplies cold so sparingly, we must act like the apothecaries, who, when they cannot obtain any simple ingredient, take a succedaneum, or quid pro quo, as they term it, such as aloes for xylobalsamum, cassia for cinnamon. In the same manner we should look diligently about us, to ascertain whether there may be any substitutes for cold, that is to say, in what other manner condensation can be effected, which is the peculiar operation of cold. Such condensations appear hitherto to be of four kinds only. 1. By simple compression, which is of little avail toward permanent condensation, on account of the elasticity of substances, but may still, however, be of some assistance. 2. By the contraction of the coarser, after the escape or departure of the finer parts of a given body; as is exemplified in induration by fire, and the repeated heating and extinguishing of metals, and the like. 3. By the cohesion of the most solid homogeneous parts of a given body, which were previously separated, and mixed with others less solid, as in the return of sublimated mercury to its simple state, in which it occupies much less space than it did in powder, and the same may be observed of the cleansing of all metals from their dross. 4. By harmony, or the application of substances which condense by some latent power. These harmonies are as yet but rarely observed, at which we cannot be surprised, since there is little to hope for from their investigation, unless the discovery of forms and confirmation be attained. With regard to animal bodies, it is not to be questioned that there are many internal and external medicines which condense by harmony, as we have before observed, but this action is rare in inanimate bodies. Written accounts, as well as report, have certainly spoken of a tree in one of the Tercera or Canary Islands (for I do not exactly recollect which) that drips perpetually, so as to supply the inhabitants, in some degree, with water; and Paracelsus says that the herb called ros solis is filled with dew at noon, while the sun gives out its greatest heat, and all other herbs around it are dry. We treat both these accounts as fables; they would, however, if true, be of the most important service, and most worthy of examination. As to the honey-dew, resembling manna, which is found in May on the leaves of the oak, we are of opinion that it is not condensed by any harmony or peculiarity of the oak leaf, but that while it falls equally upon other leaves it is retained and continues on those of the oak, because their texture is closer, and not so porous as that of most of the other leaves.[164]

With regard to heat, man possesses abundant means and power; but his observation and inquiry are defective in some respects, and those of the greatest importance, notwithstanding the boasting of quacks. For the effects of intense heat are examined and observed, while those of a more gentle degree of heat, being of the most frequent occurrence in the paths of nature, are, on that very account, least known. We see, therefore, the furnaces, which are most esteemed, employed in increasing the spirits of bodies to a great extent, as in the strong acids, and some chemical oils; while the tangible parts are hardened, and, when the volatile part has escaped, become sometimes fixed; the homogeneous parts are separated, and the heterogeneous incorporated and agglomerated in a coarse lump; and (what is chiefly worthy of remark) the junction of compound bodies, and the more delicate conformations are destroyed and confounded. But the operation of a less violent heat should be tried and investigated, by which more delicate mixtures and regular conformations may be produced and elicited, according to the example of nature, and in imitation of the effect of the sun, which we have alluded to in the aphorism on the instances of alliance. For the works of nature are carried on in much smaller portions, and in more delicate and varied positions than those of fire, as we now employ it. But man will then appear to have really augmented his power, when the works of nature can be imitated in species, perfected in power, and varied in quantity; to which should be added the acceleration in point of time. Rust, for instance, is the result of a long process, but crocus martis is obtained immediately; and the same may be observed of natural verdigris and ceruse. Crystal is formed slowly, while glass is blown immediately: stones increase slowly, while bricks are baked immediately, etc. In the meantime (with regard to our present subject) every different species of heat should, with its peculiar effects, be diligently collected and inquired into; that of the heavenly bodies, whether their rays be direct, reflected, or refracted, or condensed by a burning-glass; that of lightning, flame, and ignited charcoal; that of fire of different materials, either open or confined, straitened or overflowing, qualified by the different forms of the furnaces, excited by the bellows, or quiescent, removed to a greater or less distance, or passing through different media; moist heats, such as the balneum Mariæ, and the dunghill; the external and internal heat of animals; dry heats, such as the heat of ashes, lime, warm sand; in short, the nature of every kind of heat, and its degrees.

We should, however, particularly attend to the investigation and discovery of the effects and operations of heat, when made to approach and retire by degrees, regularly, periodically, and by proper intervals of space and time. For this systematical inequality is in truth the daughter of heaven and mother of generation, nor can any great result be expected from a vehement, precipitate, or desultory heat. For this is not only most evident in vegetables, but in the wombs of animals also there arises a great inequality of heat, from the motion, sleep, food, and passions of the female. The same inequality prevails in those subterraneous beds where metals and fossils are perpetually forming, which renders yet more remarkable the ignorance of some of the reformed alchemists, who imagined they could attain their object by the equable heat of lamps, or the like, burning uniformly. Let this suffice concerning the operation and effects of heat; nor is it time for us to investigate them thoroughly before the forms and conformations of bodies have been further examined and brought to light. When we have determined upon our models, we may seek, apply, and arrange our instruments.

4. The fourth mode of action is by continuance, the very steward and almoner, as it were, of nature. We apply the term continuance to the abandonment of a body to itself for an observable time, guarded and protected in the meanwhile from all external force. For the internal motion then commences to betray and exert itself when the external and adventitious is removed. The effects of time, however, are far more delicate than those of fire. Wine, for instance, cannot be clarified by fire as it is by continuance. Nor are the ashes produced by combustion so fine as the particles dissolved or wasted by the lapse of.ages. The incorporations and mixtures, which are hurried by fire, are very inferior to those obtained by continuance; and the various conformations assumed by bodies left to themselves, such as mouldiness, etc., are put a stop to by fire or a strong heat. It is not, in the meantime, unimportant to remark that there is a certain degree of violence in the motion of bodies entirely confined; for the confinement impedes the proper motion of the body. Continuance in an open vessel, therefore, is useful for separations, and in one hermetically sealed for mixtures, that in a vessel partly closed, but admitting the air, for putrefaction. But instances of the operation and effect of continuance must be collected diligently from every quarter.

5. The direction of motion (which is the fifth method of action) is of no small use. We adopt this term, when speaking of a body which, meeting with another, either arrests, repels, allows, or directs its original motion. This is the case principally in the figure and position of vessels. An upright cone, for instance, promotes the condensation of vapor in alembics, but when reversed, as in inverted vessels, it assists the refining of sugar. Sometimes a curved form, or one alternately contracted and dilated, is required. Strainers may be ranged under this head, where the opposed body opens a way for one portion of another substance and impedes the rest. Nor is this process or any other direction of motion carried on externally only, but sometimes by one body within another. Thus, pebbles are thrown into water to collect the muddy particles, and syrups are refined by the white of an egg, which glues the grosser particles together so as to facilitate their removal. Telesius, indeed, rashly and ignorantly enough attributes the formation of animals to this cause, by means of the channels and folds of the womb. He ought to have observed a similar formation of the young in eggs which have no wrinkles or inequalities. One may observe a real result of this direction of motion in casting and modelling.

6. The effects produced by harmony and aversion (which is the sixth method) are frequently buried in obscurity; for these occult and specific properties (as they are termed), the sympathies and antipathies, are for the most part but a corruption of philosophy. Nor can we form any great expectation of the discovery of the harmony which exists between natural objects, before that of their forms and simple conformations, for it is nothing more than the symmetry between these forms and conformations.

The greater and more universal species of harmony are not, however, so wholly obscure, and with them, therefore, we must commence. The first and principal distinction between them is this; that some bodies differ considerably in the abundance and rarity of their substance, but correspond in their conformation; others, on the contrary, correspond in the former and differ in the latter. Thus the chemists have well observed, that in their trial of first principles sulphur and mercury, as it were, pervade the universe; their reasoning about salt, however, is absurd, and merely introduced to comprise earthy dry fixed bodies. In the other two, indeed, one of the most universal species of natural harmony manifests itself. Thus there is a correspondence between sulphur, oil, greasy exhalations, flame, and, perhaps, the substance of the stars. On the other hand, there is a like correspondence between mercury, water, aqueous vapor, air, and, perhaps, pure inter-sidereal ether. Yet do these two quaternions, or great natural tribes (each within its own limits), differ immensely in quantity and density of substance, while they generally agree in conformation, as is manifest in many instances. On the other hand, the metals agree in such quantity and density (especially when compared with vegetables, etc.), but differ in many respects in conformation. Animals and vegetables, in like manner, vary in their almost infinite modes of conformation, but range within very limited degrees of quantity and density of substance.

The next most general correspondence is that between individual bodies and those which supply them by way of menstruum or support. Inquiry, therefore, must be made as to the climate, soil, and depth at which each metal is generated, and the same of gems, whether produced in rocks or mines, also as to the soil in which particular trees, shrubs, and herbs, mostly grow and, as it were, delight; and as to the best species of manure, whether dung, chalk, sea sand, or ashes, etc., and their different propriety and advantage according to the variety of soils. So also the grafting and setting of trees and plants (as regards the readiness of grafting one particular species on another) depends very much upon harmony, and it would be amusing to try an experiment I have lately heard of, in grafting forest trees (garden trees alone having hitherto been adopted), by which means the leaves and fruit are enlarged, and the trees produce more shade. The specific food of animals again should be observed, as well as that which cannot be used. Thus the carnivorous cannot be fed on herbs, for which reason the order of feuilletans, the experiment having been made, has nearly vanished; human nature being incapable of supporting their regimen, although the human will has more power over the bodily frame than that of other animals. The different kinds of putrefaction from which animals are generated should be noted.

The harmony of principal bodies with those subordinate to them (such indeed may be deemed those we have alluded to above) are sufficiently manifest, to which may be added those that exist between different bodies and their objects, and, since these latter are more apparent, they may throw great light when well observed and diligently examined upon those which are more latent.

The more internal harmony and aversion, or friendship and enmity (for superstition and folly have rendered the terms of sympathy and antipathy almost disgusting), have been either falsely assigned, or mixed with fable, or most rarely discovered from neglect. For if one were to allege that there is an enmity between the vine and the cabbage, because they will not come up well when sown together, there is a sufficient reason for it in the succulent and absorbent nature of each plant, so that the one defrauds the other. Again, if one were to say that there is a harmony and friendship between the corn and the corn-flower, or the wild poppy, because the latter seldom grow anywhere but in cultivated soils, he ought rather to say, there is an enmity between them, for the poppy and the corn-flower are produced and created by those juices which the corn has left and rejected, so that the sowing of the corn prepares the ground for their production. And there are a vast number of similar false assertions. As for fables, they must be totally exterminated. There remains, then, but a scanty supply of such species of harmony as has borne the test of experiment, such as that between the magnet and iron, gold and quicksilver, and the like. In chemical experiments on metals, however, there are some others worthy of notice, but the greatest abundance (where the whole are so few in numbers) is discovered in certain medicines, which, from their occult and specific qualities (as they are termed), affect particular limbs, humors, diseases, or constitutions. Nor should we omit the harmony between the motion and phenomena of the moon, and their effects on lower bodies, which may be brought together by an accurate and honest selection from the experiments of agriculture, navigation, and medicine, or of other sciences. By as much as these general instances, however, of more latent harmony, are rare, with so much the more diligence are they to be inquired after, through tradition, and faithful and honest reports, but without rashness and credulity, with an anxious and, as it were, hesitating degree of reliance. There remains one species of harmony which, though simple in its mode of action, is yet most valuable in its use, and must by no means be omitted, but rather diligently investigated. It is the ready or difficult coition or union of bodies in composition, or simple juxtaposition. For some bodies readily and willingly mix, and are incorporated, others tardily and perversely; thus powders mix best with water, chalk and ashes with oils, and the like. Nor are these instances of readiness and aversion to mixture to be alone collected, but others, also, of the collocation, distribution, and digestion of the parts when mingled, and the predominance after the mixture is complete.

7. Lastly, there remains the seventh, and last of the seven, modes of action; namely, that by the alternation and interchange of the other six; but of this, it will not be the right time to offer any examples, until some deeper investigation shall have taken place of each of the others. The series, or chain of this alternation, in its mode of application to separate effects, is no less powerful in its operation than difficult to be traced. But men are possessed with the most extreme impatience, both of such inquiries, and their practical application, although it be the clew of the labyrinth in all greater works. Thus far of the generally useful instances.

LI. The twenty-seventh and last place we will assign to the magical instances, a term which we apply to those where the matter or efficient agent is scanty or small, in comparison with the grandeur of the work or effect produced; so that even when common they appear miraculous, some at first sight, others even upon more attentive observation. Nature, however, of herself, supplies these but sparingly. What she will do when her whole store is thrown open, and after the discovery of forms, processes, and conformation, will appear hereafter. As far as we can yet conjecture, these magic effects are produced in three ways, either by self-multiplication, as in fire, and the poisons termed specific, and the motions transferred and multiplied from wheel to wheel; or by the excitement, or, as it were, invitation of another substance, as in the magnet, which excites innumerable needles without losing or diminishing its power; and again in leaven, and the like; or by the excess of rapidity of one species of motion over another, as has been observed in the case of gunpowder, cannon, and mines. The two former require an investigation of harmonies, the latter of a measure of motion. Whether there be any mode of changing bodies per minima (as it is termed), and transferring the delicate conformations of matter, which is of importance in all transformations of bodies, so as to enable art to effect, in a short time, that which nature works out by divers expedients, is a point of which we have as yet no indication. But, as we aspire to the extremest and highest results in that which is solid and true, so do we ever detest, and, as far as in us lies, expel all that is empty and vain.

LII. Let this suffice as to the respective dignity of prerogatives of instances. But it must be noted, that in this our organ, we treat of logic, and not of philosophy. Seeing, however, that our logic instructs and informs the understanding, in order that it may not, with the small hooks, as it were, of the mind, catch at, and grasp mere abstractions, but rather actually penetrate nature, and discover the properties and effects of bodies, and the determinate laws of their substance (so that this science of ours springs from the nature of things, as well as from that of the mind); it is not to be wondered at, if it have been continually interspersed and illustrated with natural observations and experiments, as instances of our method. The prerogative instances are, as appears from what has preceded, twenty-seven in number, and are termed, solitary instances, migrating instances, conspicuous instances, clandestine instances, constitutive instances, similar instances, singular instances, deviating instances, bordering instances, instances of power, accompanying and hostile instances, subjunctive instances, instances of alliance, instances of the cross, instances of divorce, instances of the gate, citing instances, instances of the road, supplementary instances, lancing instances, instances of the rod, instances of the course, doses of nature, wrestling instances, suggesting instances, generally useful instances, and magical instances. The advantage, by which these instances excel the more ordinary, regards specifically either theory or practice, or both. With regard to theory, they assist either the senses or the understanding; the senses, as in the five instances of the lamp; the understanding, either by expediting the exclusive mode of arriving at the form, as in solitary instances, or by confining, and more immediately indicating the affirmative, as in the migrating, conspicuous, accompanying, and subjunctive instances; or by elevating the understanding, and leading it to general and common natures, and that either immediately, as in the clandestine and singular instances, and those of alliance; or very nearly so, as in the constitutive; or still less so, as in the similar instances; or by correcting the understanding of its habits, as in the deviating instances; or by leading to the grand form or fabric of the universe, as in the bordering instances; or by guarding it from false forms and causes, as in those of the cross and of divorce. With regard to practice, they either point it out, or measure, or elevate it. They point it out, either by showing where we must commence in order not to repeat the labors of others, as in the instances of power; or by inducing us to aspire to that which may be possible, as in the suggesting instances; the four mathematical instances measure it. The generally useful and the magical elevate it.

Again, out of these twenty-seven instances, some must be collected immediately, without waiting for a particular investigation of properties. Such are the similar, singular, deviating, and bordering instances, those of power, and of the gate, and suggesting, generally useful, and magical instances; for these either assist and cure the understanding and senses, or furnish our general practice. The remainder are to be collected when we finish our synoptical tables for the work of the interpreter, upon any particular nature; for these instances, honored and gifted with such prerogatives, are like the soul amid the vulgar crowd of instances, and (as we from the first observed) a few of them are worth a multitude of the others. When, therefore, we are forming our tables they must be searched out with the greatest zeal, and placed in the table. And, since mention must be made of them in what follows, a treatise upon their nature has necessarily been prefixed. We must next, however, proceed to the supports and corrections of induction, and thence to concretes, the latent process, and latent conformations, and the other matters, which we have enumerated in their order in the twenty-first aphorism, in order that, like good and faithful guardians, we may yield up their fortune to mankind upon the emancipation and majority of their understanding; from which must necessarily follow an improvement of their estate, and an increase of their power over nature. For man, by the fall, lost at once his state of innocence, and his empire over creation, both of which can be partially recovered even in this life, the first by religion and faith, the second by the arts and sciences. For creation did not become entirely and utterly rebellious by the curse, but in consequence of the Divine decree, "in the sweat of thy brow shalt thou eat bread," she is compelled by our labors (not assuredly by our disputes or magical ceremonies), at length, to afford mankind in some degree his bread, that is to say, to supply man's daily wants.

END OF "NOVUM ORGANUM"

Footnotes:

71. Τὸ τί ἦν εἶναι, or ἦν οὐσία of Aristotle.--See lib. iii. Metap.

72. These divisions are from Aristotle's Metaphysics, where they are termed, 1. ὕλη ἢ τὸ ὑποκείμενον. 2. τὸ τί ἦν εἶναι. 3. ὅθεν ἡ ἀρχὴ τῆς κινήσεως. 4. τὸ οὗ ἕνεκεν--καὶ τὸ ἀγαθόν.

73. See Aphorism li. and second paragraph of Aphorism lxv. in the first book.

74. Bacon means, that although there exist in nature only individualities, yet a certain number of these may have common properties, and be controlled by the same laws. Now, these homogeneous qualities which distinguish them from other individuals, lead us to class them under one expression, and sometimes under a single term. Yet these classes are only pure conceptions in Bacon's opinion, and cannot be taken for distinct substances. He evidently here aims a blow at the Realists, who concluded that the essence which united individualities in a class was the only real and immutable existence in nature, inasmuch as it entered into their ideas of individual substances as a distinct and essential property, and continued in the mind as the mold, type or pattern of the class, while its individual forms were undergoing perpetual renovation and decay.--Ed.

75. Bacon's definition is obscure. All the idea we have of a law of nature consists in invariable sequence between certain classes of phenomena; but this cannot be the complete sense attached by Bacon to the term form, as he employs it in the fourth aphorism as convertible with the nature of any object; and again, in the first aphorism, as the natura naturans, or general law or condition in any substance or quality--natura naturata--which is whatever its form is, or that particular combination of forces which impresses a certain nature upon matter subject to its influence. Thus, in the Newtonian sense, the form of whiteness would be that combination of the seven primitive rays of light which give rise to that color. In combination with this word, and affording a still further insight into its meaning, we have the phrases, latens processus ad formam, et latens schematismus corporum. Now, the latens schematismus signifies the internal texture, structure, or configuration of bodies, or the result of the respective situation of all the parts of a body; while the latens processus ad formam points out the gradation of movements which takes place among the molecula of bodies when they either conserve or change their figure. Hence we may consider the form of any quality in body as something convertible with that quality, i.e., when it exists the quality is present, and vice versâ. In this sense, the form of a thing differs only from its efficient cause in being permanent, whereas we apply cause to that which exists in order of time. The latens processus and latens schematismus are subordinate to form, as concrete exemplifications of its essence. The former is the secret and invisible process by which change is effected, and involves the principle since called the law of continuity. Thus, the succession of events between the application of the match to the expulsion of the bullet is an instance of latent progress which we can now trace with some degree of accuracy. It also more directly refers to the operation by which one form or condition of being is induced upon another. For example, when the surface of iron becomes rusty, or when water is converted into steam, some change has taken place, or latent process from one form to another. Mechanics afford many exemplifications of the first latent process we have denoted, and chemistry of the second. The latens schematismus is that visible structure of bodies on which so many of their properties depend. When we inquire into the constitution of crystals, and into the internal structure of plants, we are examining into their latent schematism.--Ed.

76. By the recent discoveries in electric magnetism, copper wires, or, indeed, wires of any metal, may be transformed into magnets; the magnetic law, or form, having been to that extent discovered.

77. Haller has pursued this investigation in his "Physiology," and has left his successors little else to do than repeat his discoveries.--Ed.

78. Bacon here first seems pregnant with the important development of the higher calculus, which, in the hands of Newton and Descartes, was to effect as great a revolution in philosophy as his method.--Ed.

79. By spirit, Bacon here plainly implies material fluid too fine to be grasped by the unassisted sense, which rather operates than reasons. We sometimes adopt the same mode of expression, as in the words spirits of nitre, spirits of wine. Some such agency has been assumed by nearly all the modern physicists, a few of whom, along with Bacon, would leave us to gather from their expressions, that they believe such bodies endowed with the sentient powers of perception. As another specimen of his sentiment on this subject, we may refer to a paragraph on the decomposition of compounds, in his essay on death, beginning--"The spirit which exists in all living bodies, keeps all the parts in due subjection; when it escapes, the body decomposes, or the similar parts unite."--Ed.

80. The theory of the Epicureans and others. The atoms are supposed to be invisible, unalterable particles, endued with all the properties of the given body, and forming that body by their union. They must be separated, of course, which either takes a vacuum for granted, or introduces a tertium quid into the composition of the body.

81. Compare the three following aphorisms with the last three chapters of the third book of the "De Augmentis Scientiarum."

82. Bacon gives this unfortunate term its proper signification; μετα, in composition, with the Greeks signifying change or mutation. Most of our readers, no doubt, are aware that the obtrusion of this word into technical philosophy was purely capricious, and is of no older date than the publication of Aristotle's works by Andronicus of Rhodes, one of the learned men into whose hands the manuscripts of that philosopher fell, after they were brought by Sylla from Athens to Rome. To fourteen books in these MSS. with no distinguishing title, Andronicus is said to have prefixed the words τα μετα τα φυσικα, to denote the place which they ought to hold either in the order of Aristotle's arrangement, or in that of study. These books treat first of those subjects which are common to matter and mind; secondly, of things separate from matter, i.e. of God, and of the subordinate spirits, which were supposed by the Peripatetics to watch over particular portions of the universe. The followers of Aristotle accepted the whimsical title of Andronicus, and in their usual manner allowed a word to unite things into one science which were plainly heterogeneous. Their error was adopted by the Peripatetics of the Christian Church. The schoolmen added to it the notion of ontology, the science of the mind, or pneumatology, and as that genus of being has since become extinct with the schools, metaphysics thus in modern parlance comes to be synonymous with psychology. It were to be wished that Bacon's definition of the term had been

accepted, and mental science delivered from one of the greatest monstrosities in its nomenclature, yet Bacon whimsically enough in his De Augmentis includes mathematics in metaphysics.--Ed.

83.

"Ne tenues pluviæ, rapidive potentia solis   Acrior, aut Boreæ penetrabile frigus adurat."
--Virg. Georg. i. 92, 93.

84. This notion, which he repeats again, and particularizes in the 18th aph. of this book, is borrowed from the ancients, and we need not say is as wise as their other astronomical conjectures. The sun also approaches stars quite as large in other quarters of the zodiac, when it looks down upon the earth through the murky clouds of winter. When that luminary is in Leo, the heat of the earth is certainly greater than at any other period, but this arises from the accumulation of heat after the solstice, for the same reason that the maximum heat of the day is at two o'clock instead of noon.--Ed.

85. Bouguer, employed by Louis XIV. in philosophical researches, ascended the Andes to discover the globular form of the earth, and published an account of his passage, which verifies the statement of Bacon.

86. Montanari asserts in his book against the astrologers that he had satisfied himself by numerous and oft-repeated experiments, that the lunar rays gathered to a focus produced a sensible degree of heat. Muschenbröck, however, adopts the opposite opinion, and asserts that himself, De la Hire, Villet, and Tschirnhausen had tried with that view the strongest burning-glasses in vain. (Opera de Igne.) De la Lande makes a similar confession in his Astronomy (vol. ii. vii. § 1413). Bouguer, whom we have just quoted, demonstrated that the light of the moon was 300,000 degrees less than that of the sun; it would consequently be necessary to invent a glass with an absorbing power 300,000 degrees greater than those ordinarily in use, to try the experiment Bacon speaks of.--Ed.

87. In this thermometer, mercury was not dilated by heat or contracted by cold, as the one now in use, but a mass of air employed instead, which filled the cavity of the bulb. This being placed in an inverted position to ours, that is to say, with the bulb uppermost, pressed down the liquor when the air became dilated by heat, as ours press it upward; and when the heat diminished, the liquor rose to occupy the place vacated by the air, as the one now in use descends. It consequently was liable to be affected by a change in the temperature, as by the weight of air, and could afford only a rude standard of accuracy in scientific investigations. This thermometer was not Bacon's own contrivance, as is commonly supposed, but that of Drebbel.--Ed.

88. La Lande is indignant that the Chaldeans should have more correct notions of the nature of comets than the modern physicists, and charges Bacon with entertaining the idea that they were the mere effects of vapor and heat. This passage, with two others more positive, in the "De Aug." (cap. xl.) and the "Descript. Globi Intellect." (cap. vi.) certainly afford ground for the assertion; but if Bacon erred, he erred with Galileo, and with the foremost spirits of the times. It is true that Pythagoras and Seneca had asserted their belief in the solidity of these bodies, but the wide dominion which Aristotle subsequently exercised, threw their opinions into the shade, and made the opposite doctrine everywhere paramount.--Ed.

89. Was it a silk apron which exhibited electric sparks? Silk was then scarce.

90. The Italian fire-fly.

91. This last is found to be the real reason, air not being a good conductor, and therefore not allowing the escape of heat. The confined air is disengaged when these substances are placed under an exhausted receiver.

92. This is erroneous. Air, in fact, is one of the worst, and metals are the best conductors of heat.

93. See No. 28 in the table of the degrees of heat.

94. Bacon here mistakes sensation confined to ourselves for an internal property of distinct substances. Metals are denser than wood, and our bodies consequently coming into contact with more particles of matter when we touch them, lose a greater quantity of heat than in the case of lighter substances.--Ed.

95. This was the ancient opinion, but the moderns incline to the belief that these insects are produced by generation or fecundity from seeds deposited by their tribes in bodies on the verge of putrefaction.--Ed.

96. The correct measure of the activity of flame may be obtained by multiplying its natural force into the square of its velocity. On this account the flame of vivid lightning mentioned in No. 23 contains so much vigor, its velocity being greater than that arising from other heat.--Ed.

97. The fires supply fresh heat, the water has only a certain quantity of heat, which being diffused over a fresh supply of cooler water, must be on the whole lowered.

98. If condensation were the cause of the greater heat, Bacon concludes the centre of the flame would be the hotter part, and vice versâ. The fact is, neither of the causes assigned by Bacon is the true one; for the fire burns more quickly only because the draught of air is more rapid, the cold dense air pressing rapidly into the heated room and toward the chimney.--Ed.

99. Bacon appears to have confounded combustibility and fusibility with susceptibility of heat; for

though the metals will certainly neither dissolve as soon as ice or butter, nor be consumed as soon as wood, that only shows that different degrees of heat are required to produce similar effects on different bodies; but metals much more readily acquire and transmit the same degree of heat than any of the above substances. The rapid transmission renders them generally cold to the touch. The convenience of fixing wooden handles to vessels containing hot water illustrates these observations.

100. Another singular error, the truth being, that solid bodies are the best conductors; but of course where heat is diffused over a large mass, it is less in each part, than if that part alone absorbed the whole quantum of heat.--Ed.

101. This general law or form has been well illustrated by Newton's discovery of the decomposition of colors.

102. I.e., the common link or form which connects the various kinds of natures, such as the different hot or red natures enumerated above.--See Aphorism iii. part 2.

103. This is erroneous--all metals expand considerably when heated.

104. "Quid ipsum," the τò τì ἦν εἶναι of Aristotle.

105. To show the error of the text, we need only mention the case of water, which, when confined in corked vases, and exposed to the action of a freezing atmosphere, is sure to swell out and break those vessels which are not sufficiently large to contain its expanded volume. Megalotti narrates a hundred other instances of a similar character.--Ed.

106. Bacon's inquisition into the nature of heat, as an example of the mode of interpreting nature, cannot be looked upon otherwise than as a complete failure. Though the exact nature of this phenomenon is still an obscure and controverted matter, the science of thermotics now consists of many important truths, and to none of these truths is there so much as an approximation in Bacon's process. The steps by which this science really advanced were the discovery of a measure of a heat or temperature, the establishment of the laws of conduction and radiation, of the laws of specific heat, latent heat, and the like. Such advances have led to Ampère's hypothesis, that heat consists in the vibrations of an imponderable fluid; and to Laplace's theory, that temperature consists in the internal radiation of a similar medium. These hypotheses cannot yet be said to be even probable, but at least they are so modified as to include some of the preceding laws which are firmly established, whereas Bacon's "form," or true definition of heat, as stated in the text, includes no laws of phenomena, explains no process, and is indeed itself an example of illicit generalization.

In all the details of his example of heat he is unfortunate. He includes in his collection of instances, the hot tastes of aromatic plants, the caustic effects of acids, and many other facts which cannot be ascribed to heat without a studious laxity in the use of the word.--Ed.

107. By this term Bacon understands general phenomena, taken in order from the great mass of indiscriminative facts, which, as they lie in nature, are apt to generate confusion by their number, indistinctness and complication. Such classes of phenomena, as being peculiarly suggestive of causation, he quaintly classes under the title of prerogative inquiries, either seduced by the fanciful analogy, which such instances bore to the prerogativa centuria in the Roman Comitia, or justly considering them as Herschel supposes to hold a kind of prerogative dignity from being peculiarly suggestive of causation.

Two high authorities in physical science (v. Herschel, Nat. Phil., art. 192; Whewell's Philosophy of the Inductive Sciences, vol. ii. p. 243) pronounce these instances of little service in the task of induction, being for the most part classed not according to the ideas which they involve, or to any obvious circumstance in the facts of which they consist, but according to the extent and manner of their influence upon the inquiry in which they are employed. Thus we have solitary instances, migrating instances, ostensive instances, clandestine instances, so termed according to the degree in which they exhibit, or seem to exhibit, the property, whose nature we would examine. We have guide-post instances, crucial instances, instances of the parted road, of the doorway, of the lamp, according to the guidance they supply to our advance. Whewell remarks that such a classification is much of the same nature as if, having to teach the art of building, we were to describe tools with reference to the amount and place of the work which they must do, instead of pointing out their construction and use; as if we were to inform the pupil that we must have tools for lifting a stone up, tools for moving it sidewise, tools for laying it square, and tools for cementing it firmly. The means are thus lost in the end, and we reap the fruits of unmethodical arrangement in the confusion of cross division. In addition, all the instances are leavened with the error of confounding the laws with the causes of phenomena, and we are urged to adopt the fundamental error of seeking therein the universal agents, or general causes of phenomena, without ascending the gradual steps of intermediate laws.--Ed.

108. Of these nine general heads no more than the first is prosecuted by the author.

109. This very nearly approaches to Sir I. Newton's discovery of the decomposition of light by the prism.

110. The mineral kingdom, as displaying the same nature in all its gradations, from the shells so perfect in structure in limestone to the finer marbles in which their nature gradually disappears, is the

great theatre for instances of migration.--Ed.

111. Bacon was not aware of the fact since brought to light by Römer, that down to fourteen fathoms from the earth's mean level the thermometer remains fixed at the tenth degree, but that as the thermometer descends below that depth the heat increases in a ratio proportionate to the descent, which happens with little variation in all climates. Buffon considers this a proof of a central fire in our planet.-- Ed.

112. All the diversities of bodies depend upon two principles, i.e., the quantity and the position of the elements that enter into their composition. The primary difference is not that which depends on the greatest or least quantity of material elements, but that which depends on their position. It was the quick perception of this truth that made Leibnitz say that to complete mathematics it was necessary to join to the analysis of quantity the analysis of position.--Ed.

113. Query?

114. The real cause of this phenomenon is the attraction of the surface-water in the vessel by the sides of the bubbles. When the bubbles approach, the sides nearest each other both tend to raise the small space of water between them, and consequently less water is raised by each of these nearer sides than by the exterior part of the bubble, and the greater weight of the water raised on the exterior parts pushes the bubbles together. In the same manner a bubble near the side of a vessel is pushed toward it; the vessel and bubble both drawing the water that is between them. The latter phenomenon cannot be explained on Bacon's hypothesis.

115. Modern discoveries appear to bear out the sagacity of Bacon's remark, and the experiments of Baron Cagnard may be regarded as a first step toward its full demonstration. After the new facts elicited by that philosopher, there can be little doubt that the solid, liquid and aëriform state of bodies are merely stages in a progress of gradual transition from one extreme to the other, and that however strongly marked the distinctions between them may appear, they will ultimately turn out to be separated by no sudden or violent line of demarcation, but slide into each other by imperceptible gradations. Bacon's suggestion, however, is as old as Pythagoras, and perhaps simultaneous with the first dawn of philosophic reason. The doctrine of the reciprocal transmutation of the elements underlies all the physical systems of the ancients, and was adopted by the Epicureans as well as the Stoics. Ovid opens his last book of the Metamorphoses with the poetry of the subject, where he expressly points to the hint of Bacon:--

----"Tenuatus in auras   Aëraque humor abit, etc., etc.   *   *   *   *   *   *   Inde retro redeunt, idemque retexitur ordo."--xv. 246-249.

and Seneca, in the third book of his Natural Philosophy, quest. iv., states the opinion in more precise language than either the ancient bard or the modern philosopher.--Ed.

116. The author's own system of Memoria Technica may be found in the De Augmentis, chap. xv. We may add that, notwithstanding Bacon's assertion that he intended his method to apply to religion, politics, and morals, this is the only lengthy illustration he has adduced of any subject out of the domain of physical science.--Ed.

117. The collective instances here meant are no other than general facts or laws of some degree of generality, and are themselves the result of induction. For example, the system of Jupiter, or Saturn with its satellites, is a collective instance, and materially assisted in securing the admission of the Copernican system. We have here in miniature, and displayed at one view, a system analogous to that of the planets about the sun, of which, from the circumstance of our being involved in it, and unfavorably situated for seeing it otherwise than in detail, we are incapacitated from forming a general idea, but by slow and progressive efforts of reason.

But there is a species of collective instance which Bacon does not seem to have contemplated, in which particular phenomena are presented in such numbers at once, as to make the induction of their law a matter of ocular inspection. For example, the parabolic form assumed by a jet of water spouted out of a hole is a collective instance of the velocities and directions of the motions of all the particles which compose it seen together, and which thus leads us without trouble to recognize the law of the motion of a projectile. Again, the beautiful figures exhibited by sand strewed on regular plates of glass or metal set in vibration, are collective instances of an infinite number of points which remain at rest while the remainder of the plate vibrates, and in consequence afford us an insight into the law which regulates their arrangement and sequence throughout the whole surface. The richly colored lemniscates seen around the optic axis of crystals exposed to polarized light afford a striking instance of the same kind, pointing at once to the general mathematical expression of the law which regulates their production. Such collective instances as these lead us to a general law by an induction which offers itself spontaneously, and thus furnish advanced posts in philosophical exploration. The laws of Kepler, which Bacon ignored on account of his want of mathematical taste, may be cited as a collective instance. The first is, that the planets move in elliptical orbits, having the sun for their common focus. The second, that about this focus the radius vector of each planet describes equal areas in equal times. The third, that the squares of the periodic times of the planets are as the cubes of their mean distance from the sun. This collective instance "opened the way" to the discovery of the Newtonian law of gravitation.--Ed.

118. Is not this very hasty generalization? Do serpents move with four folds only? Observe also the motion of centipedes and other insects.

119. Shaw states another point of difference between the objects cited in the text--animals having their roots within, while plants have theirs without; for their lacteals nearly correspond with the fibres of the roots in plants; so that animals seem nourished within themselves as plants are without.--Ed.

120. Bacon falls into an error here in regarding the syllogism as something distinct from the reasoning faculty, and only one of its forms. It is not generally true that the syllogism is only a form of reasoning by which we unite ideas which accord with the middle term. This agreement is not even essential to accurate syllogisms; when the relation of the two things compared to the third is one of equality or similitude, it of course follows that the two things compared may be pronounced equal, or like to each other. But if the relation between these terms exist in a different form, then it is not true that the two extremes stand in the same relation to each other as to the middle term. For instance, if =A= is double of =B=, and =B= double of =C=, then =A= is quadruple of =C=. But then the relation of =A= to =C= is different from that of =A= to =B= and of =B= to =C=.--Ed.

121. Comparative anatomy is full of analogies of this kind. Those between natural and artificial productions are well worthy of attention, and sometimes lead to important discoveries. By observing an analogy of this kind between the plan used in hydraulic engines for preventing the counter-current of a fluid, and a similar contrivance in the blood vessels, Harvey was led to the discovery of the circulation of the blood.--Ed.

122. This is well illustrated in plants, for the gardener can produce endless varieties of any known species, but can never produce a new species itself.

123. The discoveries of Tournefort have placed moss in the class of plants. The fish alluded to below are to be found only in the tropics.--Ed.

124. There is, however, no real approximation to birds in either the flying fish or bat, any more than a man approximates to a fish because he can swim. The wings of the flying fish and bat are mere expansions of skin, bearing no resemblance whatever to those of birds.--Ed.

125. Seneca was a sounder astronomer than Bacon. He ridiculed the idea of the motion of any heavenly bodies being irregular, and predicted that the day would come, when the laws which guided the revolution of these bodies would be proved to be identical with those which controlled the motions of the planets. The anticipation, was realized by Newton.--Ed.

126. But see Bacon's own corollary at the end of the Instances of Divorce, Aphorism xxxvii. If Bacon's remark be accepted, the censure will fall upon Newton and the system so generally received at the present day. It is, however, unjust, as the centre of which Newton so often speaks is not a point with an active inherent force, but only the result of all the particular and reciprocal attractions of the different parts of the planet acting upon one spot. It is evident, that if all these forces were united in this centre, that the sum would be equal to all their partial effects.--Ed.

127. Since Newton's discovery of the law of gravitation, we find that the attractive force of the earth must extend to an infinite distance. Bacon himself alludes to the operation of this attractive force at great distances in the Instances of the Rod, Aphorism xlv.

128. Snow reflects light, but is not a source of light.

129. Bacon's sagacity here foreshadows Newton's theory of the tides.

130. The error in the text arose from Bacon's impression that the earth was immovable. It is evident, since gravitation acts at an infinite distance, that no such point could be found; and even supposing the impossible point of equilibrium discovered, the body could not maintain its position an instant, but would be hurried, at the first movement of the heavenly bodies, in the direction of the dominant gravitating power.--Ed.

131. Fly clocks are referred to in the text, not pendulum clocks, which were not known in England till 1662. The former, though clumsy and rude in their construction, still embodied sound mechanical principles. The comparison of the effect of a spring with that of a weight in producing certain motions in certain times on altitudes and in mines, has recently been tried by Professors Airy and Whewell in Dalcoath mine, by means of a pendulum, which is only a weight moved by gravity, and a chronometer balance moved and regulated by a spring. In his thirty-seventh Aphorism, Bacon also speaks of gravity as an incorporeal power, acting at a distance, and requiring time for its transmission; a consideration which occurred at a later period to Laplace in one of his most delicate investigations.

Crucial instances, as Herschel remarks, afford the readiest and securest means of eliminating extraneous causes, and deciding between the claims of rival hypotheses; especially when these, running parallel to each other, in the explanation of great classes of phenomena, at length come to be placed at issue upon a single fact. A curious example is given by M. Fresnel, as decisive in his mind of the question between the two great theories on the nature of light, which, since the time of Newton and Huyghens, have divided philosophers. When two very clean glasses are laid one on the other, if they be not perfectly flat, but one or both, in an almost imperceptible degree, convex or prominent, beautiful and vivid colors will be seen between them; and if these be viewed through a red glass, their appearance will be that of alternate dark and bright stripes. These stripes are formed between the two surfaces in apparent contact, and being applicable on both theories, are appealed to by their respective supporters as strong confirmatory facts; but there is a difference in one circumstance, according as one or other theory is employed to explain them. In the case of the Huyghenian theory, the intervals between the bright stripes ought to appear absolutely black, when a prism is used for the upper glass, in the other half bright. This curious case of difference was tried, as soon as the opposing consequences of the two theories were noted by M. Fresnel, and the result is stated by him to be decisive in favor of that theory which makes light to consist in the vibrations of an elastic medium.--Ed.

132. Bacon plainly, from this passage, was inclined to believe that the moon, like the comets, was nothing more than illuminated vapor. The Newtonian law, however, has not only established its solidity, but its density and weight. A sufficient proof of the former is afforded by the attraction of the sea, and the moon's motion round the earth.--Ed.

133. Rather the refraction; the sky or air, however, reflects the blue rays of light.

134. The polished surface of the glass causes the reflection in this case, and not the air; and a hat or other black surface put behind the window in the daytime will enable the glass to reflect distinctly for the same reason, namely, that the reflected rays are not mixed and confused with those transmitted from the other side of the window.

135. These instances, which Bacon seems to consider as a great discovery, are nothing more than disjunctive propositions combined with dilemmas. In proposing to explain an effect, we commence with the enumeration of the different causes which seem connected with its production; then with the aid of one or more dilemmas, we eliminate each of the phenomena accidental to its composition, and conclude with attributing the effect to the residue. For instance, a certain phenomenon (a) is produced either by phenomenon (=B=) or phenomenon (=C=); but =C= cannot be the cause of a, for it is found in =D=, =E=, =F=, neither of which are connected with a. Then the true cause of phenomenon (a) must be phenomenon (=B=).

This species of reasoning is liable to several paralogisms, against which Bacon has not guarded his readers, from the very fact that he stumbled into them unwittingly himself. The two principal ones are false exclusions and defective enumerations. Bacon, in his survey of the causes which are able to concur in producing the phenomena of the tides, takes no account of the periodic melting of the Polar ice, or the expansion of water by the solar heat; nor does he fare better in his exclusions. For the attraction of the planets and the progression and retrograde motion communicated by the earth's diurnal revolution, can plainly affect the sea together, and have a simultaneous influence on its surface.

Bacon is hardly just or consistent in his censure of Ramus; the end of whose dichotomy was only to render reasoning by dilemma, and crucial instances, more certain in their results, by reducing the divisions which composed their parts to two sets of contradictory propositions. The affirmative or negative of one would then necessarily have led to the acceptance or rejection of the other.--Ed.

136. Père Shenier first pointed out the spots on the sun's disk, and by the marks which they afforded him, computed its revolution to be performed in twenty-five days and some hours.--Ed.

137. Rust is now well known to be a chemical combination of oxygen with the metal, and the metal when rusty acquires additional weight. His theory as to the generation of animals, is deduced from the erroneous notion of the possibility of spontaneous generation (as it was termed). See the next paragraph but one.

138.

"*Limus ut hic durescit, et hæc ut cera liquescit   Uno eodemque igni.*"--Virg. Ecl. viii.

139. See Table of Degrees, No. 38.

140. Riccati, and all modern physicists, discover some portion of light in every body, which seems to confirm the passage in Genesis that assigns to this substance priority in creation.--Ed.

141. As instances of this kind, which the progress of science since the time of Bacon affords, we may cite the air-pump and the barometer, for manifesting the weight and elasticity of air: the measurement of the velocity of light, by means of the occultation of Jupiter's satellites and the aberration of the fixed stars: the experiments in electricity and galvanism, and in the greater part of pneumatic chemistry. In all these cases scientific facts are elicited, which sense could never have revealed to us.--Ed.

142. The itinerant instances, as well as frontier instances, are cases in which we are enabled to trace the general law of continuity which seems to pervade all nature, and which has been aptly embodied in the sentence, "natura non agit per saltum." The pursuit of this law into phenomena where its application is not at first sight obvious, has opened a mine of physical discovery, and led us to perceive an intimate connection between facts which at first seemed hostile to each other. For example,

the transparency of gold-leaf, which permits a bluish-green light to pass through it, is a frontier instance between transparent and opaque bodies, by exhibiting a body of the glass generally regarded the most opaque in nature, as still possessed of some slight degree of transparency. It thus proves that the quality of opacity is not a contrary or antagonistic quality to that of transparency, but only its extreme lowest degree.

143. Alluding to his theory of atoms.

144. Observe the approximation to Newton's theory. The same notion repeated still more clearly in the ninth motion. Newton believed that the planets might so conspire as to derange the earth's annual revolution, and to elongate the line of the apsides and ellipsis that the earth describes in its annual revolution round the sun. In the supposition that all the planets meet on the same straight line, Venus and Mercury on one side of the sun, and the earth, moon, Mars, Jupiter and Saturn on the side diametrically opposite; then Saturn would attract Jupiter, Jupiter Mars, Mars the moon, which must in its turn attract the earth in proportion to the force with which it was drawn out of its orbit. The result of this combined action on our planet would elongate its ecliptic orbit, and so far draw it from the source of heat, as to produce an intensity of cold destructive to animal life. But this movement would immediately cease with the planetary concurrence which produced it, and the earth, like a compressed spring, bound almost as near to the sun as she had been drawn from it, the reaction of the heat on its surface being about as intense as the cold caused by the first removal was severe. The earth, until it gained its regular track, would thus alternately vibrate between each side of its orbit, with successive changes in its atmosphere, proportional to the square of the variation of its distance from the sun. In no place is Bacon's genius more conspicuous than in these repeated guesses at truth. He would have been a strong Copernican, had not Gilbert defended the system.--Ed.

145. This is not true except when the projectile acquires greater velocity at every successive instant of its course, which is never the case except with falling bodies. Bacon appears to have been led into the opinion from observing that gunshots pierce many objects at a distance from which they rebound when brought within a certain proximity of contact. This apparent inconsistency, however, arises from the resistance of the parts of the object, which velocity combined with force is necessary to overcome.--Ed.

146. This passage shows that the pressure of the external atmosphere, which forces the water into the egg, was not in Bacon's time understood.--Ed.

147. We have already alluded, in a note prefixed to the same aphorism of the first book, to Newton's error of the absolute lightness of bodies. In speaking again of the volatile or spiritual substances (Aph. xl. b. ii.) which he supposed with the Platonists and some of the schoolmen to enter into the composition of every body, he ascribes to them a power of lessening the weight of the material coating in which he supposes them inclosed. It would appear from these passages and the text that Bacon had no idea of the relative density of bodies, and the capability which some have to diminish the specific gravity of the heavier substances by the dilation of their parts; or if he had, the reveries in which Aristotle indulged in treating of the soul, about the appetency of bodies to fly to kindred substances-- flame and spirit to the sky, and solid opaque substances to the earth, must have vitiated his mind.--Ed.

148. Römer, a Danish astronomer, was the first to demonstrate, by connecting the irregularities of the eclipses of Jupiter's satellites with their distances from the earth, the necessity of time for the propagation of light. The idea occurred to Dominic Cassini as well as Bacon, but both allowed the discovery to slip out of their hands.--Ed.

149. The author in the text confounds inertness, which is a simple indifference of bodies to action, with gravity, which is a force acting always in proportion to their density. He falls into the same error further on.--Ed.

150. The experiments of the last two classes of instances are considered only in relation to practice, and Bacon does not so much as mention their infinitely greater importance in the theoretical part of induction. The important law of gravitation in physical astronomy could never have been demonstrated but by such observations and experiments as assigned accurate geometrical measures to the quantities compared. It was necessary to determine with precision the demi-diameter of the earth, the velocity of falling bodies at its surface, the distance of the moon, and the speed with which she describes her orbit, before the relation could be discovered between the force which draws a stone to the ground and that which retains the moon in her sphere.

In many cases the result of a number of particular facts, or the collective instances rising out of them, can only be discovered by geometry, which so far becomes necessary to complete the work of induction. For instance, in the case of optics, when light passes from one transparent medium to another, it is refracted, and the angle which the ray of incidence makes with the superficies which bounds the two media determines that which the refracted ray makes with the same superficies. Now, all experiment can do for us in this case is, to determine for any particular angle of incidence the corresponding angle of refraction. But with respect to the general rule which in every possible case deduces one of these angles from the other, or expresses the constant and invariable relation which subsists between them, experiment gives no direct information. Geometry must, consequently, be called in, which, when a constant though unknown relation subsists between two angles, or two variable qualities of any kind, and when an indefinite number of values of those quantities are assigned, furnishes infallible means of discovering that unknown relation either accurately or by approximation. In this way it has been found, when the two media remain the same, the cosines of the above-mentioned angles have a constant ratio to each other. Hence, when the relations of the simple elements of phenomena are discovered to afford a general rule which will apply to any concrete case, the deductive method must be applied, and the elementary principles made through its agency to account for the laws of their more complex combinations. The reflection and refraction of light by the rain falling from a cloud opposite to the sun was thought, even before Newton's day, to contain the form of the rainbow. This philosopher transformed a probable conjecture into a certain fact when he deduced from the known laws of reflection and refraction the breadth of the colored arch, the diameter of the circle of which it is a part, and the relation of the latter to the place of the spectator and the sun. Doubt was at once silenced when there came out of his calculus a combination of the same laws of the simple elements of optics answering to the phenomena in nature.--Ed.

151. As far as this motion results from attraction and repulsion, it is only a simple consequence of the last two.--Ed.

152. These two cases are now resolved into the property of the capillary tubes and present only another feature of the law of attraction.--Ed.

153. This is one of the most useful practical methods in chemistry at the present day.

154. See Aphorism xxv.

155. Query?

156. Observe this approximation to Newton's theory.

157. Those differences which are generated by the masses and respective distances of bodies are only differences of quantity, and not specific; consequently those three classes are only one.--Ed.

158. See the citing instances, Aphorism xl.

159. Aristotle's doctrine, that sound takes place when bodies strike the air, which the modern science of acoustics has completely established, was rejected by Bacon in a treatise upon the same subject: "The collision or thrusting of air," he says, "which they will have to be the cause of sound, neither denotes the form nor the latent process of sound, but is a term of ignorance and of superficial contemplation." To get out of the difficulty, he betook himself to his theory of spirits, a species of phenomena which he constantly introduces to give himself the air of explaining things he could not understand, or would not admit upon the hypothesis of his opponents.--Ed.

160. The motion of trepidation, as Bacon calls it, was attributed by the ancient astronomers to the eight spheres, relative to the precession of the equinoxes. Galileo was the first to observe this kind of lunar motion.--Ed.

161. Part of the air is expanded and escapes, and part is consumed by the flame. When condensed, therefore, by the cold application, it cannot offer sufficient resistance to the external atmosphere to prevent the liquid or flesh from being forced into the glass.

162. Heat can now be abstracted by a very simple process, till the degree of cold be of almost any required intensity.--Ed.

163. It is impossible to compare a degree of heat with a degree of cold, without the assumption of some arbitrary test, to which the degrees are to be referred. In the next sentence Bacon appears to have taken the power of animal life to support heat or cold as the test, and then the comparison can only be between the degree of heat or of cold that will produce death.

The zero must be arbitrary which divides equally a certain degree of heat from a certain degree of cold.--Ed.

164. It may often be observed on the leaves of the lime and other trees.

The Essays or Counsels, Civil and Moral
OF FRANCIS Ld. VERULAM VISCOUNT ST. ALBANS

TO THE RIGHT HONORABLE MY VERY GOOD LORD THE DUKE OF BUCKINGHAM HIS GRACE, LORD HIGH ADMIRAL OF ENGLAND

EXCELLENT LORD:

SALOMON saies; A good Name is as a precious oyntment; And I assure my selfe, such wil your Graces Name bee, with Posteritie. For your Fortune, and Merit both, have been Eminent. And you have planted Things, that are like to last. I doe now publish my Essayes; which, of all my other workes, have beene most Currant: For that, as it seemes, they come home, to Mens Businesse, and Bosomes. I have enlarged them, both in Number, and Weight; So that they are indeed a New Worke. I thought it therefore agreeable, to my Affection, and Obligation to your Grace, to prefix your Name before them, both in English, and in Latine. For I doe conceive, that the Latine Volume of them, (being in the Universall Language) may last, as long as Bookes last. My Instauration, I dedicated to the King: My Historie of Henry the Seventh, (which I have now also translated into Latine) and my Portions of Naturall History, to the Prince: And these I dedicate to your Grace; Being of the best Fruits, that by the good Encrease, which God gives to my Pen and Labours, I could yeeld. God leade your Grace by the Hand. Your Graces most Obliged and faithfull Servant,

FR. ST. ALBAN

Of Truth

WHAT is truth? said jesting Pilate, and would not stay for an answer. Certainly there be, that delight in giddiness, and count it a bondage to fix a belief; affecting free-will in thinking, as well as in acting. And though the sects of philosophers of that kind be gone, yet there remain certain discoursing wits, which are of the same veins, though there be not so much blood in them, as was in those of the ancients. But it is not only the difficulty and labor, which men take in finding out of truth, nor again, that when it is found, it imposeth upon men's thoughts, that doth bring lies in favor; but a natural, though corrupt love, of the lie itself. One of the later school of the Grecians, examineth the matter, and is at a stand, to think what should be in it, that men should love lies; where neither they make for pleasure, as with poets, nor for advantage, as with the merchant; but for the lie's sake. But I cannot tell; this same truth, is a naked, and open day-light, that doth not show the masks, and mummeries, and triumphs, of the world, half so stately and daintily as candle-lights. Truth may perhaps come to the price of a pearl, that showeth best by day; but it will not rise to the price of a diamond, or carbuncle, that showeth best in varied lights. A mixture of a lie doth ever add pleasure. Doth any man doubt, that if there were taken out of men's minds, vain opinions, flattering hopes, false valuations, imaginations as one would, and the like, but it would leave the minds, of a number of men, poor shrunken things, full of melancholy and indisposition, and unpleasing to themselves?

One of the fathers, in great severity, called poesy *vinum daemonum*, because it fireth the imagination; and yet, it is but with the shadow of a lie. But it is not the lie that passeth through the mind, but the lie that sinketh in, and settleth in it, that doth the hurt; such as we spake of before. But howsoever these things are thus in men's depraved judgments, and affections, yet truth, which only doth judge itself, teacheth that the inquiry of truth, which is the love-making, or wooing of it, the knowledge of truth, which is the presence of it, and the belief of truth, which is the enjoying of it, is the sovereign good of human nature. The first creature of God, in the works of the days, was the light of the sense; the last, was the light of reason; and his sabbath work ever since, is the illumination of his Spirit. First he breathed light, upon the face of the matter or chaos; then he breathed light, into the face of man; and still he breatheth and inspireth light, into the face of his chosen. The poet, that beautified the sect, that was otherwise inferior to the rest, saith yet excellently well: It is a pleasure, to stand upon the shore, and to see ships tossed upon the sea; a pleasure, to stand in the window of a castle, and to see a battle, and the adventures thereof below: but no pleasure is comparable to the standing upon the vantage ground of truth (a hill not to be commanded, and where the air is always clear and serene), and to see the errors, and wanderings, and mists, and tempests, in the vale below; so always that this prospect be with pity, and not with swelling, or pride. Certainly, it is heaven upon earth, to have a man's mind move in charity, rest in providence, and turn upon the poles of truth.

To pass from theological, and philosophical truth, to the truth of civil business; it will be acknowledged, even by those that practise it not, that clear, and round dealing, is the honor of man's nature; and that mixture of falsehoods, is like alloy in coin of gold and silver, which may make the metal work the better, but it embaseth it. For these winding, and crooked courses, are the goings of the serpent; which goeth basely upon the belly, and not upon the feet. There is no vice, that doth so cover a man with shame, as to be found false and perfidious. And therefore Montaigne saith prettily, when he inquired the reason, why the word of the lie should be such a disgrace, and such an odious charge? Saith he, If it be well weighed, to say that a man lieth, is as much to say, as that he is brave towards God, and a coward towards men. For a lie faces God, and shrinks from man. Surely the wickedness of falsehood, and breach of faith, cannot possibly be so highly expressed, as in that it shall be the last peal, to call the judgments of God upon the generations of men; it being foretold, that when Christ cometh, he shall not find faith upon the earth.

Of Death

MEN fear death, as children fear to go in the dark; and as that natural fear in children, is increased with tales, so is the other. Certainly, the contemplation of death, as the wages of sin, and passage to another world, is holy and religious; but the fear of it, as a tribute due unto nature, is weak. Yet in religious meditations, there is sometimes mixture of vanity, and of superstition. You shall read, in some of the friars' books of mortification, that a man should think with himself, what the pain is, if he have but his finger's end pressed, or tortured, and thereby imagine, what the pains of death are, when the whole body is corrupted, and dissolved; when many times death passeth, with less pain than the torture of a limb; for the most vital parts, are not the quickest of sense. And by him that spake only as a philosopher, and natural man, it was well said, Pompa mortis magis terret, quam mors ipsa. Groans, and convulsions, and a discolored face, and friends weeping, and blacks, and obsequies, and the like, show death terrible. It is worthy the observing, that there is no passion in the mind of man, so weak, but it mates, and masters, the fear of death; and therefore, death is no such terrible enemy, when a man hath so many attendants about him, that can win the combat of him. Revenge triumphs over death; love slights it; honor aspireth to it; grief flieth to it; fear preoccupateth it; nay, we read, after Otho the emperor had slain himself, pity (which is the tenderest of affections) provoked many to die, out of mere compassion to their sovereign, and as the truest sort of followers. Nay, Seneca adds niceness and satiety: Cogita quamdiu eadem feceris; mori velle, non tantum fortis aut miser, sed etiam fastidiosus potest. A man would die, though he were neither valiant, nor miserable, only upon a weariness to do the same thing so oft, over and over. It is no less worthy, to observe, how little alteration in good spirits, the approaches of death make; for they appear to be the same men, till the last instant. Augustus Caesar died in a compliment; Livia, conjugii nostri memor, vive et vale. Tiberius in dissimulation; as Tacitus saith of him, Jam Tiberium vires et corpus, non dissimulatio, deserebant. Vespasian in a jest, sitting upon the stool; Ut puto deus fio. Galba with a sentence; Feri, si ex re sit populi Romani; holding forth his neck. Septimius Severus in despatch; Adeste si quid mihi restat agendum. And the like. Certainly the Stoics bestowed too much cost upon death, and by their great preparations, made it appear more fearful. Better saith he, qui finem vitae extremum inter munera ponat naturae. It is as natural to die, as to be born; and to a little infant, perhaps, the one is as painful, as the other. He that dies in an earnest pursuit, is like one that is wounded in hot blood; who, for the time, scarce feels the hurt; and therefore a mind fixed, and bent upon somewhat that is good, doth avert the dolors of death. But, above all, believe it, the sweetest canticle is', Nunc dimittis; when a man hath obtained worthy ends, and expectations. Death hath this also; that it openeth the gate to good fame, and extinguisheth envy.--Extinctus amabitur idem.

Of Unity In Religion

RELIGION being the chief band of human society, it is a happy thing, when itself is well contained within the true band of unity. The quarrels, and divisions about religion, were evils unknown to the heathen. The reason was, because the religion of the heathen, consisted rather in rites and ceremonies, than in any constant belief. For you may imagine, what kind of faith theirs was, when the chief doctors, and fathers of their church, were the poets. But the true God hath this attribute, that he is a jealous God; and therefore, his worship and religion, will endure no mixture, nor partner. We shall therefore speak a few words, concerning the unity of the church; what are the fruits thereof; what the bounds; and what the means.

The fruits of unity (next unto the well pleasing of God, which is all in all) are two: the one, towards those that are without the church, the other, towards those that are within. For the former; it is certain, that heresies, and schisms, are of all others the greatest scandals; yea, more than corruption of manners. For as in the natural body, a wound, or solution of continuity, is worse than a corrupt humor; so in the spiritual. So that nothing, doth so much keep men out of the church, and drive men out of the church, as breach of unity. And therefore, whensoever it cometh to that pass, that one saith, Ecce in deserto, another saith, Ecce in penetralibus; that is, when some men seek Christ, in the conventicles of heretics, and others, in an outward face of a church, that voice had need continually to sound in men's ears, Nolite exire,--Go not out. The doctor of the Gentiles (the propriety of whose vocation, drew him to have a special care of those without) saith, if an heathen come in, and hear you speak with several tongues, will he not say that you are mad? And certainly it is little better, when atheists, and profane persons, do hear of so many discordant, and contrary opinions in religion; it doth avert them from the church, and maketh them, to sit down in the chair of the scorners. It is but a light thing, to be vouched in so serious a matter, but yet it expresseth well the deformity. There is a master of scoffing, that in his catalogue of books of a feigned library, sets down this title of a book, The Morris-Dance of Heretics. For indeed, every sect of them, hath a diverse posture, or cringe by themselves, which cannot but move derision in worldlings, and depraved politics, who are apt to contemn holy things.

As for the fruit towards those that are within; it is peace; which containeth infinite blessings. It establisheth faith; it kindleth charity; the outward peace of the church, distilleth into peace of conscience; and it turneth the labors of writing, and reading of controversies, into treaties of mortification and devotion.

Concerning the bounds of unity; the true placing of them, importeth exceedingly. There appear to be two extremes. For to certain zealants, all speech of pacification is odious. Is it peace, Jehu,? What hast thou to do with peace? turn thee behind me. Peace is not the matter, but following, and party. Contrariwise, certain Laodiceans, and lukewarm persons, think they may accommodate points of religion, by middle way, and taking part of both, and witty reconcilements; as if they would make an arbitrament between God and man. Both these extremes are to be avoided; which will be done, if the league of Christians, penned by our Savior himself, were in two cross clauses thereof, soundly and plainly expounded: He that is not with us, is against us; and again, He that is not against us, is with us; that is, if the points fundamental and of substance in religion, were truly discerned and distinguished, from points not merely of faith, but of opinion, order, or good intention. This is a thing may seem to many a matter trivial, and done already. But if it were done less partially, it would be embraced more generally.

Of this I may give only this advice, according to my small model. Men ought to take heed, of rending God's church, by two kinds of controversies. The one is, when the matter of the point controverted, is too small and light, not worth the heat and strife about it, kindled only by contradiction. For, as it is noted, by one of the fathers, Christ's coat indeed had no seam, but the church's vesture was of divers colors; whereupon he saith, In veste varietas sit, scissura non sit; they be two things, unity and uniformity. The other is, when the matter of the point controverted, is great, but it is driven to an over-great subtilty, and obscurity; so that it becometh a thing rather ingenious, than substantial. A man that is of judgment and understanding, shall sometimes hear ignorant men differ, and know well within himself, that those which so differ, mean one thing, and yet they themselves would never agree. And if it come so to pass, in that distance of judgment, which is between man and man, shall we not think that God above, that knows the heart, doth not discern that frail men, in some of their contradictions, intend the same thing; and accepteth of both? The nature of such controversies is excellently expressed, by St. Paul, in the warning and precept, that he giveth concerning the same, Devita profanas vocum novitates, et oppositiones falsi nominis scientiae. Men create oppositions, which are not; and put them into new terms, so fixed, as whereas the meaning ought to govern the term, the term in effect governeth the meaning. There be also two false peaces, or unities: the one, when the peace is grounded, but upon an implicit ignorance; for all colors will agree in the dark: the other, when it is pieced up, upon a direct admission of contraries, in fundamental points. For truth and falsehood, in such things, are like the iron and clay, in the toes of Nebuchadnezzar's image; they may cleave, but they will not incorporate.

Concerning the means of procuring unity; men must beware, that in the procuring, or reuniting, of religious unity, they do not dissolve and deface the laws of charity, and of human society. There be two swords amongst Christians, the spiritual and temporal; and both have their due office and place, in the maintenance of religion. But we may not take up the third sword, which is Mahomet's sword, or like unto it; that is, to propagate religion by wars, or by sanguinary persecutions to force consciences; except it be in cases of overt scandal, blasphemy, or intermixture of practice against the state; much less to nourish seditions; to authorize conspiracies and rebellions; to put the sword into the people's hands; and the like; tending to the subversion of all government, which is the ordinance of God. For this is but to dash the first table against the second; and so to consider men as Christians, as we forget that they are men. Lucretius the poet, when he beheld the act of Agamemnon, that could endure the sacrificing of his own daughter, exclaimed: Tantum Religio potuit suadere malorum.

What would he have said, if he had known of the massacre in France, or the powder treason of England? He would have been seven times more Epicure, and atheist, than he was. For as the temporal sword is to be drawn with great circumspection in cases of religion; so it is a thing monstrous to put it into the hands of the common people. Let that be left unto the Anabaptists, and other furies. It was great blasphemy, when the devil said, I will ascend, and be like the highest; but it is greater blasphemy, to personate God, and bring him in saying, I will descend, and be like the prince of darkness; and what is it better, to make the cause of religion to descend, to the cruel and execrable actions of murthering princes, butchery of people, and subversion of states and governments? Surely this is to bring down the Holy Ghost, instead of the likeness of a dove, in the shape of a vulture or raven; and set, out of the bark of a Christian church, a flag of a bark of pirates, and assassins. Therefore it is most necessary, that the church, by doctrine and decree, princes by their sword, and all learnings, both Christian and moral, as by their Mercury rod, do damn and send to hell for ever, those facts and opinions tending to the support of the same; as hath been already in good part done. Surely in counsels concerning religion, that counsel of the apostle would be prefixed, Ira hominis non implet justitiam Dei. And it was a notable observation of a wise father, and no less ingenuously confessed; that those which held and persuaded pressure of consciences, were commonly interested therein, themselves, for their own ends.

Of Revenge

REVENGE is a kind of wild justice; which the more man's nature runs to, the more ought law to weed it out. For as for the first wrong, it doth but offend the law; but the revenge of that wrong, putteth the law out of office. Certainly, in taking revenge, a man is but even with his enemy; but in passing it over, he is superior; for it is a prince's part to pardon. And Solomon, I am sure, saith, It is the glory of a man, to pass by an offence. That which is past is gone, and irrevocable; and wise men have enough to do, with things present and to come; therefore they do but trifle with themselves, that labor in past matters. There is no man doth a wrong, for the wrong's sake; but thereby to purchase himself profit, or pleasure, or honor, or the like. Therefore why should I be angry with a man, for loving himself better than me? And if any man should do wrong, merely out of ill-nature, why, yet it is but like the thorn or briar, which prick and scratch, because they can do no other. The most tolerable sort of revenge, is for those wrongs which there is no law to remedy; but then let a man take heed, the revenge be such as there is no law to punish; else a man's enemy is still before hand, and it is two for one. Some, when they take revenge, are desirous, the party should know, whence it cometh. This is the more generous. For the delight seemeth to be, not so much in doing the hurt, as in making the party repent. But base and crafty cowards, are like the arrow that flieth in the dark. Cosmus, duke of Florence, had a desperate saying against perfidious or neglecting friends, as if those wrongs were unpardonable; You shall read (saith he) that we are commanded to forgive our enemies; but you never read, that we are commanded to forgive our friends. But yet the spirit of Job was in a better tune: Shall we (saith he) take good at God's hands, and not be content to take evil also? And so of friends in a proportion. This is certain, that a man that studieth revenge, keeps his own wounds green, which otherwise would heal, and do well. Public revenges are for the most part fortunate; as that for the death of Caesar; for the death of Pertinax; for the death of Henry the Third of France; and many more. But in private revenges, it is not so. Nay rather, vindictive persons live the life of witches; who, as they are mischievous, so end they infortunate.

Of Adversity

IT WAS an high speech of Seneca (after the manner of the Stoics), that the good things, which belong to prosperity, are to be wished; but the good things, that belong to adversity, are to be admired. Bona rerum secundarum optabilia; adversarum mirabilia. Certainly if miracles be the command over nature, they appear most in adversity. It is yet a higher speech of his, than the other (much too high for a heathen), It is true greatness, to have in one the frailty of a man, and the security of a God. Vere magnum habere fragilitatem hominis, securitatem Dei. This would have done better in poesy, where transcendences are more allowed. And the poets indeed have been busy with it; for it is in effect the thing, which figured in that strange fiction of the ancient poets, which seemeth not to be without mystery; nay, and to have some approach to the state of a Christian; that Hercules, when he went to unbind Prometheus (by whom human nature is represented), sailed the length of the great ocean, in an earthen pot or pitcher; lively describing Christian resolution, that saileth in the frail bark of the flesh, through the waves of the world. But to speak in a mean. The virtue of prosperity, is temperance; the virtue of adversity, is fortitude; which in morals is the more heroical virtue. Prosperity is the blessing of the Old Testament; adversity is the blessing of the New; which carrieth the greater benediction, and the clearer revelation of God's favor. Yet even in the Old Testament, if you listen to David's harp, you shall hear as many hearse-like airs as carols; and the pencil of the Holy Ghost hath labored more in describing the afflictions of Job, than the felicities of Solomon. Prosperity is not without many fears and distastes; and adversity is not without comforts and hopes. We see in needle-works and embroideries, it is more pleasing to have a lively work, upon a sad and solemn ground, than to have a dark and melancholy work, upon a lightsome ground: judge therefore of the pleasure of the heart, by the pleasure of the eye. Certainly virtue is like precious odors, most fragrant when they are incensed, or crushed: for prosperity doth best discover vice, but adversity doth best discover virtue.

Of Simulation And Dissimulation

DISSIMULATION is but a faint kind of policy, or wisdom; for it asketh a strong wit, and a strong heart, to know when to tell truth, and to do it. Therefore it is the weaker sort of politics, that are the great dissemblers.

Tacitus saith, Livia sorted well with the arts of her husband, and dissimulation of her son; attributing arts or policy to Augustus, and dissimulation to Tiberius. And again, when Mucianus encourageth Vespasian, to take arms against Vitellius, he saith, We rise not against the piercing judgment of Augustus, nor the extreme caution or closeness of Tiberius. These properties, of arts or policy, and dissimulation or closeness, are indeed habits and faculties several, and to be distinguished. For if a man have that penetration of judgment, as he can discern what things are to be laid open, and what to be secreted, and what to be showed at half lights, and to whom and when (which indeed are arts of state, and arts of life, as Tacitus well calleth them), to him, a habit of dissimulation is a hinderance and a poorness. But if a man cannot obtain to that judgment, then it is left to him generally, to be close, and a dissembler. For where a man cannot choose, or vary in particulars, there it is good to take the safest, and wariest way, in general; like the going softly, by one that cannot well see. Certainly the ablest men that ever were, have had all an openness, and frankness, of dealing; and a name of certainty and veracity; but then they were like horses well managed; for they could tell passing well, when to stop or turn; and at such times, when they thought the case indeed required dissimulation, if then they used it, it came to pass that the former opinion, spread abroad, of their good faith and clearness of dealing, made them almost invisible.

There be three degrees of this hiding and veiling of a man's self. The first, closeness, reservation, and secrecy; when a man leaveth himself without observation, or without hold to be taken, what he is. The second, dissimulation, in the negative; when a man lets fall signs and arguments, that he is not, that he is. And the third, simulation, in the affirmative; when a man industriously and expressly feigns and pretends to be, that he is not.

For the first of these, secrecy; it is indeed the virtue of a confessor. And assuredly, the secret man heareth many confessions. For who will open himself, to a blab or a babbler? But if a man be thought secret, it inviteth discovery; as the more close air sucketh in the more open; and as in confession, the revealing is not for worldly use, but for the ease of a man's heart, so secret men come to the knowledge of many things in that kind; while men rather discharge their minds, than impart their minds. In few words, mysteries are due to secrecy. Besides (to say truth) nakedness is uncomely, as well in mind as body; and it addeth no small reverence, to men's manners and actions, if they be not altogether open. As for talkers and futile persons, they are commonly vain and credulous withal. For he that talketh what he knoweth, will also talk what he knoweth not. Therefore set it down, that an habit of secrecy, is both politic and moral. And in this part, it is good that a man's face give his tongue leave to speak. For the discovery of a man's self, by the tracts of his countenance, is a great weakness and betraying; by how much it is many times more marked, and believed, than a man's words.

For the second, which is dissimulation; it followeth many times upon secrecy, by a necessity; so that he that will be secret, must be a dissembler in some degree. For men are too cunning, to suffer a man to keep an indifferent carriage between both, and to be secret, without swaying the balance on either side. They will so beset a man with questions, and draw him on, and pick it out of him, that, without an absurd silence, he must show an inclination one way; or if he do not, they will gather as much by his silence, as by his speech. As for equivocations, or oraculous speeches, they cannot hold out long. So that no man can be secret, except he give himself a little scope of dissimulation; which is, as it were, but the skirts or train of secrecy.

But for the third degree, which is simulation, and false profession; that I hold more culpable, and less politic; except it be in great and rare matters. And therefore a general custom of simulation (which is this last degree) is a vice, using either of a natural falseness or fearfulness, or of a mind that hath some main faults, which because a man must needs disguise, it maketh him practise simulation in other things, lest his hand should be out of use.

The great advantages of simulation and dissimulation are three. First, to lay asleep opposition, and to surprise. For where a man's intentions are published, it is an alarum, to call up all that are against them. The second is, to reserve to a man's self a fair retreat. For if a man engage himself by a manifest declaration, he must go through or take a fall. The third is, the better to discover the mind of another. For to him that opens himself, men will hardly show themselves adverse; but will fair let him go on, and turn their freedom of speech, to freedom of thought. And therefore it is a good shrewd proverb of the Spaniard, Tell a lie and find a troth. As if there were no way of discovery, but by simulation. There be also three disadvantages, to set it even. The first, that simulation and dissimulation commonly carry with them a show of fearfulness, which in any business, doth spoil the feathers, of round flying up to the mark. The second, that it puzzleth and perplexeth the conceits of many, that perhaps would otherwise co-operate with him; and makes a man walk almost alone, to his own ends. The third and greatest is, that it depriveth a man of one of the most principal instruments for action; which is trust and belief. The best composition and temperature, is to have openness in fame and opinion; secrecy in habit; dissimulation in seasonable use; and a power to feign, if there be no remedy.

Of Parents And Children

THE joys of parents are secret; and so are their griefs and fears. They cannot utter the one; nor they will not utter the other. Children sweeten labors; but they make misfortunes more bitter. They increase the cares of life; but they mitigate the remembrance of death. The perpetuity by generation is common to beasts; but memory, merit, and noble works, are proper to men. And surely a man shall see the noblest works and foundations have proceeded from childless men; which have sought to express the images of their minds, where those of their bodies have failed. So the care of posterity is most in them, that have no posterity. They that are the first raisers of their houses, are most indulgent towards their children; beholding them as the continuance, not only of their kind, but of their work; and so both children and creatures.

The difference in affection, of parents towards their several children, is many times unequal; and sometimes unworthy; especially in the mothers; as Solomon saith, A wise son rejoiceth the father, but an ungracious son shames the mother. A man shall see, where there is a house full of children, one or two of the eldest respected, and the youngest made wantons; but in the midst, some that are as it were forgotten, who many times, nevertheless, prove the best. The illiberality of parents, in allowance towards their children, is an harmful error; makes them base; acquaints them with shifts; makes them sort with mean company; and makes them surfeit more when they come to plenty. And therefore the proof is best, when men keep their authority towards the children, but not their purse. Men have a foolish manner (both parents and schoolmasters and servants) in creating and breeding an emulation between brothers, during childhood, which many times sorteth to discord when they are men, and disturbeth families. The Italians make little difference between children, and nephews or near kinsfolks; but so they be of the lump, they care not though they pass not through their own body. And, to say truth, in nature it is much a like matter; insomuch that we see a nephew sometimes resembleth an uncle, or a kinsman, more than his own parent; as the blood happens. Let parents choose betimes, the vocations and courses they mean their children should take; for then they are most flexible; and let them not too much apply themselves to the disposition of their children, as thinking they will take best to that, which they have most mind to. It is true, that if the affection or aptness of the children be extraordinary, then it is good not to cross it; but generally the precept is good, optimum elige, suave et facile illud faciet consuetudo. Younger brothers are commonly fortunate, but seldom or never where the elder are disinherited.

Of Marriage And Single Life

HE THAT hath wife and children hath given hostages to fortune; for they are impediments to great enterprises, either of virtue or mischief. Certainly the best works, and of greatest merit for the public, have proceeded from the unmarried or childless men; which both in affection and means, have married and endowed the public. Yet it were great reason that those that have children, should have greatest care of future times; unto which they know they must transmit their dearest pledges. Some there are, who though they lead a single life, yet their thoughts do end with themselves, and account future times impertinences. Nay, there are some other, that account wife and children, but as bills of charges. Nay more, there are some foolish rich covetous men, that take a pride, in having no children, because they may be thought so much the richer. For perhaps they have heard some talk, Such an one is a great rich man, and another except to it, Yea, but he hath a great charge of children; as if it were an abatement to his riches. But the most ordinary cause of a single life, is liberty, especially in certain self-pleasing and humorous minds, which are so sensible of every restraint, as they will go near to think their girdles and garters, to be bonds and shackles. Unmarried men are best friends, best masters, best servants; but not always best subjects; for they are light to run away; and almost all fugitives, are of that condition. A single life doth well with churchmen; for charity will hardly water the ground, where it must first fill a pool. It is indifferent for judges and magistrates; for if they be facile and corrupt, you shall have a servant, five times worse than a wife. For soldiers, I find the generals commonly in their hortatives, put men in mind of their wives and children; and I think the despising of marriage amongst the Turks, maketh the vulgar soldier more base. Certainly wife and children are a kind of discipline of humanity; and single men, though they may be many times more charitable, because their means are less exhaust, yet, on the other side, they are more cruel and hardhearted (good to make severe inquisitors), because their tenderness is not so oft called upon. Grave natures, led by custom, and therefore constant, are commonly loving husbands, as was said of Ulysses, *vetulam suam praetulit immortalitati*. Chaste women are often proud and froward, as presuming upon the merit of their chastity. It is one of the best bonds, both of chastity and obedience, in the wife, if she think her husband wise; which she will never do, if she find him jealous. Wives are young men's mistresses; companions for middle age; and old men's nurses. So as a man may have a quarrel to marry, when he will. But yet he was reputed one of the wise men, that made answer to the question, when a man should marry,--A young man not yet, an elder man not at all. It is often seen that bad husbands, have very good wives; whether it be, that it raiseth the price of their husband's kindness, when it comes; or that the wives take a pride in their patience. But this never fails, if the bad husbands were of their own choosing, against their friends' consent; for then they will be sure to make good their own folly.

Of Envy

THERE be none of the affections, which have been noted to fascinate or bewitch, but love and envy. They both have vehement wishes; they frame themselves readily into imaginations and suggestions; and they come easily into the eye, especially upon the present of the objects; which are the points that conduce to fascination, if any such thing there be. We see likewise, the Scripture calleth envy an evil eye; and the astrologers, call the evil influences of the stars, evil aspects; so that still there seemeth to be acknowledged, in the act of envy, an ejaculation or irradiation of the eye. Nay, some have been so curious, as to note, that the times when the stroke or percussion of an envious eye doth most hurt, are when the party envied is beheld in glory or triumph; for that sets an edge upon envy: and besides, at such times the spirits of the person envied, do come forth most into the outward parts, and so meet the blow.

But leaving these curiosities (though not unworthy to be thought on, in fit place), we will handle, what persons are apt to envy others; what persons are most subject to be envied themselves; and what is the difference between public and private envy.

A man that hath no virtue in himself, ever envieth virtue in others. For men's minds, will either feed upon their own good, or upon others' evil; and who wanteth the one, will prey upon the other; and whoso is out of hope, to attain to another's virtue, will seek to come at even hand, by depressing another's fortune.

A man that is busy, and inquisitive, is commonly envious. For to know much of other men's matters, cannot be because all that ado may concern his own estate; therefore it must needs be, that he taketh a kind of play-pleasure, in looking upon the fortunes of others. Neither can he, that mindeth but his own business, find much matter for envy. For envy is a gadding passion, and walketh the streets, and doth not keep home: Non est curiosus, quin idem sit malevolus.

Men of noble birth, are noted to be envious towards new men, when they rise. For the distance is altered, and it is like a deceit of the eye, that when others come on, they think themselves, go back.

Deformed persons, and eunuchs, and old men, and bastards, are envious. For he that cannot possibly mend his own case, will do what he can, to impair another's; except these defects light upon a very brave, and heroical nature, which thinketh to make his natural wants part of his honor; in that it should be said, that an eunuch, or a lame man, did such great matters; affecting the honor of a miracle; as it was in Narses the eunuch, and Agesilaus and Tamberlanes, that were lame men.

The same is the case of men, that rise after calamities and misfortunes. For they are as men fallen out with the times; and think other men's harms, a redemption of their own sufferings.

They that desire to excel in too many matters, out of levity and vain glory, are ever envious. For they cannot want work; it being impossible, but many, in some one of those things, should surpass them. Which was the character of Adrian the Emperor; that mortally envied poets, and painters, and artificers, in works wherein he had a vein to excel.

Lastly, near kinsfolks, and fellows in office, and those that have been bred together, are more apt to envy their equals, when they are raised. For it doth upbraid unto them their own fortunes, and pointeth at them, and cometh oftener into their remembrance, and incurreth likewise more into the note of others; and envy ever redoubleth from speech and fame. Cain's envy was the more vile and malignant, towards his brother Abel, because when his sacrifice was better accepted, there was no body to look on. Thus much for those, that are apt to envy.

Concerning those that are more or less subject to envy: First, persons of eminent virtue, when they are advanced, are less envied. For their fortune seemeth, but due unto them; and no man envieth the payment of a debt, but rewards and liberality rather. Again, envy is ever joined with the comparing of a man's self; and where there is no comparison, no envy; and therefore kings are not envied, but by kings. Nevertheless it is to be noted, that unworthy persons are most envied, at their first coming in, and afterwards overcome it better; whereas contrariwise, persons of worth and merit are most envied, when their fortune continueth long. For by that time, though their virtue be the same, yet it hath not the same lustre; for fresh men grow up that darken it.

Persons of noble blood, are less envied in their rising. For it seemeth but right done to their birth. Besides, there seemeth not much added to their fortune; and envy is as the sunbeams, that beat hotter upon a bank, or steep rising ground, than upon a flat. And for the same reason, those that are advanced by degrees, are less envied than those that are advanced suddenly and per saltum.

Those that have joined with their honor great travels, cares, or perils, are less subject to envy. For men think that they earn their honors hardly, and pity them sometimes; and pity ever healeth envy. Wherefore you shall observe, that the more deep and sober sort of politic persons, in their greatness, are ever bemoaning themselves, what a life they lead; chanting a quanta patimur! Not that they feel it so, but only to abate the edge of envy. But this is to be understood, of business that is laid upon men, and not such, as they call unto themselves. For nothing increaseth envy more, than an unnecessary and ambitious engrossing of business. And nothing doth extinguish envy more, than for a great person to preserve all other inferior officers, in their full lights and pre-eminences of their places. For by that means, there be so many screens between him and envy.

Above all, those are most subject to envy, which carry the greatness of their fortunes, in an insolent and proud manner; being never well, but while they are showing how great they are, either by outward pomp, or by triumphing over all opposition or competition; whereas wise men will rather do sacrifice to envy, in suffering themselves sometimes of purpose to be crossed, and overborne in things that do not much concern them. Notwithstanding, so much is true, that the carriage of greatness, in a plain and open manner (so it be without arrogancy and vain glory) doth draw less envy, than if it be in a more crafty and cunning fashion. For in that course, a man doth but disavow fortune; and seemeth to be conscious of his own want in worth; and doth but teach others, to envy him.

Lastly, to conclude this part; as we said in the beginning, that the act of envy had somewhat in it of witchcraft, so there is no other cure of envy, but the cure of witchcraft; and that is, to remove the lot (as they call it) and to lay it upon another. For which purpose, the wiser sort of great persons, bring in ever upon the stage somebody upon whom to derive the envy, that would come upon themselves; sometimes upon ministers and servants; sometimes upon colleagues and associates; and the like; and for that turn there are never wanting, some persons of violent and undertaking natures, who, so they may have power and business, will take it at any cost.

Now, to speak of public envy. There is yet some good in public envy, whereas in private, there is none. For public envy, is as an ostracism, that eclipseth men, when they grow too great. And therefore it is a bridle also to great ones, to keep them within bounds.

This envy, being in the Latin word invidia, goeth in the modern language, by the name of discontentment; of which we shall speak, in handling sedition. It is a disease, in a state, like to infection. For as infection spreadeth upon that which is sound, and tainteth it; so when envy is gotten once into a state, it traduceth even the best actions thereof, and turneth them into an ill odor. And therefore there is little won, by intermingling of plausible actions. For that doth argue but a weakness, and fear of envy, which hurteth so much the more, as it is likewise usual in infections; which if you fear them, you call them upon you.

This public envy, seemeth to beat chiefly upon principal officers or ministers, rather than upon kings, and estates themselves. But this is a sure rule, that if the envy upon the minister be great, when the cause of it in him is small; or if the envy be general, in a manner upon all the ministers of an estate; then the envy (though hidden) is truly upon the state itself. And so much of public envy or discontentment, and the difference thereof from private envy, which was handled in the first place.

We will add this in general, touching the affection of envy; that of all other affections, it is the most importune and continual. For of other affections, there is occasion given, but now and then; and therefore it was well said, Invidia festos dies non agit: for it is ever working upon some or other. And it is also noted, that love and envy do make a man pine, which other affections do not, because they are not so continual. It is also the vilest affection, and the most depraved; for which cause it is the proper attribute of the devil, who is called, the envious man, that soweth tares amongst the wheat by night; as it always cometh to pass, that envy worketh subtilly, and in the dark, and to the prejudice of good things, such as is the wheat.

Of Love

THE stage is more beholding to love, than the life of man. For as to the stage, love is ever matter of comedies, and now and then of tragedies; but in life it doth much mischief; sometimes like a siren, sometimes like a fury. You may observe, that amongst all the great and worthy persons (whereof the memory remaineth, either ancient or recent) there is not one, that hath been transported to the mad degree of love: which shows that great spirits, and great business, do keep out this weak passion. You must except, nevertheless, Marcus Antonius, the half partner of the empire of Rome, and Appius Claudius, the decemvir and lawgiver; whereof the former was indeed a voluptuous man, and inordinate; but the latter was an austere and wise man: and therefore it seems (though rarely) that love can find entrance, not only into an open heart, but also into a heart well fortified, if watch be not well kept. It is a poor saying of Epicurus, *Satis magnum alter alteri theatrum sumus*; as if man, made for the contemplation of heaven, and all noble objects, should do nothing but kneel before a little idol, and make himself a subject, though not of the mouth (as beasts are), yet of the eye; which was given him for higher purposes. It is a strange thing, to note the excess of this passion, and how it braves the nature, and value of things, by this; that the speaking in a perpetual hyperbole, is comely in nothing but in love. Neither is it merely in the phrase; for whereas it hath been well said, that the arch-flatterer, with whom all the petty flatterers have intelligence, is a man's self; certainly the lover is more. For there was never proud man thought so absurdly well of himself, as the lover doth of the person loved; and therefore it was well said, That it is impossible to love, and to be wise. Neither doth this weakness appear to others only, and not to the party loved; but to the loved most of all, except the love be reciproque. For it is a true rule, that love is ever rewarded, either with the reciproque, or with an inward and secret contempt. By how much the more, men ought to beware of this passion, which loseth not only other things, but itself! As for the other losses, the poet's relation doth well figure them: that he that preferred Helena, quitted the gifts of Juno and Pallas. For whosoever esteemeth too much of amorous affection, quitteth both riches and wisdom. This passion hath his floods, in very times of weakness; which are great prosperity, and great adversity; though this latter hath been less observed: both which times kindle love, and make it more fervent, and therefore show it to be the child of folly. They do best, who if they cannot but admit love, yet make it keep quarters; and sever it wholly from their serious affairs, and actions, of life; for if it check once with business, it troubleth men's fortunes, and maketh men, that they can no ways be true to their own ends. I know not how, but martial men are given to love: I think, it is but as they are given to wine; for perils commonly ask to be paid in pleasures. There is in man's nature, a secret inclination and motion, towards love of others, which if it be not spent upon some one or a few, doth naturally spread itself towards many, and maketh men become humane and charitable; as it is seen sometime in friars. Nuptial love maketh mankind; friendly love perfecteth it; but wanton love corrupteth, and embaseth it.

Of Great Place

MEN in great place are thrice servants: servants of the sovereign or state; servants of fame; and servants of business. So as they have no freedom; neither in their persons, nor in their actions, nor in their times. It is a strange desire, to seek power and to lose liberty: or to seek power over others, and to lose power over a man's self. The rising unto place is laborious; and by pains, men come to greater pains; and it is sometimes base; and by indignities, men come to dignities. The standing is slippery, and the regress is either a downfall, or at least an eclipse, which is a melancholy thing. Cum non sis qui fueris, non esse cur velis vivere. Nay, retire men cannot when they would, neither will they, when it were reason; but are impatient of privateness, even in age and sickness, which require the shadow; like old townsmen, that will be still sitting at their street door, though thereby they offer age to scorn. Certainly great persons had need to borrow other men's opinions, to think themselves happy; for if they judge by their own feeling, they cannot find it; but if they think with themselves, what other men think of them, and that other men would fain be, as they are, then they are happy, as it were, by report; when perhaps they find the contrary within. For they are the first, that find their own griefs, though they be the last, that find their own faults. Certainly men in great fortunes are strangers to themselves, and while they are in the puzzle of business, they have no time to tend their health, either of body or mind. Illi mors gravis incubat, qui notus nimis omnibus, ignotus moritur sibi. In place, there is license to do good, and evil; whereof the latter is a curse: for in evil, the best condition is not to win; the second, not to can. But power to do good, is the true and lawful end of aspiring. For good thoughts (though God accept them) yet, towards men, are little better than good dreams, except they be put in act; and that cannot be, without power and place, as the vantage, and commanding ground. Merit and good works, is the end of man's motion; and conscience of the same is the accomplishment of man's rest. For if a man can be partaker of God's theatre, he shall likewise be partaker of God's rest. Et conversus Deus, ut aspiceret opera quae fecerunt manus suae, vidit quod omnia essent bona nimis; and then the sabbath. In the discharge of thy place, set before thee the best examples; for imitation is a globe of precepts. And after a time, set before thee thine own example; and examine thyself strictly, whether thou didst not best at first. Neglect not also the examples, of those that have carried themselves ill, in the same place; not to set off thyself, by taxing their memory, but to direct thyself, what to avoid. Reform therefore, without bravery, or scandal of former times and persons; but yet set it down to thyself, as well to create good precedents, as to follow them. Reduce things to the first institution, and observe wherein, and how, they have degenerate; but yet ask counsel of both times; of the ancient time, what is best; and of the latter time, what is fittest. Seek to make thy course regular, that men may know beforehand, what they may expect; but be not too positive and peremptory; and express thyself well, when thou digressest from thy rule. Preserve the right of thy place; but stir not questions of jurisdiction; and rather assume thy right, in silence and de facto, than voice it with claims, and challenges. Preserve likewise the rights of inferior places; and think it more honor, to direct in chief, than to be busy in all. Embrace and invite helps, and advices, touching the execution of thy place; and do not drive away such, as bring thee information, as meddlers; but accept of them in good part. The vices of authority are chiefly four: delays, corruption, roughness, and facility. For delays: give easy access; keep times appointed; go through with that which is in hand, and interlace not business, but of necessity. For corruption: do not only bind thine own hands, or thy servants' hands, from taking, but bind the hands of suitors also, from offering. For integrity used doth the one; but integrity professed, and with a manifest detestation of bribery, doth the other. And avoid not only the fault, but the suspicion. Whosoever is found variable, and changeth manifestly without manifest cause, giveth suspicion of corruption. Therefore always, when thou changest thine opinion or course, profess it plainly, and declare it, together with the reasons that move thee to change; and do not think to steal it. A servant or a favorite, if he be inward, and no other apparent cause of esteem, is commonly thought, but a by-way to close corruption. For roughness: it is a needless cause of discontent: severity breedeth fear, but roughness breedeth hate. Even reproofs from authority, ought to be grave, and not taunting. As for facility: it is worse than bribery. For bribes come but now and then; but if importunity, or idle respects, lead a man, he shall never be without. As Solomon saith, To respect persons is not good; for such a man will transgress for a piece of bread. It is most true, that was anciently spoken, A place showeth the man. And it showeth some to the better, and some to the worse. Omnium consensu capax imperii, nisi imperasset, saith Tacitus of Galba; but of Vespasian he saith,

Solus imperantium, Vespasianus mutatus in melius; though the one was meant of sufficiency, the other of manners, and affection. It is an assured sign of a worthy and generous spirit, whom honor amends. For honor is, or should be, the place of virtue; and as in nature, things move violently to their place, and calmly in their place, so virtue in ambition is violent, in authority settled and calm. All rising to great place is by a winding star; and if there be factions, it is good to side a man's self, whilst he is in the rising, and to balance himself when he is placed. Use the memory of thy predecessor, fairly and tenderly; for if thou dost not, it is a debt will sure be paid when thou art gone. If thou have colleagues, respect them, and rather call them, when they look not for it, than exclude them, when they have reason to look to be called. Be not too sensible, or too remembering, of thy place in conversation, and private answers to suitors; but let it rather be said, When he sits in place, he is another man.

Of Boldness

IT IS a trivial grammar-school text, but yet worthy a wise man's consideration. Question was asked of Demosthenes, what was the chief part of an orator? he answered, action; what next? action; what next again? action. He said it, that knew it best, and had, by nature, himself no advantage in that he commended. A strange thing, that that part of an orator, which is but superficial, and rather the virtue of a player, should be placed so high, above those other noble parts, of invention, elocution, and the rest; nay, almost alone, as if it were all in all. But the reason is plain. There is in human nature generally, more of the fool than of the wise; and therefore those faculties, by which the foolish part of men's minds is taken, are most potent. Wonderful like is the case of boldness in civil business: what first? boldness; what second and third? boldness. And yet boldness is a child of ignorance and baseness, far inferior to other parts. But nevertheless it doth fascinate, and bind hand and foot, those that are either shallow in judgment, or weak in courage, which are the greatest part; yea and prevaileth with wise men at weak times. Therefore we see it hath done wonders, in popular states; but with senates, and princes less; and more ever upon the first entrance of bold persons into action, than soon after; for boldness is an ill keeper of promise. Surely, as there are mountebanks for the natural body, so are there mountebanks for the politic body; men that undertake great cures, and perhaps have been lucky, in two or three experiments, but want the grounds of science, and therefore cannot hold out. Nay, you shall see a bold fellow many times do Mahomet's miracle. Mahomet made the people believe that he would call an hill to him, and from the top of it offer up his prayers, for the observers of his law. The people assembled; Mahomet called the hill to come to him, again and again; and when the hill stood still, he was never a whit abashed, but said, If the hill will not come to Mahomet, Mahomet will go to the hill. So these men, when they have promised great matters, and failed most shamefully, yet (if they have the perfection of boldness) they will but slight it over, and make a turn, and no more ado. Certainly to men of great judgment, bold persons are a sport to behold; nay, and to the vulgar also, boldness has somewhat of the ridiculous. For if absurdity be the subject of laughter, doubt you not but great boldness is seldom without some absurdity. Especially it is a sport to see, when a bold fellow is out of countenance; for that puts his face into a most shrunken, and wooden posture; as needs it must; for in bashfulness, the spirits do a little go and come; but with bold men, upon like occasion, they stand at a stay; like a stale at chess, where it is no mate, but yet the game cannot stir. But this last were fitter for a satire than for a serious observation. This is well to be weighed; that boldness is ever blind; for it seeth not danger, and inconveniences. Therefore it is ill in counsel, good in execution; so that the right use of bold persons is, that they never command in chief, but be seconds, and under the direction of others. For in counsel, it is good to see dangers; and in execution, not to see them, except they be very great.

Of Goodness and Goodness Of Nature

I TAKE goodness in this sense, the affecting of the weal of men, which is that the Grecians call philanthropia; and the word humanity (as it is used) is a little too light to express it. Goodness I call the habit, and goodness of nature, the inclination. This of all virtues, and dignities of the mind, is the greatest; being the character of the Deity: and without it, man is a busy, mischievous, wretched thing; no better than a kind of vermin. Goodness answers to the theological virtue, charity, and admits no excess, but error. The desire of power in excess, caused the angels to fall; the desire of knowledge in excess, caused man to fall: but in charity there is no excess; neither can angel, nor man, come in danger by it. The inclination to goodness, is imprinted deeply in the nature of man; insomuch, that if it issue not towards men, it will take unto other living creatures; as it is seen in the Turks, a cruel people, who nevertheless are kind to beasts, and give alms, to dogs and birds; insomuch, as Busbechius reporteth, a Christian boy, in Constantinople, had like to have been stoned, for gagging in a waggishness a long-billed fowl. Errors indeed in this virtue of goodness, or charity, may be committed. The Italians have an ungracious proverb, Tanto buon che val niente: so good, that he is good for nothing. And one of the doctors of Italy, Nicholas Machiavel, had the confidence to put in writing, almost in plain terms, That the Christian faith, had given up good men, in prey to those that are tyrannical and unjust. Which he spake, because indeed there was never law, or sect, or opinion, did so much magnify goodness, as the Christian religion doth. Therefore, to avoid the scandal and the danger both, it is good, to take knowledge of the errors of an habit so excellent. Seek the good of other men, but be not in bondage to their faces or fancies; for that is but facility, or softness; which taketh an honest mind prisoner. Neither give thou AEsop's cock a gem, who would be better pleased, and happier, if he had had a barley-corn. The example of God, teacheth the lesson truly: He sendeth his rain, and maketh his sun to shine, upon the just and unjust; but he doth not rain wealth, nor shine honor and virtues, upon men equally. Common benefits, are to be communicate with all; but peculiar benefits, with choice. And beware how in making the portraiture, thou breakest the pattern. For divinity, maketh the love of ourselves the pattern; the love of our neighbors, but the portraiture. Sell all thou hast, and give it to the poor, and follow me: but, sell not all thou hast, except thou come and follow me; that is, except thou have a vocation, wherein thou mayest do as much good, with little means as with great; for otherwise, in feeding the streams, thou driest the fountain. Neither is there only a habit of goodness, directed by right reason; but there is in some men, even in nature, a disposition towards it; as on the other side, there is a natural malignity. For there be, that in their nature do not affect the good of others. The lighter sort of malignity, turneth but to a crassness, or frowardness, or aptness to oppose, or difficulties, or the like; but the deeper sort, to envy and mere mischief. Such men, in other men's calamities, are, as it were, in season, and are ever on the loading part: not so good as the dogs, that licked Lazarus' sores; but like flies, that are still buzzing upon any thing that is raw; misanthropi, that make it their practice, to bring men to the bough, and yet never a tree for the purpose in their gardens, as Timon had. Such dispositions, are the very errors of human nature; and yet they are the fittest timber, to make great politics of; like to knee timber, that is good for ships, that are ordained to be tossed; but not for building houses, that shall stand firm. The parts and signs of goodness, are many. If a man be gracious and courteous to strangers, it shows he is a citizen of the world, and that his heart is no island, cut off from other lands, but a continent, that joins to them. If he be compassionate towards the afflictions of others, it shows that his heart is like the noble tree, that is wounded itself, when it gives the balm. If he easily pardons, and remits offences, it shows that his mind is planted above injuries; so that he cannot be shot. If he be thankful for small benefits, it shows that he weighs men's minds, and not their trash. But above all, if he have St. Paul's perfection, that he would wish to be anathema from Christ, for the salvation of his brethren, it shows much of a divine nature, and a kind of conformity with Christ himself.

Of Nobility

WE WILL speak of nobility, first as a portion of an estate, then as a condition of particular persons. A monarchy, where there is no nobility at all, is ever a pure and absolute tyranny; as that of the Turks. For nobility attempers sovereignty, and draws the eyes of the people, somewhat aside from the line royal. But for democracies, they need it not; and they are commonly more quiet, and less subject to sedition, than where there are stirps of nobles. For men's eyes are upon the business, and not upon the persons; or if upon the persons, it is for the business' sake, as fittest, and not for flags and pedigree. We see the Switzers last well, notwithstanding their diversity of religion, and of cantons. For utility is their bond, and not respects. The united provinces of the Low Countries, in their government, excel; for where there is an equality, the consultations are more indifferent, and the payments and tributes, more cheerful. A great and potent nobility, addeth majesty to a monarch, but diminisheth power; and putteth life and spirit into the people, but presseth their fortune. It is well, when nobles are not too great for sovereignty nor for justice; and yet maintained in that height, as the insolency of inferiors may be broken upon them, before it come on too fast upon the majesty of kings. A numerous nobility causeth poverty, and inconvenience in a state; for it is a surcharge of expense; and besides, it being of necessity, that many of the nobility fall, in time, to be weak in fortune, it maketh a kind of disproportion, between honor and means.

As for nobility in particular persons; it is a reverend thing, to see an ancient castle or building, not in decay; or to see a fair timber tree, sound and perfect. How much more, to behold an ancient noble family, which has stood against the waves and weathers of time! For new nobility is but the act of power, but ancient nobility is the act of time. Those that are first raised to nobility, are commonly more virtuous, but less innocent, than their descendants; for there is rarely any rising, but by a commixture of good and evil arts. But it is reason, the memory of their virtues remain to their posterity, and their faults die with themselves. Nobility of birth commonly abateth industry; and he that is not industrious, envieth him that is. Besides, noble persons cannot go much higher; and he that standeth at a stay, when others rise, can hardly avoid motions of envy. On the other side, nobility extinguisheth the passive envy from others, towards them; because they are in possession of honor. Certainly, kings that have able men of their nobility, shall find ease in employing them, and a better slide into their business; for people naturally bend to them, as born in some sort to command.

Of Seditions And Troubles

SHEPHERDS of people, had need know the calendars of tempests in state; which are commonly greatest, when things grow to equality; as natural tempests are greatest about the Equinoctia. And as there are certain hollow blasts of wind, and secret swellings of seas before a tempest, so are there in states:

--Ille etiam caecos instare tumultus   Saepe monet, fraudesque et operta tunescere bella.

Libels and licentious discourses against the state, when they are frequent and open; and in like sort, false news often running up and down, to the disadvantage of the state, and hastily embraced; are amongst the signs of troubles. Virgil, giving the pedigree of Fame, saith, she was sister to the Giants:

Illam Terra parens, irra irritata deorum, Extremam (ut perhibent) Coeo Enceladoque sororem Progenuit.

As if fames were the relics of seditions past; but they are no less, indeed, the preludes of seditions to come. Howsoever he noteth it right, that seditious tumults, and seditious fames, differ no more but as brother and sister, masculine and feminine; especially if it come to that, that the best actions of a state, and the most plausible, and which ought to give greatest contentment, are taken in ill sense, and traduced: for that shows the envy great, as Tacitus saith; conflata magna invidia, seu bene seu male gesta premunt. Neither doth it follow, that because these fames are a sign of troubles, that the suppressing of them with too much severity, should be a remedy of troubles. For the despising of them, many times checks them best; and the going about to stop them, doth but make a wonder long-lived. Also that kind of obedience, which Tacitus speaketh of, is to be held suspected: Erant in officio, sed tamen qui mallent mandata imperantium interpretari quam exequi; disputing, excusing, cavilling upon mandates and directions, is a kind of shaking off the yoke, and assay of disobedience; especially if in those disputings, they which are for the direction, speak fearfully and tenderly, and those that are against it, audaciously.

Also, as Machiavel noteth well, when princes, that ought to be common parents, make themselves as a party, and lean to a side, it is as a boat, that is overthrown by uneven weight on the one side; as was well seen, in the time of Henry the Third of France; for first, himself entered league for the extirpation of the Protestants; and presently after, the same league was turned upon himself. For when the authority of princes, is made but an accessory to a cause, and that there be other bands, that tie faster than the band of sovereignty, kings begin to be put almost out of possession.

Also, when discords, and quarrels, and factions are carried openly and audaciously, it is a sign the reverence of government is lost. For the motions of the greatest persons in a government, ought to be as the motions of the planets under primum mobile; according to the old opinion: which is, that every of them, is carried swiftly by the highest motion, and softly in their own motion. And therefore, when great ones in their own particular motion, move violently, and, as Tacitus expresseth it well, liberius quam ut imperantium meminissent; it is a sign the orbs are out of frame. For reverence is that, wherewith princes are girt from God; who threateneth the dissolving thereof; Solvam cingula regum.

So when any of the four pillars of government, are mainly shaken, or weakened (which are religion, justice, counsel, and treasure), men had need to pray for fair weather. But let us pass from this part of predictions (concerning which, nevertheless, more light may be taken from that which followeth); and let us speak first, of the materials of seditions; then of the motives of them; and thirdly of the remedies.

Concerning the materials of seditions. It is a thing well to be considered; for the surest way to prevent seditions (if the times do bear it) is to take away the matter of them. For if there be fuel prepared, it is hard to tell, whence the spark shall come, that shall set it on fire. The matter of seditions is of two kinds: much poverty, and much discontentment. It is certain, so many overthrown estates, so many votes for troubles. Lucan noteth well the state of Rome before the Civil War,

Hinc usura vorax, rapidumque in tempore foenus,   Hinc concussa fides, et multis utile bellum.

This same multis utile bellum, is an assured and infallible sign, of a state disposed to seditions and troubles. And if this poverty and broken estate in the better sort, be joined with a want and necessity in the mean people, the danger is imminent and great. For the rebellions of the belly are the worst. As for discontentments, they are, in the politic body, like to humors in the natural, which are apt to gather a preternatural heat, and to inflame. And let no prince measure the danger of them by this, whether they be just or unjust: for that were to imagine people, to be too reasonable; who do often spurn at their own good: nor yet by this, whether the griefs whereupon they rise, be in fact great or small: for they are the most dangerous discontentments, where the fear is greater than the feeling. Dolendi modus, timendi non item. Besides, in great oppressions, the same things that provoke the patience, do withal mate the courage; but in fears it is not so. Neither let any prince, or state, be secure concerning discontentments, because they have been often, or have been long, and yet no peril hath ensued: for as it is true, that every vapor or fume doth not turn into a storm; so it is nevertheless true, that storms, though they blow over divers times, yet may fall at last; and, as the Spanish proverb noteth well, The cord breaketh at the last by the weakest pull.

The causes and motives of seditions are, innovation in religion; taxes; alteration of laws and customs; breaking of privileges; general oppression; advancement of unworthy persons; strangers; dearths; disbanded soldiers; factions grown desperate; and what soever, in offending people, joineth and knitteth them in a common cause.

For the remedies; there may be some general preservatives, whereof we will speak: as for the just cure, it must answer to the particular disease; and so be left to counsel, rather than rule.

The first remedy or prevention is to remove, by all means possible, that material cause of sedition whereof we spake; which is, want and poverty in the estate. To which purpose serveth the opening, and well-balancing of trade; the cherishing of manufactures; the banishing of idleness; the repressing of waste, and excess, by sumptuary laws; the improvement and husbanding of the soil; the regulating of prices of things vendible; the moderating of taxes and tributes; and the like. Generally, it is to be foreseen that the population of a kingdom (especially if it be not mown down by wars) do not exceed the stock of the kingdom, which should maintain them. Neither is the population to be reckoned only by number; for a smaller number, that spend more and earn less, do wear out an estate sooner, than a greater number that live lower, and gather more. Therefore the multiplying of nobility, and other degrees of quality, in an over proportion to the common people, doth speedily bring a state to necessity; and so doth likewise an overgrown clergy; for they bring nothing to the stock; and in like manner, when more are bred scholars, than preferments can take off.

It is likewise to be remembered, that forasmuch as the increase of any estate must be upon the foreigner (for whatsoever is somewhere gotten, is somewhere lost), there be but three things, which one nation selleth unto another; the commodity as nature yieldeth it; the manufacture; and the vecture, or carriage. So that if these three wheels go, wealth will flow as in a spring tide. And it cometh many times to pass, that materiam superabit opus; that the work and carriage is more worth than the material, and enricheth a state more; as is notably seen in the Low-Countrymen, who have the best mines above ground, in the world.

Above all things, good policy is to be used, that the treasure and moneys, in a state, be not gathered into few hands. For otherwise a state may have a great stock, and yet starve. And money is like muck, not good except it be spread. This is done, chiefly by suppressing, or at least keeping a strait hand, upon the devouring trades of usury, ingrossing great pasturages, and the like.

For removing discontentments, or at least the danger of them; there is in every state (as we know) two portions of subjects; the noblesse and the commonalty. When one of these is discontent, the danger is not great; for common people are of slow motion, if they be not excited by the greater sort; and the greater sort are of small strength, except the multitude be apt, and ready to move of themselves. Then is the danger, when the greater sort, do but wait for the troubling of the waters amongst the meaner, that then they may declare themselves. The poets feign, that the rest of the gods would have bound Jupiter; which he hearing of, by the counsel of Pallas, sent for Briareus, with his hundred hands, to come in to his aid. An emblem, no doubt, to show how safe it is for monarchs, to make sure of the good will of common people. To give moderate liberty for griefs and discontentments to evaporate (so it be without too great insolency or bravery), is a safe way. For he that turneth the humors back, and maketh the wound bleed inwards, endangereth malign ulcers, and pernicious imposthumations.

The part of Epimetheus mought well become Prometheus, in the case of discontentments: for there is not a better provision against them. Epimetheus, when griefs and evils flew abroad, at last shut the lid, and kept hope in the bottom of the vessel. Certainly, the politic and artificial nourishing, and entertaining of hopes, and carrying men from hopes to hopes, is one of the best antidotes against the poison of discontentments. And it is a certain sign of a wise government and proceeding, when it can hold men's hearts by hopes, when it cannot by satisfaction; and when it can handle things, in such manner, as no evil shall appear so peremptory, but that it hath some outlet of hope; which is the less hard to do, because both particular persons and factions, are apt enough to flatter themselves, or at least to brave that, which they believe not.

Also the foresight and prevention, that there be no likely or fit head, whereunto discontented persons may resort, and under whom they may join, is a known, but an excellent point of caution. I understand a fit head, to be one that hath greatness and reputation; that hath confidence with the discontented party, and upon whom they turn their eyes; and that is thought discontented, in his own particular: which kind of persons, are either to be won, and reconciled to the state, and that in a fast and true manner; or to be fronted with some other, of the same party, that may oppose them, and so divide the reputation. Generally, the dividing and breaking, of all factions and combinations that are adverse to the state, and setting them at distance, or at least distrust, amongst themselves, is not one of the worst remedies. For it is a desperate case, if those that hold with the proceeding of the state, be full of discord and faction, and those that are against it, be entire and united.

I have noted, that some witty and sharp speeches, which have fallen from princes, have given fire to seditions. Caesar did himself infinite hurt in that speech, *Sylla nescivit literas, non potuit dictare*; for it did utterly cut off that hope, which men had entertained, that he would at one time or other give over his dictatorship. Galba undid himself by that speech, *legi a se militem, non emi*; for it put the soldiers out of hope of the donative. Probus likewise, by that speech, *Si vixero, non opus erit amplius Romano imperio militibus*; a speech of great despair for the soldiers. And many the like. Surely princes had need, in tender matters and ticklish times, to beware what they say; especially in these short speeches, which fly abroad like darts, and are thought to be shot out of their secret intentions. For as for large discourses, they are flat things, and not so much noted.

Lastly, let princes, against all events, not be without some great person, one or rather more, of military valor, near unto them, for the repressing of seditions in their beginnings. For without that, there useth to be more trepidation in court upon the first breaking out of troubles, than were fit. And the state runneth the danger of that which Tacitus saith; *Atque is habitus animorum fuit, ut pessimum facinus auderent pauci, plures vellent, omnes paterentur*. But let such military persons be assured, and well reputed of, rather than factious and popular; holding also good correspondence with the other great men in the state; or else the remedy, is worse than the disease.

Of Atheism

I HAD rather believe all the fables in the Legend, and the Talmud, and the Alcoran, than that this universal frame is without a mind. And therefore, God never wrought miracle, to convince atheism, because his ordinary works convince it. It is true, that a little philosophy inclineth man's mind to atheism; but depth in philosophy bringeth men's minds about to religion. For while the mind of man looketh upon second causes scattered, it may sometimes rest in them, and go no further; but when it beholdeth the chain of them, confederate and linked together, it must needs fly to Providence and Deity. Nay, even that school which is most accused of atheism doth most demonstrate religion; that is, the school of Leucippus and Democritus and Epicurus. For it is a thousand times more credible, that four mutable elements, and one immutable fifth essence, duly and eternally placed, need no God, than that an army of infinite small portions, or seeds unplaced, should have produced this order and beauty, without a divine marshal. The Scripture saith, The fool hath said in his heart, there is no God; it is not said, The fool hath thought in his heart; so as he rather saith it, by rote to himself, as that he would have, than that he can thoroughly believe it, or be persuaded of it. For none deny, there is a God, but those, for whom it maketh that there were no God. It appeareth in nothing more, that atheism is rather in the lip, than in the heart of man, than by this; that atheists will ever be talking of that their opinion, as if they fainted in it, within themselves, and would be glad to be strengthened, by the consent of others. Nay more, you shall have atheists strive to get disciples, as it fareth with other sects. And, which is most of all, you shall have of them, that will suffer for atheism, and not recant; whereas if they did truly think, that there were no such thing as God, why should they trouble themselves? Epicurus is charged, that he did but dissemble for his credit's sake, when he affirmed there were blessed natures, but such as enjoyed themselves, without having respect to the government of the world. Wherein they say he did temporize; though in secret, he thought there was no God. But certainly he is traduced; for his words are noble and divine: Non deos vulgi negare profanum; sed vulgi opiniones diis applicare profanum. Plato could have said no more. And although he had the confidence, to deny the administration, he had not the power, to deny the nature. The Indians of the West, have names for their particular gods, though they have no name for God: as if the heathens should have had the names Jupiter, Apollo, Mars, etc., but not the word Deus; which shows that even those barbarous people have the notion, though they have not the latitude and extent of it. So that against atheists, the very savages take part, with the very subtlest philosophers. The contemplative atheist is rare: a Diagoras, a Bion, a Lucian perhaps, and some others; and yet they seem to be more than they are; for that all that impugn a received religion, or superstition, are by the adverse part branded with the name of atheists. But the great atheists, indeed are hypocrites; which are ever handling holy things, but without feeling; so as they must needs be cauterized in the end. The causes of atheism are: divisions in religion, if they be many; for any one main division, addeth zeal to both sides; but many divisions introduce atheism. Another is, scandal of priests; when it is come to that which St. Bernard saith, non est jam dicere, ut populus sic sacerdos; quia nec sic populus ut sacerdos. A third is, custom of profane scoffing in holy matters; which doth, by little and little, deface the reverence of religion. And lastly, learned times, specially with peace and prosperity; for troubles and adversities do more bow men's minds to religion. They that deny a God, destroy man's nobility; for certainly man is of kin to the beasts, by his body; and, if he be not of kin to God, by his spirit, he is a base and ignoble creature. It destroys likewise magnanimity, and the raising of human nature; for take an example of a dog, and mark what a generosity and courage he will put on, when he finds himself maintained by a man; who to him is instead of a God, or melior natura; which courage is manifestly such, as that creature, without that confidence of a better nature than his own, could never attain. So man, when he resteth and assureth himself, upon divine protection and favor, gathered a force and faith, which human nature in itself could not obtain. Therefore, as atheism is in all respects hateful, so in this, that it depriveth human nature of the means to exalt itself, above human frailty. As it is in particular persons, so it is in nations. Never was there such a state for magnanimity as Rome. Of this state hear what Cicero saith: Quam volumus licet, patres conscripti, nos amemus, tamen nec numero Hispanos, nec robore Gallos, nec calliditate Poenos, nec artibus Graecos, nec denique hoc ipso hujus gentis et terrae domestico nativoque sensu Italos ipsos et Latinos; sed pietate, ac religione, atque hac una sapientia, quod deorum immortalium numine omnia regi gubernarique perspeximus, omnes gentes nationesque superavimus.

Of Superstition

IT WERE better to have no opinion of God at all, than such an opinion, as is unworthy of him. For the one is unbelief, the other is contumely; and certainly superstition is the reproach of the Deity. Plutarch saith well to that purpose: Surely (saith he) I had rather a great deal, men should say, there was no such man at all, as Plutarch, than that they should say, that there was one Plutarch, that would eat his children as soon as they were born; as the poets speak of Saturn. And as the contumely is greater towards God, so the danger is greater towards men. Atheism leaves a man to sense, to philosophy, to natural piety, to laws, to reputation; all which may be guides to an outward moral virtue, though religion were not; but superstition dismounts all these, and erecteth an absolute monarchy, in the minds of men. Therefore atheism did never perturb states; for it makes men wary of themselves, as looking no further: and we see the times inclined to atheism (as the time of Augustus Caesar) were civil times. But superstition hath been the confusion of many states, and bringeth in a new primum mobile, that ravisheth all the spheres of government. The master of superstition, is the people; and in all superstition, wise men follow fools; and arguments are fitted to practice, in a reversed order. It was gravely said by some of the prelates in the Council of Trent, where the doctrine of the Schoolmen bare great sway, that the Schoolmen were like astronomers, which did feign eccentrics and epicycles, and such engines of orbs, to save the phenomena; though they knew there were no such things; and in like manner, that the Schoolmen had framed a number of subtle and intricate axioms, and theorems, to save the practice of the church. The causes of superstition are: pleasing and sensual rites and ceremonies; excess of outward and pharisaical holiness; overgreat reverence of traditions, which cannot but load the church; the stratagems of prelates, for their own ambition and lucre; the favoring too much of good intentions, which openeth the gate to conceits and novelties; the taking an aim at divine matters, by human, which cannot but breed mixture of imaginations: and, lastly, barbarous times, especially joined with calamities and disasters. Superstition, without a veil, is a deformed thing; for, as it addeth deformity to an ape, to be so like a man, so the similitude of superstition to religion, makes it the more deformed. And as wholesome meat corrupteth to little worms, so good forms and orders corrupt, into a number of petty observances. There is a superstition in avoiding superstition, when men think to do best, if they go furthest from the superstition, formerly received; therefore care would be had that (as it fareth in ill purgings) the good be not taken away with the bad; which commonly is done, when the people is the reformer.

Of Travel

TRAVEL, in the younger sort, is a part of education, in the elder, a part of experience. He that travelleth into a country, before he hath some entrance into the language, goeth to school, and not to travel. That young men travel under some tutor, or grave servant, I allow well; so that he be such a one that hath the language, and hath been in the country before; whereby he may be able to tell them what things are worthy to be seen, in the country where they go; what acquaintances they are to seek; what exercises, or discipline, the place yieldeth. For else, young men shall go hooded, and look abroad little. It is a strange thing, that in sea voyages, where there is nothing to be seen, but sky and sea, men should make diaries; but in land-travel, wherein so much is to be observed, for the most part they omit it; as if chance were fitter to be registered, than observation. Let diaries, therefore, be brought in use. The things to be seen and observed are: the courts of princes, especially when they give audience to ambassadors; the courts of justice, while they sit and hear causes; and so of consistories ecclesiastic; the churches and monasteries, with the monuments which are therein extant; the walls and fortifications of cities, and towns, and so the heavens and harbors; antiquities and ruins; libraries; colleges, disputations, and lectures, where any are; shipping and navies; houses and gardens of state and pleasure, near great cities; armories; arsenals; magazines; exchanges; burses; warehouses; exercises of horsemanship, fencing, training of soldiers, and the like; comedies, such whereunto the better sort of persons do resort; treasuries of jewels and robes; cabinets and rarities; and, to conclude, whatsoever is memorable, in the places where they go. After all which, the tutors, or servants, ought to make diligent inquiry. As for triumphs, masks, feasts, weddings, funerals, capital executions, and such shows, men need not to be put in mind of them; yet are they not to be neglected. If you will have a young man to put his travel into a little room, and in short time to gather much, this you must do. First, as was said, he must have some entrance into the language before he goeth. Then he must have such a servant, or tutor, as knoweth the country, as was likewise said. Let him carry with him also, some card or book, describing the country where he travelleth; which will be a good key to his inquiry. Let him keep also a diary. Let him not stay long, in one city or town; more or less as the place deserveth, but not long; nay, when he stayeth in one city or town, let him change his lodging from one end and part of the town, to another; which is a great adamant of acquaintance. Let him sequester himself, from the company of his countrymen, and diet in such places, where there is good company of the nation where he travelleth. Let him, upon his removes from one place to another, procure recommendation to some person of quality, residing in the place whither he removeth; that he may use his favor, in those things he desireth to see or know. Thus he may abridge his travel, with much profit. As for the acquaintance, which is to be sought in travel; that which is most of all profitable, is acquaintance with the secretaries and employed men of ambassadors: for so in travelling in one country, he shall suck the experience of many. Let him also see, and visit, eminent persons in all kinds, which are of great name abroad; that he may be able to tell, how the life agreeth with the fame. For quarrels, they are with care and discretion to be avoided. They are commonly for mistresses, healths, place, and words. And let a man beware, how he keepeth company with choleric and quarrelsome persons; for they will engage him into their own quarrels. When a traveller returneth home, let him not leave the countries, where he hath travelled, altogether behind him; but maintain a correspondence by letters, with those of his acquaintance, which are of most worth. And let his travel appear rather in his discourse, than his apparel or gesture; and in his discourse, let him be rather advised in his answers, than forward to tell stories; and let it appear that he doth not change his country manners, for those of foreign parts; but only prick in some flowers, of that he hath learned abroad, into the customs of his own country.

Of Empire

IT IS a miserable state of mind, to have few things to desire, and many things to fear; and yet that commonly is the case of kings; who, being at the highest, want matter of desire, which makes their minds more languishing; and have many representations of perils and shadows, which makes their minds the less clear. And this is one reason also, of that effect which the Scripture speaketh of, That the king's heart is inscrutable. For multitude of jealousies, and lack of some predominant desire, that should marshal and put in order all the rest, maketh any man's heart, hard to find or sound. Hence it comes likewise, that princes many times make themselves desires, and set their hearts upon toys; sometimes upon a building; sometimes upon erecting of an order; sometimes upon the advancing of a person; sometimes upon obtaining excellency in some art, or feat of the hand; as Nero for playing on the harp, Domitian for certainty of the hand with the arrow, Commodus for playing at fence, Caracalla for driving chariots, and the like. This seemeth incredible, unto those that know not the principle, that the mind of man, is more cheered and refreshed by profiting in small things, than by standing at a stay, in great. We see also that kings that have been fortunate conquerors, in their first years, it being not possible for them to go forward infinitely, but that they must have some check, or arrest in their fortunes, turn in their latter years to be superstitious, and melancholy; as did Alexander the Great; Diocletian; and in our memory, Charles the Fifth; and others: for he that is used to go forward, and findeth a stop, falleth out of his own favor, and is not the thing he was.

To speak now of the true temper of empire, it is a thing rare and hard to keep; for both temper, and distemper, consist of contraries. But it is one thing, to mingle contraries, another to interchange them. The answer of Apollonius to Vespasian, is full of excellent instruction. Vespasian asked him, What was Nero's overthrow? He answered, Nero could touch and tune the harp well; but in government, sometimes he used to wind the pins too high, sometimes to let them down too low. And certain it is, that nothing destroyeth authority so much, as the unequal and untimely interchange of power pressed too far, and relaxed too much.

This is true, that the wisdom of all these latter times, in princes' affairs, is rather fine deliveries, and shiftings of dangers and mischiefs, when they are near, than solid and grounded courses to keep them aloof. But this is but to try masteries with fortune. And let men beware, how they neglect and suffer matter of trouble to be prepared; for no man can forbid the spark, nor tell whence it may come. The difficulties in princes' business are many and great; but the greatest difficulty, is often in their own mind. For it is common with princes (saith Tacitus) to will contradictories, Sunt plerumque regum voluntates vehementes, et inter se contrariae. For it is the solecism of power, to think to command the end, and yet not to endure the mean.

Kings have to deal with their neighbors, their wives, their children, their prelates or clergy, their nobles, their second-nobles or gentlemen, their merchants, their commons, and their men of war; and from all these arise dangers, if care and circumspection be not used.

First for their neighbors; there can no general rule be given (for occasions are so variable), save one, which ever holdeth, which is, that princes do keep due sentinel, that none of their neighbors do ever grow so (by increase of territory, by embracing of trade, by approaches, or the like), as they become more able to annoy them, than they were. And this is generally the work of standing counsels, to foresee and to hinder it. During that triumvirate of kings, King Henry the Eighth of England, Francis the First King of France, and Charles the Fifth Emperor, there was such a watch kept, that none of the three could win a palm of ground, but the other two would straightways balance it, either by confederation, or, if need were, by a war; and would not in any wise take up peace at interest. And the like was done by that league (which Guicciardini saith was the security of Italy) made between Ferdinando King of Naples, Lorenzius Medici, and Ludovicus Sforza, potentates, the one of Florence, the other of Milan. Neither is the opinion of some of the Schoolmen, to be received, that a war cannot justly be made, but upon a precedent injury or provocation. For there is no question, but a just fear of an imminent danger, though there be no blow given, is a lawful cause of a war.

For their wives; there are cruel examples of them. Livia is infamed, for the poisoning of her husband; Roxalana, Solyman's wife, was the destruction of that renowned prince, Sultan Mustapha, and otherwise troubled his house and succession; Edward the Second of England, his queen, had the principal hand in the deposing and murder of her husband. This kind of danger, is then to be feared chiefly, when the wives have plots, for the raising of their own children; or else that they be advoutresses.

For their children; the tragedies likewise of dangers from them, have been many. And generally, the entering of fathers into suspicion of their children, hath been ever unfortunate. The destruction of Mustapha (that we named before) was so fatal to Solyman's line, as the succession of the Turks, from Solyman until this day, is suspected to be untrue, and of strange blood; for that Selymus the Second, was thought to be suppositious. The destruction of Crispus, a young prince of rare towardness, by Constantinus the Great, his father, was in like manner fatal to his house; for both Constantinus and Constance, his sons, died violent deaths; and Constantius, his other son, did little better; who died indeed of sickness, but after that Julianus had taken arms against him. The destruction of Demetrius, son to Philip the Second of Macedon, turned upon the father, who died of repentance. And many like examples there are; but few or none, where the fathers had good by such distrust; except it were, where the sons were up in open arms against them; as was Selymus the First against Bajazet; and the three sons of Henry the Second, King of England.

For their prelates; when they are proud and great, there is also danger from them; as it was in the times of Anselmus, and Thomas Becket, Archbishops of Canterbury; who, with their croziers, did almost try it with the king's sword; and yet they had to deal with stout and haughty kings, William Rufus, Henry the First, and Henry the Second. The danger is not from that state, but where it hath a dependence of foreign authority; or where the churchmen come in and are elected, not by the collation of the king, or particular patrons, but by the people.

For their nobles; to keep them at a distance, it is not amiss; but to depress them, may make a king more absolute, but less safe; and less able to perform, any thing that he desires. I have noted it, in my History of King Henry the Seventh of England, who depressed his nobility; whereupon it came to pass, that his times were full of difficulties and troubles; for the nobility, though they continued loyal unto him, yet did they not co-operate with him in his business. So that in effect, he was fain to do all things himself.

For their second-nobles; there is not much danger from them, being a body dispersed. They may sometimes discourse high, but that doth little hurt; besides, they are a counterpoise to the higher nobility, that they grow not too potent; and, lastly, being the most immediate in authority, with the common people, they do best temper popular commotions.

For their merchants; they are vena porta; and if they flourish not, a kingdom may have good limbs, but will have empty veins, and nourish little. Taxes and imposts upon them, do seldom good to the king's revenue; for that that he wins in the hundred, he leeseth in the shire; the particular rates being increased, but the total bulk of trading, rather decreased.

For their commons; there is little danger from them, except it be, where they have great and potent heads; or where you meddle with the point of religion, or their customs, or means of life.

For their men of war; it is a dangerous state, where they live and remain in a body, and are used to donatives; whereof we see examples in the janizaries, and pretorian bands of Rome; but trainings of men, and arming them in several places, and under several commanders, and without donatives, are things of defence, and no danger.

Princes are like to heavenly bodies, which cause good or evil times; and which have much veneration, but no rest. All precepts concerning kings, are in effect comprehended in those two remembrances: memento quod es homo; and memento quod es Deus, or vice Dei; the one bridleth their power, and the other their will.

Of Counsel

THE greatest trust, between man and man, is the trust of giving counsel. For in other confidences, men commit the parts of life; their lands, their goods, their children, their credit, some particular affair; but to such as they make their counsellors, they commit the whole: by how much the more, they are obliged to all faith and integrity. The wisest princes need not think it any diminution to their greatness, or derogation to their sufficiency, to rely upon counsel. God himself is not without, but hath made it one of the great names of his blessed Son: The Counsellor. Solomon hath pronounced, that in counsel is stability. Things will have their first, or second agitation: if they be not tossed upon the arguments of counsel, they will be tossed upon the waves of fortune; and be full of inconstancy, doing and undoing, like the reeling of a drunken man. Solomon's son found the force of counsel, as his father saw the necessity of it. For the beloved kingdom of God, was first rent, and broken, by ill counsel; upon which counsel, there are set for our instruction, the two marks whereby bad counsel is for ever best discerned; that it was young counsel, for the person; and violent counsel, for the matter.

The ancient times, do set forth in figure, both the incorporation, and inseparable conjunction, of counsel with kings, and the wise and politic use of counsel by kings: the one, in that they say Jupiter did marry Metis, which signifieth counsel; whereby they intend that Sovereignty, is married to Counsel: the other in that which followeth, which was thus: They say, after Jupiter was married to Metis, she conceived by him, and was with child, but Jupiter suffered her not to stay, till she brought forth, but eat her up; whereby he became himself with child, and was delivered of Pallas armed, out of his head. Which monstrous fable containeth a secret of empire; how kings are to make use of their counsel of state. That first, they ought to refer matters unto them, which is the first begetting, or impregnation; but when they are elaborate, moulded, and shaped in the womb of their counsel, and grow ripe, and ready to be brought forth, that then they suffer not their counsel to go through with the resolution and direction, as if it depended on them; but take the matter back into their own hands, and make it appear to the world, that the decrees and final directions (which, because they come forth, with prudence and power, are resembled to Pallas armed) proceeded from themselves; and not only from their authority, but (the more to add reputation to themselves) from their head and device.

Let us now speak of the inconveniences of counsel, and of the remedies. The inconveniences that have been noted, in calling and using counsel, are three. First, the revealing of affairs, whereby they become less secret. Secondly, the weakening of the authority of princes, as if they were less of themselves. Thirdly, the danger of being unfaithfully counselled, and more for the good of them that counsel, than of him that is counselled. For which inconveniences, the doctrine of Italy, and practice of France, in some kings' times, hath introduced cabinet counsels; a remedy worse than the disease.

As to secrecy; princes are not bound to communicate all matters, with all counsellors; but may extract and select. Neither is it necessary, that he that consulteth what he should do, should declare what he will do. But let princes beware, that the unsecreting of their affairs, comes not from themselves. And as for cabinet counsels, it may be their motto, plenus rimarum sum: one futile person, that maketh it his glory to tell, will do more hurt than many, that know it their duty to conceal. It is true there be some affairs, which require extreme secrecy, which will hardly go beyond one or two persons, besides the king: neither are those counsels unprosperous; for, besides the secrecy, they commonly go on constantly, in one spirit of direction, without distraction. But then it must be a prudent king, such as is able to grind with a handmill; and those inward counsellors had need also be wise men, and especially true and trusty to the king's ends; as it was with King Henry the Seventh of England, who, in his great business, imparted himself to none, except it were to Morton and Fox.

For weakening of authority; the fable showeth the remedy. Nay, the majesty of kings, is rather exalted than diminished, when they are in the chair of counsel; neither was there ever prince, bereaved of his dependences, by his counsel, except where there hath been, either an over-greatness in one counsellor, or an over-strict combination in divers; which are things soon found, and holpen.

For the last inconvenience, that men will counsel, with an eye to themselves; certainly, non inveniet fidem super terram is meant, of the nature of times, and not of all particular persons. There be, that are in nature faithful, and sincere, and plain, and direct; not crafty and involved; let princes, above all, draw to themselves such natures. Besides, counsellors are not commonly so united, but that one counsellor, keepeth sentinel over another; so that if any do counsel out of faction or private ends, it commonly comes to the king's ear. But the best remedy is, if princes know their counsellors, as well as their counsellors know them:

> *Principis est virtus maxima nosse suos.*

And on the other side, counsellors should not be too speculative into their sovereign's person. The true composition of a counsellor, is rather to be skilful in their master's business, than in his nature; for then he is like to advise him, and not feed his humor. It is of singular use to princes, if they take the opinions of their counsel, both separately and together. For private opinion is more free; but opinion before others, is more reverent. In private, men are more bold in their own humors; and in consort, men are more obnoxious to.others' humors; therefore it is good to take both; and of the inferior sort, rather in private, to preserve freedom; of the greater, rather in consort, to preserve respect. It is in vain for princes, to take counsel concerning matters, if they take no counsel likewise concerning persons; for all matters are as dead images; and the life of the execution of affairs, resteth in the good choice of persons. Neither is it enough, to consult concerning persons secundum genera, as in an idea, or mathematical description, what the kind and character of the person should be; for the greatest errors are committed, and the most judgment is shown, in the choice of individuals. It was truly said, optimi consiliarii mortui: books will speak plain, when counsellors blanch. Therefore it is good to be conversant in them, specially the books of such as themselves have been actors upon the stage.

The counsels at this day, in most places, are but familiar meetings, where matters are rather talked on, than debated. And they run too swift, to the order, or act, of counsel. It were better that in causes of weight, the matter were propounded one day, and not spoken to till the next day; in nocte consilium. So was it done in the Commission of Union, between England and Scotland; which was a grave and orderly assembly. I commend set days for petitions; for both it gives the suitors more certainty for their attendance, and it frees the meetings for matters of estate, that they may hoc agere. In choice of committees; for ripening business for the counsel, it is better to choose indifferent persons, than to make an indifferency, by putting in those, that are strong on both sides. I commend also standing commissions; as for trade, for treasure, for war, for suits, for some provinces; for where there be divers particular counsels, and but one counsel of estate (as it is in Spain), they are, in effect, no more than standing commissions: save that they have greater authority. Let such as are to inform counsels, out of their particular professions (as lawyers, seamen, mintmen, and the like) be first heard before committees; and then, as occasion serves, before the counsel. And let them not come in multitudes, or in a tribunitious manner; for that is to clamor counsels, not to inform them. A long table and a square table, or seats about the walls, seem things of form, but are things of substance; for at a long table a few at the upper end, in effect, sway all the business; but in the other form, there is more use of the counsellors' opinions, that sit lower. A king, when he presides in counsel, let him beware how he opens his own inclination too much, in that which he propoundeth; for else counsellors will but take the wind of him, and instead of giving free counsel, sing him a song of placebo.

Of Delays

FORTUNE is like the market; where many times if you can stay a little, the price will fall. Again, it is sometimes like Sibylla's offer; which at first, offereth the commodity at full, then consumeth part and part, and still holdeth up the price. For occasion (as it is in the common verse) turneth a bald noddle, after she hath presented her locks in front, and no hold taken; or at least turneth the handle of the bottle, first to be received, and after the belly, which is hard to clasp. There is surely no greater wisdom, than well to time the beginnings, and onsets, of things. Dangers are no more light, if they once seem light; and more dangers have deceived men, than forced them. Nay, it were better, to meet some dangers half way, though they come nothing near, than to keep too long a watch upon their approaches; for if a man watch too long, it is odds he will fall asleep. On the other side, to be deceived with too long shadows (as some have been, when the moon was low, and shone on their enemies' back), and so to shoot off before the time; or to teach dangers to come on, by over early buckling towards them; is another extreme. The ripeness, or unripeness, of the occasion (as we said) must ever be well weighed; and generally it is good, to commit the beginnings of all great actions to Argus, with his hundred eyes, and the ends to Briareus, with his hundred hands; first to watch, and then to speed. For the helmet of Pluto, which maketh the politic man go invisible, is secrecy in the counsel, and celerity in the execution. For when things are once come to the execution, there is no secrecy, comparable to celerity; like the motion of a bullet in the air, which flieth so swift, as it outruns the eye.

Of Cunning

WE TAKE cunning for a sinister or crooked wisdom. And certainly there is a great difference, between a cunning man, and a wise man; not only in point of honesty, but in point of ability. There be, that can pack the cards, and yet cannot play well; so there are some that are good in canvasses and factions, that are otherwise weak men. Again, it is one thing to understand persons, and another thing to understand matters; for many are perfect in men's humors, that are not greatly capable of the real part of business; which is the constitution of one that hath studied men, more than books. Such men are fitter for practice, than for counsel; and they are good, but in their own alley: turn them to new men, and they have lost their aim; so as the old rule, to know a fool from a wise man, Mitte ambos nudos ad ignotos, et videbis, doth scarce hold for them. And because these cunning men, are like haberdashers of small wares, it is not amiss to set forth their shop.

It is a point of cunning, to wait upon him with whom you speak, with your eye; as the Jesuits give it in precept: for there be many wise men, that have secret hearts, and transparent countenances. Yet this would be done with a demure abasing of your eye, sometimes, as the Jesuits also do use.

Another is, that when you have anything to obtain, of present despatch, you entertain and amuse the party, with whom you deal, with some other discourse; that he be not too much awake to make objections. I knew a counsellor and secretary, that never came to Queen Elizabeth of England, with bills to sign, but he would always first put her into some discourse of estate, that she mought the less mind the bills.

The like surprise may be made by moving things, when the party is in haste, and cannot stay to consider advisedly of that is moved.

If a man would cross a business, that he doubts some other would handsomely and effectually move, let him pretend to wish it well, and move it himself in such sort as may foil it.

The breaking off, in the midst of that one was about to say, as if he took himself up, breeds a greater appetite in him with whom you confer, to know more.

And because it works better, when anything seemeth to be gotten from you by question, than if you offer it of yourself, you may lay a bait for a question, by showing another visage, and countenance, than you are wont; to the end to give occasion, for the party to ask, what the matter is of the change? As Nehemias did; And I had not before that time, been sad before the king.

In things that are tender and unpleasing, it is good to break the ice, by some whose words are of less weight, and to reserve the more weighty voice, to come in as by chance, so that he may be asked the question upon the other's speech: as Narcissus did, relating to Claudius the marriage of Messalina and Silius.

In things that a man would not be seen in himself, it is a point of cunning, to borrow the name of the world; as to say, The world says, or There is a speech abroad.

I knew one that, when he wrote a letter, he would put that, which was most material, in the postscript, as if it had been a by-matter.

I knew another that, when he came to have speech, he would pass over that, that he intended most; and go forth, and come back again, and speak of it as of a thing, that he had almost forgot.

Some procure themselves, to be surprised, at such times as it is like the party that they work upon, will suddenly come upon them; and to be found with a letter in their hand, or doing somewhat which they are not accustomed; to the end, they may be apposed of those things, which of themselves they are desirous to utter.

It is a point of cunning, to let fall those words in a man's own name, which he would have another man learn, and use, and thereupon take advantage. I knew two, that were competitors for the secretary's place in Queen Elizabeth's time, and yet kept good quarter between themselves; and would confer, one with another, upon the business; and the one of them said, That to be a secretary, in the declination of a monarchy, was a ticklish thing, and that he did not affect it: the other straight caught up those words, and discoursed with divers of his friends, that he had no reason to desire to be secretary, in the declination of a monarchy. The first man took hold of it, and found means it was told the Queen; who, hearing of a declination of a monarchy, took it so ill, as she would never after hear of the other's suit.

There is a cunning, which we in England call, the turning of the cat in the pan; which is, when that which a man says to another, he lays it as if another had said it to him. And to say truth, it is not easy, when such a matter passed between two, to make it appear from which of them it first moved and began.

It is a way that some men have, to glance and dart at others, by justifying themselves by negatives; as to say, This I do not; as Tigellinus did towards Burrhus, Se non diversas spes, sed incolumitatem imperatoris simpliciter spectare.

Some have in readiness so many tales and stories, as there is nothing they would insinuate, but they can wrap it into a tale; which serveth both to keep themselves more in guard, and to make others carry it with more pleasure. It is a good point of cunning, for a man to shape the answer he would have, in his own words and propositions; for it makes the other party stick the less.

It is strange how long some men will lie in wait to speak somewhat they desire to say; and how far about they will fetch; and how many other matters they will beat over, to come near it. It is a thing of great patience, but yet of much use.

A sudden, bold, and unexpected question doth many times surprise a man, and lay him open. Like to him that, having changed his name, and walking in Paul's, another suddenly came behind him, and called him by his true name, whereat straightways he looked back.

But these small wares, and petty points, of cunning, are infinite; and it were a good deed to make a list of them; for that nothing doth more hurt in a state, than that cunning men pass for wise.

But certainly some there are that know the resorts and falls of business, that cannot sink into the main of it; like a house that hath convenient stairs and entries, but never a fair room. Therefore, you shall see them find out pretty looses in the conclusion, but are no ways able to examine or debate matters. And yet commonly they take advantage of their inability, and would be thought wits of direction. Some build rather upon the abusing of others, and (as we now say) putting tricks upon them, than upon soundness of their own proceedings. But Solomon saith, Prudens advertit ad gressus suos; stultus divertit ad dolos.

Of Wisdom For A Man's Self

AN ANT is a wise creature for itself, but it is a shrewd thing, in an orchard or garden. And certainly, men that are great lovers of themselves, waste the public. Divide with reason; between self-love and society; and be so true to thyself, as thou be not false to others; specially to thy king and country. It is a poor centre of a man's actions, himself. It is right earth. For that only stands fast upon his own centre; whereas all things, that have affinity with the heavens, move upon the centre of another, which they benefit. The referring of all to a man's self, is more tolerable in a sovereign prince; because themselves are not only themselves, but their good and evil is at the peril of the public fortune. But it is a desperate evil, in a servant to a prince, or a citizen in a republic. For whatsoever affairs pass such a man's hands, he crooketh them to his own ends; which must needs be often eccentric to the ends of his master, or state. Therefore, let princes, or states, choose such servants, as have not this mark; except they mean their service should be made but the accessory. That which maketh the effect more pernicious, is that all proportion is lost. It were disproportion enough, for the servant's good to be preferred before the master's; but yet it is a greater extreme, when a little good of the servant, shall carry things against a great good of the master's. And yet that is the case of bad officers, treasurers, ambassadors, generals, and other false and corrupt servants; which set a bias upon their bowl, of their own petty ends and envies, to the overthrow of their master's great and important affairs. And for the most part, the good such servants receive, is after the model of their own fortune; but the hurt they sell for that good, is after the model of their master's fortune. And certainly it is the nature of extreme self-lovers, as they will set an house on fire, and it were but to roast their eggs; and yet these men many times hold credit with their masters, because their study is but to please them, and profit themselves; and for either respect, they will abandon the good of their affairs.

Wisdom for a man's self is, in many branches thereof, a depraved thing. It is the wisdom of rats, that will be sure to leave a house, somewhat before it fall. It is the wisdom of the fox, that thrusts out the badger, who digged and made room for him. It is the wisdom of crocodiles, that shed tears when they would devour. But that which is specially to be noted is, that those which (as Cicero says of Pompey) are sui amantes, sine rivali, are many times unfortunate. And whereas they have, all their times, sacrificed to themselves, they become in the end, themselves sacrifices to the inconstancy of fortune, whose wings they thought, by their self-wisdom, to have pinioned.

Of Innovations

AS THE births of living creatures, at first are ill-shapen, so are all innovations, which are the births of time. Yet notwithstanding, as those that first bring honor into their family, are commonly more worthy than most that succeed, so the first precedent (if it be good) is seldom attained by imitation. For ill, to man's nature, as it stands perverted, hath a natural motion, strongest in continuance; but good, as a forced motion, strongest at first. Surely every medicine is an innovation; and he that will not apply new remedies, must expect new evils; for time is the greatest innovator; and if time of course alter things to the worse, and wisdom and counsel shall not alter them to the better, what shall be the end? It is true, that what is settled by custom, though it be not good, yet at least it is fit; and those things which have long gone together, are, as it were, confederate within themselves; whereas new things piece not so well; but though they help by their utility, yet they trouble by their inconformity. Besides, they are like strangers; more admired, and less favored. All this is true, if time stood still; which contrariwise moveth so round, that a froward retention of custom, is as turbulent a thing as an innovation; and they that reverence too much old times, are but a scorn to the new. It were good, therefore, that men in their innovations would follow the example of time itself; which indeed innovateth greatly, but quietly, by degrees scarce to be perceived. For otherwise, whatsoever is new unlooked for; and ever it mends some, and pairs others; and he that is holpen, takes it for a fortune, and thanks the time; and he that is hurt, for a wrong, and imputeth it to the author. It is good also, not to try experiments in states, except the necessity be urgent, or the utility evident; and well to beware, that it be the reformation, that draweth on the change, and not the desire of change, that pretendeth the reformation. And lastly, that the novelty, though it be not rejected, yet be held for a suspect; and, as the Scripture saith, that we make a stand upon the ancient way, and then look about us, and discover what is the straight and right way, and so to walk in it.

Of Dispatch

AFFECTED dispatch is one of the most dangerous things to business that can be. It is like that, which the physicians call predigestion, or hasty digestion; which is sure to fill the body full of crudities, and secret seeds of diseases. Therefore measure not dispatch, by the times of sitting, but by the advancement of the business. And as in races it is not the large stride or high lift that makes the speed; so in business, the keeping close to the matter, and not taking of it too much at once, procureth dispatch. It is the care of some, only to come off speedily for the time; or to contrive some false periods of business, because they may seem men of dispatch. But it is one thing, to abbreviate by contracting, another by cutting off. And business so handled, at several sittings or meetings, goeth commonly backward and forward in an unsteady manner. I knew a wise man that had it for a byword, when he saw men hasten to a conclusion, Stay a little, that we may make an end the sooner.

On the other side, true dispatch is a rich thing. For time is the measure of business, as money is of wares; and business is bought at a dear hand, where there is small dispatch. The Spartans and Spaniards have been noted to be of small dispatch; Mi venga la muerte de Spagna; Let my death come from Spain; for then it will be sure to be long in coming.

Give good hearing to those, that give the first information in business; and rather direct them in the beginning, than interrupt them in the continuance of their speeches; for he that is put out of his own order, will go forward and backward, and be more tedious, while he waits upon his memory, than he could have been, if he had gone on in his own course. But sometimes it is seen, that the moderator is more troublesome, than the actor.

Iterations are commonly loss of time. But there is no such gain of time, as to iterate often the state of the question; for it chaseth away many a frivolous speech, as it is coming forth. Long and curious speeches, are as fit for dispatch, as a robe or mantle, with a long train, is for race. Prefaces and passages, and excusations, and other speeches of reference to the person, are great wastes of time; and though they seem to proceed of modesty, they are bravery. Yet beware of being too material, when there is any impediment or obstruction in men's wills; for pre-occupation of mind ever requireth preface of speech; like a fomentation to make the unguent enter.

Above all things, order, and distribution, and singling out of parts, is the life of dispatch; so as the distribution be not too subtle: for he that doth not divide, will never enter well into business; and he that divideth too much, will never come out of it clearly. To choose time, is to save time; and an unseasonable motion, is but beating the air. There be three parts of business; the preparation, the debate or examination, and the perfection. Whereof, if you look for dispatch, let the middle only be the work of many, and the first and last the work of few. The proceeding upon somewhat conceived in writing, doth for the most part facilitate dispatch: for though it should be wholly rejected, yet that negative is more pregnant of direction, than an indefinite; as ashes are more generative than dust.

Of Seeming Wise

IT HATH been an opinion, that the French are wiser than they seem, and the Spaniards seem wiser than they are. But howsoever it be between nations, certainly it is so between man and man. For as the Apostle saith of godliness, Having a show of godliness, but denying the power thereof; so certainly there are, in point of wisdom and sufficiency, that do nothing or little very solemnly: magno conatu nugas. It is a ridiculous thing, and fit for a satire to persons of judgment, to see what shifts these formalists have, and what prospectives to make superficies to seem body, that hath depth and bulk. Some are so close and reserved, as they will not show their wares, but by a dark light; and seem always to keep back somewhat; and when they know within themselves, they speak of that they do not well know, would nevertheless seem to others, to know of that which they may not well speak. Some help themselves with countenance and gesture, and are wise by signs; as Cicero saith of Piso, that when he answered him, he fetched one of his brows up to his forehead, and bent the other down to his chin; Respondes, altero ad frontem sublato, altero ad mentum depresso supercilio, crudelitatem tibi non placere. Some think to bear it by speaking a great word, and being peremptory; and go on, and take by admittance, that which they cannot make good. Some, whatsoever is beyond their reach, will seem to despise, or make light of it, as impertinent or curious; and so would have their ignorance seem judgment. Some are never without a difference, and commonly by amusing men with a subtilty, blanch the matter; of whom A. Gellius saith, Hominem delirum, qui verborum minutiis rerum frangit pondera. Of which kind also, Plato, in his Protagoras, bringeth in Prodicus in scorn, and maketh him make a speech, that consisteth of distinction from the beginning to the end. Generally, such men in all deliberations find ease to be of the negative side, and affect a credit to object and foretell difficulties; for when propositions are denied, there is an end of them; but if they be allowed, it requireth a new work; which false point of wisdom is the bane of business. To conclude, there is no decaying merchant, or inward beggar, hath so many tricks to uphold the credit of their wealth, as these empty persons have, to maintain the credit of their sufficiency. Seeming wise men may make shift to get opinion; but let no man choose them for employment; for certainly you were better take for business, a man somewhat absurd, than over-formal.

Of Friendship

IT HAD been hard for him that spake it to have put more truth and untruth together in few words, than in that speech, Whatsoever is delighted in solitude, is either a wild beast or a god. For it is most true, that a natural and secret hatred, and aversation towards society, in any man, hath somewhat of the savage beast; but it is most untrue, that it should have any character at all, of the divine nature; except it proceed, not out of a pleasure in solitude, but out of a love and desire to sequester a man's self, for a higher conversation: such as is found to have been falsely and feignedly in some of the heathen; as Epimenides the Candian, Numa the Roman, Empedocles the Sicilian, and Apollonius of Tyana; and truly and really, in divers of the ancient hermits and holy fathers of the church. But little do men perceive what solitude is, and how far it extendeth. For a crowd is not company; and faces are but a gallery of pictures; and talk but a tinkling cymbal, where there is no love. The Latin adage meeteth with it a little: Magna civitas, magna solitudo; because in a great town friends are scattered; so that there is not that fellowship, for the most part, which is in less neighborhoods. But we may go further, and affirm most truly, that it is a mere and miserable solitude to want true friends; without which the world is but a wilderness; and even in this sense also of solitude, whosoever in the frame of his nature and affections, is unfit for friendship, he taketh it of the beast, and not from humanity.

A principal fruit of friendship, is the ease and discharge of the fulness and swellings of the heart, which passions of all kinds do cause and induce. We know diseases of stoppings, and suffocations, are the most dangerous in the body; and it is not much otherwise in the mind; you may take sarza to open the liver, steel to open the spleen, flowers of sulphur for the lungs, castoreum for the brain; but no receipt openeth the heart, but a true friend; to whom you may impart griefs, joys, fears, hopes, suspicions, counsels, and whatsoever lieth upon the heart to oppress it, in a kind of civil shrift or confession.

It is a strange thing to observe, how high a rate great kings and monarchs do set upon this fruit of friendship, whereof we speak: so great, as they purchase it, many times, at the hazard of their own safety and greatness. For princes, in regard of the distance of their fortune from that of their subjects and servants, cannot gather this fruit, except (to make themselves capable thereof) they raise some persons to be, as it were, companions and almost equals to themselves, which many times sorteth to inconvenience. The modern languages give unto such persons the name of favorites, or privadoes; as if it were matter of grace, or conversation. But the Roman name attaineth the true use and cause thereof, naming them participes curarum; for it is that which tieth the knot. And we see plainly that this hath been done, not by weak and passionate princes only, but by the wisest and most politic that ever reigned; who have oftentimes joined to themselves some of their servants; whom both themselves have called friends, and allowed other likewise to call them in the same manner; using the word which is received between private men.

L. Sylla, when he cómmanded Rome, raised Pompey (after surnamed the Great) to that height, that Pompey vaunted himself for Sylla's overmatch. For when he had carried the consulship for a friend of his, against the pursuit of Sylla, and that Sylla did a little resent thereat, and began to speak great, Pompey turned upon him again, and in effect bade him be quiet; for that more men adored the sun rising, than the sun setting. With Julius Caesar, Decimus Brutus had obtained that interest as he set him down in his testament, for heir in remainder, after his nephew. And this was the man that had power with him, to draw him forth to his death. For when Caesar would have discharged the senate, in regard of some ill presages, and specially a dream of Calpurnia; this man lifted him gently by the arm out of his chair, telling him he hoped he would not dismiss the senate, till his wife had dreamt a better dream. And it seemeth his favor was so great, as Antonius, in a letter which is recited verbatim in one of Cicero's Philippics, calleth him venefica, witch; as if he had enchanted Caesar. Augustus raised Agrippa (though of mean birth) to that height, as when he consulted with Maecenas, about the marriage of his daughter Julia, Maecenas took the liberty to tell him, that he must either marry his daughter to Agrippa, or take away his life; there was no third way, he had made him so great. With Tiberius Caesar, Sejanus had ascended to that height, as they two were termed, and reckoned, as a pair of friends. Tiberius in a letter to him saith, Haec pro amicitia nostra non occultavi; and the whole senate dedicated an altar to Friendship, as to a goddess, in respect of the great dearness of friendship, between them two. The like, or more, was between Septimius Severus and Plautianus. For he forced his eldest son to marry the daughter of Plautianus; and would often maintain Plautianus, in doing affronts to his son; and did write also in a letter to the senate, by these words: I love the man so well, as I wish he may over-live me. Now if these princes had been as a Trajan, or a Marcus Aurelius, a man might have thought that this had proceeded of an abundant goodness of nature; but being men so wise, of such strength and severity of mind, and so extreme lovers of themselves, as all these were, it proveth most plainly that they found their own felicity (though as great as ever happened to mortal men) but as an half piece, except they mought have a friend, to make it entire; and yet, which is more, they were princes that had wives, sons, nephews; and yet all these could not supply the comfort of friendship.

It is not to be forgotten, what Comineus observeth of his first master, Duke Charles the Hardy, namely, that he would communicate his secrets with none; and least of all, those secrets which troubled him most. Whereupon he goeth on, and saith that towards his latter time, that closeness did impair, and a little perish his understanding. Surely Comineus mought have made the same judgment also, if it had pleased him, of his second master, Lewis the Eleventh, whose closeness was indeed his tormentor. The parable of Pythagoras is dark, but true; Cor ne edito; Eat not the heart. Certainly, if a man would give it a hard phrase, those that want friends, to open themselves unto, are cannibals of their own hearts. But one thing is most admirable (wherewith I will conclude this first fruit of friendship), which is, that this communicating of a man's self to his friend, works two contrary effects; for it redoubleth joys, and cutteth griefs in halves. For there is no man, that imparteth his joys to his friend, but he joyeth the more; and no man that imparteth his griefs to his friend, but he grieveth the less. So that it is in truth, of operation upon a man's mind, of like virtue as the alchemists use to attribute to their stone, for man's body; that it worketh all contrary effects, but still to the good and benefit of nature. But yet without praying in aid of alchemists, there is a manifest image of this, in the ordinary course of nature. For in bodies, union strengtheneth and cherisheth any natural action; and on the other side, weakeneth and dulleth any violent impression: and even so it is of minds.

The second fruit of friendship, is healthful and sovereign for the understanding, as the first is for the affections. For friendship maketh indeed a fair day in the affections, from storm and tempests; but it maketh daylight in the understanding, out of darkness, and confusion of thoughts. Neither is this to be understood only of faithful counsel, which a man receiveth from his friend; but before you come to that, certain it is, that whosoever hath his mind fraught with many thoughts, his wits and understanding do clarify and break up, in the communicating and discoursing with another; he tosseth his thoughts more easily; he marshalleth them more orderly, he seeth how they look when they are turned into words: finally, he waxeth wiser than himself; and that more by an hour's discourse, than by a day's meditation. It was well said by Themistocles, to the king of Persia, That speech was like cloth of Arras, opened and put abroad; whereby the imagery doth appear in figure; whereas in thoughts they lie but as in packs. Neither is this second fruit of friendship, in opening the understanding, restrained only to such friends as are able to give a man counsel; (they indeed are best;) but even without that, a man learneth of himself, and bringeth his own thoughts to light, and whetteth his wits as against a stone, which itself cuts not. In a word, a man were better relate himself to a statua, or picture, than to suffer his thoughts to pass in smother.

Add now, to make this second fruit of friendship complete, that other point, which lieth more open, and falleth within vulgar observation; which is faithful counsel from a friend. Heraclitus saith well in one of his enigmas, Dry light is ever the best. And certain it is, that the light that a man receiveth by counsel from another, is drier and purer, than that which cometh from his own understanding and judgment; which is ever infused, and drenched, in his affections and customs. So as there is as much difference between the counsel, that a friend giveth, and that a man giveth himself, as there is between the counsel of a friend, and of a flatterer. For there is no such flatterer as is a man's self; and there is no such remedy against flattery of a man's self, as the liberty of a friend. Counsel is of two sorts: the one concerning manners, the other concerning business. For the first, the best preservative to keep the mind in health, is the faithful admonition of a friend. The calling of a man's self to a strict account, is a medicine, sometime too piercing and corrosive. Reading good books of morality, is a little flat and dead. Observing our faults in others, is sometimes improper for our case. But the best receipt (best, I say, to work, and best to take) is the admonition of a friend. It is a strange thing to behold, what gross errors and extreme absurdities many (especially of the greater sort) do commit, for want of a friend to tell them of them; to the great damage both of their fame and fortune: for, as St. James saith, they are as men that look sometimes into a glass, and presently forget their own shape and favor. As for business, a man may think, if he win, that two eyes see no more than one; or that a gamester seeth always more than a looker-on; or that a man in anger, is as wise as he that hath said over the four and twenty letters; or that a musket may be shot off as well upon the arm, as upon a rest; and such other fond and high imaginations, to think himself all in all. But when all is done, the help of good counsel, is that which setteth business straight. And if any man think that he will take counsel, but it shall be by pieces; asking counsel in one business, of one man, and in another business, of another man; it is well (that is to say, better, perhaps, than if he asked none at all); but he runneth two dangers: one, that he shall not be faithfully counselled;

for it is a rare thing, except it be from a perfect and entire friend, to have counsel given, but such as shall be bowed and crooked to some ends, which he hath, that giveth it. The other, that he shall have counsel given, hurtful and unsafe (though with good meaning), and mixed partly of mischief and partly of remedy; even as if you would call a physician, that is thought good for the cure of the disease you complain of, but is unacquainted with your body; and therefore may put you in way for a present cure, but overthroweth your health in some other kind; and so cure the disease, and kill the patient. But a friend that is wholly acquainted with a man's estate, will beware, by furthering any present business, how he dasheth upon other inconvenience. And therefore rest not upon scattered counsels; they will rather distract and mislead, than settle and direct.

After these two noble fruits of friendship (peace in the affections, and support of the judgment), followeth the last fruit; which is like the pomegranate, full of many kernels; I mean aid, and bearing a part, in all actions and occasions. Here the best way to represent to life the manifold use of friendship, is to cast and see how many things there are, which a man cannot do himself; and then it will appear, that it was a sparing speech of the ancients, to say, that a friend is another himself; for that a friend is far more than himself. Men have their time, and die many times, in desire of some things which they principally take to heart; the bestowing of a child, the finishing of a work, or the like. If a man have a true friend, he may rest almost secure that the care of those things will continue after him. So that a man hath, as it were, two lives in his desires. A man hath a body, and that body is confined to a place; but where friendship is, all offices of life are as it were granted to him, and his deputy. For he may exercise them by his friend. How many things are there which a man cannot, with any face or comeliness, say or do himself? A man can scarce allege his own merits with modesty, much less extol them; a man cannot sometimes brook to supplicate or beg; and a number of the like. But all these things are graceful, in a friend's mouth, which are blushing in a man's own. So again, a man's person hath many proper relations, which he cannot put off. A man cannot speak to his son but as a father; to his wife but as a husband; to his enemy but upon terms: whereas a friend may speak as the case requires, and not as it sorteth with the person. But to enumerate these things were endless; I have given the rule, where a man cannot fitly play his own part; if he have not a friend, he may quit the stage.

Of Expense

RICHES are for spending, and spending for honor and good actions. Therefore extraordinary expense must be limited by the worth of the occasion; for voluntary undoing, may be as well for a man's country, as for the kingdom of heaven. But ordinary expense, ought to be limited by a man's estate; and governed with such regard, as it be within his compass; and not subject to deceit and abuse of servants; and ordered to the best show, that the bills may be less than the estimation abroad. Certainly, if a man will keep but of even hand, his ordinary expenses ought to be but to the half of his receipts; and if he think to wax rich, but to the third part. It is no baseness, for the greatest to descend and look into their own estate. Some forbear it, not upon negligence alone, but doubting to bring themselves into melancholy, in respect they shall find it broken. But wounds cannot be cured without searching. He that cannot look into his own estate at all, had need both choose well those whom he employeth, and change them often; for new are more timorous and less subtle. He that can look into his estate but seldom, it behooveth him to turn all to certainties. A man had need, if he be plentiful in some kind of expense, to be as saving again in some other. As if he be plentiful in diet, to be saving in apparel; if he be plentiful in the hall, to be saving in the stable; and the like. For he that is plentiful in expenses of all kinds, will hardly be preserved from decay. In clearing of a man's estate, he may as well hurt himself in being too sudden, as in letting it run on too long. For hasty selling, is commonly as disadvantageable as interest. Besides, he that clears at once will relapse; for finding himself out of straits, he will revert to his custom: but he that cleareth by degrees, induceth a habit of frugality, and gaineth as well upon his mind, as upon his estate. Certainly, who hath a state to repair, may not despise small things; and commonly it is less dishonorable, to abridge petty charges, than to stoop to petty gettings. A man ought warily to begin charges which once begun will continue; but in matters that return not, he may be more magnificent.

Of the True Greatness Of Kingdoms And Estates

THE speech of Themistocles the Athenian, which was haughty and arrogant, in taking so much to himself, had been a grave and wise observation and censure, applied at large to others. Desired at a feast to touch a lute, he said, He could not fiddle, but yet he could make a small town, a great city. These words (holpen a little with a metaphor) may express two differing abilities, in those that deal in business of estate. For if a true survey be taken of counsellors and statesmen, there may be found (though rarely) those which can make a small state great, and yet cannot fiddle; as on the other side, there will be found a great many, that can fiddle very cunningly, but yet are so far from being able to make a small state great, as their gift lieth the other way; to bring a great and flourishing estate, to ruin and decay. And certainly whose degenerate arts and shifts, whereby many counsellors and governors gain both favor with their masters, and estimation with the vulgar, deserve no better name than fiddling; being things rather pleasing for the time, and graceful to themselves only, than tending to the weal and advancement of the state which they serve. There are also (no doubt) counsellors and governors which may be held sufficient (negotiis pares), able to manage affairs, and to keep them from precipices and manifest inconveniences; which nevertheless are far from the ability to raise and amplify an estate in power, means, and fortune. But be the workmen what they may be, let us speak of the work; that is, the true greatness of kingdoms and estates, and the means thereof. An argument fit for great and mighty princes to have in their hand; to the end that neither by over-measuring their forces, they leese themselves in vain enterprises; nor on the other side, by undervaluing them, they descend to fearful and pusillanimous counsels.

The greatness of an estate, in bulk and territory, doth fall under measure; and the greatness of finances and revenue, doth fall under computation. The population may appear by musters; and the number and greatness of cities and towns by cards and maps. But yet there is not any thing amongst civil affairs more subject to error, than the right valuation and true judgment concerning the power and forces of an estate. The kingdom of heaven is compared, not to any great kernel or nut, but to a grain of mustard-seed: which is one of the least grains, but hath in it a property and spirit hastily to get up and spread. So are there states, great in territory, and yet not apt to enlarge or command; and some that have but a small dimension of stem, and yet apt to be the foundations of great monarchies.

Walled towns, stored arsenals and armories, goodly races of horse, chariots of war, elephants, ordnance, artillery, and the like; all this is but a sheep in a lion's skin, except the breed and disposition of the people, be stout and warlike. Nay, number (itself) in armies importeth not much, where the people is of weak courage; for (as Virgil saith) It never troubles a wolf, how many the sheep be. The army of the Persians, in the plains of Arbela, was such a vast sea of people, as it did somewhat astonish the commanders in Alexander's army; who came to him therefore, and wished him to set upon them by night; and he answered, He would not pilfer the victory. And the defeat was easy. When Tigranes the Armenian, being encamped upon a hill with four hundred thousand men, discovered the army of the Romans, being not above fourteen thousand, marching towards him, he made himself merry with it, and said, Yonder men are too many for an embassage, and too few for a fight. But before the sun set, he found them enow to give him the chase with infinite slaughter. Many are the examples of the great odds, between number and courage; so that a man may truly make a judgment, that the principal point of greatness in any state, is to have a race of military men. Neither is money the sinews of war (as it is trivially said), where the sinews of men's arms, in base and effeminate people, are failing. For Solon said well to Croesus (when in ostentation he showed him his gold), Sir, if any other come, that hath better iron, than you, he will be master of all this gold. Therefore let any prince or state think solely of his forces, except his militia of natives be of good and valiant soldiers. And let princes, on the other side, that have subjects of martial disposition, know their own strength; unless they be otherwise wanting unto themselves. As for mercenary forces (which is the help in this case), all examples show, that whatsoever estate or prince doth rest upon them, he may spread his feathers for a time, but he will mew them soon after.

The blessing of Judah and Issachar will never meet; that the same people, or nation, should be both the lion's whelp and the ass between burthens; neither will it be, that a people overlaid with taxes, should ever become valiant and martial. It is true that taxes levied by consent of the estate, do abate men's courage less: as it hath been seen notably, in the excises of the Low Countries; and, in some degree, in the subsidies of England. For you must note, that we speak now of the heart, and not of the purse. So that although the same tribute and tax, laid by consent or by imposing, be all one to the purse, yet it works diversely upon the courage. So that you may conclude, that no people overcharged with tribute, is fit for empire.

Let states that aim at greatness, take heed how their nobility and gentlemen do multiply too fast. For that maketh the common subject, grow to be a peasant and base swain, driven out of heart, and in effect but the gentleman's laborer. Even as you may see in coppice woods; if you leave your staddles too thick, you shall never have clean underwood, but shrubs and bushes. So in countries, if the gentlemen be too many, the commons will be base; and you will bring it to that, that not the hundred poll, will be fit for an helmet; especially as to the infantry, which is the nerve of an army; and so there will be great population, and little strength. This which I speak of, hath been nowhere better seen, than by comparing of England and France; whereof England, though far less in territory and population, hath been (nevertheless) an overmatch; in regard the middle people of England make good soldiers, which the peasants of France do not. And herein the device of king Henry the Seventh (whereof I have spoken largely in the History of his Life) was profound and admirable; in making farms and houses of husbandry of a standard; that is, maintained with such a proportion of land unto them, as may breed a subject to live in convenient plenty and no servile condition; and to keep the plough in the hands of the owners, and not mere hirelings. And thus indeed you shall attain to Virgil's character which he gives to ancient Italy:

Terra potens armis atque ubere glebae.

Neither is that state (which, for any thing I know, is almost peculiar to England, and hardly to be found anywhere else, except it be perhaps in Poland) to be passed over; I mean the state of free servants, and attendants upon noblemen and gentlemen; which are no ways inferior unto the yeomanry for arms. And therefore out of all questions, the splendor and magnificence, and great retinues and hospitality, of noblemen and gentlemen, received into custom, doth much conduce unto martial greatness. Whereas, contrariwise, the close and reserved living of noblemen and gentlemen, causeth a penury of military forces.

By all means it is to be procured, that the trunk of Nebuchadnezzar's tree of monarchy, be great enough to bear the branches and the boughs; that is, that the natural subjects of the crown or state, bear a sufficient proportion to the stranger subjects, that they govern. Therefore all states that are liberal of naturalization towards strangers, are fit for empire. For to think that an handful of people can, with the greatest courage and policy in the world, embrace too large extent of dominion, it may hold for a time, but it will fail suddenly. The Spartans were a nice people in point of naturalization; whereby, while they kept their compass, they stood firm; but when they did spread, and their boughs were becomen too great for their stem, they became a windfall, upon the sudden. Never any state was in this point so open to receive strangers into their body, as were the Romans. Therefore it sorted with them accordingly; for they grew to the greatest monarchy. Their manner was to grant naturalization (which they called jus civitatis), and to grant it in the highest degree; that is, not only jus commercii, jus connubii, jus haereditatis; but also jus suffragii, and jus honorum. And this not to singular persons alone, but likewise to whole families; yea to cities, and sometimes to nations. Add to this their custom of plantation of colonies; whereby the Roman plant was removed into the soil of other nations. And putting both constitutions together, you will say that it was not the Romans that spread upon the world, but it was the world that spread upon the Romans; and that was the sure way of greatness. I have marvelled, sometimes, at Spain, how they clasp and contain so large dominions, with so few natural Spaniards; but sure the whole compass of Spain, is a very great body of a tree; far above Rome and Sparta at the first. And besides, though they have not had that usage, to naturalize liberally, yet they have that which is next to it; that is, to employ, almost indifferently, all nations in their militia of ordinary soldiers; yea, and sometimes in their highest commands. Nay, it seemeth at this instant they are sensible, of this want of natives; as by the Pragmatical Sanction, now published, appeareth.

It is certain that sedentary, and within-door arts, and delicate manufactures (that require rather the finger than the arm), have, in their nature, a contrariety to a military disposition. And generally, all warlike people are a little idle, and love danger better than travail. Neither must they be too much broken of it, if they shall be preserved in vigor. Therefore it was great advantage, in the ancient states of Sparta, Athens, Rome, and others, that they had the use of slaves, which commonly did rid those manufactures. But that is abolished, in greatest part, by the Christian law. That which cometh nearest to it, is to leave those arts chiefly to strangers (which, for that purpose, are the more easily to be received), and to contain the principal bulk of the vulgar natives, within those three kinds,--tillers of the ground; free servants; and handicraftsmen of strong and manly arts, as smiths, masons, carpenters, etc.; not reckoning professed soldiers.

But above all, for empire and greatness, it importeth most, that a nation do profess arms, as their principal honor, study, and occupation. For the things which we formerly have spoken of, are but habilitations towards arms; and what is habilitation without intention and act? Romulus, after his death (as they report or feign), sent a present to the Romans, that above all, they should intend arms; and then they should prove the greatest empire of the world. The fabric of the state of Sparta was wholly (though not wisely) framed and composed, to that scope and end. The Persians and Macedonians had it for a flash. The Gauls, Germans, Goths, Saxons, Normans, and others, had it for a time. The Turks have it at this day, though in great declination. Of Christian Europe, they that have it are, in effect, only the Spaniards. But it is so plain, that every man profiteth in that, he most intendeth, that it needeth not to be stood upon. It is enough to point at it; that no nation which doth not directly profess arms, may look to have greatness fall into their mouths. And on the other side, it is a most certain oracle of time, that those states that continue long in that profession (as the Romans and Turks principally have done) do wonders. And those that have professed arms but for an age, have, notwithstanding, commonly attained that greatness, in that age, which maintained them long after, when their profession and exercise of arms hath grown to decay.

Incident to this point is, for a state to have those laws or customs, which may reach forth unto them just occasions (as may be pretended) of war. For there is that justice, imprinted in the nature of men, that they enter not upon wars (whereof so many calamities do ensue) but upon some, at the least specious, grounds and quarrels. The Turk hath at hand, for cause of war, the propagation of his law or sect; a quarrel that he may always command. The Romans, though they esteemed the extending the limits of their empire, to be great honor to their generals, when it was done, yet they never rested upon that alone, to begin a war. First, therefore, let nations that pretend to greatness have this; that they be sensible of wrongs, either upon borderers, merchants, or politic ministers; and that they sit not too long upon a provocation. Secondly, let them be prest, and ready to give aids and succors, to their confederates; as it ever was with the Romans; insomuch, as if the confederate had leagues defensive, with divers other states, and, upon invasion offered, did implore their aids severally, yet the Romans would ever be the foremost, and leave it to none other to have the honor. As for the wars which were anciently made, on the behalf of a kind of party, or tacit conformity of estate, I do not see how they may be well justified: as when the Romans made a war, for the liberty of Grecia; or when the Lacedaemonians and Athenians, made wars to set up or pull down democracies and oligarchies; or when wars were made by foreigners, under the pretence of justice or protection, to deliver the subjects of others, from tyranny and oppression; and the like. Let it suffice, that no estate expect to be great, that is not awake upon any just occasion of arming.

No body can be healthful without exercise, neither natural body nor politic; and certainly to a kingdom or estate, a just and honorable war, is the true exercise. A civil war, indeed, is like the heat of a fever; but a foreign war is like the heat of exercise, and serveth to keep the body in health; for in a slothful peace, both courages will effeminate, and manners corrupt. But howsoever it be for happiness, without all question, for greatness, it maketh to be still for the most part in arms; and the strength of a veteran army (though it be a chargeable business) always on foot, is that which commonly giveth the law, or at least the reputation, amongst all neighbor states; as may well be seen in Spain, which hath had, in one part or other, a veteran army almost continually, now by the space of six score years.

To be master of the sea, is an abridgment of a monarchy. Cicero, writing to Atticus of Pompey his preparation against Caesar, saith, *Consilium Pompeii plane Themistocleum est; putat enim, qui mari potitur, eum rerum potiri.* And, without doubt, Pompey had tired out Caesar, if upon vain confidence, he had not left that way. We see the great effects of battles by sea. The battle of Actium, decided the empire of the world. The battle of Lepanto, arrested the greatness of the Turk. There be many examples, where sea-fights have been final to the war; but this is when princes or states have set up their rest, upon the battles. But thus much is certain, that he that commands the sea, is at great liberty, and may take as much, and as little, of the war as he will. Whereas those that be strongest by land, are many times nevertheless in great straits. Surely, at this day, with us of Europe, the vantage of strength at sea (which is one of the principal dowries of this kingdom of Great Britain) is great; both because most of the kingdoms of Europe, are not merely inland, but girt with the sea most part of their compass; and because the wealth of both Indies seems in great part, but an accessory to the command of the seas.

The wars of latter ages seem to be made in the dark, in respect of the glory, and honor, which reflected upon men from the wars, in ancient time. There be now, for martial encouragement, some degrees and orders of chivalry; which nevertheless are conferred promiscuously, upon soldiers and no soldiers; and some remembrance perhaps, upon the scutcheon; and some hospitals for maimed soldiers; and such like things. But in ancient times, the trophies erected upon the place of the victory; the funeral laudatives and monuments for those that died in the wars; the crowns and garlands personal; the style of emperor, which the great kings of the world after borrowed; the triumphs of the generals, upon their return; the great donatives and largesses, upon the disbanding of the armies; were things able to inflame all men's courages. But above all, that of the triumph, amongst the Romans, was not pageants or gaudery, but one of the wisest and noblest institutions, that ever was. For it contained three things: honor to the general; riches to the treasury out of the spoils; and donatives to the army. But that honor, perhaps were not fit for monarchies; except it be in the person of the monarch himself, or his sons; as it came to pass in the times of the Roman emperors, who did impropriate the actual triumphs to themselves, and their sons, for such wars as they did achieve in person; and left only, for wars achieved by subjects, some triumphal garments and ensigns to the general.

To conclude: no man can by care taking (as the Scripture saith) add a cubit to his stature, in this little model of a man's body; but in the great frame of kingdoms and commonwealths, it is in the power of princes or estates, to add amplitude and greatness to their kingdoms; for by introducing such ordinances, constitutions, and customs, as we have now touched, they may sow greatness to their posterity and succession. But these things are commonly not observed, but left to take their chance.

Of Regiment Of Health

THERE is a wisdom in this; beyond the rules of physic: a man's own observation, what he finds good of, and what he finds hurt of, is the best physic to preserve health. But it is a safer conclusion to say, This agreeth not well with me, therefore, I will not continue it; than this, I find no offence of this, therefore I may use it. For strength of nature in youth, passeth over many excesses, which are owing a man till his age. Discern of the coming on of years, and think not to do the same things still; for age will not be defied. Beware of sudden change, in any great point of diet, and, if necessity enforce it, fit the rest to it. For it is a secret both in nature and state, that it is safer to change many things, than one. Examine thy customs of diet, sleep, exercise, apparel, and the like; and try, in any thing thou shalt judge hurtful, to discontinue it, by little and little; but so, as if thou dost find any inconvenience by the change, thou come back to it again: for it is hard to distinguish that which is generally held good and wholesome, from that which is good particularly, and fit for thine own body. To be free-minded and cheerfully disposed, at hours of meat, and of sleep, and of exercise, is one of the best precepts of long lasting. As for the passions, and studies of the mind; avoid envy, anxious fears; anger fretting inwards; subtle and knotty inquisitions; joys and exhilarations in excess; sadness not communicated. Entertain hopes; mirth rather than joy; variety of delights, rather than surfeit of them; wonder and admiration, and therefore novelties; studies that fill the mind with splendid and illustrious objects, as histories, fables, and contemplations of nature. If you fly physic in health altogether, it will be too strange for your body, when you shall need it. If you make it too familiar, it will work no extraordinary effect, when sickness cometh. I commend rather some diet for certain seasons, than frequent use of physic, except it be grown into a custom. For those diets alter the body more, and trouble it less. Despise no new accident in your body, but ask opinion of it. In sickness, respect health principally; and in health, action. For those that put their bodies to endure in health, may in most sicknesses, which are not very sharp, be cured only with diet, and tendering. Celsus could never have spoken it as a physician, had he not been a wise man withal, when he giveth it for one of the great precepts of health and lasting, that a man do vary, and interchange contraries, but with an inclination to the more benign extreme: use fasting and full eating, but rather full eating; watching and sleep, but rather sleep; sitting and exercise, but rather exercise; and the like. So shall nature be cherished, and yet taught masteries. Physicians are, some of them, so pleasing and conformable to the humor of the patient, as they press not the true cure of the disease; and some other are so regular, in proceeding according to art for the disease, as they respect not sufficiently the condition of the patient. Take one of a middle temper; or if it may not be found in one man, combine two of either sort; and forget not to call as well, the best acquainted with your body, as the best reputed of for his faculty.

Of Suspicion

SUSPICIONS amongst thoughts, are like bats amongst birds, they ever fly by twilight. Certainly they are to be repressed, or at least well guarded: for they cloud the mind; they leese friends; and they check with business, whereby business cannot go on currently and constantly. They dispose kings to tyranny, husbands to jealousy, wise men to irresolution and melancholy. They are defects, not in the heart, but in the brain; for they take place in the stoutest natures; as in the example of Henry the Seventh of England. There was not a more suspicious man, nor a more stout. And in such a composition they do small hurt. For commonly they are not admitted, but with examination, whether they be likely or no. But in fearful natures they gain ground too fast. There is nothing makes a man suspect much, more than to know little; and therefore men should remedy suspicion, by procuring to know more, and not to keep their suspicions in smother. What would men have? Do they think, those they employ and deal with, are saints? Do they not think, they will have their own ends, and be truer to themselves, than to them? Therefore there is no better way, to moderate suspicions, than to account upon such suspicions as true, and yet to bridle them as false. For so far a man ought to make use of suspicions, as to provide, as if that should be true, that he suspects, yet it may do him no hurt. Suspicions that the mind of itself gathers, are but buzzes; but suspicions that are artificially nourished, and put into men's heads, by the tales and whisperings of others, have stings. Certainly, the best mean, to clear the way in this same wood of suspicions, is frankly to communicate them with the party, that he suspects; for thereby he shall be sure to know more of the truth of them, than he did before; and withal shall make that party more circumspect, not to give further cause of suspicion. But this would not be done to men of base natures; for they, if they find themselves once suspected, will never be true. The Italian says, Sospetto licentia fede; as if suspicion, did give a passport to faith; but it ought, rather, to kindle it to discharge itself.

Of Discourse

SOME, in their discourse, desire rather commendation of wit, in being able to hold all arguments, than of judgment, in discerning what is true; as if it were a praise, to know what might be said, and not, what should be thought. Some have certain common places, and themes, wherein they are good and want variety; which kind of poverty is for the most part tedious, and when it is once perceived, ridiculous. The honorablest part of talk, is to give the occasion; and again to moderate, and pass to somewhat else; for then a man leads the dance. It is good, in discourse and speech of conversation, to vary and intermingle speech of the present occasion, with arguments, tales with reasons, asking of questions, with telling of opinions, and jest with earnest: for it is a dull thing to tire, and, as we say now, to jade, any thing too far. As for jest, there be certain things, which ought to be privileged from it; namely, religion, matters of state, great persons, any man's present business of importance, and any case that deserveth pity. Yet there be some, that think their wits have been asleep, except they dart out somewhat that is piquant, and to the quick. That is a vein which would be bridled:

Parce, puer, stimulis, et fortius utere loris.

And generally, men ought to find the difference, between saltness and bitterness. Certainly, he that hath a satirical vein, as he maketh others afraid of his wit, so he had need be afraid of others' memory. He that questioneth much, shall learn much, and content much; but especially, if he apply his questions to the skill of the persons whom he asketh; for he shall give them occasion, to please themselves in speaking, and himself shall continually gather knowledge. But let his questions not be troublesome; for that is fit for a poser. And let him be sure to leave other men, their turns to speak. Nay, if there be any, that would reign and take up all the time, let him find means to take them off, and to bring others on; as musicians use to do, with those that dance too long galliards. If you dissemble, sometimes, your knowledge of that you are thought to know, you shall be thought, another time, to know that you know not. Speech of a man's self ought to be seldom, and well chosen. I knew one, was wont to say in scorn, He must needs be a wise man, he speaks so much of himself: and there is but one case, wherein a man may commend himself with good grace; and that is in commending virtue in another; especially if it be such a virtue, whereunto himself pretendeth. Speech of touch towards others, should be sparingly used; for discourse ought to be as a field, without coming home to any man. I knew two noblemen, of the west part of England, whereof the one was given to scoff, but kept ever royal cheer in his house; the other would ask, of those that had been at the other's table, Tell truly, was there never a flout or dry blow given? To which the guest would answer, Such and such a thing passed. The lord would say, I thought, he would mar a good dinner. Discretion of speech, is more than eloquence; and to speak agreeably to him, with whom we deal, is more than to speak in good words, or in good order. A good continued speech, without a good speech of interlocution, shows slowness: and a good reply or second speech, without a good settled speech, showeth shallowness and weakness. As we see in beasts, that those that are weakest in the course, are yet nimblest in the turn; as it is betwixt the greyhound and the hare. To use too many circumstances, ere one come to the matter, is wearisome; to use none at all, is blunt.

Of Plantations

PLANTATIONS are amongst ancient, primitive, and heroical works. When the world was young, it begat more children; but now it is old, it begets fewer: for I may justly account new plantations, to be the children of former kingdoms. I like a plantation in a pure soil; that is, where people are not displanted, to the end, to plant in others. For else it is rather an extirpation, than a plantation. Planting of countries, is like planting of woods; for you must make account to leese almost twenty years' profit, and expect your recompense in the end. For the principal thing, that hath been the destruction of most plantations, hath been the base and hasty drawing of profit, in the first years. It is true, speedy profit is not to be neglected, as far as may stand with the good of the plantation, but no further. It is a shameful and unblessed thing, to take the scum of people, and wicked condemned men, to be the people with whom you plant; and not only so, but it spoileth the plantation; for they will ever live like rogues, and not fall to work, but be lazy, and do mischief, and spend victuals, and be quickly weary, and then certify over to their country, to the discredit of the plantation. The people wherewith you plant ought to be gardeners, ploughmen, laborers, smiths, carpenters, joiners, fishermen, fowlers, with some few apothecaries, surgeons, cooks, and bakers. In a country of plantation, first look about, what kind of victual the country yields of itself to hand; as chestnuts, walnuts, pineapples, olives, dates, plums, cherries, wild honey, and the like; and make use of them. Then consider what victual or esculent things there are, which grow speedily, and within the year; as parsnips, carrots, turnips, onions, radish, artichokes of Hierusalem, maize, and the like. For wheat, barley, and oats, they ask too much labor; but with pease and beans you may begin, both because they ask less labor, and because they serve for meat, as well as for bread. And of rice, likewise cometh a great increase, and it is a kind of meat. Above all, there ought to be brought store of biscuit, oat-meal, flour, meal, and the like, in the beginning, till bread may be had. For beasts, or birds, take chiefly such as are least subject to diseases, and multiply fastest; as swine, goats, cocks, hens, turkeys, geese, house-doves, and the like. The victual in plantations, ought to be expended almost as in a besieged town; that is, with certain allowance. And let the main part of the ground, employed to gardens or corn, be to a common stock; and to be laid in, and stored up, and then delivered out in proportion; besides some spots of ground, that any particular person will manure for his own private. Consider likewise what commodities, the soil where the plantation is, doth naturally yield, that they may some way help to defray the charge of the plantation (so it be not, as was said, to the untimely prejudice of the main business), as it hath fared upwards with tobacco in Virginia. Wood commonly aboundeth but too much; and therefore timber is fit to be one. If there be iron ore, and streams whereupon to set the mills, iron is a brave commodity where wood aboundeth. Making of bay-salt, if the climate be proper for it, would be put in experience. Growing silk likewise, if any be, is a likely commodity. Pitch and tar, where store of firs and pines are, will not fail. So drugs and sweet woods, where they are, cannot but yield great profit. Soap-ashes likewise, and other things that may be thought of. But moil not too much under ground; for the hope of mines is very uncertain, and useth to make the planters lazy, in other things. For government; let it be in the hands of one, assisted with some counsel; and let them have commission to exercise martial laws, with some limitation. And above all, let men make that profit, of being in the wilderness, as they have God always, and his service, before their eyes. Let not the government of the plantation, depend upon too many counsellors, and undertakers, in the country that planteth, but upon a temperate number; and let those be rather noblemen and gentlemen, than merchants; for they look ever to the present gain. Let there be freedom from custom, till the plantation be of strength; and not only freedom from custom, but freedom to carry their commodities, where they may make their best of them, except there be some special cause of caution. Cram not in people, by sending too fast company after company; but rather harken how they waste, and send supplies proportionably; but so, as the number may live well in the plantation, and not by surcharge be in penury. It hath been a great endangering to the health of some plantations, that they have built along the sea and rivers, in marish and unwholesome grounds. Therefore, though you begin there, to avoid carriage and like discommodities, yet build still rather upwards from the streams, than along. It concerneth likewise the health of the plantation, that they have good store of salt with them, that they may use it in their victuals, when it shall be necessary. If you plant where savages are, do not only entertain them, with trifles and gingles, but use them justly and graciously, with sufficient guard nevertheless; and do not win their favor, by helping them to invade their enemies, but for their defence it is not amiss; and send oft of them, over to the country that plants, that they may see a better condition than their own, and commend it when they return. When the plantation grows to strength, then it is time to plant with women, as well as with men; that the plantation may spread into generations, and not be ever pieced from without. It is the sinfullest thing in the world, to forsake or destitute a plantation once in forwardness; for besides the dishonor, it is the guiltiness of blood of many commiserable persons.

Of Riches

I CANNOT call riches better than the baggage of virtue. The Roman word is better, impedimenta. For as the baggage is to an army, so is riches to virtue. It cannot be spared, nor left behind, but it hindereth the march; yea, and the care of it, sometimes loseth or disturbeth the victory. Of great riches there is no real use, except it be in the distribution; the rest is but conceit. So saith Solomon, Where much is, there are many to consume it; and what hath the owner, but the sight of it with his eyes? The personal fruition in any man, cannot reach to feel great riches: there is a custody of them; or a power of dole, and donative of them; or a fame of them; but no solid use to the owner. Do you not see what feigned prices, are set upon little stones and rarities? and what works of ostentation are undertaken, because there might seem to be some use of great riches? But then you will say, they may be of use, to buy men out of dangers or troubles. As Solomon saith, Riches are as a strong hold, in the imagination of the rich man. But this is excellently expressed, that it is in imagination, and not always in fact. For certainly great riches, have sold more men, than they have bought out. Seek not proud riches, but such as thou mayest get justly, use soberly, distribute cheerfully, and leave contentedly. Yet have no abstract nor friarly contempt of them. But distinguish, as Cicero saith well of Rabirius Posthumus, In studio rei amplificandae apparebat, non avaritiae praedam, sed instrumentum bonitati quaeri. Harken also to Solomon, and beware of hasty gathering of riches; Qui festinat ad divitias, non erit insons. The poets feign, that when Plutus (which is Riches) is sent from Jupiter, he limps and goes slowly; but when he is sent from Pluto, he runs, and is swift of foot. Meaning that riches gotten by good means, and just labor, pace slowly; but when they come by the death of others (as by the course of inheritance, testaments, and the like), they come tumbling upon a man. But it mought be applied likewise to Pluto, taking him for the devil. For when riches come from the devil (as by fraud and oppression, and unjust means), they come upon speed. The ways to enrich are many, and most of them foul. Parsimony is one of the best, and yet is not innocent; for it withholdeth men from works of liberality and charity. The improvement of the ground, is the most natural obtaining of riches; for it is our great mother's blessing, the earth's; but it is slow. And yet where men of great wealth do stoop to husbandry, it multiplieth riches exceedingly. I knew a nobleman in England, that had the greatest audits of any man in my time; a great grazier, a great sheep-master, a great timber man, a great collier, a great corn-master, a great lead-man, and so of iron, and a number of the like points of husbandry. So as the earth seemed a sea to him, in respect of the perpetual importation. It was truly observed by one, that himself came very hardly, to a little riches, and very easily, to great riches. For when a man's stock is come to that, that he can expect the prime of markets, and overcome those bargains, which for their greatness are few men's money, and be partner in the industries of younger men, he cannot but increase mainly. The gains of ordinary trades and vocations are honest; and furthered by two things chiefly: by diligence, and by a good name, for good and fair dealing. But the gains of bargains, are of a more doubtful nature; when men shall wait upon others' necessity, broke by servants and instruments to draw them on, put off others cunningly, that would be better chapmen, and the like practices, which are crafty and naught. As for the chopping of bargains, when a man buys not to hold but to sell over again, that commonly grindeth double, both upon the seller, and upon the buyer. Sharings do greatly enrich, if the hands be well chosen, that are trusted. Usury is the certainest means of gain, though one of the worst; as that whereby a man doth eat his bread, in sudore vultus alieni; and besides, doth plough upon Sundays. But yet certain though it be, it hath flaws; for that the scriveners and brokers do value unsound men, to serve their own turn. The fortune in being the first, in an invention or in a privilege, doth cause sometimes a wonderful overgrowth in riches; as it was with the first sugar man, in the Canaries. Therefore if a man can play the true logician, to have as well judgment, as invention, he may do great matters; especially if the times be fit. He that resteth upon gains certain, shall hardly grow to great riches; and he that puts all upon adventures, doth oftentimes break and come to poverty: it is good, therefore, to guard adventures with certainties, that may uphold losses. Monopolies, and coemption of wares for re-sale, where they are not restrained, are great means to enrich; especially if the party have intelligence, what things are like to come into request, and so store himself beforehand. Riches gotten by service, though it be of the best rise, yet when they are gotten by flattery, feeding humors, and other servile conditions, they may be placed amongst the worst. As for fishing for testaments and executorships (as Tacitus saith of Seneca, testamenta et orbos tamquam indagine capi), it is yet worse; by how much men submit themselves to meaner persons, than in service. Believe not much, them that seem to despise riches; for they despise them, that despair of them; and none worse, when they come to them. Be not penny-wise; riches have wings, and sometimes they fly away of themselves, sometimes they must be set flying, to bring in more. Men leave their riches, either to their kindred, or to the public; and moderate portions, prosper best in both. A great state left to an heir, is as a lure to all the birds of prey round about, to seize on him, if he be not the better stablished in years and judgment. Likewise glorious gifts and foundations, are like sacrifices without salt; and but the painted sepulchres of alms, which soon will putrefy, and corrupt inwardly. Therefore measure not thine advancements, by quantity, but frame them by measure: and defer not charities till death; for, certainly, if a man weigh it rightly, he that doth so, is rather liberal of another man's, than of his own.

Of Prophecies

I MEAN not to speak of divine prophecies; nor of heathen oracles; nor of natural predictions; but only of prophecies that have been of certain memory, and from hidden causes. Saith the Pythonissa to Saul, To-morrow thou and thy son shall be with me. Homer hath these verses:

At domus AEneae cunctis dominabitur oris, Et nati natorum, et qui nascentur ab illis.
A prophecy, as it seems, of the Roman empire. Seneca the tragedian hath these verses:

--Venient annis
Saecula seris, quibus Oceanus
Vincula rerum laxet, et ingens
Pateat Tellus, Tiphysque novos
Detegat orbes; nec sit terris
Ultima Thule:

a prophecy of the discovery of America. The daughter of Polycrates, dreamed that Jupiter bathed her father, and Apollo anointed him; and it came to pass, that he was crucified in an open place, where the sun made his body run with sweat, and the rain washed it. Philip of Macedon dreamed, he sealed up his wife's belly; whereby he did expound it, that his wife should be barren; but Aristander the soothsayer, told him his wife was with child, because men do not use to seal vessels, that are empty. A phantasm that appeared to M. Brutus, in his tent, said to him, Philippis iterum me videbis. Tiberius said to Galba, Tu quoque, Galba, degustabis imperium. In Vespasian's time, there went a prophecy in the East, that those that should come forth of Judea, should reign over the world: which though it may be was meant of our Savior; yet Tacitus expounds it of Vespasian. Domitian dreamed, the night before he was slain, that a golden head was growing, out of the nape of his neck: and indeed, the succession that followed him for many years, made golden times. Henry the Sixth of England, said of Henry the Seventh, when he was a lad, and gave him water, This is the lad that shall enjoy the crown, for which we strive. When I was in France, I heard from one Dr. Pena, that the Queen Mother, who was given to curious arts, caused the King her husband's nativity to be calculated, under a false name; and the astrologer gave a judgment, that he should be killed in a duel; at which the Queen laughed, thinking her husband to be above challenges and duels: but he was slain upon a course at tilt, the splinters of the staff of Montgomery going in at his beaver. The trivial prophecy, which I heard when I was a child, and Queen Elizabeth was in the flower of her years, was,

When hempe is spun
England's done:

whereby it was generally conceived, that after the princes had reigned, which had the principal letters of that word hempe (which were Henry, Edward, Mary, Philip, and Elizabeth), England should come to utter confusion; which, thanks be to God, is verified only in the change of the name; for that the King's style, is now no more of England but of Britain. There was also another prophecy, before the year of '88, which I do not well understand.

There shall be seen upon a day,
Between the Baugh and the May,
The black fleet of Norway.
When that that is come and gone,
England build houses of lime and stone,
For after wars shall you have none.

It was generally conceived to be meant, of the Spanish fleet that came in '88: for that the king of Spain's surname, as they say, is Norway. The prediction of Regiomontanus,

Octogesimus octavus mirabilis annus,

was thought likewise accomplished in the sending of that great fleet, being the greatest in strength, though not in number, of all that ever swam upon the sea. As for Cleon's dream, I think it was a jest. It was, that he was devoured of a long dragon; and it was expounded of a maker of sausages, that troubled him exceedingly. There are numbers of the like kind; especially if you include dreams, and predictions of astrology. But I have set down these few only, of certain credit, for example. My judgment is, that they ought all to be despised; and ought to serve but for winter talk by the fireside. Though when I say despised, I mean it as for belief; for otherwise, the spreading, or publishing, of them, is in no sort to be despised. For they have done much mischief; and I see many severe laws made, to suppress them. That that hath given them grace, and some credit, consisteth in three things. First, that men mark when they hit, and never mark when they miss; as they do generally also of dreams. The second is, that probable conjectures, or obscure traditions, many times turn themselves into prophecies; while the nature of man, which coveteth divination, thinks it no peril to foretell that which indeed they do but collect. As that of Seneca's verse. For so much was then subject to demonstration, that the globe of the earth had great parts beyond the Atlantic, which mought be probably conceived not to be all sea: and adding thereto the tradition in Plato's Timaeus, and his Atlanticus, it mought encourage one to turn it to a prediction. The third and last (which is the great one) is, that almost all of them, being infinite in number, have been impostures, and by idle and crafty brains merely contrived and feigned, after the event past.

Of Ambition

AMBITION is like choler; which is an humor that maketh men active, earnest, full of alacrity, and stirring, if it be not stopped. But if it be stopped, and cannot have his way, it becometh adust, and thereby malign and venomous. So ambitious men, if they find the way open for their rising, and still get forward, they are rather busy than dangerous; but if they be checked in their desires, they become secretly discontent, and look upon men and matters with an evil eye, and are best pleased, when things go backward; which is the worst property in a servant of a prince, or state. Therefore it is good for princes, if they use ambitious men, to handle it, so as they be still progressive and not retrograde; which, because it cannot be without inconvenience, it is good not to use such natures at all. For if they rise not with their service, they will take order, to make their service fall with them. But since we have said, it were good not to use men of ambitious natures, except it be upon necessity, it is fit we speak, in what cases they are of necessity. Good commanders in the wars must be taken, be they never so ambitious; for the use of their service, dispenseth with the rest; and to take a soldier without ambition, is to pull off his spurs. There is also great use of ambitious men, in being screens to princes in matters of danger and envy; for no man will take that part, except he be like a seeled dove, that mounts and mounts, because he cannot see about him. There is use also of ambitious men, in pulling down the greatness of any subject that overtops; as Tiberius used Marco, in the pulling down of Sejanus. Since, therefore, they must be used in such cases, there resteth to speak, how they are to be bridled, that they may be less dangerous. There is less danger of them, if they be of mean birth, than if they be noble; and if they be rather harsh of nature, than gracious and popular: and if they be rather new raised, than grown cunning, and fortified, in their greatness. It is counted by some, a weakness in princes, to have favorites; but it is, of all others, the best remedy against ambitious great-ones. For when the way of pleasuring, and displeasuring, lieth by the favorite, it is impossible any other should be overgreat. Another means to curb them, is to balance them by others, as proud as they. But then there must be some middle counsellors, to keep things steady; for without that ballast, the ship will roll too much. At the least, a prince may animate and inure some meaner persons, to be as it were scourges, to ambitious men. As for the having of them obnoxious to ruin; if they be of fearful natures, it may do well; but if they be stout and daring, it may precipitate their designs, and prove dangerous. As for the pulling of them down, if the affairs require it, and that it may not be done with safety suddenly, the only way is the interchange, continually, of favors and disgraces; whereby they may not know what to expect, and be, as it were, in a wood. Of ambitions, it is less harmful, the ambition to prevail in great things, than that other, to appear in every thing; for that breeds confusion, and mars business. But yet it is less danger, to have an ambitious man stirring in business, than great in dependences. He that seeketh to be eminent amongst able men, hath a great task; but that is ever good for the public. But he, that plots to be the only figure amongst ciphers, is the decay of a whole age. Honor hath three things in it: the vantage ground to do good; the approach to kings and principal persons; and the raising of a man's own fortunes. He that hath the best of these intentions, when he aspireth, is an honest man; and that prince, that can discern of these intentions in another that aspireth, is a wise prince. Generally, let princes and states choose such ministers, as are more sensible of duty than of using; and such as love business rather upon conscience, than upon bravery, and let them discern a busy nature, from a willing mind.

Of Masques And Triumphs

THESE things are but toys, to come amongst such serious observations. But yet, since princes will have such things, it is better they should be graced with elegancy, than daubed with cost. Dancing to song, is a thing of great state and pleasure. I understand it, that the song be in quire, placed aloft, and accompanied with some broken music; and the ditty fitted to the device. Acting in song, especially in dialogues, hath an extreme good grace; I say acting, not dancing (for that is a mean and vulgar thing); and the voices of the dialogue would be strong and manly (a base and a tenor; no treble); and the ditty high and tragical; not nice or dainty. Several quires, placed one over against another, and taking the voice by catches, anthem-wise, give great pleasure. Turning dances into figure, is a childish curiosity. And generally let it be noted, that those things which I here set down, are such as do naturally take the sense, and not respect petty wonderments. It is true, the alterations of scenes, so it be quietly and without noise, are things of great beauty and pleasure; for they feed and relieve the eye, before it be full of the same object. Let the scenes abound with light, specially colored and varied; and let the masquers, or any other, that are to come down from the scene, have some motions upon the scene itself, before their coming down; for it draws the eye strangely, and makes it, with great pleasure, to desire to see, that it cannot perfectly discern. Let the songs be loud and cheerful, and not chirpings or pulings. Let the music likewise be sharp and loud, and well placed. The colors that show best by candle-light are white, carnation, and a kind of sea-water-green; and oes, or spangs, as they are of no great cost, so they are of most glory. As for rich embroidery, it is lost and not discerned. Let the suits of the masquers be graceful, and such as become the person, when the vizors are off; not after examples of known attires; Turke, soldiers, mariners', and the like. Let anti-masques not be long; they have been commonly of fools, satyrs, baboons, wild-men, antics, beasts, sprites, witches, Ethiops, pigmies, turquets, nymphs, rustics, Cupids, statuas moving, and the like. As for angels, it is not comical enough, to put them in anti-masques; and anything that is hideous, as devils, giants, is on the other side as unfit. But chiefly, let the music of them be recreative, and with some strange changes. Some sweet odors suddenly coming forth, without any drops falling, are, in such a company as there is steam and heat, things of great pleasure and refreshment. Double masques, one of men, another of ladies, addeth state and variety. But all is nothing except the room be kept clear and neat.

For justs, and tourneys, and barriers; the glories of them are chiefly in the chariots, wherein the challengers make their entry; especially if they be drawn with strange beasts: as lions, bears, camels, and the like; or in the devices of their entrance; or in the bravery of their liveries; or in the goodly furniture of their horses and armor. But enough of these toys.

Of Nature In Men

NATURE is often hidden; sometimes overcome; seldom extinguished. Force, maketh nature more violent in the return; doctrine and discourse, maketh nature less importune; but custom only doth alter and subdue nature. He that seeketh victory over his nature, let him not set himself too great, nor too small tasks; for the first will make him dejected by often failings; and the second will make him a small proceeder, though by often prevailings. And at the first let him practise with helps, as swimmers do with bladders or rushes; but after a time let him practise with disadvantages, as dancers do with thick shoes. For it breeds great perfection, if the practice be harder than the use. Where nature is mighty, and therefore the victory hard, the degrees had need be, first to stay and arrest nature in time; like to him that would say over the four and twenty letters when he was angry; then to go less in quantity; as if one should, in forbearing wine, come from drinking healths, to a draught at a meal; and lastly, to discontinue altogether. But if a man have the fortitude, and resolution, to enfranchise himself at once, that is the best:

Optimus ille animi vindex laedentia pectus
Vincula qui rupit, dedoluitque semel.

Neither is the ancient rule amiss, to bend nature, as a wand, to a contrary extreme, whereby to set it right, understanding it, where the contrary extreme is no vice. Let not a man force a habit upon himself, with a perpetual continuance, but with some intermission. For both the pause reinforceth the new onset; and if a man that is not perfect, be ever in practice, he shall as well practise his errors, as his abilities, and induce one habit of both; and there is no means to help this, but by seasonable intermissions. But let not a man trust his victory over his nature, too far; for nature will lay buried a great time, and yet revive, upon the occasion or temptation. Like as it was with AEsop's damsel, turned from a cat to a woman, who sat very demurely at the board's end, till a mouse ran before her. Therefore, let a man either avoid the occasion altogether; or put himself often to it, that he may be little moved with it. A man's nature is best perceived in privateness, for there is no affectation; in passion, for that putteth a man out of his precepts; and in a new case or experiment, for there custom leaveth him. They are happy men, whose natures sort with their vocations; otherwise they may say, multum incola fuit anima mea; when they converse in those things, they do not affect. In studies, whatsoever a man commandeth upon himself, let him set hours for it; but whatsoever is agreeable to his nature, let him take no care for any set times; for his thoughts will fly to it, of themselves; so as the spaces of other business, or studies, will suffice. A man's nature, runs either to herbs or weeds; therefore let him seasonably water the one, and destroy the other.

Of Custom And Education

MEN'S thoughts, are much according to their inclination; their discourse and speeches, according to their learning and infused opinions; but their deeds, are after as they have been accustomed. And therefore, as Machiavel well noteth (though in an evil-favored instance), there is no trusting to the force of nature, nor to the bravery of words, except it be corroborate by custom. His instance is, that for the achieving of a desperate conspiracy, a man should not rest upon the fierceness of any man's nature, or his resolute undertakings; but take such an one, as hath had his hands formerly in blood. But Machiavel knew not of a Friar Clement, nor a Ravillac, nor a Jaureguy, nor a Baltazar Gerard; yet his rule holdeth still, that nature, nor the engagement of words, are not so forcible, as custom. Only superstition is now so well advanced, that men of the first blood, are as firm as butchers by occupation; and votary resolution, is made equipollent to custom, even in matter of blood. In other things, the predominancy of custom is everywhere visible; insomuch as a man would wonder, to hear men profess, protest, engage, give great words, and then do, just as they have done before; as if they were dead images, and engines moved only by the wheels of custom. We see also the reign or tyranny of custom, what it is. The Indians (I mean the sect of their wise men) lay themselves quietly upon a stock of wood, and so sacrifice themselves by fire. Nay, the wives strive to be burned, with the corpses of their husbands. The lads of Sparta, of ancient time, were wont to be scourged upon the altar of Diana, without so much as queching. I remember, in the beginning of Queen Elizabeth's time of England, an Irish rebel condemned, put up a petition to the deputy, that he might be hanged in a withe, and not in an halter; because it had been so used, with former rebels. There be monks in Russia, for penance, that will sit a whole night in a vessel of water, till they be engaged with hard ice. Many examples may be put of the force of custom, both upon mind and body. Therefore, since custom is the principal magistrate of man's life, let men by all means endeavor, to obtain good customs. Certainly custom is most perfect, when it beginneth in young years: this we call education; which is, in effect, but an early custom. So we see, in languages, the tongue is more pliant to all expressions and sounds, the joints are more supple, to all feats of activity and motions, in youth than afterwards. For it is true, that late learners cannot so well take the ply; except it be in some minds, that have not suffered themselves to fix, but have kept themselves open, and prepared to receive continual amendment, which is exceeding rare. But if the force of custom simple and separate, be great, the force of custom copulate and conjoined and collegiate, is far greater. For there example teacheth, company comforteth, emulation quickeneth, glory raiseth: so as in such places the force of custom is in his exaltation. Certainly the great multiplication of virtues upon human nature, resteth upon societies well ordained and disciplined. For commonwealths, and good governments, do nourish virtue grown but do not much mend the deeds. But the misery is, that the most effectual means, are now applied to the ends, least to be desired.

Of Fortune

IT CANNOT be denied, but outward accidents conduce much to fortune; favor, opportunity, death of others, occasion fitting virtue. But chiefly, the mould of a man's fortune is in his own hands. Faber quisque fortunae suae, saith the poet. And the most frequent of external causes is, that the folly of one man, is the fortune of another. For no man prospers so suddenly, as by others' errors. Serpens nisi serpentem comederit non fit draco. Overt and apparent virtues, bring forth praise; but there be secret and hidden virtues, that bring forth fortune; certain deliveries of a man's self, which have no name. The Spanish name, desemboltura, partly expresseth them; when there be not stonds nor restiveness in a man's nature; but that the wheels of his mind, keep way with the wheels of his fortune. For so Livy (after he had described Cato Major in these words, In illo viro tantum robur corporis et animi fuit, ut quocunque loco natus esset, fortunam sibi facturus videretur) falleth upon that, that he had versatile ingenium. Therefore if a man look sharply and attentively, he shall see Fortune: for though she be blind, yet she is not invisible. The way of fortune, is like the Milken Way in the sky; which is a meeting or knot of a number of small stars; not seen asunder, but giving light together. So are there a number of little, and scarce discerned virtues, or rather faculties and customs, that make men fortunate. The Italians note some of them, such as a man would little think. When they speak of one that cannot do amiss, they will throw in, into his other conditions, that he hath Poco di matto. And certainly there be not two more fortunate properties, than to have a little of the fool, and not too much of the honest. Therefore extreme lovers of their country or masters, were never fortunate, neither can they be. For when a man placeth his thoughts without himself, he goeth not his own way. An hasty fortune maketh an enterpriser and remover (the French hath it better, entreprenant, or remuant); but the exercised fortune maketh the able man. Fortune is to be honored and respected, and it be but for her daughters, Confidence and Reputation. For those two, Felicity breedeth; the first within a man's self, the latter in others towards him. All wise men, to decline the envy of their own virtues, use to ascribe them to Providence and Fortune; for so they may the better assume them: and, besides, it is greatness in a man, to be the care of the higher powers. So Caesar said to the pilot in the tempest, Caesarem portas, et fortunam ejus. So Sylla chose the name of Felix, and not of Magnus. And it hath been noted, that those who ascribe openly too much to their own wisdom and policy, end infortunate. It is written that Timotheus the Athenian, after he had, in the account he gave to the state of his government, often interlaced this speech, and in this, Fortune had no part, never prospered in anything, he undertook afterwards. Certainly there be, whose fortunes are like Homer's verses, that have a slide and easiness more than the verses of other poets; as Plutarch saith of Timoleon's fortune, in respect of that of Agesilaus or Epaminondas. And that this should be, no doubt it is much, in a man's self.

Of Usury

MANY have made witty invectives against usury. They say that it is a pity, the devil should have God's part, which is the tithe. That the usurer is the greatest Sabbath-breaker, because his plough goeth every Sunday. That the usurer is the drone, that Virgil speaketh of;

Ignavum fucos pecus a praesepibus arcent.

That the usurer breaketh the first law, that was made for mankind after the fall, which was, in sudore vultus tui comedes panem tuum; not, in sudore vultus alieni. That usurers should have orange-tawny bonnets, because they do judaize. That it is against nature for money to beget money; and the like. I say this only, that usury is a concessum propter duritiem cordis; for since there must be borrowing and lending, and men are so hard of heart, as they will not lend freely, usury must be permitted. Some others, have made suspicious and cunning propositions of banks, discovery of men's estates, and other inventions. But few have spoken of usury usefully. It is good to set before us, the incommodities and commodities of usury, that the good, may be either weighed out or culled out; and warily to provide, that while we make forth to that which is better, we meet not with that which is worse.

The discommodities of usury are, First, that it makes fewer merchants. For were it not for this lazy trade of usury, money would not be still, but would in great part be employed upon merchandizing; which is the vena porta of wealth in a state. The second, that it makes poor merchants. For, as a farmer cannot husband his ground so well, if he sit at a great rent; so the merchant cannot drive his trade so well, if he sit at great usury. The third is incident to the other two; and that is the decay of customs of kings or states, which ebb or flow, with merchandizing. The fourth, that it bringeth the treasure of a realm, or state, into a few hands. For the usurer being at certainties, and others at uncertainties, at the end of the game, most of the money will be in the box; and ever a state flourisheth, when wealth is more equally spread. The fifth, that it beats down the price of land; for the employment of money, is chiefly either merchandizing or purchasing; and usury waylays both. The sixth, that it doth dull and damp all industries, improvements, and new inventions, wherein money would be stirring, if it were not for this slug. The last, that it is the canker and ruin of many men's estates; which, in process of time, breeds a public poverty.

On the other side, the commodities of usury are, first, that howsoever usury in some respect hindereth merchandizing, yet in some other it advanceth it; for it is certain that the greatest part of trade is driven by young merchants, upon borrowing at interest; so as if the usurer either call in, or keep back, his money, there will ensue, presently, a great stand of trade. The second is, that were it not for this easy borrowing upon interest, men's necessities would draw upon them a most sudden undoing; in that they would be forced to sell their means (be it lands or goods) far under foot; and so, whereas usury doth but gnaw upon them, bad markets would swallow them quite up. As for mortgaging or pawning, it will little mend the matter: for either men will not take pawns without use; or if they do, they will look precisely for the forfeiture. I remember a cruel moneyed man in the country, that would say, The devil take this usury, it keeps us from forfeitures, of mortgages and bonds. The third and last is, that it is a vanity to conceive, that there would be ordinary borrowing without profit; and it is impossible to conceive, the number of inconveniences that will ensue, if borrowing be cramped. Therefore to speak of the abolishing of usury is idle. All states have ever had it, in one kind or rate, or other. So as that opinion must be sent to Utopia.

To speak now of the reformation, and reiglement, of usury; how the discommodities of it may be best avoided, and the commodities retained. It appears, by the balance of commodities and discommodities of usury, two things are to be reconciled. The one, that the tooth of usury be grinded, that it bite not too much; the other, that there be left open a means, to invite moneyed men to lend to the merchants, for the continuing and quickening of trade. This cannot be done, except you introduce two several sorts of usury, a less and a greater. For if you reduce usury to one low rate, it will ease the common borrower, but the merchant will be to seek for money. And it is to be noted, that the trade of merchandize, being the most lucrative, may bear usury at a good rate; other contracts not so.

To serve both intentions, the way would be briefly thus. That there be two rates of usury: the one free, and general for all; the other under license only, to certain persons, and in certain places of merchandizing. First, therefore, let usury in general, be reduced to five in the hundred; and let that rate be proclaimed, to be free and current; and let the state shut itself out, to take any penalty for the same. This will preserve borrowing, from any general stop or dryness. This will ease infinite borrowers in the country. This will, in good part, raise the price of land, because land purchased at sixteen years' purchase will yield six in the hundred, and somewhat more; whereas this rate of interest, yields but five. This by like reason will encourage, and edge, industrious and profitable improvements; because many will rather venture in that kind, than take five in the hundred, especially having been used to greater profit. Secondly, let there be certain persons licensed, to lend to known merchants, upon usury at a higher rate; and let it be with the cautions following. Let the rate be, even with the merchant himself, somewhat more easy than that he used formerly to pay; for by that means, all borrowers, shall have some ease by this reformation, be he merchant, or whosoever. Let it be no bank or common stock, but every man be master of his own money. Not that I altogether mislike banks, but they will hardly be brooked, in regard of certain suspicions. Let the state be answered some small matter for the license, and the rest left to the lender; for if the abatement be but small, it will no whit discourage the lender. For he, for example, that took before ten or nine in the hundred, will sooner descend to eight in the hundred than give over his trade of usury, and go from certain gains, to gains of hazard. Let these licensed lenders be in number indefinite, but restrained to certain principal cities and towns of merchandizing; for then they will be hardly able to color other men's moneys in the country: so as the license of nine will not suck away the current rate of five; for no man will send his moneys far off, nor put them into unknown hands.

If it be objected that this doth in a sort authorize usury, which before, was in some places but permissive; the answer is, that it is better to mitigate usury, by declaration, than to suffer it to rage, by connivance.

Of Youth And Age

A MAN that is young in years, may be old in hours, if he have lost no time. But that happeneth rarely. Generally, youth is like the first cogitations, not so wise as the second. For there is a youth in thoughts, as well as in ages. And yet the invention of young men, is more lively than that of old; and imaginations stream into their minds better, and, as it were, more divinely. Natures that have much heat, and great and violent desires and perturbations, are not ripe for action, till they have passed the meridian of their years; as it was with Julius Caesar and Septimius Severus. Of the latter, of whom it is said, Juventutem egit erroribus, imo furoribus, plenam. And yet he was the ablest emperor, almost, of all the list. But reposed natures may do well in youth. As it is seen in Augustus Caesar, Cosmus Duke of Florence, Gaston de Foix, and others. On the other side, heat and vivacity in age, is an excellent composition for business. Young men are fitter to invent, than to judge; fitter for execution, than for counsel; and fitter for new projects, than for settled business. For the experience of age, in things that fall within the compass of it, directeth them; but in new things, abuseth them.

The errors of young men, are the ruin of business; but the errors of aged men, amount but to this, that more might have been done, or sooner. Young men, in the conduct and manage of actions, embrace more than they can hold; stir more than they can quiet; fly to the end, without consideration of the means and degrees; pursue some few principles, which they have chanced upon absurdly; care not to innovate, which draws unknown inconveniences; use extreme remedies at first; and, that which doubleth all errors, will not acknowledge or retract them; like an unready horse, that will neither stop nor turn. Men of age object too much, consult too long, adventure too little, repent too soon, and seldom drive business home to the full period, but content themselves with a mediocrity of success. Certainly it is good to compound employments of both; for that will be good for the present, because the virtues of either age, may correct the defects of both; and good for succession, that young men may be learners, while men in age are actors; and, lastly, good for extern accidents, because authority followeth old men, and favor and popularity, youth. But for the moral part, perhaps youth will have the pre-eminence, as age hath for the politic. A certain rabbin, upon the text, Your young men shall see visions, and your old men shall dream dreams, inferreth that young men, are admitted nearer to God than old, because vision, is a clearer revelation, than a dream. And certainly, the more a man drinketh of the world, the more it intoxicateth; and age doth profit rather in the powers of understanding, than in the virtues of the will and affections. There be some, have an over-early ripeness in their years, which fadeth betimes. These are, first, such as have brittle wits, the edge whereof is soon turned; such as was Hermogenes the rhetorician, whose books are exceeding subtle; who afterwards waxed stupid. A second sort, is of those that have some natural dispositions which have better grace in youth, than in age; such as is a fluent and luxuriant speech; which becomes youth well, but not age: so Tully saith of Hortensius, Idem manebat, neque idem decebat. The third is of such, as take too high a strain at the first, and are magnanimous, more than tract of years can uphold. As was Scipio Africanus, of whom Livy saith in effect, Ultima primis cedebant.

Of Beauty

VIRTUE is like a rich stone, best plain set; and surely virtue is best, in a body that is comely, though not of delicate features; and that hath rather dignity of presence, than beauty of aspect. Neither is it almost seen, that very beautiful persons are otherwise of great virtue; as if nature were rather busy, not to err, than in labor to produce excellency. And therefore they prove accomplished, but not of great spirit; and study rather behavior, than virtue. But this holds not always: for Augustus Caesar, Titus Vespasianus, Philip le Belle of France, Edward the Fourth of England, Alcibiades of Athens, Ismael the Sophy of Persia, were all high and great spirits; and yet the most beautiful men of their times. In beauty, that of favor, is more than that of color; and that of decent and gracious motion, more than that of favor. That is the best part of beauty, which a picture cannot express; no, nor the first sight of the life. There is no excellent beauty, that hath not some strangeness in the proportion. A man cannot tell whether Apelles, or Albert Durer, were the more trifler; whereof the one, would make a personage by geometrical proportions; the other, by taking the best parts out of divers faces, to make one excellent. Such personages, I think, would please nobody, but the painter that made them. Not but I think a painter may make a better face than ever was; but he must do it by a kind of felicity (as a musician that maketh an excellent air in music), and not by rule. A man shall see faces, that if you examine them part by part, you shall find never a good; and yet altogether do well. If it be true that the principal part of beauty is in decent motion, certainly it is no marvel, though persons in years seem many times more amiable; pulchrorum autumnus pulcher; for no youth can be comely but by pardon, and considering the youth, as to make up the comeliness. Beauty is as summer fruits, which are easy to corrupt, and cannot last; and for the most part it makes a dissolute youth, and an age a little out of countenance; but yet certainly again, if it light well, it maketh virtue shine, and vices blush.

Of Deformity

DEFORMED persons are commonly even with nature; for as nature hath done ill by them, so do they by nature; being for the most part (as the Scripture saith) void of natural affection; and so they have their revenge of nature. Certainly there is a consent, between the body and the mind; and where nature erreth in the one, she ventureth in the other. Ubi peccat in uno, periclitatur in altero. But because there is, in man, an election touching the frame of his mind, and a necessity in the frame of his body, the stars of natural inclination are sometimes obscured, by the sun of discipline and virtue. Therefore it is good to consider of deformity, not as a sign, which is more deceivable; but as a cause, which seldom faileth of the effect. Whosoever hath anything fixed in his person, that doth induce contempt, hath also a perpetual spur in himself, to rescue and deliver himself from scorn. Therefore all deformed persons, are extreme bold. First, as in their own defence, as being exposed to scorn; but in process of time, by a general habit. Also it stirreth in them industry, and especially of this kind, to watch and observe the weakness of others, that they may have somewhat to repay. Again, in their superiors, it quencheth jealousy towards them, as persons that they think they may, at pleasure, despise: and it layeth their competitors and emulators asleep; as never believing they should be in possibility of advancement, till they see them in possession. So that upon the matter, in a great wit, deformity is an advantage to rising. Kings in ancient times (and at this present in some countries) were wont to put great trust in eunuchs; because they that are envious towards all are more obnoxious and officious, towards one. But yet their trust towards them, hath rather been as to good spials, and good whisperers, than good magistrates and officers. And much like is the reason of deformed persons. Still the ground is, they will, if they be of spirit, seek to free themselves from scorn; which must be either by virtue or malice; and therefore let it not be marvelled, if sometimes they prove excellent persons; as was Agesilaus, Zanger the son of Solyman, AEsop, Gasca, President of Peru; and Socrates may go likewise amongst them; with others.

Of Building

HOUSES are built to live in, and not to look on; therefore let use be preferred before uniformity, except where both may be had. Leave the goodly fabrics of houses, for beauty only, to the enchanted palaces of the poets; who build them with small cost. He that builds a fair house, upon an ill seat, committeth himself to prison. Neither do I reckon it an ill seat, only where the air is unwholesome; but likewise where the air is unequal; as you shall see many fine seats set upon a knap of ground, environed with higher hills round about it; whereby the heat of the sun is pent in, and the wind gathereth as in troughs; so as you shall have, and that suddenly, as great diversity of heat and cold as if you dwelt in several places. Neither is it ill air only that maketh an ill seat, but ill ways, ill markets; and, if you will consult with Momus, ill neighbors. I speak not of many more; want of water; want of wood, shade, and shelter; want of fruitfulness, and mixture of grounds of several natures; want of prospect; want of level grounds; want of places at some near distance for sports of hunting, hawking, and races; too near the sea, too remote; having the commodity of navigable rivers, or the discommodity of their overflowing; too far off from great cities, which may hinder business, or too near them, which lurcheth all provisions, and maketh everything dear; where a man hath a great living laid together, and where he is scanted: all which, as it is impossible perhaps to find together, so it is good to know them, and think of them, that a man may take as many as he can; and if he have several dwellings, that he sort them so that what he wanteth in the one, he may find in the other. Lucullus answered Pompey well; who, when he saw his stately galleries, and rooms so large and lightsome, in one of his houses, said, Surely an excellent place for summer, but how do you in winter? Lucullus answered, Why, do you not think me as wise as some fowl are, that ever change their abode towards the winter?

To pass from the seat, to the house itself; we will do as Cicero doth in the orator's art; who writes books De Oratore, and a book he entitles Orator; whereof the former, delivers the precepts of the art, and the latter, the perfection. We will therefore describe a princely palace, making a brief model thereof. For it is strange to see, now in Europe, such huge buildings as the Vatican and Escurial and some others be, and yet scarce a very fair room in them.

First, therefore, I say you cannot have a perfect palace except you have two several sides; a side for the banquet, as it is spoken of in the book of Hester, and a side for the household; the one for feasts and triumphs, and the other for dwelling. I understand both these sides to be not only returns, but parts of the front; and to be uniform without, though severally partitioned within; and to be on both sides of a great and stately tower, in the midst of the front, that, as it were, joineth them together on either hand. I would have on the side of the banquet, in front, one only goodly room above stairs, of some forty foot high; and under it a room for a dressing, or preparing place, at times of triumphs. On the other side, which is the household side, I wish it divided at the first, into a hall and a chapel (with a partition between); both of good state and bigness; and those not to go all the length, but to have at the further end, a winter and a summer parlor, both fair. And under these rooms, a fair and large cellar, sunk under ground; and likewise some privy kitchens, with butteries and pantries, and the like. As for the tower, I would have it two stories, of eighteen foot high apiece, above the two wings; and a goodly leads upon the top, railed with statuas interposed; and the same tower to be divided into rooms, as shall be thought fit. The stairs likewise to the upper rooms, let them be upon a fair open newel, and finely railed in, with images of wood, cast into a brass color; and a very fair landing-place at the top. But this to be, if you do not point any of the lower rooms, for a dining place of servants. For otherwise, you shall have the servants' dinner after your own: for the steam of it, will come up as in a tunnel. And so much for the front. Only I understand the height of the first stairs to be sixteen foot, which is the height of the lower room.

Beyond this front, is there to be a fair court, but three sides of it, of a far lower building than the front. And in all the four corners of that court, fair staircases, cast into turrets, on the outside, and not within the row of buildings themselves. But those towers, are not to be of the height of the front, but rather proportionable to the lower building. Let the court not be paved, for that striketh up a great heat in summer, and much cold in winter. But only some side alleys, with a cross, and the quarters to graze, being kept shorn, but not too near shorn. The row of return on the banquet side, let it be all stately galleries: in which galleries let there be three, or five, fine cupolas in the length of it, placed at equal distance; and fine colored windows of several works. On the household side, chambers of presence and ordinary entertainments, with some bed-chambers; and let all three sides be a double house, without thorough lights on the sides, that you may have rooms from the sun, both for forenoon and afternoon. Cast it also, that you may have rooms, both for summer and winter; shady for summer, and warm for winter. You shall have sometimes fair houses so full of glass, that one cannot tell where to become, to be out of the sun or cold. For inbowed windows, I hold them of good use (in cities, indeed, upright do better, in respect of the uniformity towards the street); for they be pretty retiring places for conference; and besides, they keep both the wind and sun off; for that which would strike almost through the room, doth scarce pass the window. But let them be but few, four in the court, on the sides only.

Beyond this court, let there be an inward court, of the same square and height; which is to be environed with the garden on all sides; and in the inside, cloistered on all sides, upon decent and beautiful arches, as high as the first story. On the under story, towards the garden, let it be turned to a grotto, or a place of shade, or estivation. And only have opening and windows towards the garden; and be level upon the floor, no whit sunken under ground, to avoid all dampishness. And let there be a fountain, or some fair work of statuas, in the midst of this court; and to be paved as the other court was. These buildings to be for privy lodgings on both sides; and the end for privy galleries. Whereof you must foresee that one of them be for an infirmary, if the prince or any special person should be sick, with chambers, bed-chamber, antecamera, and recamera joining to it. This upon the second story. Upon the ground story, a fair gallery, open, upon pillars; and upon the third story likewise, an open gallery, upon pillars, to take the prospect and freshness of the garden. At both corners of the further side, by way of return, let there be two delicate or rich cabinets, daintily paved, richly hanged, glazed with crystalline glass, and a rich cupola in the midst; and all other elegancy that may be thought upon. In the upper gallery too, I wish that there may be, if the place will yield it, some fountains running in divers places from the wall, with some fine avoidances. And thus much for the model of the palace; save that you must have, before you come to the front, three courts. A green court plain, with a wall about it; a second court of the same, but more garnished, with little turrets, or rather embellishments, upon the wall; and a third court, to make a square with the front, but not to be built, nor yet enclosed with a naked wall, but enclosed with terraces, leaded aloft, and fairly garnished, on the three sides; and cloistered on the inside, with pillars, and not with arches below. As for offices, let them stand at distance, with some low galleries, to pass from them to the palace itself.

Of Gardens

GOD Almighty first planted a garden. And indeed it is the purest of human pleasures. It is the greatest refreshment to the spirits of man; without which, buildings and palaces are but gross handiworks; and a man shall ever see, that when ages grow to civility and elegancy, men come to build stately sooner than to garden finely; as if gardening were the greater perfection. I do hold it, in the royal ordering of gardens, there ought to be gardens, for all the months in the year; in which severally things of beauty may be then in season. For December, and January, and the latter part of November, you must take such things as are green all winter: holly; ivy; bays; juniper; cypress-trees; yew; pine-apple-trees; fir-trees; rosemary; lavender; periwinkle, the white, the purple, and the blue; germander; flags; orange-trees; lemon-trees; and myrtles, if they be stoved; and sweet marjoram, warm set. There followeth, for the latter part of January and February, the mezereon-tree, which then blossoms; crocus vernus, both the yellow and the grey; primroses, anemones; the early tulippa; hyacinthus orientalis; chamairis; fritellaria. For March, there come violets, specially the single blue, which are the earliest; the yellow daffodil; the daisy; the almond-tree in blossom; the peach-tree in blossom; the cornelian-tree in blossom; sweet-briar. In April follow the double white violet; the wallflower; the stock-gilliflower; the cowslip; flowerdelices, and lilies of all natures; rosemary-flowers; the tulippa; the double peony; the pale daffodil; the French honeysuckle; the cherry-tree in blossom; the damson and plum-trees in blossom; the white thorn in leaf; the lilac-tree. In May and June come pinks of all sorts, specially the blushpink; roses of all kinds, except the musk, which comes later; honeysuckles; strawberries; bugloss; columbine; the French marigold, flos Africanus; cherry-tree in fruit; ribes; figs in fruit; rasps; vineflowers; lavender in flowers; the sweet satyrian, with the white flower; herba muscaria; lilium convallium; the apple-tree in blossom. In July come gilliflowers of all varieties; musk-roses; the lime-tree in blossom; early pears and plums in fruit; jennetings, codlins. In August come plums of all sorts in fruit; pears; apricocks; berberries; filberds; musk-melons; monks-hoods, of all colors. In September come grapes; apples; poppies of all colors; peaches; melocotones; nectarines; cornelians; wardens; quinces. In October and the beginning of November come services; medlars; bullaces; roses cut or removed to come late; hollyhocks; and such like. These particulars are for the climate of London; but my meaning is perceived, that you may have ver perpetuum, as the place affords.

And because the breath of flowers is far sweeter in the air (where it comes and goes like the warbling of music) than in the hand, therefore nothing is more fit for that delight, than to know what be the flowers and plants that do best perfume the air. Roses, damask and red, are fast flowers of their smells; so that you may walk by a whole row of them, and find nothing of their sweetness; yea though it be in a morning's dew. Bays likewise yield no smell as they grow. Rosemary little; nor sweet marjoram. That which above all others yields the sweetest smell in the air is the violet, specially the white double violet, which comes twice a year; about the middle of April, and about Bartholomew-tide. Next to that is the musk-rose. Then the strawberry-leaves dying, which yield a most excellent cordial smell. Then the flower of vines; it is a little dust, like the dust of a bent, which grows upon the cluster in the first coming forth. Then sweet-briar. Then wall-flowers, which are very delightful to be set under a parlor or lower chamber window. Then pinks and gilliflowers, especially the matted pink and clove gilliflower. Then the flowers of the lime-tree. Then the honeysuckles, so they be somewhat afar off. Of beanflowers I speak not, because they are field flowers. But those which perfume the air most delightfully, not passed by as the rest, but being trodden upon and crushed, are three; that is, burnet, wildthyme, and watermints. Therefore you are to set whole alleys of them, to have the pleasure when you walk or tread.

For gardens (speaking of those which are indeed princelike, as we have done of buildings), the contents ought not well to be under thirty acres of ground; and to be divided into three parts; a green in the entrance; a heath or desert in the going forth; and the main garden in the midst; besides alleys on both sides. And I like well that four acres of ground be assigned to the green; six to the heath; four and four to either side; and twelve to the main garden. The green hath two pleasures: the one, because nothing is more pleasant to the eye than green grass kept finely shorn; the other, because it will give you a fair alley in the midst, by which you may go in front upon a stately hedge, which is to enclose the garden. But because the alley will be long, and, in great heat of the year or day, you ought not to buy the shade in the garden, by going in the sun through the green, therefore you are, of either side the green, to plant a covert alley upon carpenter's work, about twelve foot in height, by which you may go in shade into the garden. As for the making of knots or figures, with divers colored earths, that they may lie under the windows of the house on that side which the garden stands, they be but toys; you may see as good sights, many times, in tarts. The garden is best to be square, encompassed on all the four sides with a stately arched hedge. The arches to be upon pillars of carpenter's work, of some ten foot high, and six foot broad; and the spaces between of the same dimension with the breadth of the arch. Over the arches let there be an entire hedge of some four foot high, framed also upon carpenter's work; and upon the upper hedge, over every arch, a little turret, with a belly, enough to receive a cage of birds: and over every space between the arches some other little figure, with broad plates of round colored glass gilt, for the sun to play upon. But this hedge I intend to be raised upon a bank, not steep, but gently slope, of some six foot, set all with flowers. Also I understand, that this square of the garden, should not be the whole breadth of the ground, but to leave on either side, ground enough for diversity of side alleys; unto which the two covert alleys of the green, may deliver you. But there must be no alleys with hedges, at either end of this great enclosure; not at the hither end, for letting your prospect upon this fair hedge from the green; nor at the further end, for letting your prospect from the hedge, through the arches upon the heath.

For the ordering of the ground, within the great hedge, I leave it to variety of device; advising nevertheless, that whatsoever form you cast it into, first, it be not too busy, or full of work. Wherein I, for my part, do not like images cut out in juniper or other garden stuff; they be for children. Little low hedges, round, like welts, with some pretty pyramids, I like well; and in some places, fair columns upon frames of carpenter's work. I would also have the alleys, spacious and fair. You may have closer alleys, upon the side grounds, but none in the main garden. I wish also, in the very middle, a fair mount, with three ascents, and alleys, enough for four to walk abreast; which I would have to be perfect circles, without any bulwarks or embossments; and the whole mount to be thirty foot high; and some fine banqueting-house, with some chimneys neatly cast, and without too much glass.

For fountains, they are a great beauty and refreshment; but pools mar all, and make the garden unwholesome, and full of flies and frogs. Fountains I intend to be of two natures: the one that sprinkleth or spouteth water; the other a fair receipt of water, of some thirty or forty foot square, but without fish, or slime, or mud. For the first, the ornaments of images gilt, or of marble, which are in use, do well: but the main matter is so to convey the water, as it never stay, either in the bowls or in the cistern; that the water be never by rest discolored, green or red or the like; or gather any mossiness or putrefaction. Besides that, it is to be cleansed every day by the hand. Also some steps up to it, and some fine pavement about it, doth well. As for the other kind of fountain, which we may call a bathing pool, it may admit much curiosity and beauty; wherewith we will not trouble ourselves: as, that the bottom be finely paved, and with images; the sides likewise; and withal embellished with colored glass, and such things of lustre; encompassed also with fine rails of low statuas. But the main point is the same which we mentioned in the former kind of fountain; which is, that the water be in perpetual motion, fed by a water higher than the pool, and delivered into it by fair spouts, and then discharged away under ground, by some equality of bores, that it stay little. And for fine devices, of arching water without spilling, and making it rise in several forms (of feathers, drinking glasses, canopies, and the like), they be pretty things to look on, but nothing to health and sweetness.

For the heath, which was the third part of our plot, I wish it to be framed, as much as may be, to a natural wildness. Trees I would have none in it, but some thickets made only of sweet-briar and honeysuckle, and some wild vine amongst; and the ground set with violets, strawberries, and primroses. For these are sweet, and prosper in the shade. And these to be in the heath, here and there, not in any order. I like also little heaps, in the nature of mole-hills (such as are in wild heaths), to be set, some with wild thyme; some with pinks; some with germander, that gives a good flower to the eye; some with periwinkle; some with violets; some with strawberries; some with cowslips; some with daisies; some with red roses; some with lilium convallium; some with sweet-williams red; some with bear's-foot: and the like low flowers, being withal sweet and sightly. Part of which heaps, are to be with standards of little bushes pricked upon their top, and part without. The standards to be roses; juniper; holly; berberries (but here and there, because of the smell of their blossoms); red currants; gooseberries; rosemary; bays; sweetbriar; and such like. But these standards to be kept with cutting, that they grow not out of course.

For the side grounds, you are to fill them with variety of alleys, private, to give a full shade, some of them, wheresoever the sun be. You are to frame some of them, likewise, for shelter, that when the wind blows sharp you may walk as in a gallery. And those alleys must be likewise hedged at both ends, to keep out the wind; and these closer alleys must be ever finely gravelled, and no grass, because of going wet. In many of these alleys, likewise, you are to set fruit-trees of all sorts; as well upon the walls, as in ranges. And this would be generally observed, that the borders wherein you plant your fruit-trees, be fair and large, and low, and not steep; and set with fine flowers, but thin and sparingly, lest they deceive the trees. At the end of both the side grounds, I would have a mount of some pretty height, leaving the wall of the enclosure breast high, to look abroad into the fields.

For the main garden, I do not deny, but there should be some fair alleys ranged on both sides, with fruit-trees; and some pretty tufts of fruit-trees, and arbors with seats, set in some decent order; but these to be by no means set too thick; but to leave the main garden so as it be not close, but the air open and free. For as for shade, I would have you rest upon the alleys of the side grounds, there to walk, if you be disposed, in the heat of the year or day; but to make account, that the main garden is for the more temperate parts of the year; and in the heat of summer, for the morning and the evening, or overcast days.

For aviaries, I like them not, except they be of that largeness as they may be turfed, and have living plants and bushes set in them; that the birds may have more scope, and natural nesting, and that no foulness appear in the floor of the aviary. So I have made a platform of a princely garden, partly by precept, partly by drawing, not a model, but some general lines of it; and in this I have spared for no cost. But it is nothing for great princes, that for the most part taking advice with workmen, with no less cost set their things together; and sometimes add statuas and such things for state and magnificence, but nothing to the true pleasure of a garden.

Of Negotiating

IT IS generally better to deal by speech than by letter; and by the mediation of a third than by a man's self. Letters are good, when a man would draw an answer by letter back again; or when it may serve for a man's justification afterwards to produce his own letter; or where it may be danger to be interrupted, or heard by pieces. To deal in person is good, when a man's face breedeth regard, as commonly with inferiors; or in tender cases, where a man's eye, upon the countenance of him with whom he speaketh, may give him a direction how far to go; and generally, where a man will reserve to himself liberty, either to disavow or to expound. In choice of instruments, it is better to choose men of a plainer sort, that are like to do that, that is committed to them, and to report back again faithfully the success, than those that are cunning, to contrive, out of other men's business, somewhat to grace themselves, and will help the matter in report for satisfaction's sake. Use also such persons as affect the business, wherein they are employed; for that quickeneth much; and such, as are fit for the matter; as bold men for expostulation, fair-spoken men for persuasion, crafty men for inquiry and observation, froward, and absurd men, for business that doth not well bear out itself. Use also such as have been lucky, and prevailed before, in things wherein you have employed them; for that breeds confidence, and they will strive to maintain their prescription. It is better to sound a person, with whom one deals afar off, than to fall upon the point at first; except you mean to surprise him by some short question. It is better dealing with men in appetite, than with those that are where they would be. If a man deal with another upon conditions, the start or first performance is all; which a man cannot reasonably demand, except either the nature of the thing be such, which must go before; or else a man can persuade the other party, that he shall still need him in some other thing; or else that he be counted the honester man. All practice is to discover, or to work. Men discover themselves in trust, in passion, at unawares, and of necessity, when they would have somewhat done, and cannot find an apt pretext. If you would work any man, you must either know his nature and fashions, and so lead him; or his ends, and so persuade him; or his weakness and disadvantages, and so awe him; or those that have interest in him, and so govern him. In dealing with cunning persons, we must ever consider their ends, to interpret their speeches; and it is good to say little to them, and that which they least look for. In all negotiations of difficulty, a man may not look to sow and reap at once; but must prepare business, and so ripen it by degrees.

Of Followers And Friends

COSTLY followers are not to be liked; lest while a man maketh his train longer, he make his wings shorter. I reckon to be costly, not them alone which charge the purse, but which are wearisome, and importune in suits. Ordinary followers ought to challenge no higher conditions, than countenance, recommendation, and protection from wrongs. Factious followers are worse to be liked, which follow not upon affection to him, with whom they range themselves, but upon discontentment conceived against some other; whereupon commonly ensueth that ill intelligence, that we many times see between great personages. Likewise glorious followers, who make themselves as trumpets of the commendation of those they follow, are full of inconvenience; for they taint business through want of secrecy; and they export honor from a man, and make him a return in envy. There is a kind of followers likewise, which are dangerous, being indeed espials; which inquire the secrets of the house, and bear tales of them, to others. Yet such men, many times, are in great favor; for they are officious, and commonly exchange tales. The following by certain estates of men, answerable to that, which a great person himself professeth (as of soldiers, to him that hath been employed in the wars, and the like), hath ever been a thing civil, and well taken, even in monarchies; so it be without too much pomp or popularity. But the most honorable kind of following, is to be followed as one, that apprehendeth to advance virtue, and desert, in all sorts of persons. And yet, where there is no eminent odds in sufficiency, it is better to take with the more passable, than with the more able. And besides, to speak truth, in base times, active men are of more use than virtuous. It is true that in government, it is good to use men of one rank equally: for to countenance some extraordinarily, is to make them insolent, and the rest discontent; because they may claim a due. But contrariwise, in favor, to use men with much difference and election is good; for it maketh the persons preferred more thankful, and the rest more officious: because all is of favor. It is good discretion, not to make too much of any man at the first; because one cannot hold out that proportion. To be governed (as we call it) by one is not safe; for it shows softness, and gives a freedom, to scandal and disreputation; for those, that would not censure or speak ill of a man immediately, will talk more boldly of those that are so great with them, and thereby wound their honor. Yet to be distracted with many is worse; for it makes men to be of the last impression, and full of change. To take advice of some few friends, is ever honorable; for lookers-on many times see more than gamesters; and the vale best discovereth the hill. There is little friendship in the world, and least of all between equals, which was wont to be magnified. That that is, is between superior and inferior, whose fortunes may comprehend the one the other.

Of Suitors

MANY ill matters and projects are undertaken; and private suits do putrefy the public good. Many good matters, are undertaken with bad minds; I mean not only corrupt minds, but crafty minds, that intend not performance. Some embrace suits, which never mean to deal effectually in them; but if they see there may be life in the matter, by some other mean, they will be content to win a thank, or take a second reward, or at least to make use, in the meantime, of the suitor's hopes. Some take hold of suits, only for an occasion to cross some other; or to make an information, whereof they could not otherwise have apt pretext; without care what become of the suit, when that turn is served; or, generally, to make other men's business a kind of entertainment, to bring in their own. Nay, some undertake suits, with a full purpose to let them fall; to the end to gratify the adverse party, or competitor. Surely there is in some sort a right in every suit; either a right of equity, if it be a suit of controversy; or a right of desert, if it be a suit of petition. If affection lead a man to favor the wrong side in justice, let him rather use his countenance to compound the matter, than to carry it. If affection lead a man to favor the less worthy in desert, let him do it, without depraving or disabling the better deserver. In suits which a man doth not well understand, it is good to refer them to some friend of trust and judgment, that may report, whether he may deal in them with honor: but let him choose well his referendaries, for else he may be led by the nose. Suitors are so distasted with delays and abuses, that plain dealing, in denying to deal in suits at first, and reporting the success barely, and in challenging no more thanks than one hath deserved, is grown not only honorable, but also gracious. In suits of favor, the first coming ought to take little place: so far forth, consideration may be had of his trust, that if intelligence of the matter could not otherwise have been had, but by him, advantage be not taken of the note, but the party left to his other means; and in some sort recompensed, for his discovery. To be ignorant of the value of a suit, is simplicity; as well as to be ignorant of the right thereof, is want of conscience. Secrecy in suits, is a great mean of obtaining; for voicing them to be in forwardness, may discourage some kind of suitors, but doth quicken and awake others. But timing of the suit is the principal. Timing, I say, not only in respect of the person that should grant it, but in respect of those, which are like to cross it. Let a man, in the choice of his mean, rather choose the fittest mean, than the greatest mean; and rather them that deal in certain things, than those that are general. The reparation of a denial, is sometimes equal to the first grant; if a man show himself neither dejected nor discontented. Iniquum petas ut aequum feras is a good rule, where a man hath strength of favor: but otherwise, a man were better rise in his suit; for he, that would have ventured at first to have lost the suitor, will not in the conclusion lose both the suitor, and his own former favor. Nothing is thought so easy a request to a great person, as his letter; and yet, if it be not in a good cause, it is so much out of his reputation. There are no worse instruments, than these general contrivers of suits; for they are but a kind of poison, and infection, to public proceedings.

Of Studies

STUDIES serve for delight, for ornament, and for ability. Their chief use for delight, is in privateness and retiring; for ornament, is in discourse; and for ability, is in the judgment, and disposition of business. For expert men can execute, and perhaps judge of particulars, one by one; but the general counsels, and the plots and marshalling of affairs, come best, from those that are learned. To spend too much time in studies is sloth; to use them too much for ornament, is affectation; to make judgment wholly by their rules, is the humor of a scholar. They perfect nature, and are perfected by experience: for natural abilities are like natural plants, that need proyning, by study; and studies themselves, do give forth directions too much at large, except they be bounded in by experience. Crafty men contemn studies, simple men admire them, and wise men use them; for they teach not their own use; but that is a wisdom without them, and above them, won by observation. Read not to contradict and confute; nor to believe and take for granted; nor to find talk and discourse; but to weigh and consider. Some books are to be tasted, others to be swallowed, and some few to be chewed and digested; that is, some books are to be read only in parts; others to be read, but not curiously; and some few to be read wholly, and with diligence and attention. Some books also may be read by deputy, and extracts made of them by others; but that would be only in the less important arguments, and the meaner sort of books, else distilled books are like common distilled waters, flashy things. Reading maketh a full man; conference a ready man; and writing an exact man. And therefore, if a man write little, he had need have a great memory; if he confer little, he had need have a present wit: and if he read little, he had need have much cunning, to seem to know, that he doth not. Histories make men wise; poets witty; the mathematics subtile; natural philosophy deep; moral grave; logic and rhetoric able to contend. Abeunt studia in mores. Nay, there is no stond or impediment in the wit, but may be wrought out by fit studies; like as diseases of the body, may have appropriate exercises. Bowling is good for the stone and reins; shooting for the lungs and breast; gentle walking for the stomach; riding for the head; and the like. So if a man's wit be wandering, let him study the mathematics; for in demonstrations, if his wit be called away never so little, he must begin again. If his wit be not apt to distinguish or find differences, let him study the Schoolmen; for they are cymini sectores. If he be not apt to beat over matters, and to call up one thing to prove and illustrate another, let him study the lawyers' cases. So every defect of the mind, may have a special receipt.

Of Faction

MANY have an opinion not wise, that for a prince to govern his estate, or for a great person to govern his proceedings, according to the respect of factions, is a principal part of policy; whereas contrariwise, the chiefest wisdom, is either in ordering those things which are general, and wherein men of several factions do nevertheless agree; or in dealing with correspondence to particular persons, one by one. But I say not that the considerations of factions, is to be neglected. Mean men, in their rising, must adhere; but great men, that have strength in themselves, were better to maintain themselves indifferent, and neutral. Yet even in beginners, to adhere so moderately, as he be a man of the one faction, which is most passable with the other, commonly giveth best way. The lower and weaker faction, is the firmer in conjunction; and it is often seen, that a few that are stiff, do tire out a greater number, that are more moderate. When one of the factions is extinguished, the remaining subdivideth; as the faction between Lucullus, and the rest of the nobles of the senate (which they called Optimates) held out awhile, against the faction of Pompey and Caesar; but when the senate's authority was pulled down, Caesar and Pompey soon after brake. The faction or party of Antonius and Octavianus Caesar, against Brutus and Cassius, held out likewise for a time; but when Brutus and Cassius were overthrown, then soon after, Antonius and Octavianus brake and subdivided. These examples are of wars, but the same holdeth in private factions. And therefore, those that are seconds in factions, do many times, when the faction subdivideth, prove principals; but many times also, they prove ciphers and cashiered; for many a man's strength is in opposition; and when that faileth, he groweth out of use. It is commonly seen, that men, once placed, take in with the contrary faction, to that by which they enter: thinking belike, that they have the first sure, and now are ready for a new purchase. The traitor in faction, lightly goeth away with it; for when matters have stuck long in balancing, the winning of some one man casteth them, and he getteth all the thanks. The even carriage between two factions, proceedeth not always of moderation, but of a trueness to a man's self, with end to make use of both. Certainly in Italy, they hold it a little suspect in popes, when they have often in their mouth Padre commune: and take it to be a sign of one, that meaneth to refer all to the greatness of his own house. Kings had need beware, how they side themselves, and make themselves as of a faction or party; for leagues within the state, are ever pernicious to monarchies: for they raise an obligation, paramount to obligation of sovereignty, and make the king tanquam unus ex nobis; as was to be seen in the League of France. When factions are carried too high and too violently, it is a sign of weakness in princes; and much to the prejudice, both of their authority and business. The motions of factions under kings ought to be, like the motions (as the astronomers speak) of the inferior orbs, which may have their proper motions, but yet still are quietly carried, by the higher motion of primum mobile.

Of Ceremonies, And Respects

HE THAT is only real, had need have exceeding great parts of virtue; as the stone had need to be rich, that is set without foil. But if a man mark it well, it is, in praise and commendation of men, as it is in gettings and gains: for the proverb is true, That light gains make heavy purses; for light gains come thick, whereas great, come but now and then. So it is true, that small matters win great commendation, because they are continually in use and in note: whereas the occasion of any great virtue, cometh but on festivals. Therefore it doth much add to a man's reputation, and is (as Queen Isabella said) like perpetual letters commendatory, to have good forms. To attain them, it almost sufficeth not to despise them; for so shall a man observe them in others; and let him trust himself with the rest. For if he labor too much to express them, he shall lose their grace; which is to be natural and unaffected. Some men's behavior is like a verse, wherein every syllable is measured; how can a man comprehend great matters, that breaketh his mind too much, to small observations? Not to use ceremonies at all, is to teach others not to use them again; and so diminisheth respect to himself; especially they be not to be omitted, to strangers and formal natures; but the dwelling upon them, and exalting them above the moon, is not only tedious, but doth diminish the faith and credit of him that speaks. And certainly, there is a kind of conveying, of effectual and imprinting passages amongst compliments, which is of singular use, if a man can hit upon it. Amongst a man's peers, a man shall be sure of familiarity; and therefore it is good, a little to keep state. Amongst a man's inferiors one shall be sure of reverence; and therefore it is good, a little to be familiar. He that is too much in anything, so that he giveth another occasion of satiety, maketh himself cheap. To apply one's self to others, is good; so it be with demonstration, that a man doth it upon regard, and not upon facility. It is a good precept generally, in seconding another, yet to add somewhat of one's own: as if you will grant his opinion, let it be with some distinction; if you will follow his motion, let it be with condition; if you allow his counsel, let it be with alleging further reason. Men had need beware, how they be too perfect in compliments; for be they never so sufficient otherwise, their enviers will be sure to give them that attribute, to the disadvantage of their greater virtues. It is loss also in business, to be too full of respects, or to be curious, in observing times and opportunities. Solomon saith, He that considereth the wind, shall not sow, and he that looketh to the clouds, shall not reap. A wise man will make more opportunities, than he finds. Men's behavior should be, like their apparel, not too strait or point device, but free for exercise or motion.

Of Praise

PRAISE is the reflection of virtue; but it is as the glass or body, which giveth the reflection. If it be from the common people, it is commonly false and naught; and rather followeth vain persons, than virtuous. For the common people understand not many excellent virtues. The lowest virtues draw praise from them; the middle virtues work in them astonishment or admiration; but of the highest virtues, they have no sense of perceiving at all. But shows, and species virtutibus similes, serve best with them. Certainly fame is like a river, that beareth up things light and swoln, and drowns things weighty and solid. But if persons of quality and judgment concur, then it is (as the Scripture saith) nomen bonum instar unguenti fragrantis. It fireth all round about, and will not easily away. For the odors of ointments are more durable, than those of flowers. There be so many false points of praise, that a man may justly hold it a suspect. Some praises proceed merely of flattery; and if he be an ordinary flatterer, he will have certain common attributes, which may serve every man; if he be a cunning flatterer, he will follow the archflatterer, which is a man's self; and wherein a man thinketh best of himself, therein the flatterer will uphold him most: but if he be an impudent flatterer, look wherein a man is conscious to himself, that he is most defective, and is most out of countenance in himself, that will the flatterer entitle him to perforce, spreta conscientia. Some praises come of good wishes and respects, which is a form due, in civility, to kings and great persons, laudando praecipere, when by telling men what they are, they represent to them, what they should be. Some men are praised maliciously, to their hurt, thereby to stir envy and jealousy towards them: pessimum genus inimicorum laudantium; insomuch as it was a proverb, amongst the Grecians, that he that was praised to his hurt, should have a push rise upon his nose; as we say, that a blister will rise upon one's tongue, that tells a lie. Certainly moderate praise, used with opportunity, and not vulgar, is that which doth the good. Solomon saith, He that praiseth his friend aloud, rising early, it shall be to him no better than a curse. Too much magnifying of man or matter, doth irritate contradiction, and procure envy and scorn. To praise a man's self, cannot be decent, except it be in rare cases; but to praise a man's office or profession, he may do it with good grace, and with a kind of magnanimity. The cardinals of Rome, which are theologues, and friars, and Schoolmen, have a phrase of notable contempt and scorn towards civil business: for they call all temporal business of wars, embassages, judicature, and other employments, sbirrerie, which is under-sheriffries; as if they were but matters, for under-sheriffs and catchpoles: though many times those under-sheriffries do more good, than their high speculations. St. Paul, when he boasts of himself, he doth oft interlace, I speak like a fool; but speaking of his calling, he saith, magnificabo apostolatum meum.

Of Vain-glory

IT WAS prettily devised of AEsop, The fly sat upon the axle-tree of the chariot wheel, and said, What a dust do I raise! So are there some vain persons, that whatsoever goeth alone, or moveth upon greater means, if they have never so little hand in it, they think it is they that carry it. They that are glorious, must needs be factious; for all bravery stands upon comparisons. They must needs be violent, to make good their own vaunts. Neither can they be secret, and therefore not effectual; but according to the French proverb, Beaucoup de bruit, peu de fruit; Much bruit little fruit. Yet certainly, there is use of this quality in civil affairs. Where there is an opinion and fame to be created, either of virtue or greatness, these men are good trumpeters. Again, as Titus Livius noteth, in the case of Antiochus and the AEtolians, There are sometimes great effects, of cross lies; as if a man, that negotiates between two princes, to draw them to join in a war against the third, doth extol the forces of either of them, above measure, the one to the other: and sometimes he that deals between man and man, raiseth his own credit with both, by pretending greater interest than he hath in either. And in these and the like kinds, it often falls out, that somewhat is produced of nothing; for lies are sufficient to breed opinion, and opinion brings on substance. In militar commanders and soldiers, vain-glory is an essential point; for as iron sharpens iron, so by glory, one courage sharpeneth another. In cases of great enterprise upon charge and adventure, a composition of glorious natures, doth put life into business; and those that are of solid and sober natures, have more of the ballast, than of the sail. In fame of learning, the flight will be slow without some feathers of ostentation. Qui de contemnenda gloria libros scribunt, nomen, suum inscribunt. Socrates, Aristotle, Galen, were men full of ostentation. Certainly vain-glory helpeth to perpetuate a man's memory; and virtue was never so beholding to human nature, as it received his due at the second hand. Neither had the fame of Cicero, Seneca, Plinius Secundus, borne her age so well, if it had not been joined with some vanity in themselves; like unto varnish, that makes ceilings not only shine but last. But all this while, when I speak of vain-glory, I mean not of that property, that Tacitus doth attribute to Mucianus; Omnium quae dixerat feceratque arte quadam ostentator: for that proceeds not of vanity, but of natural magnanimity and discretion; and in some persons, is not only comely, but gracious. For excusations, cessions, modesty itself well governed, are but arts of ostentation. And amongst those arts, there is none better than that which Plinius Secundus speaketh of, which is to be liberal of praise and commendation to others, in that, wherein a man's self hath any perfection. For saith Pliny, very wittily, In commending another, you do yourself right; for he that you commend, is either superior to you in that you commend, or inferior. If he be inferior, if he be to be commended, you much more; if he be superior, if he be not to be commended, you much less. Glorious men are the scorn of wise men, the admiration of fools, the idols of parasites, and the slaves of their own vaunts.

Of Honor And Reputation

THE winning of honor, is but the revealing of a man's virtue and worth, without disadvantage. For some in their actions, do woo and effect honor and reputation, which sort of men, are commonly much talked of, but inwardly little admired. And some, contrariwise, darken their virtue in the show of it; so as they be undervalued in opinion. If a man perform that, which hath not been attempted before; or attempted and given over; or hath been achieved, but not with so good circumstance; he shall purchase more honor, than by effecting a matter of greater difficulty or virtue, wherein he is but a follower. If a man so temper his actions, as in some one of them he doth content every faction, or combination of people, the music will be the fuller. A man is an ill husband of his honor, that entereth into any action, the failing wherein may disgrace him, more than the carrying of it through, can honor him. Honor that is gained and broken upon another, hath the quickest reflection, like diamonds cut with facets. And therefore, let a man contend to excel any competitors of his in honor, in outshooting them, if he can, in their own bow. Discreet followers and servants, help much to reputation. Omnis fama a domesticis emanat. Envy, which is the canker of honor, is best extinguished by declaring a man's self in his ends, rather to seek merit than fame; and by attributing a man's successes, rather to divine Providence and felicity, than to his own virtue or policy.

The true marshalling of the degrees of sovereign honor, are these: In the first place are conditores imperiorum, founders of states and commonwealths; such as were Romulus, Cyrus, Caesar, Ottoman, Ismael. In the second place are legislatores, lawgivers; which are also called second founders, or perpetui principes, because they govern by their ordinances after they are gone; such were Lycurgus, Solon, Justinian, Eadgar, Alphonsus of Castile, the Wise, that made the Siete Partidas. In the third place are liberatores, or salvatores, such as compound the long miseries of civil wars, or deliver their countries from servitude of strangers or tyrants; as Augustus Caesar, Vespasianus, Aurelianus, Theodoricus, King Henry the Seventh of England, King Henry the Fourth of France. In the fourth place are propagatores or propugnatores imperii; such as in honorable wars enlarge their territories, or make noble defence against invaders. And in the last place are patres patriae; which reign justly, and make the times good wherein they live. Both which last kinds need no examples, they are in such number. Degrees of honor, in subjects, are, first participes curarum, those upon whom, princes do discharge the greatest weight of their affairs; their right hands, as we call them. The next are duces belli, great leaders in war; such as are princes' lieutenants, and do them notable services in the wars. The third are gratiosi, favorites; such as exceed not this scantling, to be solace to the sovereign, and harmless to the people. And the fourth, negotiis pares; such as have great places under princes, and execute their places, with sufficiency. There is an honor, likewise, which may be ranked amongst the greatest, which happeneth rarely; that is, of such as sacrifice themselves to death or danger for the good of their country; as was M. Regulus, and the two Decii.

Of Judicature

JUDGES ought to remember, that their office is jus dicere, and not jus dare; to interpret law, and not to make law, or give law. Else will it be like the authority, claimed by the Church of Rome, which under pretext of exposition of Scripture, doth not stick to add and alter; and to pronounce that which they do not find; and by show of antiquity, to introduce novelty. Judges ought to be more learned, than witty, more reverend, than plausible, and more advised, than confident. Above all things, integrity is their portion and proper virtue. Cursed (saith the law) is he that removeth the landmark. The mislayer of a mere-stone is to blame. But it is the unjust judge, that is the capital remover of landmarks, when he defineth amiss, of lands and property. One foul sentence doth more hurt, than many foul examples. For these do but corrupt the stream, the other corrupteth the fountain. So with Solomon, Fons turbatus, et vena corrupta, est justus cadens in causa sua coram adversario. The office of judges may have reference unto the parties that use, unto the advocates that plead, unto the clerks and ministers of justice underneath them, and to the sovereign or state above them.

First, for the causes or parties that sue. There be (saith the Scripture) that turn judgment, into wormwood; and surely there be also, that turn it into vinegar; for injustice maketh it bitter, and delays make it sour. The principal duty of a judge, is to suppress force and fraud; whereof force is the more pernicious, when it is open, and fraud, when it is close and disguised. Add thereto contentious suits, which ought to be spewed out, as the surfeit of courts. A judge ought to prepare his way to a just sentence, as God useth to prepare his way, by raising valleys and taking down hills: so when there appeareth on either side an high hand, violent prosecution, cunning advantages taken, combination, power, great counsel, then is the virtue of a judge seen, to make inequality equal; that he may plant his judgment as upon an even ground. Qui fortiter emungit, elicit sanguinem; and where the wine-press is hard wrought, it yields a harsh wine, that tastes of the grape-stone. Judges must beware of hard constructions, and strained inferences; for there is no worse torture, than the torture of laws. Specially in case of laws penal, they ought to have care, that that which was meant for terror, be not turned into rigor; and that they bring not upon the people, that shower whereof the Scripture speaketh, Pluet super eos laqueos; for penal laws pressed, are a shower of snares upon the people. Therefore let penal laws, if they have been sleepers of long, or if they be grown unfit for the present time, be by wise judges confined in the execution: Judicis officium est, ut res, ita tempora rerum, etc. In causes of life and death, judges ought (as far as the law permitteth) in justice to remember mercy; and to cast a severe eye upon the example, but a merciful eye upon the person.

Secondly, for the advocates and counsel that plead. Patience and gravity of hearing, is an essential part of justice; and an overspeaking judge is no well-tuned cymbal. It is no grace to a judge, first to find that, which he might have heard in due time from the bar; or to show quickness of conceit, in cutting off evidence or counsel too short; or to prevent information by questions, though pertinent. The parts of a judge in hearing, are four: to direct the evidence; to moderate length, repetition, or impertinency of speech; to recapitulate, select, and collate the material points, of that which hath been said; and to give the rule or sentence. Whatsoever is above these is too much; and proceedeth either of glory, and willingness to speak, or of impatience to hear, or of shortness of memory, or of want of a staid and equal attention. It is a strange thing to see, that the boldness of advocates should prevail with judges; whereas they should imitate God, in whose seat they sit; who represseth the presumptuous, and giveth grace to the modest. But it is more strange, that judges should have noted favorites; which cannot but cause multiplication of fees, and suspicion of by-ways. There is due from the judge to the advocate, some commendation and gracing, where causes are well handled and fair pleaded; especially towards the side which obtaineth not; for that upholds in the client, the reputation of his counsel, and beats down in him the conceit of his cause. There is likewise due to the public, a civil reprehension of advocates, where there appeareth cunning counsel, gross neglect, slight information, indiscreet pressing, or an overbold defence. And let not the counsel at the bar, chop with the judge, nor wind himself into the handling of the cause anew, after the judge hath declared his sentence; but, on the other side, let not the judge meet the cause half way, nor give occasion to the party, to say his counsel or proofs were not heard.

Thirdly, for that that concerns clerks and ministers. The place of justice is an hallowed place; and therefore not only the bench, but the foot-place; and precincts and purprise thereof, ought to be preserved without scandal and corruption. For certainly grapes (as the Scripture saith) will not be gathered of thorns or thistles; neither can justice yield her fruit with sweetness, amongst the briars and brambles of catching and polling clerks, and ministers. The attendance of courts, is subject to four bad instruments. First, certain persons that are sowers of suits; which make the court swell, and the country pine. The second sort is of those, that engage courts in quarrels of jurisdiction, and are not truly amici curiae, but parasiti curiae, in puffing a court up beyond her bounds, for their own scraps and advantage. The third sort, is of those that may be accounted the left hands of courts; persons that are full of nimble and sinister tricks and shifts, whereby they pervert the plain and direct courses of courts, and bring justice into oblique lines and labyrinths. And the fourth, is the poller and exacter of fees; which justifies the common resemblance of the courts of justice, to the bush whereunto, while the sheep flies for defence in weather, he is sure to lose part of his fleece. On the other side, an ancient clerk, skilful in precedents, wary in proceeding, and understanding in the business of the court, is an excellent finger of a court; and doth many times point the way to the judge himself.

Fourthly, for that which may concern the sovereign and estate. Judges ought above all to remember the conclusion of the Roman Twelve Tables; Salus populi suprema lex; and to know that laws, except they be in order to that end, are but things captious, and oracles not well inspired. Therefore it is an happy thing in a state, when kings and states do often consult with judges; and again, when judges do often consult with the king and state: the one, when there is matter of law, intervenient in business of state; the other, when there is some consideration of state, intervenient in matter of law. For many times the things deduced to judgment may be meum and tuum, when the reason and consequence thereof may trench to point of estate: I call matter of estate, not only the parts of sovereignty, but whatsoever introduceth any great alteration, or dangerous precedent; or concerneth manifestly any great portion of people. And let no man weakly conceive, that just laws and true policy have any antipathy; for they are like the spirits and sinews, that one moves with the other. Let judges also remember, that Solomon's throne was supported by lions on both sides: let them be lions, but yet lions under the throne; being circumspect that they do not check or oppose any points of sovereignty. Let not judges also be ignorant of their own right, as to think there is not left to them, as a principal part of their office, a wise use and application of laws. For they may remember, what the apostle saith of a greater law than theirs; Nos scimus quia lex bona est, modo quis ea utatur legitime.

Of Anger

TO SEEK to extinguish anger utterly, is but a bravery of the Stoics. We have better oracles: Be angry, but sin not. Let not the sun go down upon your anger. Anger must be limited and confined, both in race and in time. We will first speak how the natural inclination and habit to be angry, may be attempted and calmed. Secondly, how the particular motions of anger may be repressed, or at least refrained from doing mischief. Thirdly, how to raise anger, or appease anger in another.

For the first; there is no other way but to meditate, and ruminate well upon the effects of anger, how it troubles man's life. And the best time to do this, is to look back upon anger, when the fit is thoroughly over. Seneca saith well, That anger is like ruin, which breaks itself upon that it falls. The Scripture exhorteth us to possess our souls in patience. Whosoever is out of patience, is out of possession of his soul. Men must not turn bees;

... animasque in vulnere ponunt.

Anger is certainly a kind of baseness; as it appears well in the weakness of those subjects in whom it reigns; children, women, old folks, sick folks. Only men must beware, that they carry their anger rather with scorn, than with fear; so that they may seem rather to be above the injury, than below it; which is a thing easily done, if a man will give law to himself in it.

For the second point; the causes and motives of anger, are chiefly three. First, to be too sensible of hurt; for no man is angry, that feels not himself hurt; and therefore tender and delicate persons must needs be oft angry; they have so many things to trouble them, which more robust natures have little sense of. The next is, the apprehension and construction of the injury offered, to be, in the circumstances thereof, full of contempt: for contempt is that, which putteth an edge upon anger, as much or more than the hurt itself. And therefore, when men are ingenious in picking out circumstances of contempt, they do kindle their anger much. Lastly, opinion of the touch of a man's reputation, doth multiply and sharpen anger. Wherein the remedy is, that a man should have, as Consalvo was wont to say, telam honoris crassiorem. But in all refrainings of anger, it is the best remedy to win time; and to make a man's self believe, that the opportunity of his revenge is not yet come, but that he foresees a time for it; and so to still himself in the meantime, and reserve it.

To contain anger from mischief, though it take hold of a man, there be two things, whereof you must have special caution. The one, of extreme bitterness of words, especially if they be aculeate and proper; for cummunia maledicta are nothing so much; and again, that in anger a man reveal no secrets; for that, makes him not fit for society. The other, that you do not peremptorily break off, in any business, in a fit of anger; but howsoever you show bitterness, do not act anything, that is not revocable.

For raising and appeasing anger in another; it is done chiefly by choosing of times, when men are frowardest and worst disposed, to incense them. Again, by gathering (as was touched before) all that you can find out, to aggravate the contempt. And the two remedies are by the contraries. The former to take good times, when first to relate to a man an angry business; for the first impression is much; and the other is, to sever, as much as may be, the construction of the injury from the point of contempt; imputing it to misunderstanding, fear, passion, or what you will.

Of Vicissitude Of Things

SOLOMON saith, There is no new thing upon the earth. So that as Plato had an imagination, That all knowledge was but remembrance; so Solomon giveth his sentence, That all novelty is but oblivion. Whereby you may see, that the river of Lethe runneth as well above ground as below. There is an abstruse astrologer that saith, If it were not for two things that are constant (the one is, that the fixed stars ever stand a like distance one from another, and never come nearer together, nor go further asunder; the other, that the diurnal motion perpetually keepeth time), no individual would last one moment. Certain it is, that the matter is in a perpetual flux, and never at a stay. The great winding-sheets, that bury all things in oblivion, are two; deluges and earthquakes. As for conflagrations and great droughts, they do not merely dispeople and destroy. Phaeton's car went but a day. And the three years' drought in the time of Elias, was but particular, and left people alive. As for the great burnings by lightnings, which are often in the West Indies, they are but narrow. But in the other two destructions, by deluge and earthquake, it is further to be noted, that the remnant of people which hap to be reserved, are commonly ignorant and mountainous people, that can give no account of the time past; so that the oblivion is all one, as if none had been left. If you consider well of the people of the West Indies, it is very probable that they are a newer or a younger people, than the people of the Old World. And it is much more likely, that the destruction that hath heretofore been there, was not by earthquakes (as the Egyptian priest told Solon concerning the island of Atlantis, that it was swallowed by an earthquake), but rather that it was desolated by a particular deluge. For earthquakes are seldom in those parts. But on the other side, they have such pouring rivers, as the rivers of Asia and Africk and Europe, are but brooks to them. Their Andes, likewise, or mountains, are far higher than those with us; whereby it seems, that the remnants of generation of men, were in such a particular deluge saved. As for the observation that Machiavel hath, that the jealousy of sects, doth much extinguish the memory of things; traducing Gregory the Great, that he did what in him lay, to extinguish all heathen antiquities; I do not find that those zeals do any great effects, nor last long; as it appeared in the succession of Sabinian, who did revive the former antiquities.

The vicissitude of mutations in the superior globe, are no fit matter for this present argument. It may be, Plato's great year, if the world should last so long, would have some effect; not in renewing the state of like individuals (for that is the fume of those, that conceive the celestial bodies have more accurate influences upon these things below, than indeed they have), but in gross. Comets, out of question, have likewise power and effect, over the gross and mass of things; but they are rather gazed upon, and waited upon in their journey, than wisely observed in their effects; specially in, their respective effects; that is, what kind of comet, for magnitude, color, version of the beams, placing in the region of heaven, or lasting, produceth what kind of effects.

There is a toy which I have heard, and I would not have it given over, but waited upon a little. They say it is observed in the Low Countries (I know not in what part) that every five and thirty years, the same kind and suit of years and weathers come about again; as great frosts, great wet, great droughts, warm winters, summers with little heat, and the like; and they call it the Prime. It is a thing I do the rather mention, because, computing backwards, I have found some concurrence.

But to leave these points of nature, and to come to men. The greatest vicissitude of things amongst men, is the vicissitude of sects and religions. For those orbs rule in men's minds most. The true religion is built upon the rock; the rest are tossed, upon the waves of time. To speak, therefore, of the causes of new sects; and to give some counsel concerning them, as far as the weakness of human judgment can give stay, to so great revolutions. When the religion formerly received, is rent by discords; and when the holiness of the professors of religion, is decayed and full of scandal; and withal the times be stupid, ignorant, and barbarous; you may doubt the springing up of a new sect; if then also, there should arise any extravagant and strange spirit, to make himself author thereof. All which points held, when Mahomet published his law. If a new sect have not two properties, fear it not; for it will not spread. The one is the supplanting, or the opposing, of authority established; for nothing is more popular than that. The other is the giving license to pleasures, and a voluptuous life. For as for speculative heresies (such as were in ancient times the Arians, and now the Arminians), though they work mightily upon men's wits, yet they do not produce any great alterations in states; except it be by the help of civil occasions. There be three manner of plantations of new sects. By the power of signs and miracles; by the eloquence, and wisdom, of speech and persuasion; and by the sword. For martyrdoms, I reckon them amongst miracles; because they seem to exceed the strength of human nature: and I may do the like, of superlative and admirable holiness of life. Surely there is no better way, to stop the rising of new sects and schisms, than to reform abuses; to compound the smaller differences; to proceed mildly, and not with sanguinary persecutions; and rather to take off the principal authors by winning and advancing them, than to enrage them by violence and bitterness.

The changes and vicissitude in wars are many; but chiefly in three things; in the seats or stages of the war; in the weapons; and in the manner of the conduct. Wars, in ancient time, seemed more to move from east to west; for the Persians, Assyrians, Arabians, Tartars (which were the invaders) were all eastern people. It is true, the Gauls were western; but we read but of two incursions of theirs: the one to Gallo-Grecia, the other to Rome. But east and west have no certain points of heaven; and no more have the wars, either from the east or west, any certainty of observation. But north and south are fixed; and it hath seldom or never been seen that the far southern people have invaded the northern, but contrariwise. Whereby it is manifest that the northern tract of the world, is in nature the more martial region: be it in respect of the stars of that hemisphere; or of the great continents that are upon the north, whereas the south part, for aught that is known, is almost all sea; or (which is most apparent) of the cold of the northern parts, which is that which, without aid of discipline, doth make the bodies hardest, and the courages warmest.

Upon the breaking and shivering of a great state and empire, you may be sure to have wars. For great empires, while they stand, do enervate and destroy the forces of the natives which they have subdued, resting upon their own protecting forces; and then when they fail also, all goes to ruin, and they become a prey. So was it in the decay of the Roman empire; and likewise in the empire of Almaigne, after Charles the Great, every bird taking a feather; and were not unlike to befall to Spain, if it should break. The great accessions and unions of kingdoms, do likewise stir up wars; for when a state grows to an over-power, it is like a great flood, that will be sure to overflow. As it hath been seen in the states of Rome, Turkey, Spain, and others. Look when the world hath fewest barbarous peoples, but such as commonly will not marry or generate, except they know means to live (as it is almost everywhere at this day, except Tartary), there is no danger of inundations of people; but when there be great shoals of people, which go on to populate, without foreseeing means of life and sustentation, it is of necessity that once in an age or two, they discharge a portion of their people upon other nations; which the ancient northern people were wont to do by lot; casting lots what part should stay at home, and what should seek their fortunes. When a warlike state grows soft and effeminate, they may be sure of a war. For commonly such states are grown rich in the time of their degenerating; and so the prey inviteth, and their decay in valor, encourageth a war.

As for the weapons, it hardly falleth under rule and observation: yet we see even they, have returns and vicissitudes. For certain it is, that ordnance was known in the city of the Oxidrakes in India; and was that, which the Macedonians called thunder and lightning, and magic. And it is well known that the use of ordnance, hath been in China above two thousand years. The conditions of weapons, and their improvement, are; First, the fetching afar off; for that outruns the danger; as it is seen in ordnance and muskets. Secondly, the strength of the percussion; wherein likewise ordnance do exceed all arietations and ancient inventions. The third is, the commodious use of them; as that they may serve in all weathers; that the carriage may be light and manageable; and the like.

For the conduct of the war: at the first, men rested extremely upon number: they did put the wars likewise upon main force and valor; pointing days for pitched fields, and so trying it out upon an even match and they were more ignorant in ranging and arraying their battles. After, they grew to rest upon number rather competent, than vast; they grew to advantages of place, cunning diversions, and the like: and they grew more skilful in the ordering of their battles.

In the youth of a state, arms do flourish; in the middle age of a state, learning; and then both of them together for a time; in the declining age of a state, mechanical arts and merchandize. Learning hath his infancy, when it is but beginning and almost childish; then his youth, when it is luxuriant and juvenile; then his strength of years, when it is solid and reduced; and lastly, his old age, when it waxeth dry and exhaust. But it is not good to look too long upon these turning wheels of vicissitude, lest we become giddy. As for the philology of them, that is but a circle of tales, and therefore not fit for this writing.

Of Fame

THE poets make Fame a monster. They describe her in part finely and elegantly, and in part gravely and sententiously. They say, look how many feathers she hath, so many eyes she hath underneath; so many tongues; so many voices; she pricks up so many ears.

This is a flourish. There follow excellent parables; as that, she gathereth strength in going; that she goeth upon the ground, and yet hideth her head in the clouds; that in the daytime she sitteth in a watch tower, and flieth most by night; that she mingleth things done, with things not done; and that she is a terror to great cities. But that which passeth all the rest is: They do recount that the Earth, mother of the giants that made war against Jupiter, and were by him destroyed, thereupon in an anger brought forth Fame. For certain it is, that rebels, figured by the giants, and seditious fames and libels, are but brothers and sisters, masculine and feminine. But now, if a man can tame this monster, and bring her to feed at the hand, and govern her, and with her fly other ravening fowl and kill them, it is somewhat worth. But we are infected with the style of the poets. To speak now in a sad and serious manner: There is not, in all the politics, a place less handled and more worthy to be handled, than this of fame. We will therefore speak of these points: What are false fames; and what are true fames; and how they may be best discerned; how fames may be sown, and raised; how they may be spread, and multiplied; and how they may be checked, and laid dead. And other things concerning the nature of fame. Fame is of that force, as there is scarcely any great action, wherein it hath not a great part; especially in the war. Mucianus undid Vitellius, by a fame that he scattered, that Vitellius had in purpose to remove the legions of Syria into Germany, and the legions of Germany into Syria; whereupon the legions of Syria were infinitely inflamed. Julius Caesar took Pompey unprovided, and laid asleep his industry and preparations, by a fame that he cunningly gave out: Caesar's own soldiers loved him not, and being wearied with the wars, and laden with the spoils of Gaul, would forsake him, as soon as he came into Italy. Livia settled all things for the succession of her son Tiberius, by continual giving out, that her husband Augustus was upon recovery and amendment, and it is an usual thing with the pashas, to conceal the death of the Great Turk from the janizaries and men of war, to save the sacking of Constantinople and other towns, as their manner is. Themistocles made Xerxes, king of Persia, post apace out of Grecia, by giving out, that the Grecians had a purpose to break his bridge of ships, which he had made athwart Hellespont. There be a thousand such like examples; and the more they are, the less they need to be repeated; because a man meeteth with them everywhere. Therefore let all wise governors have as great a watch and care over fames, as they have of the actions and designs themselves.

[This essay was not finished]

A Glossary Of Archaic Words And Phrases

Abridgment: miniature
Absurd: stupid, unpolished
Abuse: cheat, deceive
Aculeate: stinging
Adamant: loadstone
Adust: scorched
Advoutress: adulteress
Affect: like, desire
Antic: clown
Appose: question
Arietation: battering-ram
Audit: revenue
Avoidance: secret outlet
Battle: battalion
Bestow: settle in life
Blanch: flatter, evade
Brave: boastful
Bravery: boast, ostentation

Broke: deal in brokerage
Broken: shine by comparison
Broken music: part music
Cabinet: secret
Calendar: weather forecast
Card: chart, map
Care not to: are reckless
Cast: plan
Cat: cate, cake
Charge and adventure: cost and risk
Check with: interfere
Chop: bandy words
Civil: peaceful
Close: secret, secretive
Collect: infer
Compound: compromise
Consent: agreement
Curious: elaborate
Custom: import duties
Deceive: rob
Derive: divert
Difficileness: moroseness
Discover: reveal
Donative: money gift
Doubt: fear
Equipollent: equally powerful
Espial: spy
Estate: state
Facility: of easy persuasion
Fair: rather
Fame: rumor
Favor: feature
Flashy: insipid
Foot-pace: lobby
Foreseen: guarded against
Froward: stubborn
Futile: babbling
Globe: complete body
Glorious: showy, boastful
Humorous: capricious
Hundred poll: hundredth head
Impertinent: irrelevant
Implicit: entangled
In a mean: in moderation
In smother: suppressed
Indifferent: impartial
Intend: attend to
Knap: knoll
Leese: lose
Let: hinder
Loose: shot
Lot: spell
Lurch: intercept
Make: profit, get
Manage: train
Mate: conquer
Material: business-like
Mere-stone: boundary stone
Muniting: fortifying
Nerve: sinew
Obnoxious: subservient, liable

Oes: round spangles
Pair: impair
Pardon: allowance
Passable: mediocre
Pine-apple-tree: pine
Plantation: colony
Platform: plan
Plausible: praiseworthy
Point device: excessively precise
Politic: politician
Poll: extort
Poser: examiner
Practice: plotting
Preoccupate: anticipate
Prest: prepared
Prick: plant
Proper: personal
Prospective: stereoscope
Proyne: prune
Purprise: enclosure
Push: pimple
Quarrel: pretext
Quech: flinch
Reason: principle
Recamera: retiring-room
Return: reaction
Return: wing running back
Rise: dignity
Round: straight
Save: account for
Scantling: measure
Seel: blind
Shrewd: mischievous
Sort: associate
Spial: spy
Staddle: sapling
Steal: do secretly
Stirp: family
Stond: stop, stand
Stoved: hot-housed
Style: title
Success: outcome
Sumptuary law: law against
extravagance
Superior globe: the heavens
Temper: proportion
Tendering: nursing
Tract: line, trait
Travel: travail, labor
Treaties: treatises
Trench to: touch
Trivial: common
Turquet: Turkish dwarf
Under foot: below value
Unready: untrained
Usury: interest
Value: certify
Virtuous: able
Votary: vowed
Wanton: spoiled
Wood: maze
Work: manage, utilize

The Advancement of Learning

Originally Published By:
Cassell & Company, Limited: London, Paris & Melbourne. 1893.

INTRODUCTION

"THE TVVOO Bookes of Francis Bacon. Of the proficience and aduancement of Learning, divine and humane. To the King. At London. Printed for Henrie Tomes, and are to be sould at his shop at Graies Inne Gate in Holborne. 1605." That was the original title-page of the book now in the reader's hand—a living book that led the way to a new world of thought. It was the book in which Bacon, early in the reign of James the First, prepared the way for a full setting forth of his New Organon, or instrument of knowledge.

The Organon of Aristotle was a set of treatises in which Aristotle had written the doctrine of propositions. Study of these treatises was a chief occupation of young men when they passed from school to college, and proceeded from Grammar to Logic, the second of the Seven Sciences. Francis Bacon as a youth of sixteen, at Trinity College, Cambridge, felt the unfruitfulness of this method of search after truth. He was the son of Sir Nicholas Bacon, Queen Elizabeth's Lord Keeper, and was born at York House, in the Strand, on the 22nd of January, 1561. His mother was the Lord Keeper's second wife, one of two sisters, of whom the other married Sir William Cecil, afterwards Lord Burleigh. Sir Nicholas Bacon had six children by his former marriage, and by his second wife two sons, Antony and Francis, of whom Antony was about two years the elder. The family home was at York Place, and at Gorhambury, near St. Albans, from which town, in its ancient and its modern style, Bacon afterwards took his titles of Verulam and St. Albans.

Antony and Francis Bacon went together to Trinity College, Cambridge, when Antony was fourteen years old and Francis twelve. Francis remained at Cambridge only until his sixteenth year; and Dr. Rawley, his chaplain in after-years, reports of him that "whilst he was commorant in the University, about sixteen years of age (as his lordship hath been pleased to impart unto myself), he first fell into dislike of the philosophy of Aristotle; not for the worthlessness of the author, to whom he would ascribe all high attributes, but for the unfruitfulness of the way, being a philosophy (as his lordship used to say) only strong for disputatious and contentions, but barren of the production of works for the benefit of the life of man; in which mind he continued to his dying day." Bacon was sent as a youth of sixteen to Paris with the ambassador Sir Amyas Paulet, to begin his training for the public service; but his father's death, in February, 1579, before he had completed the provision he was making for his youngest children, obliged him to return to London, and, at the age of eighteen, to settle down at Gray's Inn to the study of law as a profession. He was admitted to the outer bar in June, 1582, and about that time, at the age of twenty-one, wrote a sketch of his conception of a New Organon that should lead man to more fruitful knowledge, in a little Latin tract, which he called "Temporis Partus Maximus" ("The Greatest Birth of Time").

In November, 1584, Bacon took his seat in the House of Commons as member for Melcombe Regis, in Dorsetshire. In October, 1586, he sat for Taunton. He was member afterwards for Liverpool; and he was one of those who petitioned for the speedy execution of Mary Queen of Scots. In October, 1589, he obtained the reversion of the office of Clerk of the Council in the Star Chamber, which was worth £1,600 or £2,000 a year; but for the succession to this office he had to wait until 1608. It had not yet fallen to him when he wrote his "Two Books of the Advancement of Learning." In the Parliament that met in February, 1593, Bacon sat as member for Middlesex. He raised difficulties of procedure in the way of the grant of a treble subsidy, by just objection to the joining of the Lords with the Commons in a money grant, and a desire to extend the time allowed for payment from three years to six; it was, in fact, extended to four years. The Queen was offended. Francis Bacon and his brother Antony had attached themselves to the young Earl of Essex, who was their friend and patron. The office of Attorney-General became vacant. Essex asked the Queen to appoint Francis Bacon. The Queen gave the office to Sir Edward Coke, who was already Solicitor-General, and by nine years Bacon's senior. The office of Solicitor-General thus became vacant, and that was sought for Francis Bacon. The Queen, after delay and hesitation, gave it, in November, 1595, to Serjeant Fleming. The Earl of Essex consoled his friend by giving him "a piece of land"—Twickenham Park—which Bacon afterwards sold for £1,800—equal, say, to £12,000 in present buying power. In 1597 Bacon was returned to Parliament as member for Ipswich, and in that year he was hoping to marry the rich widow of Sir William Hatton, Essex helping; but the lady married, in the next year, Sir Edward Coke. It was in 1597 that Bacon published the First Edition of his Essays. That was a little book containing only ten essays in English, with twelve "Meditationes Sacræ," which were essays in Latin on religious subjects. From 1597 onward to the end of his life, Bacon's Essays were subject to continuous addition and revision. The author's Second Edition, in which the number of the Essays was increased from ten to thirty-eight, did not appear until November or December, 1612, seven years later than these two books on the "Advancement of Learning;" and the final edition of the Essays, in which their number was increased from thirty-eight to fifty-eight, appeared only in 1625; and Bacon died on the 9th of April, 1626. The edition of the Essays published in 1597, under Elizabeth, marked only the beginning of a course of thought that afterwards flowed in one stream with his teachings in philosophy.

In February, 1601, there was the rebellion of Essex. Francis Bacon had separated himself from his patron after giving him advice that was disregarded. Bacon, now Queen's Counsel, not only appeared against his old friend, but with excess of zeal, by which, perhaps, he hoped to win back the Queen's favour, he twice obtruded violent attacks upon Essex when he was not called upon to speak. On the 25th of February, 1601, Essex was beheaded. The genius of Bacon was next employed to justify that act by "A Declaration of the Practices and Treasons attempted and committed by Robert late Earle of Essex and his Complices." But James of Scotland, on whose behalf Essex had intervened, came to the throne by the death of Elizabeth on the 24th of March, 1603. Bacon was among the crowd of men who were made knights by James I., and he had to justify himself under the new order of things by writing "Sir Francis Bacon his Apologie in certain Imputations concerning the late Earle of Essex." He was returned to the first Parliament of James I. by Ipswich and St. Albans, and he was confirmed in his office of King's Counsel in August, 1604; but he was not appointed to the office of Solicitor-General when it became vacant in that year.

That was the position of Francis Bacon in 1605, when he published this work, where in his First Book he pointed out the discredits of learning from human defects of the learned, and emptiness of many of the studies chosen, or the way of dealing with them. This came, he said, especially by the mistaking or misplacing of the last or furthest end of knowledge, as if there were sought in it "a couch whereupon to rest a searching and restless spirit; or a terrace for a wandering and variable mind to walk up and down with a fair prospect; or a tower of state for a proud mind to raise itself upon; or a fort or commanding ground for strife and contention; or a shop for profit or sale; and not a rich storehouse for the glory of the Creator and the relief of man's estate." The rest of the First Book was given to an argument upon the Dignity of Learning; and the Second Book, on the Advancement of Learning, is, as Bacon himself described it, "a general and faithful perambulation of learning, with an inquiry what parts thereof lie fresh and waste, and not improved and converted by the industry of man; to the end that such a plot made and recorded to memory may both minister light to any public designation and also serve to excite voluntary endeavours." Bacon makes, by a sort of exhaustive analysis, a ground-plan of all subjects of study, as an intellectual map, helping the right inquirer in his search for the right path. The right path is that by which he has the best chance of adding to the stock of knowledge in the world something worth labouring for; and the true worth is in labour for "the glory of the Creator and the relief of man's estate."

H. M.

THE FIRST BOOK OF FRANCIS BACON; OF THE PROFICIENCE AND ADVANCEMENT OF LEARNING, DIVINE AND HUMAN.

To the King.

THERE were under the law, excellent King, both daily sacrifices and freewill offerings; the one proceeding upon ordinary observance, the other upon a devout cheerfulness: in like manner there belongeth to kings from their servants both tribute of duty and presents of affection. In the former of these I hope I shall not live to be wanting, according to my most humble duty and the good pleasure of your Majesty's employments: for the latter, I thought it more respective to make choice of some oblation which might rather refer to the propriety and excellency of your individual person, than to the business of your crown and state.

Wherefore, representing your Majesty many times unto my mind, and beholding you not with the inquisitive eye of presumption, to discover that which the Scripture telleth me is inscrutable, but with the observant eye of duty and admiration, leaving aside the other parts of your virtue and fortune, I have been touched—yea, and possessed—with an extreme wonder at those your virtues and faculties, which the philosophers call intellectual; the largeness of your capacity, the faithfulness of your memory, the swiftness of your apprehension, the penetration of your judgment, and the facility and order of your elocution: and I have often thought that of all the persons living that I have known, your Majesty were the best instance to make a man of Plato's opinion, that all knowledge is but remembrance, and that the mind of man by Nature knoweth all things, and hath but her own native and original notions (which by the strangeness and darkness of this tabernacle of the body are sequestered) again revived and restored: such a light of Nature I have observed in your Majesty, and such a readiness to take flame and blaze from the least occasion presented, or the least spark of another's knowledge delivered. And as the Scripture saith of the wisest king, "That his heart was as the sands of the sea;" which, though it be one of the largest bodies, yet it consisteth of the smallest and finest portions; so hath God given your Majesty a composition of understanding admirable, being able to compass and comprehend the greatest matters, and nevertheless to touch and apprehend the least; whereas it should seem an impossibility in Nature for the same instrument to make itself fit for great and small works. And for your gift of speech, I call to mind what Cornelius Tacitus saith of Augustus Cæsar: Augusto profluens, et quæ principem deceret, eloquentia fuit. For if we note it well, speech that is uttered with labour and difficulty, or speech that savoureth of the affectation of art and precepts, or speech that is framed after the imitation of some pattern of eloquence, though never so excellent; all this hath somewhat servile, and holding of the subject. But your Majesty's manner of speech is, indeed, prince-like, flowing as from a fountain, and yet streaming and branching itself into Nature's order, full of facility and felicity, imitating none, and inimitable by any. And as in your civil estate there appeareth to be an emulation and contention of your Majesty's virtue with your fortune; a virtuous disposition with a fortunate regiment; a virtuous expectation (when time was) of your greater fortune, with a prosperous possession thereof in the due time; a virtuous observation of the laws of marriage, with most blessed and happy fruit of marriage; a virtuous and most Christian desire of peace, with a fortunate inclination in your neighbour princes thereunto: so likewise in these intellectual matters there seemeth to be no less contention between the excellency of your Majesty's gifts of Nature and the universality and perfection of your learning. For I am well assured that this which I shall say is no amplification at all, but a positive and measured truth; which is, that there hath not been since Christ's time any king or temporal monarch which hath been so learned in all literature and erudition, divine and human. For let a man seriously and diligently revolve and peruse the succession of the Emperors of Rome, of which Cæsar the Dictator (who lived some years before Christ) and Marcus Antoninus were the best learned, and so descend to the Emperors of Græcia, or of the West, and then to the lines of France, Spain, England, Scotland, and the rest, and he shall find this judgment is truly made. For it seemeth much in a king if, by the compendious extractions of other men's wits and labours, he can take hold of any superficial ornaments and shows of learning, or if he countenance and prefer learning and learned men; but to drink, indeed, of the true fountains of learning—nay, to have such a fountain of learning in himself, in a king, and in a king born—is almost a miracle. And the more, because there is met in your Majesty a rare conjunction, as well of divine and sacred literature as of profane and human; so as your Majesty standeth invested of that triplicity, which in great veneration was ascribed to the ancient Hermes: the power and fortune of a king, the knowledge and illumination of a priest, and the learning and universality of a philosopher. This propriety inherent and individual attribute in your Majesty deserveth to be expressed not only in the fame and admiration of the present time, nor in the history or tradition of the ages succeeding, but also in some solid work, fixed memorial, and immortal monument, bearing a character or signature both of the power of a king and the difference and perfection of such a king.

Therefore I did conclude with myself that I could not make unto your Majesty a better oblation than of some treatise tending to that end, whereof the sum will consist of these two parts: the former concerning the excellency of learning and knowledge, and the excellency of the merit and true glory in the augmentation and propagation thereof; the latter, what the particular acts and works are which have been embraced and undertaken for the advancement of learning; and again, what defects and undervalues I find in such particular acts: to the end that though I cannot positively or affirmatively advise your Majesty, or propound unto you framed particulars, yet I may excite your princely cogitations to visit the excellent treasure of your own mind, and thence to extract particulars for this purpose agreeable to your magnanimity and wisdom.

I

(1) In the entrance to the former of these—to clear the way and, as it were, to make silence, to have the true testimonies concerning the dignity of learning to be better heard, without the interruption of tacit objections—I think good to deliver it from the discredits and disgraces which it hath received, all from ignorance, but ignorance severally disguised; appearing sometimes in the zeal and jealousy of divines, sometimes in the severity and arrogancy of politics, and sometimes in the errors and imperfections of learned men themselves.

(2) I hear the former sort say that knowledge is of those things which are to be accepted of with great limitation and caution; that the aspiring to overmuch knowledge was the original temptation and sin whereupon ensued the fall of man; that knowledge hath in it somewhat of the serpent, and, therefore, where it entereth into a man it makes him swell; *Scientia inflat*; that Solomon gives a censure, "That there is no end of making books, and that much reading is weariness of the flesh;" and again in another place, "That in spacious knowledge there is much contristation, and that he that increaseth knowledge increaseth anxiety;" that Saint Paul gives a caveat, "That we be not spoiled through vain philosophy;" that experience demonstrates how learned men have been arch-heretics, how learned times have been inclined to atheism, and how the contemplation of second causes doth derogate from our dependence upon God, who is the first cause.

(3) To discover, then, the ignorance and error of this opinion, and the misunderstanding in the grounds thereof, it may well appear these men do not observe or consider that it was not the pure knowledge of Nature and universality, a knowledge by the light whereof man did give names unto other creatures in Paradise as they were brought before him according unto their proprieties, which gave the occasion to the fall; but it was the proud knowledge of good and evil, with an intent in man to give law unto himself, and to depend no more upon God's commandments, which was the form of the temptation. Neither is it any quantity of knowledge, how great soever, that can make the mind of man to swell; for nothing can fill, much less extend the soul of man, but God and the contemplation of God; and, therefore, Solomon, speaking of the two principal senses of inquisition, the eye and the ear, affirmeth that the eye is never satisfied with seeing, nor the ear with hearing; and if there be no fulness, then is the continent greater than the content: so of knowledge itself and the mind of man, whereto the senses are but reporters, he defineth likewise in these words, placed after that calendar or ephemerides which he maketh of the diversities of times and seasons for all actions and purposes, and concludeth thus: "God hath made all things beautiful, or decent, in the true return of their seasons. Also He hath placed the world in man's heart, yet cannot man find out the work which God worketh from the beginning to the end"—declaring not obscurely that God hath framed the mind of man as a mirror or glass, capable of the image of the universal world, and joyful to receive the impression thereof, as the eye joyeth to receive light; and not only delighted in beholding the variety of things and vicissitude of times, but raised also to find out and discern the ordinances and decrees which throughout all those changes are infallibly observed. And although he doth insinuate that the supreme or summary law of Nature (which he calleth "the work which God worketh from the beginning to the end") is not possible to be found out by man, yet that doth not derogate from the capacity of the mind; but may be referred to the impediments, as of shortness of life, ill conjunction of labours, ill tradition of knowledge over from hand to hand, and many other inconveniences, whereunto the condition of man is subject. For that nothing parcel of the world is denied to man's inquiry and invention, he doth in another place rule over,

when he saith, "The spirit of man is as the lamp of God, wherewith He searcheth the inwardness of all secrets." If, then, such be the capacity and receipt of the mind of man, it is manifest that there is no danger at all in the proportion or quantity of knowledge, how large soever, lest it should make it swell or out-compass itself; no, but it is merely the quality of knowledge, which, be it in quantity more or less, if it be taken without the true corrective thereof, hath in it some nature of venom or malignity, and some effects of that venom, which is ventosity or swelling. This corrective spice, the mixture whereof maketh knowledge so sovereign, is charity, which the Apostle immediately addeth to the former clause; for so he saith, "Knowledge bloweth up, but charity buildeth up;" not unlike unto that which he deilvereth in another place: "If I spake," saith he, "with the tongues of men and angels, and had not charity, it were but as a tinkling cymbal." Not but that it is an excellent thing to speak with the tongues of men and angels, but because, if it be severed from charity, and not referred to the good of men and mankind, it hath rather a sounding and unworthy glory than a meriting and substantial virtue. And as for that censure of Solomon concerning the excess of writing and reading books, and the anxiety of spirit which redoundeth from knowledge, and that admonition of St. Paul, "That we be not seduced by vain philosophy," let those places be rightly understood; and they do, indeed, excellently set forth the true bounds and limitations whereby human knowledge is confined and circumscribed, and yet without any such contracting or coarctation, but that it may comprehend all the universal nature of things; for these limitations are three: the first, "That we do not so place our felicity in knowledge, as we forget our mortality;" the second, "That we make application of our knowledge, to give ourselves repose and contentment, and not distaste or repining;" the third, "That we do not presume by the contemplation of Nature to attain to the mysteries of God." For as touching the first of these, Solomon doth excellently expound himself in another place of the same book, where he saith: "I saw well that knowledge recedeth as far from ignorance as light doth from darkness; and that the wise man's eyes keep watch in his head, whereas this fool roundeth about in darkness: but withal I learned that the same mortality involveth them both." And for the second, certain it is there is no vexation or anxiety of mind which resulteth from knowledge otherwise than merely by accident; for all knowledge and wonder (which is the seed of knowledge) is an impression of pleasure in itself; but when men fall to framing conclusions out of their knowledge, applying it to their particular, and ministering to themselves thereby weak fears or vast desires, there groweth that carefulness and trouble of mind which is spoken of; for then knowledge is no more Lumen siccum, whereof Heraclitus the profound said, Lumen siccum optima anima; but it becometh Lumen madidum, or maceratum, being steeped and infused in the humours of the affections. And as for the third point, it deserveth to be a little stood upon, and not to be lightly passed over; for if any man shall think by view and inquiry into these sensible and material things to attain that light, whereby he may reveal unto himself the nature or will of God, then, indeed, is he spoiled by vain philosophy; for the contemplation of God's creatures and works produceth (having regard to the works and creatures themselves) knowledge, but having regard to God no perfect knowledge, but wonder, which is broken knowledge. And, therefore, it was most aptly said by one of Plato's school, "That the sense of man carrieth a resemblance with the sun, which (as we see) openeth and revealeth all the terrestrial globe; but then, again, it obscureth and concealeth the stars and celestial globe: so doth the sense discover natural things, but it darkeneth and shutteth up divine." And hence it is true that it hath proceeded, that divers great learned men have been heretical, whilst they have sought to fly up to the secrets of the Deity by this waxen wings of the senses. And as for the conceit that too much knowledge should incline a man to atheism, and that the ignorance of second causes should make a more devout dependence upon God, which is the first cause; first, it is good to ask the question which Job asked of his friends: "Will you lie for God, as one man will lie for another, to gratify him?" For certain it is that God worketh nothing in Nature but by second causes; and if they would have it otherwise believed, it is mere imposture, as it were in favour towards God, and nothing else but to offer to the Author of truth the unclean sacrifice of a lie. But further, it is an assured truth, and a conclusion of experience, that a little or superficial knowledge of philosophy may incline the mind of men to atheism, but a further proceeding therein doth bring the mind back again to religion. For in the entrance of philosophy, when the second causes, which are next unto the senses, do offer themselves to the mind of man, if it dwell and stay there it may induce some oblivion of the highest cause; but when a man passeth on further and seeth the dependence of causes and the works of Providence; then, according to the allegory of the poets, he will easily believe that the highest link of Nature's chain must needs he tied to the foot of Jupiter's chair. To conclude, therefore, let no man upon a weak conceit of sobriety or an ill-applied moderation think or maintain that a man can search too far, or be too well studied in the book of God's word, or in the book of God's works, divinity or philosophy; but rather let men endeavour an endless progress or proficience in both; only let men beware that they apply both to charity, and not to swelling; to use, and not to ostentation; and again, that they do not unwisely mingle or confound these learnings together.

II

(1) And as for the disgraces which learning receiveth from politics, they be of this nature: that learning doth soften men's minds, and makes them more unapt for the honour and exercise of arms; that it doth mar and pervert men's dispositions for matter of government and policy, in making them too curious and irresolute by variety of reading, or too peremptory or positive by strictness of rules and axioms, or too immoderate and overweening by reason of the greatness of examples, or too incompatible and differing from the times by reason of the dissimilitude of examples; or at least, that it doth divert men's travails from action and business, and bringeth them to a love of leisure and privateness; and that it doth bring into states a relaxation of discipline, whilst every man is more ready to argue than to obey and execute. Out of this conceit Cato, surnamed the Censor, one of the wisest men indeed that ever lived, when Carneades the philosopher came in embassage to Rome, and that the young men of Rome began to flock about him, being allured with the sweetness and majesty of his eloquence and learning, gave counsel in open senate that they should give him his despatch with all speed, lest he should infect and enchant the minds and affections of the youth, and at unawares bring in an alteration of the manners and customs of the state. Out of the same conceit or humour did Virgil, turning his pen to the advantage of his country and the disadvantage of his own profession, make a kind of separation between policy and government, and between arts and sciences, in the verses so much renowned, attributing and challenging the one to the Romans, and leaving and yielding the other to the Grecians: Tu regere imperio popules, Romane, memento, Hæ tibi erunt artes, &c. So likewise we see that Anytus, the accuser of Socrates, laid it as an article of charge and accusation against him, that he did, with the variety and power of his discourses and disputatious, withdraw young men from due reverence to the laws and customs of their country, and that he did profess a dangerous and pernicious science, which was to make the worse matter seem the better, and to suppress truth by force of eloquence and speech.

(2) But these and the like imputations have rather a countenance of gravity than any ground of justice: for experience doth warrant that, both in persons and in times, there hath been a meeting and concurrence in learning and arms, flourishing and excelling in the same men and the same ages. For as 'for men, there cannot be a better nor the hike instance as of that pair, Alexander the Great and Julius Cæsar, the Dictator; whereof the one was Aristotle's scholar in philosophy, and the other was Cicero's rival in eloquence; or if any man had rather call for scholars that were great generals, than generals that were great scholars, let him take Epaminondas the Theban, or Xenophon the Athenian; whereof the one was the first that abated the power of Sparta, and the other was the first that made way to the overthrow of the monarchy of Persia. And this concurrence is yet more visible in times than in persons, by how much an age is greater object than a man. For both in Egypt, Assyria, Persia, Græcia, and Rome, the same times that are most renowned for arms are, likewise, most admired for learning, so that the greatest authors and philosophers, and the greatest captains and governors, have lived in the same ages. Neither can it otherwise he: for as in man the ripeness of strength of the body and mind cometh much about an age, save that the strength of the body cometh somewhat the more early, so in states, arms and learning, whereof the one correspondeth to the body, the other to the soul of man, have a concurrence or near sequence in times.

(3) And for matter of policy and government, that learning, should rather hurt, than enable thereunto, is a thing very improbable; we see it is accounted an error to commit a natural body to empiric physicians, which commonly have a few pleasing receipts whereupon they are confident and adventurous, but know neither the causes of diseases, nor the complexions of patients, nor peril of accidents, nor the true method of cures; we see it is a like error to rely upon advocates or lawyers which are only men of practice, and not grounded in their books, who are many times easily surprised when matter falleth out besides their experience, to the prejudice of the causes they handle: so by like reason it cannot be but a matter of doubtful consequence if states be managed by empiric statesmen, not well mingled with men grounded in learning. But contrariwise, it is almost without instance contradictory that ever any government was disastrous that was in the hands of learned governors. For howsoever it hath been ordinary with politic men to extenuate and disable learned men by the names of pedantes; yet in the records of time it appeareth in many particulars that the governments of princes in minority (notwithstanding the infinite disadvantage of that kind of state)—have nevertheless excelled the government of princes of mature age, even for that reason which they seek to traduce, which is that by that occasion the state hath been in the hands of pedantes: for so was the state of Rome for the first five years, which are so much magnified, during the minority of Nero, in the hands of Seneca, a pedenti; so it was again, for ten years' space or more, during the minority of Gordianus the younger, with great applause and contentation in the hands of Misitheus, a pedanti: so was it before that, in the minority of Alexander Severus, in like happiness, in hands not much unlike, by reason of the rule of the women, who were aided by the teachers and preceptors. Nay, let a man look into the government of the Bishops

of Rome, as by name, into the government of Pius Quintus and Sextus Quintus in our times, who were both at their entrance esteemed but as pedantical friars, and he shall find that such Popes do greater things, and proceed upon truer principles of state, than those which have ascended to the papacy from an education and breeding in affairs of state and courts of princes; for although men bred in learning are perhaps to seek in points of convenience and accommodating for the present, which the Italians call ragioni di stato, whereof the same Pius Quintus could not hear spoken with patience, terming them inventions against religion and the moral virtues; yet on the other side, to recompense that, they are perfect in those same plain grounds of religion, justice, honour, and moral virtue, which if they be well and watchfully pursued, there will be seldom use of those other, no more than of physic in a sound or well-dieted body. Neither can the experience of one man's life furnish examples and precedents for the event of one man's life. For as it happeneth sometimes that the grandchild, or other descendant, resembleth the ancestor more than the son; so many times occurrences of present times may sort better with ancient examples than with those of the later or immediate times; and lastly, the wit of one man can no more countervail learning than one man's means can hold way with a common purse.

(4) And as for those particular seducements or indispositions of the mind for policy and government, which learning is pretended to insinuate; if it be granted that any such thing be, it must be remembered withal that learning ministereth in every of them greater strength of medicine or remedy than it offereth cause of indisposition or infirmity. For if by a secret operation it make men perplexed and irresolute, on the other side by plain precept it teacheth them when and upon what ground to resolve; yea, and how to carry things in suspense, without prejudice, till they resolve. If it make men positive and regular, it teacheth them what things are in their nature demonstrative, and what are conjectural, and as well the use of distinctions and exceptions, as the latitude of principles and rules. If it mislead by disproportion or dissimilitude of examples, it teacheth men the force of circumstances, the errors of comparisons, and all the cautions of application; so that in all these it doth rectify more effactually than it can pervert. And these medicines it conveyeth into men's minds much more forcibly by the quickness and penetration of examples. For let a man look into the errors of Clement VII., so lively described by Guicciardini, who served under him, or into the errors of Cicero, painted out by his own pencil in his Epistles to Atticus, and he will fly apace from being irresolute. Let him look into the errors of Phocion, and he will beware how he be obstinate or inflexible. Let him but read the fable of Ixion, and it will hold him from being vaporous or imaginative. Let him look into the errors of Cato II., and he will never be one of the Antipodes, to tread opposite to the present world.

(5) And for the conceit that learning should dispose men to leisure and privateness, and make men slothful: it were a strange thing if that which accustometh the mind to a perpetual motion and agitation should induce slothfulness, whereas, contrariwise, it may be truly affirmed that no kind of men love business for itself but those that are learned; for other persons love it for profit, as a hireling that loves the work for the wages; or for honour, as because it beareth them up in the eyes of men, and refresheth their reputation, which otherwise would wear; or because it putteth them in mind of their fortune, and giveth them occasion to pleasure and displeasure; or because it exerciseth some faculty wherein they take pride, and so entertaineth them in good-humour and pleasing conceits towards themselves; or because it advanceth any other their ends. So that as it is said of untrue valours, that some men's valours are in the eyes of them that look on, so such men's industries are in the eyes of others, or, at least, in regard of their own designments; only learned men love business as an action according to nature, as agreeable to health of mind as exercise is to health of body, taking pleasure in the action itself, and not in the purchase, so that of all men they are the most indefatigable, if it be towards any business which can hold òr detain their mind.

(6) And if any man be laborious in reading and study, and yet idle in business and action, it groweth from some weakness of body or softness of spirit, such as Seneca speaketh of: Quidam tam sunt umbratiles, ut putent in turbido esse quicquid in luce est; and not of learning: well may it be that such a point of a man's nature may make him give himself to learning, but it is not learning that breedeth any such point in his nature.

(7) And that learning should take up too much time or leisure: I answer, the most active or busy man that hath been or can be, hath (no question) many vacant times of leisure while he expecteth the tides and returns of business (except he be either tedious and of no despatch, or lightly and unworthily ambitious to meddle in things that may be better done by others), and then the question is but how those spaces and times of leisure shall be filled and spent; whether in pleasure or in studies; as was well answered by Demosthenes to his adversary Æschines, that was a man given to pleasure, and told him "That his orations did smell of the lamp." "Indeed," said Demosthenes, "there is a great difference between the things that you and I do by lamp-light." So as no man need doubt that learning will expel business, but rather it will keep and defend the possession of the mind against idleness and pleasure, which otherwise at unawares may enter to the prejudice of both.

(8) Again, for that other conceit that learning should undermine the reverence of laws and government, it is assuredly a mere depravation and calumny, without all shadow of truth. For to say

that a blind custom of obedience should be a surer obligation than duty taught and understood, it is to affirm that a blind man may tread surer by a guide than a seeing man can by a light. And it is without all controversy that learning doth make the minds of men gentle, generous, manageable, and pliant to government; whereas ignorance makes them churlish, thwart, and mutinous: and the evidence of time doth clear this assertion, considering that the most barbarous, rude, and unlearned times have been most subject to tumults, seditious, and changes.

(9) And as to the judgment of Cato the Censor, he was well punished for his blasphemy against learning, in the same kind wherein he offended; for when he was past threescore years old, he was taken with an extreme desire to go to school again, and to learn the Greek tongue, to the end to peruse the Greek authors; which doth well demonstrate that his former censure of the Grecian learning was rather an affected gravity, than according to the inward sense of his own opinion. And as for Virgil's verses, though it pleased him to brave the world in taking to the Romans the art of empire, and leaving to others the arts of subjects, yet so much is manifest—that the Romans never ascended to that height of empire till the time they had ascended to the height of other arts. For in the time of the two first Cæsars, which had the art of government in greatest perfection, there lived the best poet, Virgilius Maro; the best historiographer, Titus Livius; the best antiquary, Marcus Varro; and the best or second orator, Marcus Cicero, that to the memory of man are known. As for the accusation of Socrates, the time must be remembered when it was prosecuted; which was under the Thirty Tyrants, the most base, bloody, and envious persons that have governed; which revolution of state was no sooner over but Socrates, whom they had made a person criminal, was made a person heroical, and his memory accumulate with honours divine and human; and those discourses of his which were then termed corrupting of manners, were after acknowledged for sovereign medicines of the mind and manners, and so have been received ever since till this day. Let this, therefore, serve for answer to politiques, which in their humorous severity, or in their feigned gravity, have presumed to throw imputations upon learning; which redargution nevertheless (save that we know not whether our labours may extend to other ages) were not needful for the present, in regard of the love and reverence towards learning which the example and countenance of two so learned princes, Queen Elizabeth and your Majesty, being as Castor and Pollux, *lucida sidera*, stars of excellent light and most benign influence, hath wrought in all men of place and authority in our nation.

III

(1) Now therefore we come to that third sort of discredit or diminution of credit that groweth unto learning from learned men themselves, which commonly cleaveth fastest: it is either from their fortune, or from their manners, or from the nature of their studies. For the first, it is not in their power; and the second is accidental; the third only is proper to be handled: but because we are not in hand with true measure, but with popular estimation and conceit, it is not amiss to speak somewhat of the two former. The derogations therefore which grow to learning from the fortune or condition of learned men, are either in respect of scarcity of means, or in respect of privateness of life and meanness of employments.

(2) Concerning want, and that it is the case of learned men usually to begin with little, and not to grow rich so fast as other men, by reason they convert not their labours chiefly to lucre and increase, it were good to leave the commonplace in commendation of povery to some friar to handle, to whom much was attributed by Machiavel in this point when he said, "That the kingdom of the clergy had been long before at an end, if the reputation and reverence towards the poverty of friars had not borne out the scandal of the superfluities and excesses of bishops and prelates." So a man might say that the felicity and delicacy of princes and great persons had long since turned to rudeness and barbarism, if the poverty of learning had not kept up civility and honour of life; but without any such advantages, it is worthy the observation what a reverent and honoured thing poverty of fortune was for some ages in the Roman state, which nevertheless was a state without paradoxes. For we see what Titus Livius saith in his introduction: *Cæterum aut me amor negotii suscepti fallit aut nulla unquam respublica nec major, nec sanctior, nec bonis exemplis ditior fuit; nec in quam tam sero avaritia luxuriaeque immigraverint; nec ubi tantus ac tam diu paupertati ac parsimoniæ honos fuerit.* We see likewise, after that the state of Rome was not itself, but did degenerate, how that person that took upon him to be counsellor to Julius Cæsar after his victory where to begin his restoration of the state, maketh it of all points the most summary to take away the estimation of wealth: *Verum hæc et omnia mala pariter cum honore pecuniæ desinent; si neque magistratus, neque alia vulgo cupienda, venalia erunt.* To conclude this point: as it was truly said that *Paupertas est virtutis fortuna*, though sometimes it come from vice, so it may be fitly said that, though some times it may proceed from misgovernment and accident. Surely Solomon hath pronounced it both in censure, *Qui festinat ad divitias non erit insons*; and in precept, "Buy the truth, and sell it not; and so of wisdom and knowledge;" judging that means were to be spent upon learning, and not learning to be applied to means. And as for the privateness or obscureness (as it may be in

vulgar estimation accounted) of life of contemplative men, it is a theme so common to extol a private life, not taxed with sensuality and sloth, in comparison and to the disadvantage of a civil life, for safety, liberty, pleasure, and dignity, or at least freedom from indignity, as no man handleth it but handleth it well; such a consonancy it hath to men's conceits in the expressing, and to men's consents in the allowing. This only I will add, that learned men forgotten in states and not living in the eyes of men, are like the images of Cassius and Brutus in the funeral of Junia, of which, not being represented as many others were, Tacitus saith, Eo ipso præfulgebant quod non visebantur.

(3) And for meanness of employment, that which is most traduced to contempt is that the government of youth is commonly allotted to them; which age, because it is the age of least authority, it is transferred to the disesteeming of those employments wherein youth is conversant, and which are conversant about youth. But how unjust this traducement is (if you will reduce things from popularity of opinion to measure of reason) may appear in that we see men are more curious what they put into a new vessel than into a vessel seasoned; and what mould they lay about a young plant than about a plant corroborate; so as this weakest terms and times of all things use to have the best applications and helps. And will you hearken to the Hebrew rabbins? "Your young men shall see visions, and your old men shall dream dreams:" say they, youth is the worthier age, for that visions are nearer apparitions of God than dreams? And let it be noted that howsoever the condition of life of pedantes hath been scorned upon theatres, as the ape of tyranny; and that the modern looseness or negligence hath taken no due regard to the choice of schoolmasters and tutors; yet the ancient wisdom of the best times did always make a just complaint, that states were too busy with their laws and too negligent in point of education: which excellent part of ancient discipline hath been in some sort revived of late times by the colleges of the Jesuits; of whom, although in regard of their superstition I may say, Quo meliores, eo deteriores; yet in regard of this, and some other points concerning human learning and moral matters, I may say, as Agesilaus said to his enemy Pharnabazus, Talis quum sis, utunam noster esses. And that much touching the discredits drawn from the fortunes of learned men.

(4) As touching the manners of learned men, it is a thing personal and individual: and no doubt there be amongst them, as in other professions, of all temperatures: but yet so as it is not without truth which is said, that Abeunt studua in mores, studies have an influence and operation upon the manners of those that are conversant in them.

(5) But upon an attentive and indifferent review, I for my part cannot find any disgrace to learning can proceed from the manners of learned men; not inherent to them as they are learned; except it be a fault (which was the supposed fault of Demosthenes, Cicero, Cato II., Seneca, and many more) that because the times they read of are commonly better than the times they live in, and the duties taught better than the duties practised, they contend sometimes too far to bring things to perfection, and to reduce the corruption of manners to honesty of precepts or examples of too great height. And yet hereof they have caveats enough in their own walks. For Solon, when he was asked whether he had given his citizens the best laws, answered wisely, "Yea, of such as they would receive:" and Plato, finding that his own heart could not agree with the corrupt manners of his country, refused to bear place or office, saying, "That a man's country was to be used as his parents were, that is, with humble persuasions, and not with contestations." And Cæsar's counsellor put in the same caveat, Non ad vetera instituta revocans quæ jampridem corruptis moribus ludibrio sunt; and Cicero noteth this error directly in Cato II. when he writes to his friend Atticus, Cato optime sentit, sed nocet interdum reipublicæ; loquitur enim tanquam in republicâ Platonis, non tanquam in fæce Romuli. And the same Cicero doth excuse and expound the philosophers for going too far and being too exact in their prescripts when he saith, Isti ipse præceptores virtutis et mágistri videntur fines officiorum paulo longius quam natura vellet protulisse, ut cum ad ultimum animo contendissemus, ibi tamen, ubi oportet, consisteremus: and yet himself might have said, Monitis sum minor ipse meis; for it was his own fault, though not in so extreme a degree.

(6) Another fault likewise much of this kind hath been incident to learned men, which is, that they have esteemed the preservation, good, and honour of their countries or masters before their own fortunes or safeties. For so saith Demosthenes unto the Athenians: "If it please you to note it, my counsels unto you are not such whereby I should grow great amongst you, and you become little amongst the Grecians; but they be of that nature as they are sometimes not good for me to give, but are always good for you to follow." And so Seneca, after he had consecrated that Quinquennium Neronis to the eternal glory of learned governors, held on his honest and loyal course of good and free counsel after his master grew extremely corrupt in his government. Neither can this point otherwise be, for learning endueth men's minds with a true sense of the frailty of their persons, the casuality of their fortunes, and the dignity of their soul and vocation, so that it is impossible for them to esteem that any greatness of their own fortune can be a true or worthy end of their being and ordainment, and therefore are desirous to give their account to God, and so likewise to their masters under God (as kings and the states that they serve) in those words, Ecce tibi lucrefeci, and not Ecce mihi lucrefeci; whereas the corrupter sort of mere politiques, that have not their thoughts established by learning in the love and apprehension of duty, nor never look abroad into universality, do refer all things to themselves, and thrust themselves

into the centre of the world, as if all lines should meet in them and their fortunes, never caring in all tempests what becomes of the ship of state, so they may save themselves in the cockboat of their own fortune; whereas men that feel the weight of duty and know the limits of self-love use to make good their places and duties, though with peril; and if they stand in seditious and violent alterations, it is rather the reverence which many times both adverse parts do give to honesty, than any versatile advantage of their own carriage. But for this point of tender sense and fast obligation of duty which learning doth endue the mind withal, howsoever fortune may tax it, and many in the depth of their corrupt principles may despise it, yet it will receive an open allowance, and therefore needs the less disproof or excuse.

(7) Another fault incident commonly to learned men, which may be more properly defended than truly denied, is that they fail sometimes in applying themselves to particular persons, which want of exact application ariseth from two causes—the one, because the largeness of their mind can hardly confine itself to dwell in the exquisite observation or examination of the nature and customs of one person, for it is a speech for a lover, and not for a wise man, Satis magnum alter alteri theatrum sumus. Nevertheless I shall yield that he that cannot contract the sight of his mind as well as disperse and dilate it, wanteth a great faculty. But there is a second cause, which is no inability, but a rejection upon choice and judgment. For the honest and just bounds of observation by one person upon another extend no further but to understand him sufficiently, whereby not to give him offence, or whereby to be able to give him faithful counsel, or whereby to stand upon reasonable guard and caution in respect of a man's self. But to be speculative into another man to the end to know how to work him, or wind him, or govern him, proceedeth from a heart that is double and cloven, and not entire and ingenuous; which as in friendship it is want of integrity, so towards princes or superiors is want of duty. For the custom of the Levant, which is that subjects do forbear to gaze or fix their eyes upon princes, is in the outward ceremony barbarous, but the moral is good; for men ought not, by cunning and bent observations, to pierce and penetrate into the hearts of kings, which the Scripture hath declared to be inscrutable.

(8) There is yet another fault (with which I will conclude this part) which is often noted in learned men, that they do many times fail to observe decency and discretion in their behaviour and carriage, and commit errors in small and ordinary points of action, so as the vulgar sort of capacities do make a judgment of them in greater matters by that which they find wanting in them in smaller. But this consequence doth oft deceive men, for which I do refer them over to that which was said by Themistocles, arrogantly and uncivilly being applied to himself out of his own mouth, but, being applied to the general state of this question, pertinently and justly, when, being invited to touch a lute, he said, "He could not fiddle, but he could make a small town a great state." So no doubt many may be well seen in the passages of government and policy which are to seek in little and punctual occasions. I refer them also to that which Plato said of his master Socrates, whom he compared to the gallipots of apothecaries, which on the outside had apes and owls and antiques, but contained within sovereign and precious liquors and confections; acknowledging that, to an external report, he was not without superficial levities and deformities, but was inwardly replenished with excellent virtues and powers. And so much touching the point of manners of learned men.

(9) But in the meantime I have no purpose to give allowance to some conditions and courses base and unworthy, wherein divers professors of learning have wronged themselves and gone too far; such as were those trencher philosophers which in the later age of the Roman state were usually in the houses of great persons, being little better than solemn parasites, of which kind, Lucian maketh a merry description of the philosopher that the great lady took to ride with her in her coach, and would needs have him carry her little dog, which he doing officiously and yet uncomely, the page scoffed and said, "That he doubted the philosopher of a Stoic would turn to be a Cynic." But, above all the rest, this gross and palpable flattery whereunto many not unlearned have abased and abused their wits and pens, turning (as Du Bartas saith) Hecuba into Helena, and Faustina into Lucretia, hath most diminished the price and estimation of learning. Neither is the modern dedication of books and writings, as to patrons, to be commended, for that books (such as are worthy the name of books) ought to have no patrons but truth and reason. And the ancient custom was to dedicate them only to private and equal friends, or to entitle the books with their names; or if to kings and great persons, it was to some such as the argument of the book was fit and proper for; but these and the like courses may deserve rather reprehension than defence.

(10) Not that I can tax or condemn the morigeration or application of learned men to men in fortune. For the answer was good that Diogenes made to one that asked him in mockery, "How it came to pass that philosophers were the followers of rich men, and not rich men of philosophers?" He answered soberly, and yet sharply, "Because the one sort knew what they had need of, and the other did not." And of the like nature was the answer which Aristippus made, when having a petition to Dionysius, and no ear given to him, he fell down at his feet, whereupon Dionysius stayed and gave him the hearing, and granted it; and afterwards some person, tender on the behalf of philosophy, reproved Aristippus that he would offer the profession of philosophy such an indignity as for a private suit to fall

at a tyrant's feet; but he answered, "It was not his fault, but it was the fault of Dionysius, that had his ears in his feet." Neither was it accounted weakness, but discretion, in him that would not dispute his best with Adrianus Cæsar, excusing himself, "That it was reason to yield to him that commanded thirty legions." These and the like, applications, and stooping to points of necessity and convenience, cannot be disallowed; for though they may have some outward baseness, yet in a judgment truly made they are to be accounted submissions to the occasion and not to the person.

IV

(1) Now I proceed to those errors and vanities which have intervened amongst the studies themselves of the learned, which is that which is principal and proper to the present argument; wherein my purpose is not to make a justification of the errors, but by a censure and separation of the errors to make a justification of that which is good and sound, and to deliver that from the aspersion of the other. For we see that it is the manner of men to scandalise and deprave that which retaineth the state and virtue, by taking advantage upon that which is corrupt and degenerate, as the heathens in the primitive Church used to blemish and taint the Christians with the faults and corruptions of heretics. But nevertheless I have no meaning at this time to make any exact animadversion of the errors and impediments in matters of learning, which are more secret and remote from vulgar opinion, but only to speak unto such as do fall under or near unto a popular observation.

(2) There be therefore chiefly three vanities in studies, whereby learning hath been most traduced. For those things we do esteem vain which are either false or frivolous, those which either have no truth or no use; and those persons we esteem vain which are either credulous or curious; and curiosity is either in matter or words: so that in reason as well as in experience there fall out to be these three distempers (as I may term them) of learning—the first, fantastical learning; the second, contentious learning; and the last, delicate learning; vain imaginations, vain altercations, and vain affectations; and with the last I will begin. Martin Luther, conducted, no doubt, by a higher Providence, but in discourse of reason, finding what a province he had undertaken against the Bishop of Rome and the degenerate traditions of the Church, and finding his own solitude, being in nowise aided by the opinions of his own time, was enforced to awake all antiquity, and to call former times to his succours to make a party against the present time. So that the ancient authors, both in divinity and in humanity, which had long time slept in libraries, began generally to be read and revolved. This, by consequence, did draw on a necessity of a more exquisite travail in the languages original, wherein those authors did write, for the better understanding of those authors, and the better advantage of pressing and applying their words. And thereof grew, again, a delight in their manner of style and phrase, and an admiration of that kind of writing, which was much furthered and precipitated by the enmity and opposition that the propounders of those primitive but seeming new opinions had against the schoolmen, who were generally of the contrary part, and whose writings were altogether in a differing style and form; taking liberty to coin and frame new terms of art to express their own sense, and to avoid circuit of speech, without regard to the pureness, pleasantness, and (as I may call it) lawfulness of the phrase or word. And again, because the great labour then was with the people (of whom the Pharisees were wont to say, Execrabilis ista turba, quæ non novit legem), for the winning and persuading of them, there grew of necessity in chief price and request eloquence and variety of discourse, as the fittest and forciblest access into the capacity of the vulgar sort; so that these four causes concurring—the admiration of ancient authors, the hate of the schoolmen, the exact study of languages, and the efficacy of preaching—did bring in an affectionate study of eloquence and copy of speech, which then began to flourish. This grew speedily to an excess; for men began to hunt more after words than matter—more after the choiceness of the phrase, and the round and clean composition of the sentence, and the sweet falling of the clauses, and the varying and illustration of their works with tropes and figures, than after the weight of matter, worth of subject, soundness of argument, life of invention, or depth of judgment. Then grew the flowing and watery vein of Osorius, the Portugal bishop, to be in price. Then did Sturmius spend such infinite and curious pains upon Cicero the Orator and Hermogenes the Rhetorician, besides his own books of Periods and Imitation, and the like. Then did Car of Cambridge and Ascham with their lectures and writings almost deify Cicero and Demosthenes, and allure all young men that were studious unto that delicate and polished kind of learning. Then did Erasmus take occasion to make the scoffing echo, Decem annos consuumpsi in legendo Cicerone; and the echo answered in Greek, One, Asine. Then grew the learning of the schoolmen to be utterly despised as barbarous. In sum, the whole inclination and bent of those times was rather towards copy than weight.

(3) Here therefore [is] the first distemper of learning, when men study words and not matter; whereof, though I have represented an example of late times, yet it hath been and will be secundum majus et minus in all time. And how is it possible but this should have an operation to discredit learning, even with vulgar capacities, when they see learned men's works like the first letter of a patent

or limited book, which though it hath large flourishes, yet it is but a letter? It seems to me that Pygmalion's frenzy is a good emblem or portraiture of this vanity; for words are but the images of matter, and except they have life of reason and invention, to fall in love with them is all one as to fall in love with a picture.

(4) But yet notwithstanding it is a thing not hastily to be condemned, to clothe and adorn the obscurity even of philosophy itself with sensible and plausible elocution. For hereof we have great examples in Xenophon, Cicero, Seneca, Plutarch, and of Plato also in some degree; and hereof likewise there is great use, for surely, to the severe inquisition of truth and the deep progress into philosophy, it is some hindrance because it is too early satisfactory to the mind of man, and quencheth the desire of further search before we come to a just period. But then if a man be to have any use of such knowledge in civil occasions, of conference, counsel, persuasion, discourse, or the like, then shall he find it prepared to his hands in those authors which write in that manner. But the excess of this is so justly contemptible, that as Hercules, when he saw the image of Adonis, Venus' minion, in a temple, said in disdain, Nil sacri es; so there is none of Hercules' followers in learning—that is, the more severe and laborious sort of inquirers into truth—but will despise those delicacies and affectations, as indeed capable of no divineness. And thus much of the first disease or distemper of learning.

(5) The second which followeth is in nature worse than the former: for as substance of matter is better than beauty of words, so contrariwise vain matter is worse than vain words: wherein it seemeth the reprehension of St. Paul was not only proper for those times, but prophetical for the times following; and not only respective to divinity, but extensive to all knowledge: Devita profanas vocum novitates, et oppositiones falsi nominis scientiæ. For he assigneth two marks and badges of suspected and falsified science: the one, the novelty and strangeness of terms; the other, the strictness of positions, which of necessity doth induce oppositions, and so questions and altercations. Surely, like as many substances in nature which are solid do putrefy and corrupt into worms;—so it is the property of good and sound knowledge to putrefy and dissolve into a number of subtle, idle, unwholesome, and (as I may term them) vermiculate questions, which have indeed a kind of quickness and life of spirit, but no soundness of matter or goodness of quality. This kind of degenerate learning did chiefly reign amongst the schoolmen, who having sharp and strong wits, and abundance of leisure, and small variety of reading, but their wits being shut up in the cells of a few authors (chiefly Aristotle their dictator) as their persons were shut up in the cells of monasteries and colleges, and knowing little history, either of nature or time, did out of no great quantity of matter and infinite agitation of wit spin out unto us those laborious webs of learning which are extant in their books. For the wit and mind of man, if it work upon matter, which is the contemplation of the creatures of God, worketh according to the stuff and is limited thereby; but if it work upon itself, as the spider worketh his web, then it is endless, and brings forth indeed cobwebs of learning, admirable for the fineness of thread and work, but of no substance or profit.

(6) This same unprofitable subtility or curiosity is of two sorts: either in the subject itself that they handle, when it is a fruitless speculation or controversy (whereof there are no small number both in divinity and philosophy), or in the manner or method of handling of a knowledge, which amongst them was this—upon every particular position or assertion to frame objections, and to those objections, solutions; which solutions were for the most part not confutations, but distinctions: whereas indeed the strength of all sciences is, as the strength of the old man's faggot, in the bond. For the harmony of a science, supporting each part the other, is and ought to be the true and brief confutation and suppression of all the smaller sort of objections. But, on the other side, if you take out every axiom, as the sticks of the faggot, one by one, you may quarrel with them and bend them and break them at your pleasure: so that, as was said of Seneca, Verborum minutiis rerum frangit pondera, so a man may truly say of the schoolmen, Quæstionum minutiis scientiarum frangunt soliditatem. For were it not better for a man in fair room to set up one great light, or branching candlestick of lights, than to go about with a small watch-candle into every corner? And such is their method, that rests not so much upon evidence of truth proved by arguments, authorities, similitudes, examples, as upon particular confutations and solutions of every scruple, cavillation, and objection; breeding for the most part one question as fast as it solveth another; even as in the former resemblance, when you carry the light into one corner, you darken the rest; so that the fable and fiction of Scylla seemeth to be a lively image of this kind of philosophy or knowledge; which was transformed into a comely virgin for the upper parts; but then Candida succinctam latrantibus inguina monstris: so the generalities of the schoolmen are for a while good and proportionable; but then when you descend into their distinctions and decisions, instead of a fruitful womb for the use and benefit of man's life, they end in monstrous altercations and barking questions. So as it is not possible but this quality of knowledge must fall under popular contempt, the people being apt to contemn truths upon occasion of controversies and altercations, and to think they are all out of their way which never meet; and when they see such digladiation about subtleties, and matters of no use or moment, they easily fall upon that judgment of Dionysius of Syracusa, Verba ista sunt senum otiosorum.

(7) Notwithstanding, certain it is that if those schoolmen to their great thirst of truth and unwearied

travail of wit had joined variety and universality of reading and contemplation, they had proved excellent lights, to the great advancement of all learning and knowledge; but as they are, they are great undertakers indeed, and fierce with dark keeping. But as in the inquiry of the divine truth, their pride inclined to leave the oracle of God's word, and to vanish in the mixture of their own inventions; so in the inquisition of nature, they ever left the oracle of God's works, and adored the deceiving and deformed images which the unequal mirror of their own minds, or a few received authors or principles, did represent unto them. And thus much for the second disease of learning.

(8) For the third vice or disease of learning, which concerneth deceit or untruth, it is of all the rest the foulest; as that which doth destroy the essential form of knowledge, which is nothing but a representation of truth: for the truth of being and the truth of knowing are one, differing no more than the direct beam and the beam reflected. This vice therefore brancheth itself into two sorts; delight in deceiving and aptness to be deceived; imposture and credulity; which, although they appear to be of a diverse nature, the one seeming to proceed of cunning and the other of simplicity, yet certainly they do for the most part concur: for, as the verse noteth—

"Percontatorem fugito, nam garrulus idem est,"

an inquisitive man is a prattler; so upon the like reason a credulous man is a deceiver: as we see it in fame, that he that will easily believe rumours will as easily augment rumours and add somewhat to them of his own; which Tacitus wisely noteth, when he saith, Fingunt simul creduntque: so great an affinity hath fiction and belief.

(9) This facility of credit and accepting or admitting things weakly authorised or warranted is of two kinds according to the subject: for it is either a belief of history, or, as the lawyers speak, matter of fact; or else of matter of art and opinion. As to the former, we see the experience and inconvenience of this error in ecclesiastical history; which hath too easily received and registered reports and narrations of miracles wrought by martyrs, hermits, or monks of the desert, and other holy men, and their relics, shrines, chapels and images: which though they had a passage for a time by the ignorance of the people, the superstitious simplicity of some and the politic toleration of others holding them but as divine poesies, yet after a period of time, when the mist began to clear up, they grew to be esteemed but as old wives' fables, impostures of the clergy, illusions of spirits, and badges of Antichrist, to the great scandal and detriment of religion.

(10) So in natural history, we see there hath not been that choice and judgment used as ought to have been; as may appear in the writings of Plinius, Cardanus, Albertus, and divers of the Arabians, being fraught with much fabulous matter, a great part not only untried, but notoriously untrue, to the great derogation of the credit of natural philosophy with the grave and sober kind of wits: wherein the wisdom and integrity of Aristotle is worthy to be observed, that, having made so diligent and exquisite a history of living creatures, hath mingled it sparingly with any vain or feigned matter; and yet on the other side hath cast all prodigious narrations, which he thought worthy the recording, into one book, excellently discerning that matter of manifest truth, such whereupon observation and rule was to be built, was not to be mingled or weakened with matter of doubtful credit; and yet again, that rarities and reports that seem uncredible are not to be suppressed or denied to the memory of men.

(11) And as for the facility of credit which is yielded to arts and opinions, it is likewise of two kinds; either when too much belief is attributed to the arts themselves, or to certain authors in any art. The sciences themselves, which have had better intelligence and confederacy with the imagination of man than with his reason, are three in number: astrology, natural magic, and alchemy; of which sciences, nevertheless, the ends or pretences are noble. For astrology pretendeth to discover that correspondence or concatenation which is between the superior globe and the inferior; natural magic pretendeth to call and reduce natural philosophy from variety of speculations to the magnitude of works; and alchemy pretendeth to make separation of all the unlike parts of bodies which in mixtures of natures are incorporate. But the derivations and prosecutions to these ends, both in the theories and in the practices, are full of error and vanity; which the great professors themselves have sought to veil over and conceal by enigmatical writings, and referring themselves to auricular traditions and such other devices, to save the credit of impostures. And yet surely to alchemy this right is due, that it may be compared to the husbandman whereof Æsop makes the fable; that, when he died, told his sons that he had left unto them gold buried underground in his vineyard; and they digged over all the ground, and gold they found none; but by reason of their stirring and digging the mould about the roots of their vines, they had a great vintage the year following: so assuredly the search and stir to make gold hath brought to light a great number of good and fruitful inventions and experiments, as well for the disclosing of nature as for the use of man's life.

(12) And as for the overmuch credit that hath been given unto authors in sciences, in making them dictators, that their words should stand, and not consuls, to give advice; the damage is infinite that sciences have received thereby, as the principal cause that hath kept them low at a stay without growth

or advancement. For hence it hath come, that in arts mechanical the first deviser comes shortest, and time addeth and perfecteth; but in sciences the first author goeth furthest, and time leeseth and corrupteth. So we see artillery, sailing, printing, and the like, were grossly managed at the first, and by time accommodated and refined; but contrariwise, the philosophies and sciences of Aristotle, Plato, Democritus, Hippocrates, Euclides, Archimedes, of most vigour at the first, and by time degenerate and imbased: whereof the reason is no other, but that in the former many wits and industries have contributed in one; and in the latter many wits and industries have been spent about the wit of some one, whom many times they have rather depraved than illustrated; for, as water will not ascend higher than the level of the first spring-head from whence it descendeth, so knowledge derived from Aristotle, and exempted from liberty of examination, will not rise again higher than the knowledge of Aristotle. And, therefore, although the position be good, Oportet discentem credere, yet it must be coupled with this, Oportet edoctum judicare; for disciples do owe unto masters only a temporary belief and a suspension of their own judgment till they be fully instructed, and not an absolute resignation or perpetual captivity; and therefore, to conclude this point, I will say no more, but so let great authors have their due, as time, which is the author of authors, be not deprived of his due—which is, further and further to discover truth. Thus have I gone over these three diseases of learning; besides the which there are some other rather peccant humours than formed diseases, which, nevertheless, are not so secret and intrinsic, but that they fall under a popular observation and traducement, and, therefore, are not to be passed over.

V

(1) The first of these is the extreme affecting of two extremities: the one antiquity, the other novelty; wherein it seemeth the children of time do take after the nature and malice of the father. For as he devoureth his children, so one of them seeketh to devour and suppress the other; while antiquity envieth there should be new additions, and novelty cannot be content to add but it must deface; surely the advice of the prophet is the true direction in this matter, State super vias antiquas, et videte quænam sit via recta et bona et ambulate in ea. Antiquity deserveth that reverence, that men should make a stand thereupon and discover what is the best way; but when the discovery is well taken, then to make progression. And to speak truly, Antiquitas sæculi juventus mundi. These times are the ancient times, when the world is ancient, and not those which we account ancient ordine retrogrado, by a computation backward from ourselves.

(2) Another error induced by the former is a distrust that anything should be now to be found out, which the world should have missed and passed over so long time: as if the same objection were to be made to time that Lucian maketh to Jupiter and other the heathen gods; of which he wondereth that they begot so many children in old time, and begot none in his time; and asketh whether they were become septuagenary, or whether the law Papia, made against old men's marriages, had restrained them. So it seemeth men doubt lest time is become past children and generation; wherein contrariwise we see commonly the levity and unconstancy of men's judgments, which, till a matter be done, wonder that it can be done; and as soon as it is done, wonder again that it was no sooner done: as we see in the expedition of Alexander into Asia, which at first was prejudged as a vast and impossible enterprise; and yet afterwards it pleaseth Livy to make no more of it than this, Nil aliud quàm bene ausus vana contemnere. And the same happened to Columbus in the western navigation. But in intellectual matters it is much more common, as may be seen in most of the propositions of Euclid; which till they be demonstrate, they seem strange to our assent; but being demonstrate, our mind accepteth of them by a kind of relation (as the lawyers speak), as if we had known them before.

(3) Another error, that hath also some affinity with the former, is a conceit that of former opinions or sects after variety and examination the best hath still prevailed and suppressed the rest; so as if a man should begin the labour of a new search, he were but like to light upon somewhat formerly rejected, and by rejection brought into oblivion; as if the multitude, or the wisest for the multitude's sake, were not ready to give passage rather to that which is popular and superficial than to that which is substantial and profound for the truth is, that time seemeth to be of the nature of a river or stream, which carrieth down to us that which is light and blown up, and sinketh and drowneth that which is weighty and solid.

(4) Another error, of a diverse nature from all the former, is the over-early and peremptory reduction of knowledge into arts and methods; from which time commonly sciences receive small or no augmentation. But as young men, when they knit and shape perfectly, do seldom grow to a further stature, so knowledge, while it is in aphorisms and observations, it is in growth; but when it once is comprehended in exact methods, it may, perchance, be further polished, and illustrate and accommodated for use and practice, but it increaseth no more in bulk and substance.

(5) Another error which doth succeed that which we last mentioned is, that after the distribution of particular arts and sciences, men have abandoned universality, or philosophia prima, which cannot but cease and stop all progression. For no perfect discovery can be made upon a flat or a level; neither is it

possible to discover the more remote and deeper parts of any science if you stand but upon the level of the same science, and ascend not to a higher science.

(6) Another error hath proceeded from too great a reverence, and a kind of adoration of the mind and understanding of man; by means whereof, men have withdrawn themselves too much from the contemplation of nature, and the observations of experience, and have tumbled up and down in their own reason and conceits. Upon these intellectualists, which are notwithstanding commonly taken for the most sublime and divine philosophers, Heraclitus gave a just censure, saying:—"Men sought truth in their own little worlds, and not in the great and common world;" for they disdain to spell, and so by degrees to read in the volume of God's works; and contrariwise by continual meditation and agitation of wit do urge and, as it were, invocate their own spirits to divine and give oracles unto them, whereby they are deservedly deluded.

(7) Another error that hath some connection with this latter is, that men have used to infect their meditations, opinions, and doctrines with some conceits which they have most admired, or some sciences which they have most applied, and given all things else a tincture according to them, utterly untrue and improper. So hath Plato intermingled his philosophy with theology, and Aristotle with logic; and the second school of Plato, Proclus and the rest, with the mathematics; for these were the arts which had a kind of primogeniture with them severally. So have the alchemists made a philosophy out of a few experiments of the furnace; and Gilbertus our countryman hath made a philosophy out of the observations of a loadstone. So Cicero, when reciting the several opinions of the nature of the soul, he found a musician that held the soul was but a harmony, saith pleasantly, Hic ab arte sua non recessit, &c. But of these conceits Aristotle speaketh seriously and wisely when he saith, Qui respiciunt ad pauca de facili pronunciant.

(8) Another error is an impatience of doubt, and haste to assertion without due and mature suspension of judgment. For the two ways of contemplation are not unlike the two ways of action commonly spoken of by the ancients: the one plain and smooth in the beginning, and in the end impassable; the other rough and troublesome in the entrance, but after a while fair and even. So it is in contemplation: if a man will begin with certainties, he shall end in doubts; but if he will be content to begin with doubts, he shall end in certainties.

(9) Another error is in the manner of the tradition and delivery of knowledge, which is for the most part magistral and peremptory, and not ingenuous and faithful; in a sort as may be soonest believed, and not easiest examined. It is true, that in compendious treatises for practice that form is not to be disallowed; but in the true handling of knowledge men ought not to fall either on the one side into the vein of Velleius the Epicurean, Nil tam metuens quam ne dubitare aliqua de revideretur: nor, on the other side, into Socrates, his ironical doubting of all things; but to propound things sincerely with more or less asseveration, as they stand in a man's own judgment proved more or less.

(10) Other errors there are in the scope that men propound to themselves, whereunto they bend their endeavours; for, whereas the more constant and devote kind of professors of any science ought to propound to themselves to make some additions to their science, they convert their labours to aspire to certain second prizes: as to be a profound interpreter or commentor, to be a sharp champion or defender, to be a methodical compounder or abridger, and so the patrimony of knowledge cometh to be sometimes improved, but seldom augmented.

(11) But the greatest error of all the rest is the mistaking or misplacing of the last or furthest end of knowledge. For men have entered into a desire of learning and knowledge, sometimes upon a natural curiosity and inquisitive appetite; sometimes to entertain their minds with variety and delight; sometimes for ornament and reputation; and sometimes to enable them to victory of wit and contradiction; and most times for lucre and profession; and seldom sincerely to give a true account of their gift of reason to the benefit and use of men: as if there were sought in knowledge a couch whereupon to rest a searching and restless spirit; or a terrace for a wandering and variable mind to walk up and down with a fair prospect; or a tower of state, for a proud mind to raise itself upon; or a fort or commanding ground, for strife and contention; or a shop, for profit or sale; and not a rich storehouse for the glory of the Creator and the relief of man's estate. But this is that which will indeed dignify and exalt knowledge, if contemplation and action may be more nearly and straitly conjoined and united together than they have been: a conjunction like unto that of the two highest planets, Saturn, the planet of rest and contemplation; and Jupiter, the planet of civil society and action, howbeit, I do not mean, when I speak of use and action, that end before-mentioned of the applying of knowledge to lucre and profession; for I am not ignorant how much that diverteth and interrupteth the prosecution and advancement of knowledge, like unto the golden ball thrown before Atalanta, which, while she goeth aside and stoopeth to take up, the race is hindered,

"Declinat cursus, aurumque volubile tollit." [1]

Neither is my meaning, as was spoken of Socrates, to call philosophy down from heaven to converse upon the earth—that is, to leave natural philosophy aside, and to apply knowledge only to manners and policy. But as both heaven and earth do conspire and contribute to the use and benefit of man, so the end ought to be, from both philosophies to separate and reject vain speculations, and whatsoever is empty and void, and to preserve and augment whatsoever is solid and fruitful; that knowledge may not be as a courtesan, for pleasure and vanity only, or as a bond-woman, to acquire and gain to her master's use; but as a spouse, for generation, fruit, and comfort.

(12) Thus have I described and opened, as by a kind of dissection, those peccant humours (the principal of them) which have not only given impediment to the proficience of learning, but have given also occasion to the traducement thereof: wherein, if I have been too plain, it must be remembered, *fidelia vulnera amantis, sed dolosa oscula malignantis*. This I think I have gained, that I ought to be the better believed in that which I shall say pertaining to commendation; because I have proceeded so freely in that which concerneth censure. And yet I have no purpose to enter into a laudative of learning, or to make a hymn to the Muses (though I am of opinion that it is long since their rites were duly celebrated), but my intent is, without varnish or amplification justly to weigh the dignity of knowledge in the balance with other things, and to take the true value thereof by testimonies and arguments, divine and human.

VI

(1) First, therefore, let us seek the dignity of knowledge in the archetype or first platform, which is in the attributes and acts of God, as far as they are revealed to man and may be observed with sobriety; wherein we may not seek it by the name of learning, for all learning is knowledge acquired, and all knowledge in God is original, and therefore we must look for it by another name, that of wisdom or sapience, as the Scriptures call it.

(2) It is so, then, that in the work of the creation we see a double emanation of virtue from God; the one referring more properly to power, the other to wisdom; the one expressed in making the subsistence of the matter, and the other in disposing the beauty of the form. This being supposed, it is to be observed that for anything which appeareth in the history of the creation, the confused mass and matter of heaven and earth was made in a moment, and the order and disposition of that chaos or mass was the work of six days; such a note of difference it pleased God to put upon the works of power, and the works of wisdom; wherewith concurreth, that in the former it is not set down that God said, "Let there be heaven and earth," as it is set down of the works following; but actually, that God made heaven and earth: the one carrying the style of a manufacture, and the other of a law, decree, or counsel.

(3) To proceed, to that which is next in order from God, to spirits: we find, as far as credit is to be given to the celestial hierarchy of that supposed Dionysius, the senator of Athens, the first place or degree is given to the angels of love, which are termed seraphim; the second to the angels of light, which are termed cherubim; and the third, and so following places, to thrones, principalities, and the rest, which are all angels of power and ministry; so as this angels of knowledge and illumination are placed before the angels of office and domination.

(4) To descend from spirits and intellectual forms to sensible and material forms, we read the first form that was created was light, which hath a relation and correspondence in nature and corporal things to knowledge in spirits and incorporal things.

(5) So in the distribution of days we see the day wherein God did rest and contemplate His own works was blessed above all the days wherein He did effect and accomplish them.

(6) After the creation was finished, it is set down unto us that man was placed in the garden to work therein; which work, so appointed to him, could be no other than work of contemplation; that is, when the end of work is but for exercise and experiment, not for necessity; for there being then no reluctation of the creature, nor sweat of the brow, man's employment must of consequence have been matter of delight in the experiment, and not matter of labour for the use. Again, the first acts which man performed in Paradise consisted of the two summary parts of knowledge; the view of creatures, and the imposition of names. As for the knowledge which induced the fall, it was, as was touched before, not the natural knowledge of creatures, but the moral knowledge of good and evil; wherein the supposition was, that God's commandments or prohibitions were not the originals of good and evil, but that they had other beginnings, which man aspired to know, to the end to make a total defection from God and to depend wholly upon himself.

(7) To pass on: in the first event or occurrence after the fall of man, we see (as the Scriptures have infinite mysteries, not violating at all the truth of this story or letter) an image of the two estates, the contemplative state and the active state, figured in the two persons of Abel and Cain, and in the two simplest and most primitive trades of life; that of the shepherd (who, by reason of his leisure, rest in a place, and lying in view of heaven, is a lively image of a contemplative life), and that of the husbandman, where we see again the favour and election of God went to the shepherd, and not to the

tiller of the ground.

(8) So in the age before the flood, the holy records within those few memorials which are there entered and registered have vouchsafed to mention and honour the name of the inventors and authors of music and works in metal. In the age after the flood, the first great judgment of God upon the ambition of man was the confusion of tongues; whereby the open trade and intercourse of learning and knowledge was chiefly imbarred.

(9) To descend to Moses the lawgiver, and God's first pen: he is adorned by the Scriptures with this addition and commendation, "That he was seen in all the learning of the Egyptians," which nation we know was one of the most ancient schools of the world: for so Plato brings in the Egyptian priest saying unto Solon, "You Grecians are ever children; you have no knowledge of antiquity, nor antiquity of knowledge." Take a view of the ceremonial law of Moses; you shall find, besides the prefiguration of Christ, the badge or difference of the people of God, the exercise and impression of obedience, and other divine uses and fruits thereof, that some of the most learned Rabbins have travailed profitably and profoundly to observe, some of them a natural, some of them a moral sense, or reduction of many of the ceremonies and ordinances. As in the law of the leprosy, where it is said, "If the whiteness have overspread the flesh, the patient may pass abroad for clean; but if there be any whole flesh remaining, he is to be shut up for unclean;" one of them noteth a principle of nature, that putrefaction is more contagious before maturity than after; and another noteth a position of moral philosophy, that men abandoned to vice do not so much corrupt manners, as those that are half good and half evil. So in this and very many other places in that law, there is to be found, besides the theological sense, much aspersion of philosophy.

(10) So likewise in that excellent hook of Job, if it be revolved with diligence, it will be found pregnant and swelling with natural philosophy; as for example, cosmography, and the roundness of the world, Qui extendit aquilonem super vacuum, et appendit terram super nihilum; wherein the pensileness of the earth, the pole of the north, and the finiteness or convexity of heaven are manifestly touched. So again, matter of astronomy: Spiritus ejus ornavit cælos, et obstetricante manu ejus eductus est Coluber tortuoses. And in another place, Nunquid conjungere valebis micantes stellas Pleiadas, aut gyrum Arcturi poteris dissipare? Where the fixing of the stars, ever standing at equal distance, is with great elegancy noted. And in another place, Qui facit Arcturum, et Oriona, et Hyadas, et interiora Austri; where again he takes knowledge of the depression of the southern pole, calling it the secrets of the south, because the southern stars were in that climate unseen. Matter of generation: Annon sicut lac mulsisti me, et sicut caseum coagulasti me? &c. Matter of minerals: Habet argentum venarum suarum principia; et auro locus est in quo conflatur, ferrum de terra tollitur, et lapis solutus calore in æs vertitur; and so forwards in that chapter.

(11) So likewise in the person of Solomon the king, we see the gift or endowment of wisdom and learning, both in Solomon's petition and in God's assent thereunto, preferred before all other terrene and temporal felicity. By virtue of which grant or donative of God Solomon became enabled not only to write those excellent parables or aphorisms concerning divine and moral philosophy, but also to compile a natural history of all verdure, from the cedar upon the mountain to the moss upon the wall (which is but a rudiment between putrefaction and an herb), and also of all things that breathe or move. Nay, the same Solomon the king, although he excelled in the glory of treasure and magnificent buildings, of shipping and navigation, of service and attendance, of fame and renown, and the like, yet he maketh no claim to any of those glories, but only to the glory of inquisition of truth; for so he saith expressly, "The glory of God is to conceal a thing, but the glory of the king is to find it out;" as if, according to the innocent play of children, the Divine Majesty took delight to hide His works, to the end to have them found out; and as if kings could not obtain a greater honour than to be God's playfellows in that game; considering the great commandment of wits and means, whereby nothing needeth to be hidden from them.

(12) Neither did the dispensation of God vary in the times after our Saviour came into the world; for our Saviour himself did first show His power to subdue ignorance, by His conference with the priests and doctors of the law, before He showed His power to subdue nature by His miracles. And the coming of this Holy Spirit was chiefly figured and expressed in the similitude and gift of tongues, which are but vehicula scientiæ.

(13) So in the election of those instruments, which it pleased God to use for the plantation of the faith, notwithstanding that at the first He did employ persons altogether unlearned, otherwise than by inspiration, more evidently to declare His immediate working, and to abase all human wisdom or knowledge; yet nevertheless that counsel of His was no sooner performed, but in the next vicissitude and succession He did send His divine truth into the world, waited on with other learnings, as with servants or handmaids: for so we see St. Paul, who was only learned amongst the Apostles, had his pen most used in the Scriptures of the New Testament.

(14) So again we find that many of the ancient bishops and fathers of the Church were excellently read and studied in all the learning of this heathen; insomuch that the edict of the Emperor Julianus

(whereby it was interdicted unto Christians to be admitted into schools, lectures, or exercises of learning) was esteemed and accounted a more pernicious engine and machination against the Christian Faith than were all the sanguinary prosecutions of his predecessors; neither could the emulation and jealousy of Gregory, the first of that name, Bishop of Rome, ever obtain the opinion of piety or devotion; but contrariwise received the censure of humour, malignity, and pusillanimity, even amongst holy men; in that he designed to obliterate and extinguish the memory of heathen antiquity and authors. But contrariwise it was the Christian Church, which, amidst the inundations of the Scythians on the one side from the north-west, and the Saracens from the east, did preserve in the sacred lap and bosom thereof the precious relics even of heathen learning, which otherwise had been extinguished, as if no such thing had ever been.

(15) And we see before our eyes, that in the age of ourselves and our fathers, when it pleased God to call the Church of Rome to account for their degenerate manners and ceremonies, and sundry doctrines obnoxious and framed to uphold the same abuses; at one and the same time it was ordained by the Divine Providence that there should attend withal a renovation and new spring of all other knowledges. And on the other side we see the Jesuits, who partly in themselves, and partly by the emulation and provocation of their example, have much quickened and strengthened the state of learning; we see (I say) what notable service and reparation they have done to the Roman see.

(16) Wherefore, to conclude this part, let it be observed, that there be two principal duties and services, besides ornament and illustration, which philosophy and human learning do perform to faith and religion. The one, because they are an effectual inducement to the exaltation of the glory of God. For as the Psalms and other Scriptures do often invite us to consider and magnify the great and wonderful works of God, so if we should rest only in the contemplation of the exterior of them as they first offer themselves to our senses, we should do a like injury unto the majesty of God, as if we should judge or construe of the store of some excellent jeweller by that only which is set out toward the street in his shop. The other, because they minister a singular help and preservative against unbelief and error. For our Saviour saith, "You err, not knowing the Scriptures, nor the power of God;" laying before us two books or volumes to study, if we will be secured from error: first the Scriptures, revealing the will of God, and then the creatures expressing His power; whereof the latter is a key unto the former: not only opening our understanding to conceive the true sense of the Scriptures by the general notions of reason and rules of speech, but chiefly opening our belief, in drawing us into a due meditation of the omnipotency of God, which is chiefly signed and engraven upon His works. Thus much therefore for divine testimony and evidence concerning the true dignity and value of learning.

VII

(1) As for human proofs, it is so large a field, as in a discourse of this nature and brevity it is fit rather to use choice of those things which we shall produce, than to embrace the variety of them. First, therefore, in the degrees of human honour amongst the heathen, it was the highest to obtain to a veneration and adoration as a God. This unto the Christians is as the forbidden fruit. But we speak now separately of human testimony, according to which—that which the Grecians call apotheosis, and the Latins relatio inter divos—was the supreme honour which man could attribute unto man, specially when it was given, not by a formal decree or act of state (as it was used among the Roman Emperors), but by an inward assent and belief. Which honour, being so high, had also a degree or middle term; for there were reckoned above human honours, honours heroical and divine: in the attribution and distribution of which honours we see antiquity made this difference; that whereas founders and uniters of states and cities, lawgivers, extirpers of tyrants, fathers of the people, and other eminent persons in civil merit, were honoured but with the titles of worthies or demigods, such as were Hercules, Theseus, Minus, Romulus, and the like; on the other side, such as were inventors and authors of new arts, endowments, and commodities towards man's life, were ever consecrated amongst the gods themselves, as was Ceres, Bacchus, Mercurius, Apollo, and others. And justly; for the merit of the former is confined within the circle of an age or a nation, and is like fruitful showers, which though they be profitable and good, yet serve but for that season, and for a latitude of ground where they fall; but the other is, indeed, like the benefits of heaven, which are permanent and universal. The former again is mixed with strife and perturbation, but the latter hath the true character of Divine Presence, coming in aura leni, without noise or agitation.

(2) Neither is certainly that other merit of learning, in repressing the inconveniences which grow from man to man, much inferior to the former, of relieving the necessities which arise from nature, which merit was lively set forth by the ancients in that feigned relation of Orpheus' theatre, where all beasts and birds assembled, and, forgetting their several appetites—some of prey, some of game, some of quarrel—stood all sociably together listening unto the airs and accords of the harp, the sound whereof no sooner ceased, or was drowned by some louder noise, but every beast returned to his own nature;

wherein is aptly described the nature and condition of men, who are full of savage and unreclaimed desires, of profit, of lust, of revenge; which as long as they give ear to precepts, to laws, to religion, sweetly touched with eloquence and persuasion of books, of sermons, of harangues, so long is society and peace maintained; but if these instruments be silent, or that sedition and tumult make them not audible, all things dissolve into anarchy and confusion.

(3) But this appeareth more manifestly when kings themselves, or persons of authority under them, or other governors in commonwealths and popular estates, are endued with learning. For although he might be thought partial to his own profession that said "Then should people and estates be happy when either kings were philosophers, or philosophers kings;" yet so much is verified by experience, that under learned princes and governors there have been ever the best times: for howsoever kings may have their imperfections in their passions and customs, yet, if they be illuminate by learning, they have those notions of religion, policy, and morality, which do preserve them and refrain them from all ruinous and peremptory errors and excesses, whispering evermore in their ears, when counsellors and servants stand mute and silent. And senators or counsellors, likewise, which be learned, to proceed upon more safe and substantial principles, than counsellors which are only men of experience; the one sort keeping dangers afar off, whereas the other discover them not till they come near hand, and then trust to the agility of their wit to ward or avoid them.

(4) Which felicity of times under learned princes (to keep still the law of brevity, by using the most eminent and selected examples) doth best appear in the age which passed from the death of Domitianus the emperor until the reign of Commodus; comprehending a succession of six princes, all learned, or singular favourers and advancers of learning, which age for temporal respects was the most happy and flourishing that ever the Roman Empire (which then was a model of the world) enjoyed—a matter revealed and prefigured unto Domitian in a dream the night before he was slain: for he thought there was grown behind upon his shoulders a neck and a head of gold, which came accordingly to pass in those golden times which succeeded; of which princes we will make some commemoration; wherein, although the matter will be vulgar, and may be thought fitter for a declamation than agreeable to a treatise infolded as this is, yet, because it is pertinent to the point in hand—Neque semper arcum tendit Apollo—and to name them only were too naked and cursory, I will not omit it altogether. The first was Nerva, the excellent temper of whose government is by a glance in Cornelius Tacitus touched to the life: Postquam divus Nerva res oluim insociabiles miscuisset, imperium et libertatem. And in token of his learning, the last act of his short reign left to memory was a missive to his adopted son, Trajan, proceeding upon some inward discontent at the ingratitude of the times, comprehended in a verse of Homer's—

"Telis, Phœbe, tuis, lacrymas ulciscere nostras."

(5) Trajan, who succeeded, was for his person not learned; but if we will hearken to the speech of our Saviour, that saith, "He that receiveth a prophet in the name of a prophet shall have a prophet's reward," he deserveth to be placed amongst the most learned princes; for there was not a greater admirer of learning or benefactor of learning, a founder of famous libraries, a perpetual advancer of learned men to office, and familiar converser with learned professors and preceptors who were noted to have then most credit in court. On the other side how much Trajan's virtue and government was admired and renowned, surely no testimony of grave and faithful history doth more lively set forth than that legend tale of Gregorius Magnum, Bishop of Rome, who was noted for the extreme envy he bare towards all heathen excellency; and yet he is reported, out of the love and estimation of Trajan's moral virtues, to have made unto God passionate and fervent prayers for the delivery of his soul out of hell, and to have obtained it, with a caveat that he should make no more such petitions. In this prince's time also the persecutions against the Christians received intermission upon the certificate of Plinius Secundus, a man of excellent learning and by Trajan advanced.

(6) Adrian, his successor, was the most curious man that lived, and the most universal inquirer: insomuch as it was noted for an error in his mind that he desired to comprehend all things, and not to reserve himself for the worthiest things, falling into the like humour that was long before noted in Philip of Macedon, who, when he would needs overrule and put down an excellent musician in an argument touching music, was well answered by him again—"God forbid, sir," saith he, "that your fortune should be so bad as to know these things better than I." It pleased God likewise to use the curiosity of this emperor as an inducement to the peace of His Church in those days; for having Christ in veneration, not as a God or Saviour, but as a wonder or novelty, and having his picture in his gallery matched with Apollonius (with whom in his vain imagination he thought its had some conformity), yet it served the turn to allay the bitter hatred of those times against the Christian name, so as the Church had peace during his time. And for his government civil, although he did not attain to that of Trajan's in glory of arms or perfection of justice, yet in deserving of the weal of the subject he did exceed him. For Trajan erected many famous monuments and buildings, insomuch as Constantine the Great in emulation was

wont to call him Parietaria, "wall-flower," because his name was upon so many walls; but his buildings and works were more of glory and triumph than use and necessity. But Adrian spent his whole reign, which was peaceable, in a perambulation or survey of the Roman Empire, giving order and making assignation where he went for re-edifying of cities, towns, and forts decayed, and for cutting of rivers and streams, and for making bridges and passages, and for policing of cities and commonalties with new ordinances and constitutions, and granting new franchises and incorporations; so that his whole time was a very restoration of all the lapses and decays of former times.

(7) Antoninus Pius, who succeeded him, was a prince excellently learned, and had the patient and subtle wit of a schoolman, insomuch as in common speech (which leaves no virtue untaxed) he was called Cymini Sector, a carver or a divider of cummin seed, which is one of the least seeds. Such a patience he had and settled spirit to enter into the least and most exact differences of causes, a fruit no doubt of the exceeding tranquillity and serenity of his mind, which being no ways charged or encumbered, either with fears, remorses, or scruples, but having been noted for a man of the purest goodness, without all fiction or affectation, that hath reigned or lived, made his mind continually present and entire. He likewise approached a degree nearer unto Christianity, and became, as Agrippa said unto St. Paul, "half a Christian," holding their religion and law in good opinion, and not only ceasing persecution, but giving way to the advancement of Christians.

(5) There succeeded him the first Divi fratres, the two adoptive brethren—Lucius Commodus Verus, son to Ælius Verus, who delighted much in the softer kind of learning, and was wont to call the poet Martial his Virgil; and Marcus Aurelius Antoninus: whereof the latter, who obscured his colleague and survived him long, was named the "Philosopher," who, as he excelled all the rest in learning, so he excelled them likewise in perfection of all royal virtues; insomuch as Julianus the emperor, in his book entitled Cærsares, being as a pasquil or satire to deride all his predecessors, feigned that they were all invited to a banquet of the gods, and Silenus the jester sat at the nether end of the table and bestowed a scoff on everyone as they came in; but when Marcus Philosophus came in, Silenus was gravelled and out of countenance, not knowing where to carp at him, save at the last he gave a glance at his patience towards his wife. And the virtue of this prince, continued with that of his predecessor, made the name of Antoninus so sacred in the world, that though it were extremely dishonoured in Commodus, Caracalla, and Heliogabalus, who all bare the name, yet, when Alexander Severus refused the name because he was a stranger to the family, the Senate with one acclamation said, Quomodo Augustus, sic et Antoninus. In such renown and veneration was the name of these two princes in those days, that they would have had it as a perpetual addition in all the emperors' style. In this emperor's time also the Church for the most part was in peace; so as in this sequence of six princes we do see the blessed effects of learning in sovereignty, painted forth in the greatest table of the world.

(9) But for a tablet or picture of smaller volume (not presuming to speak of your Majesty that liveth), in my judgment the most excellent is that of Queen Elizabeth, your immediate predecessor in this part of Britain; a prince that, if Plutarch were now alive to write lives by parallels, would trouble him, I think, to find for her a parallel amongst women. This lady was endued with learning in her sex singular, and rare even amongst masculine princes—whether we speak of learning, of language, or of science, modern or ancient, divinity or humanity—and unto the very last year of her life she accustomed to appoint set hours for reading, scarcely any young student in a university more daily or more duly. As for her government, I assure myself (I shall not exceed if I do affirm) that this part of the island never had forty-five years of better tines, and yet not through the calmness of the season, but through the wisdom of her regiment. For if there be considered, of the one side, the truth of religion established, the constant peace and security, the good administration of justice, the temperate use of the prerogative, not slackened, nor much strained; the flourishing state of learning, sortable to so excellent a patroness; the convenient estate of wealth and means, both of crown and subject; the habit of obedience, and the moderation of discontents; and there be considered, on the other side, the differences of religion, the troubles of neighbour countries, the ambition of Spain, and opposition of Rome, and then that she was solitary and of herself; these things, I say, considered, as I could not have chosen an instance so recent and so proper, so I suppose I could not have chosen one more remarkable or eminent to the purpose now in hand, which is concerning the conjunction of learning in the prince with felicity in the people.

(10) Neither hath learning an influence and operation only upon civil merit and moral virtue, and the arts or temperature of peace and peaceable government; but likewise it hath no less power and efficacy in enablement towards martial and military virtue and prowess, as may be notably represented in the examples of Alexander the Great and Cæsar the Dictator (mentioned before, but now in fit place to be resumed), of whose virtues and acts in war there needs no note or recital, having been the wonders of time in that kind; but of their affections towards learning and perfections in learning it is pertinent to say somewhat.

(11) Alexander was bred and taught under Aristotle, the great philosopher, who dedicated divers of his books of philosophy unto him; he was attended with Callisthenes and divers other learned persons, that followed him in camp, throughout his journeys and conquests. What price and estimation

he had learning in doth notably appear in these three particulars: first, in the envy he used to express that he bare towards Achilles, in this, that he had so good a trumpet of his praises as Homer's verses; secondly, in the judgment or solution he gave touching that precious cabinet of Darius, which was found among his jewels (whereof question was made what thing was worthy to be put into it, and he gave his opinion for Homer's works); thirdly, in his letter to Aristotle, after he had set forth his books of nature, wherein he expostulateth with him for publishing the secrets or mysteries of philosophy; and gave him to understand that himself esteemed it more to excel other men in learning and knowledge than in power and empire. And what use he had of learning doth appear, or rather shine, in all his speeches and answers, being full of science and use of science, and that in all variety.

(12) And herein again it may seem a thing scholastical, and somewhat idle to recite things that every man knoweth; but yet, since the argument I handle leadeth me thereunto, I am glad that men shall perceive I am as willing to flatter (if they will so call it) an Alexander, or a Cæsar, or an Antoninus, that are dead many hundred years since, as any that now liveth; for it is the displaying of the glory of learning in sovereignty that I propound to myself, and not a humour of declaiming in any man's praises. Observe, then, the speech he used of Diogenes, and see if it tend not to the true state of one of the greatest questions of moral philosophy: whether the enjoying of outward things, or the contemning of them, be the greatest happiness; for when he saw Diogenes so perfectly contented with so little, he said to those that mocked at his condition, "were I not Alexander, I would wish to be Diogenes." But Seneca inverteth it, and saith, "Plus erat, quod hic nollet accipere, quàm quod ille posset dare." There were more things which Diogenes would have refused than those were which Alexander could have given or enjoyed.

(13) Observe, again, that speech which was usual with him,—"That he felt his mortality chiefly in two things, sleep and lust;" and see if it were not a speech extracted out of the depth of natural philosophy, and liker to have come out of the mouth of Aristotle or Democritus than from Alexander.

(14) See, again, that speech of humanity and poesy, when, upon the bleeding of his wounds, he called unto him one of his flatterers, that was wont to ascribe to him divine honour, and said, "Look, this is very blood; this is not such a liquor as Homer speaketh of, which ran from Venus' hand when it was pierced by Diomedes."

(15) See likewise his readiness in reprehension of logic in the speech he used to Cassander, upon a complaint that was made against his father Antipater; for when Alexander happened to say, "Do you think these men would have come from so far to complain except they had just cause of grief?" and Cassander answered, "Yea, that was the matter, because they thought they should not be disproved;" said Alexander, laughing, "See the subtleties of Aristotle, to take a matter both ways, pro et contra, &c."

(16) But note, again, how well he could use the same art which he reprehended to serve his own humour: when bearing a secret grudge to Callisthenes, because he was against the new ceremony of his adoration, feasting one night where the same Callisthenes was at the table, it was moved by some after supper, for entertainment sake, that Callisthenes, who was an eloquent man, might speak of some theme or purpose at his own choice; which Callisthenes did, choosing the praise of the Macedonian nation for his discourse, and performing the same with so good manner as the hearers were much ravished; whereupon Alexander, nothing pleased, said, "It was easy to be eloquent upon so good a subject; but," saith he, "turn your style, and let us hear what you can say against us;" which Callisthenes presently undertook, and did with that sting and life that Alexander interrupted him, and said, "The goodness of the cause made him eloquent before, and despite made him eloquent then again."

(17) Consider further, for tropes of rhetoric, that excellent use of a metaphor or translation, wherewith he taxeth Antipater, who was an imperious and tyrannous governor; for when one of Antipater's friends commended him to Alexander for his moderation, that he did not degenerate as his other lieutenants did into the Persian pride, in uses of purple, but kept the ancient habit of Macedon, of black. "True," saith Alexander; "but Antipater is all purple within." Or that other, when Parmenio came to him in the plain of Arbela and showed him the innumerable multitude of his enemies, specially as they appeared by the infinite number of lights as it had been a new firmament of stars, and thereupon advised him to assail them by night; whereupon he answered, "That he would not steal the victory."

(18) For matter of policy, weigh that significant distinction, so much in all ages embraced, that he made between his two friends Hephæstion and Craterus, when he said, "That the one loved Alexander, and the other loved the king:" describing the principal difference of princes' best servants, that some in affection love their person, and other in duty love their crown.

(19) Weigh also that excellent taxation of an error, ordinary with counsellors of princes, that they counsel their masters according to the model of their own mind and fortune, and not of their masters. When upon Darius' great offers Parmenio had said, "Surely I would accept these offers were I as Alexander;" saith Alexander, "So would I were I as Parmenio."

(20) Lastly, weigh that quick and acute reply which he made when he gave so large gifts to his friends and servants, and was asked what he did reserve for himself, and he answered, "Hope." Weigh, I say, whether he had not cast up his account aright, because hope must be the portion of all that resolve

upon great enterprises; for this was Cæsar's portion when he went first into Gaul, his estate being then utterly overthrown with largesses. And this was likewise the portion of that noble prince, howsoever transported with ambition, Henry Duke of Guise, of whom it was usually said that he was the greatest usurer in France, because he had turned all his estate into obligations.

(21) To conclude, therefore, as certain critics are used to say hyperbolically, "That if all sciences were lost they might be found in Virgil," so certainly this may be said truly, there are the prints and footsteps of learning in those few speeches which are reported of this prince, the admiration of whom, when I consider him not as Alexander the Great, but as Aristotle's scholar, hath carried me too far.

(22) As for Julius Cæsar, the excellency of his learning needeth not to be argued from his education, or his company, or his speeches; but in a further degree doth declare itself in his writings and works: whereof some are extant and permanent, and some unfortunately perished. For first, we see there is left unto us that excellent history of his own wars, which he entitled only a Commentary, wherein all succeeding times have admired the solid weight of matter, and the real passages and lively images of actions and persons, expressed in the greatest propriety of words and perspicuity of narration that ever was; which that it was not the effect of a natural gift, but of learning and precept, is well witnessed by that work of his entitled De Analogia, being a grammatical philosophy, wherein he did labour to make this same Vox ad placitum to become Vox ad licitum, and to reduce custom of speech to congruity of speech; and took as it were the pictures of words from the life of reason.

(23) So we receive from him, as a monument both of his power and learning, the then reformed computation of the year; well expressing that he took it to be as great a glory to himself to observe and know the law of the heavens, as to give law to men upon the earth.

(24) So likewise in that book of his, Anti-Cato, it may easily appear that he did aspire as well to victory of wit as victory of war: undertaking therein a conflict against the greatest champion with the pen that then lived, Cicero the orator.

(25) So, again, in his book of Apophthegms, which he collected, we see that he esteemed it more honour to make himself but a pair of tables, to take the wise and pithy words of others, than to have every word of his own to be made an apophthegm or an oracle, as vain princes, by custom of flattery, pretend to do. And yet if I should enumerate divers of his speeches, as I did those of Alexander, they are truly such as Solomon noteth, when he saith, Verba sapientum tanquam aculei, et tanquam clavi in altum defixi: whereof I will only recite three, not so delectable for elegancy, but admirable for vigour and efficacy.

(26) As first, it is reason he be thought a master of words, that could with one word appease a mutiny in his army, which was thus: The Romans, when their generals did speak to their army, did use the word Milites, but when the magistrates spake to the people they did use the word Quirites. The soldiers were in tumult, and seditiously prayed to be cashiered; not that they so meant, but by expostulation thereof to draw Cæsar to other conditions; wherein he being resolute not to give way, after some silence, he began his speech, Ego Quirites, which did admit them already cashiered— wherewith they were so surprised, crossed, and confused, as they would not suffer him to go on in his speech, but relinquished their demands, and made it their suit to be again called by the name of Milites.

(27) The second speech was thus: Cæsar did extremely affect the name of king; and some were set on as he passed by in popular acclamation to salute him king. Whereupon, finding the cry weak and poor, he put it off thus, in a kind of jest, as if they had mistaken his surname: Non Rex sum, sed Cæsar; a speech that, if it be searched, the life and fulness of it can scarce be expressed. For, first, it was a refusal of the name, but yet not serious; again, it did signify an infinite confidence and magnanimity, as if he presumed Cæsar was the greater title, as by his worthiness it is come to pass till this day. But chiefly it was a speech of great allurement toward his own purpose, as if the state did strive with him but for a name, whereof mean families were vested; for Rex was a surname with the Romans, as well as King is with us.

(28) The last speech which I will mention was used to Metellus, when Cæsar, after war declared, did possess himself of this city of Rome; at which time, entering into the inner treasury to take the money there accumulate, Metellus, being tribune, forbade him. Whereto Cæsar said, "That if he did not desist, he would lay him dead in the place." And presently taking himself up, he added, "Young man, it is harder for me to speak it than to do it—Adolescens, durius est mihi hoc dicere quàm facere." A speech compounded of the greatest terror and greatest clemency that could proceed out of the mouth of man.

(29) But to return and conclude with him, it is evident himself knew well his own perfection in learning, and took it upon him, as appeared when upon occasion that some spake what a strange resolution it was in Lucius Sylla to resign his dictators, he, scoffing at him to his own advantage, answered, "That Sylla could not skill of letters, and therefore knew not how to dictate."

(30) And here it were fit to leave this point, touching the concurrence of military virtue and learning (for what example should come with any grace after those two of Alexander and Cæsar?), were it not in regard of the rareness of circumstance, that I find in one other particular, as that which did so

suddenly pass from extreme scorn to extreme wonder: and it is of Xenophon the philosopher, who went from Socrates' school into Asia in the expedition of Cyrus the younger against King Artaxerxes. This Xenophon at that time was very young, and never had seen the wars before, neither had any command in the army, but only followed the war as a voluntary, for the love and conversation of Proxenus, his friend. He was present when Falinus came in message from the great king to the Grecians, after that Cyrus was slain in the field, and they, a handful of men, left to themselves in the midst of the king's territories, cut off from their country by many navigable rivers and many hundred miles. The message imported that they should deliver up their arms and submit themselves to the king's mercy. To which message, before answer was made, divers of the army conferred familiarly with Falinus; and amongst the rest Xenophon happened to say, "Why, Falinus, we have now but these two things left, our arms and our virtue; and if we yield up our arms, how shall we make use of our virtue?" Whereto Falinus, smiling on him, said, "If I be not deceived, young gentleman, you are an Athenian, and I believe you study philosophy, and it is pretty that you say; but you are much abused if you think your virtue can withstand the king's power." Here was the scorn; the wonder followed: which was that this young scholar or philosopher, after all the captains were murdered in parley by treason, conducted those ten thousand foot, through the heart of all the king's high countries, from Babylon to Græcia in safety, in despite of all the king's forces, to the astonishment of the world, and the encouragement of the Grecians in times succeeding to make invasion upon the kings of Persia, as was after purposed by Jason the Thessalian, attempted by Agesilaus the Spartan, and achieved by Alexander the Macedonian, all upon the ground of the act of that young scholar.

VIII

(1) To proceed now from imperial and military virtue to moral and private virtue; first, it is an assured truth, which is contained in the verses:—

> *"Scilicet ingenuas didicisse fideliter artes*
> *Emollit mores nec sinit esse feros."*

It taketh away the wildness and barbarism and fierceness of men's minds; but indeed the accent had need be upon *fideliter*; for a little superficial learning doth rather work a contrary effect. It taketh away all levity, temerity, and insolency, by copious suggestion of all doubts and difficulties, and acquainting the mind to balance reasons on both sides, and to turn back the first offers and conceits of the mind, and to accept of nothing but examined and tried. It taketh away vain admiration of anything, which is the root of all weakness. For all things are admired, either because they are new, or because they are great. For novelty, no man that wadeth in learning or contemplation thoroughly but will find that printed in his heart, *Nil novi super terram*. Neither can any man marvel at the play of puppets, that goeth behind the curtain, and adviseth well of the motion. And for magnitude, as Alexander the Great, after that he was used to great armies, and the great conquests of the spacious provinces in Asia, when he received letters out of Greece, of some fights and services there, which were commonly for a passage or a fort, or some walled town at the most, he said:—"It seemed to him that he was advertised of the battles of the frogs and the mice, that the old tales went of." So certainly, if a man meditate much upon the universal frame of nature, the earth with men upon it (the divineness of souls except) will not seem much other than an ant-hill, whereas some ants carry corn, and some carry their young, and some go empty, and all to and fro a little heap of dust. It taketh away or mitigateth fear of death or adverse fortune, which is one of the greatest impediments of virtue and imperfections of manners. For if a man's mind be deeply seasoned with the consideration of the mortality and corruptible nature of things, he will easily concur with Epictetus, who went forth one day and saw a woman weeping for her pitcher of earth that was broken, and went forth the next day and saw a woman weeping for her son that was dead, and thereupon said, "Heri vidi fragilem frangi, hodie vidi mortalem mori." And, therefore, Virgil did excellently and profoundly couple the knowledge of causes and the conquest of all fears together, as *concomitantia*.

> *"Felix, qui potuit rerum cognoscere causas,*
> *Quique metus omnes, et inexorabile fatum*
> *Subjecit pedibus, strepitumque Acherontis avari."*

(2) It were too long to go over the particular remedies which learning doth minister to all the diseases of the mind: sometimes purging the ill humours, sometimes opening the obstructions, sometimes helping digestion, sometimes increasing appetite, sometimes healing the wounds and exulcerations thereof, and the like; and, therefore, I will conclude with that which hath *rationem totius*—which is, that it disposeth the constitution of the mind not to be fixed or settled in the defects

thereof, but still to be capable and susceptible of growth and reformation. For the unlearned man knows not what it is to descend into himself, or to call himself to account, nor the pleasure of that *suavissima vita, indies sentire se fieri meliorem.* The good parts he hath he will learn to show to the full, and use them dexterously, but not much to increase them. The faults he hath he will learn how to hide and colour them, but not much to amend them; like an ill mower, that mows on still, and never whets his scythe. Whereas with the learned man it fares otherwise, that he doth ever intermix the correction and amendment of his mind with the use and employment thereof. Nay, further, in general and in sum, certain it is that Veritas and Bonitas differ but as the seal and the print; for truth prints goodness, and they be the clouds of error which descend in the storms of passions and perturbations.

(3) From moral virtue let us pass on to matter of power and commandment, and consider whether in right reason there be any comparable with that wherewith knowledge investeth and crowneth man's nature. We see the dignity of the commandment is according to the dignity of the commanded; to have commandment over beasts as herdmen have, is a thing contemptible; to have commandment over children as schoolmasters have, is a matter of small honour; to have commandment over galley-slaves is a disparagement rather than an honour. Neither is the commandment of tyrants much better, over people which have put off the generosity of their minds; and, therefore, it was ever holden that honours in free monarchies and commonwealths had a sweetness more than in tyrannies, because the commandment extendeth more over the wills of men, and not only over their deeds and services. And therefore, when Virgil putteth himself forth to attribute to Augustus Cæsar the best of human honours, he doth it in these words:—

"Victorque volentes
Per populos dat jura, viamque affectat Olympo."

But yet the commandment of knowledge is yet higher than the commandment over the will; for it is a commandment over the reason, belief, and understanding of man, which is the highest part of the mind, and giveth law to the will itself. For there is no power on earth which setteth up a throne or chair of estate in the spirits and souls of men, and in their cogitations, imaginations, opinions, and beliefs, but knowledge and learning. And therefore we see the detestable and extreme pleasure that arch-heretics, and false prophets, and impostors are transported with, when they once find in themselves that they have a superiority in the faith and conscience of men; so great as if they have once tasted of it, it is seldom seen that any torture or persecution can make them relinquish or abandon it. But as this is that which the author of the Revelation calleth the depth or profoundness of Satan, so by argument of contraries, the just and lawful sovereignty over men's understanding, by force of truth rightly interpreted, is that which approacheth nearest to the similitude of the divine rule.

(4) As for fortune and advancement, the beneficence of learning is not so confined to give fortune only to states and commonwealths, as it doth not likewise give fortune to particular persons. For it was well noted long ago, that Homer hath given more men their livings, than either Sylla, or Cæsar, or Augustus ever did, notwithstanding their great largesses and donatives, and distributions of lands to so many legions. And no doubt it is hard to say whether arms or learning have advanced greater numbers. And in case of sovereignty we see, that if arms or descent have carried away the kingdom, yet learning hath carried the priesthood, which ever hath been in some competition with empire.

(5) Again, for the pleasure and delight of knowledge and learning, it far surpasseth all other in nature. For, shall the pleasures of the affections so exceed the pleasure of the sense, as much as the obtaining of desire or victory exceedeth a song or a dinner? and must not of consequence the pleasures of the intellect or understanding exceed the pleasures of the affections? We see in all other pleasures there is satiety, and after they be used, their verdure departeth, which showeth well they be but deceits of pleasure, and not pleasures; and that it was the novelty which pleased, and not the quality. And, therefore, we see that voluptuous men turn friars, and ambitions princes turn melancholy. But of knowledge there is no satiety, but satisfaction and appetite are perpetually interchangeable; and, therefore, appeareth to be good in itself simply, without fallacy or accident. Neither is that pleasure of small efficacy and contentment to the mind of man, which the poet Lucretius describeth elegantly:—

"Suave mari magno, turbantibus æquora ventis, &c."

"It is a view of delight," saith he, "to stand or walk upon the shore side, and to see a ship tossed with tempest upon the sea; or to be in a fortified tower, and to see two battles join upon a plain. But it is a pleasure incomparable, for the mind of man to be settled, landed, and fortified in the certainty of truth; and from thence to descry and behold the errors, perturbations, labours, and wanderings up and down of other men."

(6) Lastly, leaving the vulgar arguments, that by learning man excelleth man in that wherein man excelleth beasts; that by learning man ascendeth to the heavens and their motions, where in body he

cannot come; and the like: let us conclude with the dignity and excellency of knowledge and learning in that whereunto man's nature doth most aspire, which is immortality, or continuance; for to this tendeth generation, and raising of houses and families; to this tend buildings, foundations, and monuments; to this tendeth the desire of memory, fame, and celebration; and in effect the strength of all other human desires. We see then how far the monuments of wit and learning are more durable than the monuments of power or of the hands. For have not the verses of Homer continued twenty-five hundred years, or more, without the loss of a syllable or letter; during which the infinite palaces, temples, castles, cities, have been decayed and demolished? It is not possible to have the true pictures or statues of Cyrus, Alexander, Cæsar, no nor of the kings or great personages of much later years; for the originals cannot last, and the copies cannot but leese of the life and truth. But the images of men's wits and knowledges remain in books, exempted from the wrong of time and capable of perpetual renovation. Neither are they fitly to be called images, because they generate still, and cast their seeds in the minds of others, provoking and causing infinite actions and opinions in succeeding ages. So that if the invention of the ship was thought so noble, which carrieth riches and commodities from place to place, and consociateth the most remote regions in participation of their fruits, how much more are letters to be magnified, which as ships pass through the vast seas of time, and make ages so distant to participate of the wisdom, illuminations, and inventions, the one of the other? Nay, further, we see some of the philosophers which were least divine, and most immersed in the senses, and denied generally the immortality of the soul, yet came to this point, that whatsoever motions the spirit of man could act and perform without the organs of the body, they thought might remain after death, which were only those of the understanding and not of the affection; so immortal and incorruptible a thing did knowledge seem unto them to be. But we, that know by divine revelation that not only the understanding but the affections purified, not only the spirit but the body changed, shall be advanced to immortality, do disclaim in these rudiments of the senses. But it must be remembered, both in this last point, and so it may likewise be needful in other places, that in probation of the dignity of knowledge or learning, I did in the beginning separate divine testimony from human, which method I have pursued, and so handled them both apart.

(7) Nevertheless I do not pretend, and I know it will be impossible for me, by any pleading of mine, to reverse the judgment, either of Æsop's cock, that preferred the barleycorn before the gem; or of Midas, that being chosen judge between Apollo, president of the Muses, and Pan, god of the flocks, judged for plenty; or of Paris, that judged for beauty and love against wisdom and power; or of Agrippina, *occidat matrem, modo imperet*, that preferred empire with any condition never so detestable; or of Ulysses, *qui vetulam prætulit immortalitati*, being a figure of those which prefer custom and habit before all excellency, or of a number of the like popular judgments. For these things must continue as they have been; but so will that also continue whereupon learning hath ever relied, and which faileth not: *Justificata est sapientia a filiis suis*.

THE SECOND BOOK

To the King.

1. IT might seem to have more convenience, though it come often otherwise to pass (excellent King), that those which are fruitful in their generations, and have in themselves the foresight of immortality in their descendants, should likewise be more careful of the good estate of future times, unto which they know they must transmit and commend over their dearest pledges. Queen Elizabeth was a sojourner in the world in respect of her unmarried life, and was a blessing to her own times; and yet so as the impression of her good government, besides her happy memory, is not without some effect which doth survive her. But to your Majesty, whom God hath already blessed with so much royal issue, worthy to continue and represent you for ever, and whose youthful and fruitful bed doth yet promise many the like renovations, it is proper and agreeable to be conversant not only in the transitory parts of good government, but in those acts also which are in their nature permanent and perpetual. Amongst the which (if affection do not transport me) there is not any more worthy than the further endowment of the world with sound and fruitful knowledge. For why should a few received authors stand up like Hercules' columns, beyond which there should be no sailing or discovering, since we have so bright and benign a star as your Majesty to conduct and prosper us? To return therefore where we left, it remaineth to consider of what kind those acts are which have been undertaken and performed by kings and others for the increase and advancement of learning, wherein I purpose to speak actively, without digressing or dilating.

2. Let this ground therefore be laid, that all works are over common by amplitude of reward, by soundness of direction, and by the conjunction of labours. The first multiplieth endeavour, the second preventeth error, and the third supplieth the frailty of man. But the principal of these is direction, for claudus in via antevertit cursorem extra viam; and Solomon excellently setteth it down, "If the iron be not sharp, it requireth more strength, but wisdom is that which prevaileth," signifying that the invention or election of the mean is more effectual than any enforcement or accumulation of endeavours. This I am induced to speak, for that (not derogating from the noble intention of any that have been deservers towards the state of learning), I do observe nevertheless that their works and acts are rather matters of magnificence and memory than of progression and proficience, and tend rather to augment the mass of learning in the multitude of learned men than to rectify or raise the sciences themselves.

3. The works or acts of merit towards learning are conversant about three objects—the places of learning, the books of learning, and the persons of the learned. For as water, whether it be the dew of heaven or the springs of the earth, doth scatter and leese itself in the ground, except it be collected into some receptacle where it may by union comfort and sustain itself; and for that cause the industry of man hath made and framed springheads, conduits, cisterns, and pools, which men have accustomed likewise to beautify and adorn with accomplishments of magnificence and state, as well as of use and necessity; so this excellent liquor of knowledge, whether it descend from divine inspiration, or spring from human sense, would soon perish and vanish to oblivion, if it were not preserved in books, traditions, conferences, and places appointed, as universities, colleges, and schools, for the receipt and comforting of the same.

4. The works which concern the seats and places of learning are four—foundations and buildings, endowments with revenues, endowments with franchises and privileges, institutions and ordinances for government—all tending to quietness and privateness of life, and discharge of cares and troubles; much like the stations which Virgil prescribeth for the hiving of bees:

> *"Principio sedes apibus statioque petenda,*
> *Quo neque sit ventis aditus, &c."*

5. The works touching books are two—first, libraries, which are as the shrines where all the relics of the ancient saints, full of true virtue, and that without delusion or imposture, are preserved and reposed; secondly, new editions of authors, with more correct impressions, more faithful translations, more profitable glosses, more diligent annotations, and the like.

6. The works pertaining to the persons of learned men (besides the advancement and countenancing of them in general) are two—the reward and designation of readers in sciences already

extant and invented; and the reward and designation of writers and inquirers concerning any parts of learning not sufficiently laboured and prosecuted.

7. These are summarily the works and acts wherein the merits of many excellent princes and other worthy personages, have been conversant. As for any particular commemorations, I call to mind what Cicero said when he gave general thanks, Difficile non aliquem, ingratum quenquam præterire. Let us rather, according to the Scriptures, look unto that part of the race which is before us, than look back to that which is already attained.

8. First, therefore, amongst so many great foundations of colleges in Europe, I find strange that they are all dedicated to professions, and none left free to arts and sciences at large. For if men judge that learning should be referred to action, they judge well; but in this they fall into the error described in the ancient fable, in which the other parts of the body did suppose the stomach had been idle, because it neither performed the office of motion, as the limbs do, nor of sense, as the head doth; but yet notwithstanding it is the stomach that digesteth and distributeth to all the rest. So if any man think philosophy and universality to be idle studies, he doth not consider that all professions are from thence served and supplied. And this I take to be a great cause that hath hindered the progression of learning, because these fundamental knowledges have been studied but in passage. For if you will have a tree bear more fruit than it hath used to do, it is not anything you can do to the boughs, but it is the stirring of the earth and putting new mould about thee roots that must work it. Neither is it to be forgotten, that this dedicating of foundations and dotations to professory learning hath not only had a malign aspect and influence upon the growth of sciences, but hath also been prejudicial to states, and governments. For hence it proceedeth that princes find a solitude in regard of able men to serve them in causes of estate, because there is no education collegiate which is free, where such as were so disposed might give themselves in histories, modern languages, books of policy and civil discourse, and other the like enablements unto service of estate.

9. And because founders of colleges do plant, and founders of lectures do water, it followeth well in order to speak of the defect which is in public lectures; namely, in the smallness, and meanness of the salary or reward which in most places is assigned unto them, whether they be lectures of arts, or of professions. For it is necessary to the progression of sciences that readers be of the most able and sufficient men; as those which are ordained for generating and propagating of sciences, and not for transitory use. This cannot be, except their condition and endowment be such as may content the ablest man to appropriate his whole labour and continue his whole age in that function and attendance; and therefore must have a proportion answerable to that mediocrity or competency of advancement, which may be expected from a profession or the practice of a profession. So as, if you will have sciences flourish, you must observe David's military law, which was, "That those which stayed with the carriage should have equal part with those which were in the action;" else will the carriages be ill attended. So readers in sciences are indeed the guardians of the stores and provisions of sciences, whence men in active courses are furnished, and therefore ought to have equal entertainment with them; otherwise if the fathers in sciences be of the weakest sort or be ill maintained,

"Et patrum invalidi referent jejunia nati."

10. Another defect I note, wherein I shall need some alchemist to help me, who call upon men to sell their books, and to build furnaces; quitting and forsaking Minerva and the Muses as barren virgins, and relying upon Vulcan. But certain it is, that unto the deep, fruitful, and operative study of many sciences, specialty natural philosophy and physic, books be not only the instrumentals; wherein also the beneficence of men hath not been altogether wanting. For we see spheres, globes, astrolabes, maps, and the like, have been provided as appurtenances to astronomy and cosmography, as well as books. We see likewise that some places instituted for physic have annexed the commodity of gardens for simples of all sorts, and do likewise command the use of dead bodies for anatomies. But these do respect but a few things. In general, there will hardly be any main proficience in the disclosing of nature, except there be some allowance for expenses about experiments; whether they be experiments appertaining to Vulcanus or Dædalus, furnace or engine, or any other kind. And therefore as secretaries and spials of princes and states bring in bills for intelligence, so you must allow the spials and intelligencers of nature to bring in their bills; or else you shall be ill advertised.

11. And if Alexander made such a liberal assignation to Aristotle of treasure for the allowance of hunters, fowlers, fishers, and the like, that he might compile a history of nature, much better do they deserve it that travail in arts of nature.

12. Another defect which I note is an intermission or neglect in those which are governors in universities, of consultation, and in princes or superior persons, of visitation: to enter into account and consideration, whether the readings, exercises, and other customs appertaining unto learning, anciently begun and since continued, be well instituted or no; and thereupon to ground an amendment or reformation in that which shall be found inconvenient. For it is one of your Majesty's own most wise

and princely maxims, "That in all usages and precedents, the times be considered wherein they first began; which if they were weak or ignorant, it derogateth from the authority of the usage, and leaveth it for suspect." And therefore inasmuch as most of the usages and orders of the universities were derived from more obscure times, it is the more requisite they be re-examined. In this kind I will give an instance or two, for example sake, of things that are the most obvious and familiar. The one is a matter, which though it be ancient and general, yet I hold to be an error; which is, that scholars in universities come too soon and too unripe to logic and rhetoric, arts fitter for graduates than children and novices. For these two, rightly taken, are the gravest of sciences, being the arts of arts; the one for judgment, the other for ornament. And they be the rules and directions how to set forth and dispose matter: and therefore for minds empty and unfraught with matter, and which have not gathered that which Cicero calleth sylva and supellex, stuff and variety, to begin with those arts (as if one should learn to weigh, or to measure, or to paint the wind) doth work but this effect, that the wisdom of those arts, which is great and universal, is almost made contemptible, and is degenerate into childish sophistry and ridiculous affectation. And further, the untimely learning of them hath drawn on by consequence the superficial and unprofitable teaching and writing of them, as fitteth indeed to the capacity of children. Another is a lack I find in the exercises used in the universities, which do snake too great a divorce between invention and memory. For their speeches are either premeditate, in verbis conceptis, where nothing is left to invention, or merely extemporal, where little is left to memory. Whereas in life and action there is least use of either of these, but rather of intermixtures of premeditation and invention, notes and memory. So as the exercise fitteth not the practice, nor the image the life; and it is ever a true rule in exercises, that they be framed as near as may be to the life of practice; for otherwise they do pervert the motions and faculties of the mind, and not prepare them. The truth whereof is not obscure, when scholars come to the practices of professions, or other actions of civil life; which when they set into, this want is soon found by themselves, and sooner by others. But this part, touching the amendment of the institutions and orders of universities, I will conclude with the clause of Cæsar's letter to Oppius and Balbes, Hoc quemadmodum fieri possit, nonnulla mihi in mentem veniunt, et multa reperiri possunt: de iis rebus rgo vos ut cogitationem suscipiatis.

13. Another defect which I note ascendeth a little higher than the precedent. For as the proficience of learning consisteth much in the orders and institutions of universities in the same states and kingdoms, so it would be yet more advanced, if there were more intelligence mutual between the universities of Europe than now there is. We see there be many orders and foundations, which though they be divided under several sovereignties and territories, yet they take themselves to have a kind of contract, fraternity, and correspondence one with the other, insomuch as they have provincials and generals. And surely as nature createth brotherhood in families, and arts mechanical contract brotherhoods in communalties, and the anointment of God superinduceth a brotherhood in kings and bishops, so in like manner there cannot but be a fraternity in learning and illumination, relating to that paternity which is attributed to God, who is called the Father of illuminations or lights.

14. The last defect which I will note is, that there hath not been, or very rarely been, any public designation of writers or inquirers concerning such parts of knowledge as may appear not to have been already sufficiently laboured or undertaken; unto which point it is an inducement to enter into a view and examination what parts of learning have been prosecuted, and what omitted. For the opinion of plenty is amongst the causes of want, and the great quantity of books maketh a show rather of superfluity than lack; which surcharge nevertheless is not to be remedied by making no more books, but by making more good books, which, as the serpent of Moses, might devour the serpents of the enchanters.

15. The removing of all the defects formerly enumerate, except the last, and of the active part also of the last (which is the designation of writers), are opera basilica; towards which the endeavours of a private man may be but as an image in a crossway, that may point at the way, but cannot go it. But the inducing part of the latter (which is the survey of learning) may be set forward by private travail. Wherefore I will now attempt to make a general and faithful perambulation of learning, with an inquiry what parts thereof lie fresh and waste, and not improved and converted by the industry of man, to the end that such a plot made and recorded to memory may both minister light to any public designation, and, also serve to excite voluntary endeavours. Wherein, nevertheless, my purpose is at this time to note only omissions and deficiences, and not to make any redargution of errors or incomplete prosecutions. For it is one thing to set forth what ground lieth unmanured, and another thing to correct ill husbandry in that which is manured.

In the handling and undertaking of which work I am not ignorant what it is that I do now move and attempt, nor insensible of mine own weakness to sustain my purpose. But my hope is, that if my extreme love to learning carry me too far, I may obtain the excuse of affection; for that "It is not granted to man to love and to be wise." But I know well I can use no other liberty of judgment than I must leave to others; and I for my part shall be indifferently glad either to perform myself, or accept from another, that duty of humanity—Nam qui erranti comiter monstrat viam, &c. I do foresee likewise that of those things which I shall enter and register as deficiences and omissions, many will conceive and censure that some of them are already done and extant; others to be but curiosities, and things of no great use; and others to be of too great difficulty, and almost impossibility to be compassed and effected. But for the two first, I refer myself to the particulars. For the last, touching impossibility, I take it those things are to be held possible which may be done by some person, though not by every one; and which may be done by many, though not by any one; and which may be done in the succession of ages, though not within the hourglass of one man's life; and which may be done by public designation, though not by private endeavour. But, notwithstanding, if any man will take to himself rather that of Solomon, "Dicit piger, Leo est in via," than that of Virgil, "Possunt quia posse videntur," I shall be content that my labours be esteemed but as the better sort of wishes; for as it asketh some knowledge to demand a question not impertinent, so it requireth some sense to make a wish not absurd.

I

(1) The parts of human learning have reference to the three parts of man's understanding, which is the seat of learning: history to his memory, poesy to his imagination, and philosophy to his reason. Divine learning receiveth the same distribution; for, the spirit of man is the same, though the revelation of oracle and sense be diverse. So as theology consisteth also of history of the Church; of parables, which is divine poesy; and of holy doctrine or precept. For as for that part which seemeth supernumerary, which is prophecy, it is but divine history, which hath that prerogative over human, as the narration may be before the fact as well as after.

(2) History is natural, civil, ecclesiastical, and literary; whereof the first three I allow as extant, the fourth I note as deficient. For no man hath propounded to himself the general state of learning to be described and represented from age to age, as many have done the works of Nature, and the state, civil and ecclesiastical; without which the history of the world seemeth to me to be as the statue of Polyphemus with his eye out, that part being wanting which doth most show the spirit and life of the person. And yet I am not ignorant that in divers particular sciences, as of the jurisconsults, the mathematicians, the rhetoricians, the philosophers, there are set down some small memorials of the schools, authors, and books; and so likewise some barren relations touching the invention of arts or usages. But a just story of learning, containing the antiquities and originals of knowledges and their sects, their inventions, their traditions, their diverse administrations and managings, their flourishings, their oppositions, decays, depressions, oblivions, removes, with the causes and occasions of them, and all other events concerning learning, throughout the ages of the world, I may truly affirm to be wanting; the use and end of which work I do not so much design for curiosity or satisfaction of those that are the lovers of learning, but chiefly for a more serious and grave purpose, which is this in few words, that it will make learned men wise in the use and administration of learning. For it is not Saint Augustine's nor Saint Ambrose's works that will make so wise a divine as ecclesiastical history thoroughly read and observed, and the same reason is of learning.

(3) History of Nature is of three sorts; of Nature in course, of Nature erring or varying, and of Nature altered or wrought; that is, history of creatures, history of marvels, and history of arts. The first of these no doubt is extant, and that in good perfection; the two latter are bandied so weakly and unprofitably as I am moved to note them as deficient. For I find no sufficient or competent collection of the works of Nature which have a digression and deflexion from the ordinary course of generations, productions, and motions; whether they be singularities of place and region, or the strange events of time and chance, or the effects of yet unknown properties, or the instances of exception to general kinds. It is true I find a number of books of fabulous experiments and secrets, and frivolous impostures for pleasure and strangeness; but a substantial and severe collection of the heteroclites or irregulars of Nature, well examined and described, I find not, specially not with due rejection of fables and popular errors. For as things now are, if an untruth in Nature be once on foot, what by reason of the neglect of examination, and countenance of antiquity, and what by reason of the use of the opinion in similitudes and ornaments of speech, it is never called down.

(4) The use of this work, honoured with a precedent in Aristotle, is nothing less than to give contentment to the appetite of curious and vain wits, as the manner of Mirabilaries is to do; but for two reasons, both of great weight: the one to correct the partiality of axioms and opinions, which are commonly framed only upon common and familiar examples; the other because from the wonders of Nature is the nearest intelligence and passage towards the wonders of art, for it is no more but by following and, as it were, hounding Nature in her wanderings, to be able to lead her afterwards to the same place again. Neither am I of opinion, in this history of marvels, that superstitious narrations of sorceries, witchcrafts, dreams, divinations, and the like, where there is an assurance and clear evidence of the fact, be altogether excluded. For it is not yet known in what cases and how far effects attributed to superstition do participate of natural causes; and, therefore, howsoever the practice of such things is

to be condemned, yet from the speculation and consideration of them light may be taken, not only for the discerning of the offences, but for the further disclosing of Nature. Neither ought a man to make scruple of entering into these things for inquisition of truth, as your Majesty hath showed in your own example, who, with the two clear eyes of religion and natural philosophy, have looked deeply and wisely into these shadows, and yet proved yourself to be of the nature of the sun, which passeth through pollutions and itself remains as pure as before. But this I hold fit, that these narrations, which have mixture with superstition, be sorted by themselves, and not to be mingled with the narrations which are merely and sincerely natural. But as for the narrations touching the prodigies and miracles of religions, they are either not true or not natural; and, therefore, impertinent for the story of Nature.

(5) For history of Nature, wrought or mechanical, I find some collections made of agriculture, and likewise of manual arts; but commonly with a rejection of experiments familiar and vulgar; for it is esteemed a kind of dishonour unto learning to descend to inquiry or meditation upon matters mechanical, except they be such as may be thought secrets, rarities, and special subtleties; which humour of vain and supercilious arrogancy is justly derided in Plato, where he brings in Hippias, a vaunting sophist, disputing with Socrates, a true and unfeigned inquisitor of truth; where, the subject being touching beauty, Socrates, after his wandering manner of inductions, put first an example of a fair virgin, and then of a fair horse, and then of a fair pot well glazed, whereat Hippias was offended, and said, "More than for courtesy's sake, he did think much to dispute with any that did allege such base and sordid instances." Whereunto Socrates answereth, "You have reason, and it becomes you well, being a man so trim in your vestments," &c., and so goeth on in an irony. But the truth is, they be not the highest instances that give the securest information, as may be well expressed in the tale so common of the philosopher that, while he gazed upwards to the stars, fell into the water; for if he had looked down he might have seen the stars in the water, but looking aloft he could not see the water in the stars. So it cometh often to pass that mean and small things discover great, better than great can discover the small; and therefore Aristotle noteth well, "That the nature of everything is best seen in his smallest portions." And for that cause he inquireth the nature of a commonwealth, first in a family, and the simple conjugations of man and wife, parent and child, master and servant, which are in every cottage. Even so likewise the nature of this great city of the world, and the policy thereof, must be first sought in mean concordances and small portions. So we see how that secret of Nature, of the turning of iron touched with the loadstone towards the north, was found out in needles of iron, not in bars of iron.

(6) But if my judgment be of any weight, the use of history mechanical is of all others the most radical and fundamental towards natural philosophy; such natural philosophy as shall not vanish in the fume of subtle, sublime, or delectable speculation, but such as shall be operative to the endowment and benefit of man's life. For it will not only minister and suggest for the present many ingenious practices in all trades, by a connection and transferring of the observations of one art to the use of another, when the experiences of several mysteries shall fall under the consideration of one man's mind; but further, it will give a more true and real illumination concerning causes and axioms than is hitherto attained. For like as a man's disposition is never well known till he be crossed, nor Proteus ever changed shapes till he was straitened and held fast; so the passages and variations of nature cannot appear so fully in the liberty of nature as in the trials and vexations of art.

II

(1) For civil history, it is of three kinds; not unfitly to be compared with the three kinds of pictures or images. For of pictures or images we see some are unfinished, some are perfect, and some are defaced. So of histories we may find three kinds: memorials, perfect histories, and antiquities; for memorials are history unfinished, or the first or rough drafts of history; and antiquities are history defaced, or some remnants of history which have casually escaped the shipwreck of time.

(2) Memorials, or preparatory history, are of two sorts; whereof the one may be termed commentaries, and the other registers. Commentaries are they which set down a continuance of the naked events and actions, without the motives or designs, the counsels, the speeches, the pretexts, the occasions, and other passages of action. For this is the true nature of a commentary (though Cæsar, in modesty mixed with greatness, did for his pleasure apply the name of a commentary to the best history of the world). Registers are collections of public acts, as decrees of council, judicial proceedings, declarations and letters of estate, orations, and the like, without a perfect continuance or contexture of the thread of the narration.

(3) Antiquities, or remnants of history, are, as was said, tanquam tabula naufragii: when industrious persons, by an exact and scrupulous diligence and observation, out of monuments, names, words, proverbs, traditions, private records and evidences, fragments of stories, passages of books that concern not story, and the like, do save and recover somewhat from the deluge of time.

(4) In these kinds of unperfect histories I do assign no deficience, for they are tanquam imperfecte

mista; and therefore any deficience in them is but their nature. As for the corruptions and moths of history, which are epitomes, the use of them deserveth to be banished, as all men of sound judgment have confessed, as those that have fretted and corroded the sound bodies of many excellent histories, and wrought them into base and unprofitable dregs.

(5) History, which may be called just and perfect history, is of three kinds, according to the object which it propoundeth, or pretendeth to represent: for it either representeth a time, or a person, or an action. The first we call chronicles, the second lives, and the third narrations or relations. Of these, although the first be the most complete and absolute kind of history, and hath most estimation and glory, yet the second excelleth it in profit and use, and the third in verity and sincerity. For history of times representeth the magnitude of actions, and the public faces and deportments of persons, and passeth over in silence the smaller passages and motions of men and matters. But such being the workmanship of God, as He doth hang the greatest weight upon the smallest wires, maxima è minimis, suspendens, it comes therefore to pass, that such histories do rather set forth the pomp of business than the true and inward resorts thereof. But lives, if they be well written, propounding to themselves a person to represent, in whom actions, both greater and smaller, public and private, have a commixture, must of necessity contain a more true, native, and lively representation. So again narrations and relations of actions, as the war of Peloponnesus, the expedition of Cyrus Minor, the conspiracy of Catiline, cannot but be more purely and exactly true than histories of times, because they may choose an argument comprehensible within the notice and instructions of the writer: whereas he that undertaketh the story of a time, specially of any length, cannot but meet with many blanks and spaces, which he must be forced to fill up out of his own wit and conjecture.

(6) For the history of times, I mean of civil history, the providence of God hath made the distribution. For it hath pleased God to ordain and illustrate two exemplar states of the world for arms, learning, moral virtue, policy, and laws; the state of Græcia and the state of Rome; the histories whereof occupying the middle part of time, have more ancient to them histories which may by one common name be termed the antiquities of the world; and after them, histories which may be likewise called by the name of modern history.

(7) Now to speak of the deficiences. As to the heathen antiquities of the world it is in vain to note them for deficient. Deficient they are no doubt, consisting most of fables and fragments; but the deficience cannot be holpen; for antiquity is like fame, caput inter nubila condit, her head is muffled from our sight. For the history of the exemplar states, it is extant in good perfection. Not but I could wish there were a perfect course of history for Græcia, from Theseus to Philopœmen (what time the affairs of Græcia drowned and extinguished in the affairs of Rome), and for Rome from Romulus to Justinianus, who may be truly said to be ultimus Romanorum. In which sequences of story the text of Thucydides and Xenophon in the one, and the texts of Livius, Polybius, Sallustius, Cæsar, Appianus, Tacitus, Herodianus in the other, to be kept entire, without any diminution at all, and only to be supplied and continued. But this is a matter of magnificence, rather to be commended than required; and we speak now of parts of learning supplemental, and not of supererogation.

(8) But for modern histories, whereof there are some few very worthy, but the greater part beneath mediocrity, leaving the care of foreign stories to foreign states, because I will not be curiosus in aliena republica, I cannot fail to represent to your Majesty the unworthiness of the history of England in the main continuance thereof, and the partiality and obliquity of that of Scotland in the latest and largest author that I have seen: supposing that it would be honour for your Majesty, and a work very memorable, if this island of Great Britain, as it is now joined in monarchy for the ages to come, so were joined in one history for the times passed, after the manner of the sacred history, which draweth down the story of the ten tribes and of the two tribes as twins together. And if it shall seem that the greatness of this work may make it less exactly performed, there is an excellent period of a much smaller compass of time, as to the story of England; that is to say, from the uniting of the Roses to the uniting of the kingdoms; a portion of time wherein, to my understanding, there hath been the rarest varieties that in like number of successions of any hereditary monarchy hath been known. For it beginneth with the mixed adoption of a crown by arms and title; an entry by battle, an establishment by marriage; and therefore times answerable, like waters after a tempest, full of working and swelling, though without extremity of storm; but well passed through by the wisdom of the pilot, being one of the most sufficient kings of all the number. Then followeth the reign of a king, whose actions, howsoever conducted, had much intermixture with the affairs of Europe, balancing and inclining them variably; in whose time also began that great alteration in the state ecclesiastical, an action which seldom cometh upon the stage. Then the reign of a minor; then an offer of a usurpation (though it was but as febris ephemera). Then the reign of a queen matched with a foreigner; then of a queen that lived solitary and unmarried, and yet her government so masculine, as it had greater impression and operation upon the states abroad than it any ways received from thence. And now last, this most happy and glorious event, that this island of Britain, divided from all the world, should be united in itself, and that oracle of rest given to Æneas, antiquam exquirite matrem, should now be performed and fulfilled upon the nations of England and

Scotland, being now reunited in the ancient mother name of Britain, as a full period of all instability and peregrinations. So that as it cometh to pass in massive bodies, that they have certain trepidations and waverings before they fix and settle, so it seemeth that by the providence of God this monarchy, before it was to settle in your majesty and your generations (in which I hope it is now established for ever), it had these prelusive changes and varieties.

(9) For lives, I do find strange that these times have so little esteemed the virtues of the times, as that the writings of lives should be no more frequent. For although there be not many sovereign princes or absolute commanders, and that states are most collected into monarchies, yet are there many worthy personages that deserve better than dispersed report or barren eulogies. For herein the invention of one of the late poets is proper, and doth well enrich the ancient fiction. For he feigneth that at the end of the thread or web of every man's life there was a little medal containing the person's name, and that Time waited upon the shears, and as soon as the thread was cut caught the medals, and carried them to the river of Lathe; and about the bank there were many birds flying up and down, that would get the medals and carry them in their beak a little while, and then let them fall into the river. Only there were a few swans, which if they got a name would carry it to a temple where it was consecrate. And although many men, more mortal in their affections than in their bodies, do esteem desire of name and memory but as a vanity and ventosity,

"Animi nil magnæ laudis egentes;"

which opinion cometh from that root, Non prius laudes contempsimus, quam laudanda facere desivimus: yet that will not alter Solomon's judgment, Memoria justi cum laudibus, at impiorum nomen putrescet: the one flourisheth, the other either consumeth to present oblivion, or turneth to an ill odour. And therefore in that style or addition, which is and hath been long well received and brought in use, felicis memoriæ, piæ memoriæ, bonæ memoriæ, we do acknowledge that which Cicero saith, borrowing it from Demosthenes, that bona fama propria possessio defunctorum; which possession I cannot but note that in our times it lieth much waste, and that therein there is a deficience.

(10) For narrations and relations of particular actions, there were also to be wished a greater diligence therein; for there is no great action but hath some good pen which attends it. And because it is an ability not common to write a good history, as may well appear by the small number of them; yet if particularity of actions memorable were but tolerably reported as they pass, the compiling of a complete history of times might be the better expected, when a writer should arise that were fit for it: for the collection of such relations might be as a nursery garden, whereby to plant a fair and stately garden when time should serve.

(11) There is yet another partition of history which Cornelius Tacitus maketh, which is not to be forgotten, specially with that application which he accoupleth it withal, annals and journals: appropriating to the former matters of estate, and to the latter acts and accidents of a meaner nature. For giving but a touch of certain magnificent buildings, he addeth, Cum ex dignitate populi Romani repertum sit, res illustres annalibus, talia diurnis urbis actis mandare. So as there is a kind of contemplative heraldry, as well as civil. And as nothing doth derogate from the dignity of a state more than confusion of degrees, so it doth not a little imbase the authority of a history to intermingle matters of triumph, or matters of ceremony, or matters of novelty, with matters of state. But the use of a journal hath not only been in the history of time, but likewise in the history of persons, and chiefly of actions; for princes in ancient time had, upon point of honour and policy both, journals kept, what passed day by day. For we see the chronicle which was read before Ahasuerus, when he could not take rest, contained matter of affairs, indeed, but such as had passed in his own time and very lately before. But the journal of Alexander's house expressed every small particularity, even concerning his person and court; and it is yet a use well received in enterprises memorable, as expeditions of war, navigations, and the like, to keep diaries of that which passeth continually.

(12) I cannot likewise be ignorant of a form of writing which some grave and wise men have used, containing a scattered history of those actions which they have thought worthy of memory, with politic discourse and observation thereupon: not incorporate into the history, but separately, and as the more principal in their intention; which kind of ruminated history I think more fit to place amongst books of policy, whereof we shall hereafter speak, than amongst books of history. For it is the true office of history to represent the events themselves together with the counsels, and to leave the observations and conclusions thereupon to the liberty and faculty of every man's judgment. But mixtures are things irregular, whereof no man can define.

(13) So also is there another kind of history manifoldly mixed, and that is history of cosmography: being compounded of natural history, in respect of the regions themselves; of history civil, in respect of the habitations, regiments, and manners of the people; and the mathematics, in respect of the climates and configurations towards the heavens: which part of learning of all others in this latter time hath obtained most proficience. For it may be truly affirmed to the honour of these times, and in a virtuous

emulation with antiquity, that this great building of the world had never through-lights made in it, till the age of us and our fathers. For although they had knowledge of the antipodes,

> *"Nosque ubi primus equis Oriens afflavit anhelis,*
> *Illic sera rubens accendit lumina Vesper,"*

yet that might be by demonstration, and not in fact; and if by travel, it requireth the voyage but of half the globe. But to circle the earth, as the heavenly bodies do, was not done nor enterprised till these later times: and therefore these times may justly bear in their word, not only plus ultra, in precedence of the ancient non ultra, and imitabile fulmen, in precedence of the ancient non imitabile fulmen,

> *"Demens qui nimbos et non imitabile fulmen,"* &c.

but likewise imitabile cælum; in respect of the many memorable voyages after the manner of heaven about the globe of the earth.

(14) And this proficience in navigation and discoveries may plant also an expectation of the further proficience and augmentation of all sciences; because it may seem they are ordained by God to be coevals, that is, to meet in one age. For so the prophet Daniel speaking of the latter times foretelleth, Plurimi pertransibunt, et multiplex erit scientia: as if the openness and through-passage of the world and the increase of knowledge were appointed to be in the same ages; as we see it is already performed in great part: the learning of these later times not much giving place to the former two periods or returns of learning, the one of the Grecians, the other of the Romans.

III

(1) History ecclesiastical receiveth the same divisions with history civil: but further in the propriety thereof may be divided into the history of the Church, by a general name; history of prophecy; and history of providence. The first describeth the times of the militant Church, whether it be fluctuant, as the ark of Noah, or movable, as the ark in the wilderness, or at rest, as the ark in the Temple: that is, the state of the Church in persecution, in remove, and in peace. This part I ought in no sort to note as deficient; only I would that the virtue and sincerity of it were according to the mass and quantity. But I am not now in hand with censures, but with omissions.

(2) The second, which is history of prophecy, consisteth of two relatives—the prophecy and the accomplishment; and, therefore, the nature of such a work ought to be, that every prophecy of the Scripture be sorted with the event fulfilling the same throughout the ages of the world, both for the better confirmation of faith and for the better illumination of the Church touching those parts of prophecies which are yet unfulfilled: allowing, nevertheless, that latitude which is agreeable and familiar unto divine prophecies, being of the nature of their Author, with whom a thousand years are but as one day, and therefore are not fulfilled punctually at once, but have springing and germinant accomplishment throughout many ages, though the height or fulness of them may refer to some one age. This is a work which I find deficient, but is to be done with wisdom, sobriety, and reverence, or not at all.

(3) The third, which is history of Providence, containeth that excellent correspondence which is between God's revealed will and His secret will; which though it be so obscure, as for the most part it is not legible to the natural man—no, nor many times to those that behold it from the tabernacle—yet, at some times it pleaseth God, for our better establishment and the confuting of those which are as without God in the world, to write it in such text and capital letters, that, as the prophet saith, "He that runneth by may read it"—that is, mere sensual persons, which hasten by God's judgments, and never bend or fix their cogitations upon them, are nevertheless in their passage and race urged to discern it. Such are the notable events and examples of God's judgments, chastisements, deliverances, and blessings; and this is a work which has passed through the labour of many, and therefore I cannot present as omitted.

(4) There are also other parts of learning which are appendices to history. For all the exterior proceedings of man consist of words and deeds, whereof history doth properly receive and retain in memory the deeds; and if words, yet but as inducements and passages to deeds; so are there other books and writings which are appropriate to the custody and receipt of words only, which likewise are of three sorts—orations, letters, and brief speeches or sayings. Orations are pleadings, speeches of counsel, laudatives, invectives, apologies, reprehensions, orations of formality or ceremony, and the like. Letters are according to all the variety of occasions, advertisements, advises, directions, propositions, petitions, commendatory, expostulatory, satisfactory, of compliment, of pleasure, of discourse, and all other passages of action. And such as are written from wise men, are of all the words of man, in my judgment, the best; for they are more natural than orations and public speeches, and more advised than

conferences or present speeches. So again letters of affairs from such as manage them, or are privy to them, are of all others the best instructions for history, and to a diligent reader the best histories in themselves. For apophthegms, it is a great loss of that book of Cæsar's; for as his history, and those few letters of his which we have, and those apophthegms which were of his own, excel all men's else, so I suppose would his collection of apophthegms have done; for as for those which are collected by others, either I have no taste in such matters or else their choice hath not been happy. But upon these three kinds of writings I do not insist, because I have no deficiences to propound concerning them.

(5) Thus much therefore concerning history, which is that part of learning which answereth to one of the cells, domiciles, or offices of the mind of man, which is that of the memory.

IV

(1) Poesy is a part of learning in measure of words, for the most part restrained, but in all other points extremely licensed, and doth truly refer to the imagination; which, being not tied to the laws of matter, may at pleasure join that which nature hath severed, and sever that which nature hath joined, and so make unlawful matches and divorces of things—Pictoribus atque poetis, &c. It is taken in two senses in respect of words or matter. In the first sense, it is but a character of style, and belongeth to arts of speech, and is not pertinent for the present. In the latter, it is—as hath been said—one of the principal portions of learning, and is nothing else but feigned history, which may be styled as well in prose as in verse.

(2) The use of this feigned history hath been to give some shadow of satisfaction to the mind of man in those points wherein the nature of things doth deny it, the world being in proportion inferior to the soul; by reason whereof there is, agreeable to the spirit of man, a more ample greatness, a more exact goodness, and a more absolute variety, than can be found in the nature of things. Therefore, because the acts or events of true history have not that magnitude which satisfieth the mind of man, poesy feigneth acts and events greater and more heroical. Because true history propoundeth the successes and issues of actions not so agreeable to the merits of virtue and vice, therefore poesy feigns them more just in retribution, and more according to revealed Providence. Because true history representeth actions and events more ordinary and less interchanged, therefore poesy endueth them with more rareness and more unexpected and alternative variations. So as it appeareth that poesy serveth and conferreth to magnanimity, morality and to delectation. And therefore, it was ever thought to have some participation of divineness, because it doth raise and erect the mind, by submitting the shows of things to the desires of the mind; whereas reason doth buckle and bow the mind unto the nature of things. And we see that by these insinuations and congruities with man's nature and pleasure, joined also with the agreement and consort it hath with music, it hath had access and estimation in rude times and barbarous regions, where other learning stood excluded.

(3) The division of poesy which is aptest in the propriety thereof (besides those divisions which are common unto it with history, as feigned chronicles, feigned lives, and the appendices of history, as feigned epistles, feigned orations, and the rest) is into poesy narrative, representative, and allusive. The narrative is a mere imitation of history, with the excesses before remembered, choosing for subjects commonly wars and love, rarely state, and sometimes pleasure or mirth. Representative is as a visible history, and is an image of actions as if they were present, as history is of actions in nature as they are (that is) past. Allusive, or parabolical, is a narration applied only to express some special purpose or conceit; which latter kind of parabolical wisdom was much more in use in the ancient times, as by the fables of Æsop, and the brief sentences of the seven, and the use of hieroglyphics may appear. And the cause was (for that it was then of necessity to express any point of reason which was more sharp or subtle than the vulgar in that manner) because men in those times wanted both variety of examples and subtlety of conceit. And as hieroglyphics were before letters, so parables were before arguments; and nevertheless now and at all times they do retain much life and rigour, because reason cannot be so sensible nor examples so fit.

(4) But there remaineth yet another use of poesy parabolical, opposite to that which we last mentioned; for that tendeth to demonstrate and illustrate that which is taught or delivered, and this other to retire and obscure it—that is, when the secrets and mysteries of religion, policy, or philosophy, are involved in fables or parables. Of this in divine poesy we see the use is authorised. In heathen poesy we see the exposition of fables doth fall out sometimes with great felicity: as in the fable that the giants being overthrown in their war against the gods, the earth their mother in revenge thereof brought forth Fame:

"*Illam terra parens, ira irritat Deorum,*
Extremam, ut perhibent, Cœo Enceladoque soroem,
Progenuit."

Expounded that when princes and monarchs have suppressed actual and open rebels, then the malignity of people (which is the mother of rebellion) doth bring forth libels and slanders, and taxations of the states, which is of the same kind with rebellion but more feminine. So in the fable that the rest of the gods having conspired to bind Jupiter, Pallas called Briareus with his hundred hands to his aid: expounded that monarchies need not fear any curbing of their absoluteness by mighty subjects, as long as by wisdom they keep the hearts of the people, who will be sure to come in on their side. So in the fable that Achilles was brought up under Chiron, the centaur, who was part a man and part a beast, expounded ingeniously but corruptly by Machiavel, that it belongeth to the education and discipline of princes to know as well how to play the part of a lion in violence, and the fox in guile, as of the man in virtue and justice. Nevertheless, in many the like encounters, I do rather think that the fable was first, and the exposition devised, than that the moral was first, and thereupon the fable framed; for I find it was an ancient vanity in Chrysippus, that troubled himself with great contention to fasten the assertions of the Stoics upon the fictions of the ancient poets; but yet that all the fables and fictions of the poets were but pleasure and not figure, I interpose no opinion. Surely of these poets which are now extant, even Homer himself (notwithstanding he was made a kind of scripture by the later schools of the Grecians), yet I should without any difficulty pronounce that his fables had no such inwardness in his own meaning. But what they might have upon a more original tradition is not easy to affirm, for he was not the inventor of many of them.

(5) In this third part of learning, which is poesy, I can report no deficience; for being as a plant that cometh of the lust of the earth, without a formal seed, it hath sprung up and spread abroad more than any other kind. But to ascribe unto it that which is due, for the expressing of affections, passions, corruptions, and customs, we are beholding to poets more than to the philosophers' works; and for wit and eloquence, not much less than to orators' harangues. But it is not good to stay too long in the theatre. Let us now pass on to the judicial place or palace of the mind, which we are to approach and view with more reverence and attention.

V

(1) The knowledge of man is as the waters, some descending from above, and some springing from beneath: the one informed by the light of nature, the other inspired by divine revelation. The light of nature consisteth in the notions of the mind and the reports of the senses; for as for knowledge which man receiveth by teaching, it is cumulative and not original, as in a water that besides his own spring-head is fed with other springs and streams. So then, according to these two differing illuminations or originals, knowledge is first of all divided into divinity and philosophy.

(2) In philosophy the contemplations of man do either penetrate unto God, or are circumferred to nature, or are reflected or reverted upon himself. Out of which several inquiries there do arise three knowledges—divine philosophy, natural philosophy, and human philosophy or humanity. For all things are marked and stamped with this triple character—the power of God, the difference of nature and the use of man. But because the distributions and partitions of knowledge are not like several lines that meet in one angle, and so touch but in a point, but are like branches of a tree that meet in a stem, which hath a dimension and quantity of entireness and continuance before it come to discontinue and break itself into arms and boughs; therefore it is good, before we enter into the former distribution, to erect and constitute one universal science, by the name of philosophia prima, primitive or summary philosophy, as the main and common way, before we come where the ways part and divide themselves; which science whether I should report as deficient or no, I stand doubtful. For I find a certain rhapsody of natural theology, and of divers parts of logic; and of that part of natural philosophy which concerneth the principles, and of that other part of natural philosophy which concerneth the soul or spirit—all these strangely commixed and confused; but being examined, it seemeth to me rather a depredation of other sciences, advanced and exalted unto some height of terms, than anything solid or substantive of itself. Nevertheless I cannot be ignorant of the distinction which is current, that the same things are handled but in several respects. As for example, that logic considereth of many things as they are in notion, and this philosophy as they are in nature—the one in appearance, the other in existence; but I find this difference better made than pursued. For if they had considered quantity, similitude, diversity, and the rest of those extern characters of things, as philosophers, and in nature, their inquiries must of force have been of a far other kind than they are. For doth any of them, in handling quantity, speak of the force of union, how and how far it multiplieth virtue? Doth any give the reason why some things in nature are so common, and in so great mass, and others so rare, and in so small quantity? Doth any, in handling similitude and diversity, assign the cause why iron should not move to iron, which is more like, but move to the loadstone, which is less like? Why in all diversities of things there should be certain participles in nature which are almost ambiguous to which kind they should be referred? But

there is a mere and deep silence touching the nature and operation of those common adjuncts of things, as in nature; and only a resuming and repeating of the force and use of them in speech or argument. Therefore, because in a writing of this nature I avoid all subtlety, my meaning touching this original or universal philosophy is thus, in a plain and gross description by negative: "That it be a receptacle for all such profitable observations and axioms as fall not within the compass of any of the special parts of philosophy or sciences, but are more common and of a higher stage."

(3) Now that there are many of that kind need not be doubted. For example: Is not the rule, Si inœqualibus æqualia addas, omnia erunt inæqualia, an axiom as well of justice as of the mathematics? and is there not a true coincidence between commutative and distributive justice, and arithmetical and geometrical proportion? Is not that other rule, Quæ in eodem tertio conveniunt, et inter se conveniunt, a rule taken from the mathematics, but so potent in logic as all syllogisms are built upon it? Is not the observation, Omnia mutantur, nil interit, a contemplation in philosophy thus, that the quantum of nature is eternal? in natural theology thus, that it requireth the same omnipotency to make somewhat nothing, which at the first made nothing somewhat? according to the Scripture, Didici quod omnia opera, quœ fecit Deus, perseverent in perpetuum; non possumus eis quicquam addere nec auferre. Is not the ground, which Machiavel wisely and largely discourseth concerning governments, that the way to establish and preserve them is to reduce them ad principia—a rule in religion and nature, as well as in civil administration? Was not the Persian magic a reduction or correspondence of the principles and architectures of nature to the rules and policy of governments? Is not the precept of a musician, to fall from a discord or harsh accord upon a concord or sweet accord, alike true in affection? Is not the trope of music, to avoid or slide from the close or cadence, common with the trope of rhetoric of deceiving expectation? Is not the delight of the quavering upon a stop in music the same with the playing of light upon the water?

"Splendet tremulo sub lumine pontus."

Are not the organs of the senses of one kind with the organs of reflection, the eye with a glass, the ear with a cave or strait, determined and bounded? Neither are these only similitudes, as men of narrow observation may conceive them to be, but the same footsteps of nature, treading or printing upon several subjects or matters. This science therefore (as I understand it) I may justly report as deficient; for I see sometimes the profounder sort of wits, in handling some particular argument, will now and then draw a bucket of water out of this well for their present use; but the spring-head thereof seemeth to me not to have been visited, being of so excellent use both for the disclosing of nature and the abridgment of art.

VI

(1) This science being therefore first placed as a common parent like unto Berecynthia, which had so much heavenly issue, omnes cœlicolas, omnes supera alta tenetes; we may return to the former distribution of the three philosophies—divine, natural, and human. And as concerning divine philosophy or natural theology, it is that knowledge or rudiment of knowledge concerning God which may be obtained by the contemplation of His creatures; which knowledge may be truly termed divine in respect of the object, and natural in respect of the light. The bounds of this knowledge are, that it sufficeth to convince atheism, but not to inform religion; and therefore there was never miracle wrought by God to convert an atheist, because the light of nature might have led him to confess a God; but miracles have been wrought to convert idolaters and the superstitious, because no light of nature extendeth to declare the will and true worship of God. For as all works do show forth the power and skill of the workman, and not his image, so it is of the works of God, which do show the omnipotency and wisdom of the Maker, but not His image. And therefore therein the heathen opinion differeth from the sacred truth: for they supposed the world to be the image of God, and man to be an extract or compendious image of the world; but the Scriptures never vouchsafe to attribute to the world that honour, as to be the image of God, but only the work of His hands; neither do they speak of any other image of God but man. Wherefore by the contemplation of nature to induce and enforce the acknowledgment of God, and to demonstrate His power, providence, and goodness, is an excellent argument, and hath been excellently handled by divers, but on the other side, out of the contemplation of nature, or ground of human knowledges, to induce any verity or persuasion concerning the points of faith, is in my judgment not safe; Da fidei quæ fidei sunt. For the heathen themselves conclude as much in that excellent and divine fable of the golden chain, "That men and gods were not able to draw Jupiter down to the earth; but, contrariwise, Jupiter was able to draw them up to heaven." So as we ought not to attempt to draw down or submit the mysteries of God to our reason, but contrariwise to raise and advance our reason to the divine truth. So as in this part of knowledge, touching divine philosophy, I am so far from noting any deficience, as I rather note an excess; whereunto I have digressed because of

the extreme prejudice which both religion and philosophy hath received and may receive by being commixed together; as that which undoubtedly will make an heretical religion, and an imaginary and fabulous philosophy.

(2) Otherwise it is of the nature of angels and spirits, which is an appendix of theology, both divine and natural, and is neither inscrutable nor interdicted. For although the Scripture saith, "Let no man deceive you in sublime discourse touching the worship of angels, pressing into that he knoweth not," &c., yet notwithstanding if you observe well that precept, it may appear thereby that there be two things only forbidden—adoration of them, and opinion fantastical of them, either to extol them further than appertaineth to the degree of a creature, or to extol a man's knowledge of them further than he hath ground. But the sober.and grounded inquiry, which may arise out of the passages of Holy Scriptures, or out of the gradations of nature, is not restrained. So of degenerate and revolted spirits, the conversing with them or the employment of them is prohibited, much more any veneration towards them; but the contemplation or science of their nature, their power, their illusions, either by Scripture or reason, is a part of spiritual wisdom. For so the apostle saith, "We are not ignorant of his stratagems." And it is no more unlawful to inquire the nature of evil spirits, than to inquire the force of poisons in nature, or the nature of sin and vice in morality. But this part touching angels and spirits I cannot note as deficient, for many have occupied themselves in it; I may rather challenge it, in many of the writers thereof, as fabulous and fantastical.

VII

(1) Leaving therefore divine philosophy or natural theology (not divinity or inspired theology, which we reserve for the last of all as the haven and sabbath of all man's contemplations) we will now proceed to natural philosophy. If then it be true that Democritus said, "That the truth of nature lieth hid in certain deep mines and caves;" and if it be true likewise that the alchemists do so much inculcate, that Vulcan is a second nature, and imitateth that dexterously and compendiously, which nature worketh by ambages and length of time, it were good to divide natural philosophy into the mine and the furnace, and to make two professions or occupations of natural philosophers—some to be pioneers and some smiths; some to dig, and some to refine and hammer. And surely I do best allow of a division of that kind, though in more familiar and scholastical terms: namely, that these be the two parts of natural philosophy—the inquisition of causes, and the production of effects; speculative and operative; natural science, and natural prudence. For as in civil matters there is a wisdom of discourse, and a wisdom of direction; so is it in natural. And here I will make a request, that for the latter (or at least for a part thereof) I may revive and reintegrate the misapplied and abused name of natural magic, which in the true sense is but natural wisdom, or natural prudence; taken according to the ancient acception, purged from vanity and superstition. Now although it be true, and I know it well, that there is an intercourse between causes and effects, so as both these knowledges, speculative and operative, have a great connection between themselves; yet because all true and fruitful natural philosophy hath a double scale or ladder, ascendent and descendent, ascending from experiments to the invention of causes, and descending from causes to the invention of new experiments; therefore I judge it most requisite that these two parts be severally considered and handled.

(2) Natural science or theory is divided into physic and metaphysic; wherein I desire it may be conceived that I use the word metaphysic in a differing sense from that that is received. And in like manner, I doubt not but it will easily appear to men of judgment, that in this and other particulars, wheresoever my conception and notion may differ from the ancient, yet I am studious to keep the ancient terms. For hoping well to deliver myself from mistaking, by the order and perspicuous expressing of that I do propound, I am otherwise zealous and affectionate to recede as little from antiquity, either in terms or opinions, as may stand with truth and the proficience of knowledge. And herein I cannot a little marvel at the philosopher Aristotle, that did proceed in such a spirit of difference and contradiction towards all antiquity; undertaking not only to frame new words of science at pleasure, but to confound and extinguish all ancient wisdom; insomuch as he never nameth or mentioneth an ancient author or opinion, but to confute and reprove; wherein for glory, and drawing followers and disciples, he took the right course. For certainly there cometh to pass, and hath place in human truth, that which was noted and pronounced in the highest truth:—Veni in nomine partis, nec recipits me; si quis venerit in nomine suo eum recipietis. But in this divine aphorism (considering to whom it was applied, namely, to antichrist, the highest deceiver), we may discern well that the coming in a man's own name, without regard of antiquity or paternity, is no good sign of truth, although it be joined with the fortune and success of an eum recipietis. But for this excellent person Aristotle, I will think of him that he learned that humour of his scholar, with whom it seemeth he did emulate; the one to conquer all opinions, as the other to conquer all nations. Wherein, nevertheless, it may be, he may at some men's hands, that are of a bitter disposition, get a like title as his scholar did:—

"Felix terrarum prædo, non utile mundo
Editus exemplum, &c."
So,

"Felix doctrinæ prædo."

But to me, on the other side, that do desire as much as lieth in my pen to ground a sociable intercourse between antiquity and proficience, it seemeth best to keep way with antiquity usque ad aras; and, therefore, to retain the ancient terms, though I sometimes alter the uses and definitions, according to the moderate proceeding in civil government; where, although there be some alteration, yet that holdeth which Tacitus wisely noteth, eadem magistratuum vocabula.

(3) To return, therefore, to the use and acception of the term metaphysic as I do now understand the word; it appeareth, by that which hath been already said, that I intend philosophia prima, summary philosophy and metaphysic, which heretofore have been confounded as one, to be two distinct things. For the one I have made as a parent or common ancestor to all knowledge; and the other I have now brought in as a branch or descendant of natural science. It appeareth likewise that I have assigned to summary philosophy the common principles and axioms which are promiscuous and indifferent to several sciences; I have assigned unto it likewise the inquiry touching the operation or the relative and adventive characters of essences, as quantity, similitude, diversity, possibility, and the rest, with this distinction and provision; that they be handled as they have efficacy in nature, and not logically. It appeareth likewise that natural theology, which heretofore hath been handled confusedly with metaphysic, I have enclosed and bounded by itself. It is therefore now a question what is left remaining for metaphysic; wherein I may without prejudice preserve thus much of the conceit of antiquity, that physic should contemplate that which is inherent in matter, and therefore transitory; and metaphysic that which is abstracted and fixed. And again, that physic should handle that which supposeth in nature only a being and moving; and metaphysic should handle that which supposeth further in nature a reason, understanding, and platform. But the difference, perspicuously expressed, is most familiar and sensible. For as we divided natural philosophy in general into the inquiry of causes and productions of effects, so that part which concerneth the inquiry of causes we do subdivide according to the received and sound division of causes. The one part, which is physic, inquireth and handleth the material and efficient causes; and the other, which is metaphysic, handleth the formal and final causes.

(4) Physic (taking it according to the derivation, and not according to our idiom for medicine) is situate in a middle term or distance between natural history and metaphysic. For natural history describeth the variety of things; physic the causes, but variable or respective causes; and metaphysic the fixed and constant causes.

"Limus ut hic durescit, et hæc ut cera liquescit,
Uno eodemque igni."

Fire is the cause of induration, but respective to clay; fire is the cause of colliquation, but respective to wax. But fire is no constant cause either of induration or colliquation; so then the physical causes are but the efficient and the matter. Physic hath three parts, whereof two respect nature united or collected, the third contemplateth nature diffused or distributed. Nature is collected either into one entire total, or else into the same principles or seeds. So as the first doctrine is touching the contexture or configuration of things, as de mundo, de universitate rerum. The second is the doctrine concerning the principles or originals of things. The third is the doctrine concerning all variety and particularity of things; whether it be of the differing substances, or their differing qualities and natures; whereof there needeth no enumeration, this part being but as a gloss or paraphrase that attendeth upon the text of natural history. Of these three I cannot report any as deficient. In what truth or perfection they are handled, I make not now any judgment; but they are parts of knowledge not deserted by the labour of man.

(5) For metaphysic, we have assigned unto it the inquiry of formal and final causes; which assignation, as to the former of them, may seem to be nugatory and void, because of the received and inveterate opinion, that the inquisition of man is not competent to find out essential forms or true differences; of which opinion we will take this hold, that the invention of forms is of all other parts of knowledge the worthiest to be sought, if it be possible to be found. As for the possibility, they are ill discoverers that think there is no land, when they can see nothing but sea. But it is manifest that Plato, in his opinion of ideas, as one that had a wit of elevation situate as upon a cliff, did descry that forms were the true object of knowledge; but lost the real fruit of his opinion, by considering of forms as absolutely abstracted from matter, and not confined and determined by matter; and so turning his opinion upon theology, wherewith all his natural philosophy is infected. But if any man shall keep a

continual watchful and severe eye upon action, operation, and the use of knowledge, he may advise and take notice what are the forms, the disclosures whereof are fruitful and important to the state of man. For as to the forms of substances (man only except, of whom it is said, Formavit hominem de limo terræ, et spiravit in faciem ejus spiraculum vitæ, and not as of all other creatures, Producant aquæ, producat terra), the forms of substances I say (as they are now by compounding and transplanting multiplied) are so perplexed, as they are not to be inquired; no more than it were either possible or to purpose to seek in gross the forms of those sounds which make words, which by composition and transposition of letters are infinite. But, on the other side, to inquire the form of those sounds or voices which make simple letters is easily comprehensible; and being known induceth and manifesteth the forms of all words, which consist and are compounded of them. In the same manner to inquire the form of a lion, of an oak, of gold; nay, of water, of air, is a vain pursuit; but to inquire the forms of sense, of voluntary motion, of vegetation, of colours, of gravity and levity, of density, of tenuity, of heat, of cold, and all other natures and qualities, which, like an alphabet, are not many, and of which the essences (upheld by matter) of all creatures do consist; to inquire, I say, the true forms of these, is that part of metaphysic which we now define of. Not but that physic doth make inquiry and take consideration of the same natures; but how? Only as to the material and efficient causes of them, and not as to the forms. For example, if the cause of whiteness in snow or froth be inquired, and it be rendered thus, that the subtle intermixture of air and water is the cause, it is well rendered; but, nevertheless, is this the form of whiteness? No; but it is the efficient, which is ever but vehiculum formæ. This part of metaphysic I do not find laboured and performed; whereat I marvel not; because I hold it not possible to be invented by that course of invention which hath been used; in regard that men (which is the root of all error) have made too untimely a departure, and too remote a recess from particulars.

(6) But the use of this part of metaphysic, which I report as deficient, is of the rest the most excellent in two respects: the one, because it is the duty and virtue of all knowledge to abridge the infinity of individual experience, as much as the conception of truth will permit, and to remedy the complaint of vita brevis, ars longa; which is performed by uniting the notions and conceptions of sciences. For knowledges are as pyramids, whereof history is the basis. So of natural philosophy, the basis is natural history; the stage next the basis is physic; the stage next the vertical point is metaphysic. As for the vertical point, opus quod operatur Deus à principio usque ad finem, the summary law of nature, we know not whether man's inquiry can attain unto it. But these three be the true stages of knowledge, and are to them that are depraved no better than the giants' hills:—

> *"Ter sunt conati imponere Pelio Ossam,*
> *Scilicet atque Ossæ frondsum involvere Olympum."*

But to those which refer all things to the glory of God, they are as the three acclamations, Sante, sancte, sancte! holy in the description or dilatation of His works; holy in the connection or concatenation of them; and holy in the union of them in a perpetual and uniform law. And, therefore, the speculation was excellent in Parmenides and Plato, although but a speculation in them, that all things by scale did ascend to unity. So then always that knowledge is worthiest which is charged with least multiplicity, which appeareth to be metaphysic; as that which considereth the simple forms or differences of things, which are few in number, and the degrees and co-ordinations whereof make all this variety. The second respect, which valueth and commendeth this part of metaphysic, is that it doth enfranchise the power of man unto the greatest liberty and possibility of works and effects. For physic carrieth men in narrow and restrained ways, subject to many accidents and impediments, imitating the ordinary flexuous courses of nature. But latæ undique sunt sapientibus viæ; to sapience (which was anciently defined to be rerum divinarum et humanarum scientia) there is ever a choice of means. For physical causes give light to new invention in simili materia. But whosoever knoweth any form, knoweth the utmost possibility of superinducing that nature upon any variety of matter; and so is less restrained in operation, either to the basis of the matter, or the condition of the efficient; which kind of knowledge Solomon likewise, though in a more divine sense, elegantly describeth: non arctabuntur gressus tui, et currens non habebis offendiculum. The ways of sapience are not much liable either to particularity or chance.

(7) The second part of metaphysic is the inquiry of final causes, which I am moved to report not as omitted, but as misplaced. And yet if it were but a fault in order, I would not speak of it; for order is matter of illustration, but pertaineth not to the substance of sciences. But this misplacing hath caused a deficience, or at least a great improficience in the sciences themselves. For the handling of final causes, mixed with the rest in physical inquiries, hath intercepted the severe and diligent inquiry of all real and physical causes, and given men the occasion to stay upon these satisfactory and specious causes, to the great arrest and prejudice of further discovery. For this I find done not only by Plato, who ever anchoreth upon that shore, but by Aristotle, Galen, and others which do usually likewise fall upon these flats of discoursing causes. For to say that "the hairs of the eyelids are for a quickset and fence about the

sight;" or that "the firmness of the skins and hides of living creatures is to defend them from the extremities of heat or cold;" or that "the bones are for the columns or beams, whereupon the frames of the bodies of living creatures are built;" or that "the leaves of trees are for protecting of the fruit;" or that "the clouds are for watering of the earth;" or that "the solidness of the earth is for the station and mansion of living creatures;" and the like, is well inquired and collected in metaphysic, but in physic they are impertinent. Nay, they are, indeed, but remoras and hindrances to stay and slug the ship from further sailing; and have brought this to pass, that the search of the physical causes hath been neglected and passed in silence. And, therefore, the natural philosophy of Democritus and some others, who did not suppose a mind or reason in the frame of things, but attributed the form thereof able to maintain itself to infinite essays or proofs of Nature, which they term fortune, seemeth to me (as far as I can judge by the recital and fragments which remain unto us) in particularities of physical causes more real and better inquired than that of Aristotle and Plato; whereof both intermingled final causes, the one as a part of theology, and the other as a part of logic, which were the favourite studies respectively of both those persons; not because those final causes are not true and worthy to be inquired, being kept within their own province, but because their excursions into the limits of physical causes hath bred a vastness and solitude in that tract. For otherwise, keeping their precincts and borders, men are extremely deceived if they think there is an enmity or repugnancy at all between them. For the cause rendered, that "the hairs about the eyelids are for the safeguard of the sight," doth not impugn the cause rendered, that "pilosity is incident to orifices of moisture—*muscosi fontes*, &c." Nor the cause rendered, that "the firmness of hides is for the armour of the body against extremities of heat or cold," doth not impugn the cause rendered, that "contraction of pores is incident to the outwardest parts, in regard of their adjacence to foreign or unlike bodies;" and so of the rest, both causes being true and compatible, the one declaring an intention, the other a consequence only. Neither doth this call in question or derogate from Divine Providence, but highly confirm and exalt it. For as in civil actions he is the greater and deeper politique that can make other men the instruments of his will and ends, and yet never acquaint them with his purpose, so as they shall do it and yet not know what they do, than he that imparteth his meaning to those he employeth; so is the wisdom of God more admirable, when Nature intendeth one thing and Providence draweth forth another, than if He had communicated to particular creatures and motions the characters and impressions of His Providence. And thus much for metaphysic; the latter part whereof I allow as extant, but wish it confined to his proper place.

VIII

(1) Nevertheless, there remaineth yet another part of natural philosophy, which is commonly made a principal part, and holdeth rank with physic special and metaphysic, which is mathematic; but I think it more agreeable to the nature of things, and to the light of order, to place it as a branch of metaphysic. For the subject of it being quantity, not quantity indefinite, which is but a relative, and belongeth to philosophia prima (as hath been said), but quantity determined or proportionable, it appeareth to be one of the essential forms of things, as that that is causative in Nature of a number of effects; insomuch as we see in the schools both of Democritus and of Pythagoras that the one did ascribe figure to the first seeds of things, and the other did suppose numbers to be the principles and originals of things. And it is true also that of all other forms (as we understand forms) it is the most abstracted and separable from matter, and therefore most proper to metaphysic; which hath likewise been the cause why it hath been better laboured and inquired than any of the other forms, which are more immersed in matter. For it being the nature of the mind of man (to the extreme prejudice of knowledge) to delight in the spacious liberty of generalities, as in a champaign region, and not in the inclosures of particularity, the mathematics of all other knowledge were the goodliest fields to satisfy that appetite. But for the placing of this science, it is not much material: only we have endeavoured in these our partitions to observe a kind of perspective, that one part may cast light upon another.

(2) The mathematics are either pure or mixed. To the pure mathematics are those sciences belonging which handle quantity determinate, merely severed from any axioms of natural philosophy; and these are two, geometry and arithmetic, the one handling quantity continued, and the other dissevered. Mixed hath for subject some axioms or parts of natural philosophy, and considereth quantity determined, as it is auxiliary and incident unto them. For many parts of Nature can neither be invented with sufficient subtlety, nor demonstrated with sufficient perspicuity, nor accommodated unto use with sufficient dexterity, without the aid and intervening of the mathematics, of which sort are perspective, music, astronomy, cosmography, architecture, engineery, and divers others. In the mathematics I can report no deficience, except it be that men do not sufficiently understand this excellent use of the pure mathematics, in that they do remedy and cure many defects in the wit and faculties intellectual. For if the wit be too dull, they sharpen it; if too wandering, they fix it; if too inherent in the sense, they abstract it. So that as tennis is a game of no use in itself, but of great use in

respect it maketh a quick eye and a body ready to put itself into all postures, so in the mathematics that use which is collateral and intervenient is no less worthy than that which is principal and intended. And as for the mixed mathematics, I may only make this prediction, that there cannot fail to be more kinds of them as Nature grows further disclosed. Thus much of natural science, or the part of Nature speculative.

(3) For natural prudence, or the part operative of natural philosophy, we will divide it into three parts—experimental, philosophical, and magical; which three parts active have a correspondence and analogy with the three parts speculative, natural history, physic, and metaphysic. For many operations have been invented, sometimes by a casual incidence and occurrence, sometimes by a purposed experiment; and of those which have been found by an intentional experiment, some have been found out by varying or extending the same experiment, some by transferring and compounding divers experiments the one into the other, which kind of invention an empiric may manage. Again, by the knowledge of physical causes there cannot fail to follow many indications and designations of new particulars, if men in their speculation will keep one eye upon use and practice. But these are but coastings along the shore, *premendo littus iniquum*; for it seemeth to me there can hardly be discovered any radical or fundamental alterations and innovations in Nature, either by the fortune and essays of experiments, or by the light and direction of physical causes. If, therefore, we have reported metaphysic deficient, it must follow that we do the like of natural magic, which hath relation thereunto. For as for the natural magic whereof now there is mention in books, containing certain credulous and superstitious conceits and observations of sympathies and antipathies, and hidden proprieties, and some frivolous experiments, strange rather by disguisement than in themselves, it is as far differing in truth of Nature from such a knowledge as we require as the story of King Arthur of Britain, or Hugh of Bourdeaux, differs from Cæsar's Commentaries in truth of story; for it is manifest that Cæsar did greater things de vero than those imaginary heroes were feigned to do. But he did them not in that fabulous manner. Of this kind of learning the fable of Ixion was a figure, who designed to enjoy Juno, the goddess of power, and instead of her had copulation with a cloud, of which mixture were begotten centaurs and chimeras. So whosoever shall entertain high and vaporous imaginations, instead of a laborious and sober inquiry of truth, shall beget hopes and beliefs of strange and impossible shapes. And, therefore, we may note in these sciences which hold so much of imagination and belief, as this degenerate natural magic, alchemy, astrology, and the like, that in their propositions the description of the means is ever more monstrous than the pretence or end. For it is a thing more probable that he that knoweth well the natures of weight, of colour, of pliant and fragile in respect of the hammer, of volatile and fixed in respect of the fire, and the rest, may superinduce upon some metal the nature and form of gold by such mechanic as longeth to the production of the natures afore rehearsed, than that some grains of the medicine projected should in a few moments of time turn a sea of quicksilver or other material into gold. So it is more probable that he that knoweth the nature of arefaction, the nature of assimilation of nourishment to the thing nourished, the manner of increase and clearing of spirits, the manner of the depredations which spirits make upon the humours and solid parts, shall by ambages of diets, bathings, anointings, medicines, motions, and the like, prolong life, or restore some degree of youth or vivacity, than that it can be done with the use of a few drops or scruples of a liquor or receipt. To conclude, therefore, the true natural magic, which is that great liberty and latitude of operation which dependeth upon the knowledge of forms, I may report deficient, as the relative thereof is. To which part, if we be serious and incline not to vanities and plausible discourse, besides the deriving and deducing the operations themselves from metaphysic, there are pertinent two points of much purpose, the one by way of preparation, the other by way of caution. The first is, that there be made a calendar, resembling an inventory of the estate of man, containing all the inventions (being the works or fruits of Nature or art) which are now extant, and whereof man is already possessed; out of which doth naturally result a note what things are yet held impossible, or not invented, which calendar will be the more artificial and serviceable if to every reputed impossibility you add what thing is extant which cometh the nearest in degree to that impossibility; to the end that by these optatives and potentials man's inquiry may be the more awake in deducing direction of works from the speculation of causes. And secondly, that these experiments be not only esteemed which have an immediate and present use, but those principally which are of most universal consequence for invention of other experiments, and those which give most light to the invention of causes; for the invention of the mariner's needle, which giveth the direction, is of no less benefit for navigation than the invention of the sails which give the motion.

(4) Thus have I passed through natural philosophy and the deficiences thereof; wherein if I have differed from the ancient and received doctrines, and thereby shall move contradiction, for my part, as I affect not to dissent, so I purpose not to contend. If it be truth,

"Non canimus surdis, respondent omnia sylvæ,"

the voice of Nature will consent, whether the voice of man do or no. And as Alexander Borgia was wont to say of the expedition of the French for Naples, that they came with chalk in their hands to mark up their lodgings, and not with weapons to fight; so I like better that entry of truth which cometh peaceably with chalk to mark up those minds which are capable to lodge and harbour it, than that which cometh with pugnacity and contention.

(5) But there remaineth a division of natural philosophy according to the report of the inquiry, and nothing concerning the matter or subject: and that is positive and considerative, when the inquiry reporteth either an assertion or a doubt. These doubts or non liquets are of two sorts, particular and total. For the first, we see a good example thereof in Aristotle's Problems which deserved to have had a better continuance; but so nevertheless as there is one point whereof warning is to be given and taken. The registering of doubts hath two excellent uses: the one, that it saveth philosophy from errors and falsehoods; when that which is not fully appearing is not collected into assertion, whereby error might draw error, but reserved in doubt; the other, that the entry of doubts are as so many suckers or sponges to draw use of knowledge; insomuch as that which if doubts had not preceded, a man should never have advised, but passed it over without note, by the suggestion and solicitation of doubts is made to be attended and applied. But both these commodities do scarcely countervail and inconvenience, which will intrude itself if it be not debarred; which is, that when a doubt is once received, men labour rather how to keep it a doubt still, than how to solve it, and accordingly bend their wits. Of this we see the familiar example in lawyers and scholars, both which, if they have once admitted a doubt, it goeth ever after authorised for a doubt. But that use of wit and knowledge is to be allowed, which laboureth to make doubtful things certain, and not those which labour to make certain things doubtful. Therefore these calendars of doubts I commend as excellent things; so that there be this caution used, that when they be thoroughly sifted and brought to resolution, they be from thenceforth omitted, discarded, and not continued to cherish and encourage men in doubting. To which calendar of doubts or problems I advise be annexed another calendar, as much or more material which is a calendar of popular errors: I mean chiefly in natural history, such as pass in speech and conceit, and are nevertheless apparently detected and convicted of untruth, that man's knowledge be not weakened nor embased by such dross and vanity. As for the doubts or non liquets general or in total, I understand those differences of opinions touching the principles of nature, and the fundamental points of the same, which have caused the diversity of sects, schools, and philosophies, as that of Empedocles, Pythagoras, Democritus, Parmenides, and the rest. For although Aristotle, as though he had been of the race of the Ottomans, thought he could not reign except the first thing he did he killed all his brethren; yet to those that seek truth and not magistrality, it cannot but seem a matter of great profit, to see before them the several opinions touching the foundations of nature. Not for any exact truth that can be expected in those theories; for as the same phenomena in astronomy are satisfied by this received astronomy of the diurnal motion, and the proper motions of the planets, with their eccentrics and epicycles, and likewise by the theory of Copernicus, who supposed the earth to move, and the calculations are indifferently agreeable to both, so the ordinary face and view of experience is many times satisfied by several theories and philosophies; whereas to find the real truth requireth another manner of severity and attention. For as Aristotle saith, that children at the first will call every woman mother, but afterward they come to distinguish according to truth, so experience, if it be in childhood, will call every philosophy mother, but when it cometh to ripeness it will discern the true mother. So as in the meantime it is good to see the several glosses and opinions upon Nature, whereof it may be everyone in some one point hath seen clearer than his fellows, therefore I wish some collection to be made painfully and understandingly de antiquis philosophiis, out of all the possible light which remaineth to us of them: which kind of work I find deficient. But here I must give warning, that it be done distinctly and severedly; the philosophies of everyone throughout by themselves, and not by titles packed and faggoted up together, as hath been done by Plutarch. For it is the harmony of a philosophy in itself, which giveth it light and credence; whereas if it be singled and broken, it will seem more foreign and dissonant. For as when I read in Tacitus the actions of Nero or Claudius, with circumstances of times, inducements, and occasions, I find them not so strange; but when I read them in Suetonius Tranquillus, gathered into titles and bundles and not in order of time, they seem more monstrous and incredible: so is it of any philosophy reported entire, and dismembered by articles. Neither do I exclude opinions of latter times to be likewise represented in this calendar of sects of philosophy, as that of Theophrastus Paracelsus, eloquently reduced into an harmony by the pen of Severinus the Dane; and that of Tilesius, and his scholar Donius, being as a pastoral philosophy, full of sense, but of no great depth; and that of Fracastorius, who, though he pretended not to make any new philosophy, yet did use the absoluteness of his own sense upon the old; and that of Gilbertus our countryman, who revived, with some alterations and demonstrations, the opinions of Xenophanes; and any other worthy to be admitted.

(6) Thus have we now dealt with two of the three beams of man's knowledge; that is radius directus, which is referred to nature, radius refractus, which is referred to God, and cannot report truly because of the inequality of the medium. There resteth radius reflexus, whereby man beholdeth and contemplateth himself.

IX

(1) We come therefore now to that knowledge whereunto the ancient oracle directeth us, which is the knowledge of ourselves; which deserveth the more accurate handling, by how much it toucheth us more nearly. This knowledge, as it is the end and term of natural philosophy in the intention of man, so notwithstanding it is but a portion of natural philosophy in the continent of Nature. And generally let this be a rule, that all partitions of knowledges be accepted rather for lines and veins than for sections and separations; and that the continuance and entireness of knowledge be preserved. For the contrary hereof hath made particular sciences to become barren, shallow, and erroneous, while they have not been nourished and maintained from the common fountain. So we see Cicero, the orator, complained of Socrates and his school, that he was the first that separated philosophy and rhetoric; whereupon rhetoric became an empty and verbal art. So we may see that the opinion of Copernicus, touching the rotation of the earth, which astronomy itself cannot correct, because it is not repugnant to any of the phenomena, yet natural philosophy may correct. So we see also that the science of medicine if it be destituted and forsaken by natural philosophy, it is not much better than an empirical practice. With this reservation, therefore, we proceed to human philosophy or humanity, which hath two parts: the one considereth man segregate or distributively, the other congregate or in society; so as human philosophy is either simple and particular, or conjugate and civil. Humanity particular consisteth of the same parts whereof man consisteth: that is, of knowledges which respect the body, and of knowledges that respect the mind. But before we distribute so far, it is good to constitute. For I do take the consideration in general, and at large, of human nature to be fit to be emancipate and made a knowledge by itself, not so much in regard of those delightful and elegant discourses which have been made of the dignity of man, of his miseries, of his state and life, and the like adjuncts of his common and undivided nature; but chiefly in regard of the knowledge concerning the sympathies and concordances between the mind and body, which being mixed cannot be properly assigned to the sciences of either.

(2) This knowledge hath two branches: for as all leagues and amities consist of mutual intelligence and mutual offices, so this league of mind and body hath these two parts: how the one discloseth the other, and how the one worketh upon the other; discovery and impression. The former of these hath begotten two arts, both of prediction or prenotion; whereof the one is honoured with the inquiry of Aristotle, and the other of Hippocrates. And although they have of later time been used to be coupled with superstitions and fantastical arts, yet being purged and restored to their true state, they have both of them a solid ground in Nature, and a profitable use in life. The first is physiognomy, which discovereth the disposition of the mind by the lineaments of the body. The second is the exposition of natural dreams, which discovereth the state of the body by the imaginations of the mind. In the former of these I note a deficience. For Aristotle hath very ingeniously and diligently handled the factures of the body, but not the gestures of the body, which are no less comprehensible by art, and of greater use and advantage. For the lineaments of the body do disclose the disposition and inclination of the mind in general; but the motions of the countenance and parts do not only so, but do further disclose the present humour and state of the mind and will. For as your majesty saith most aptly and elegantly, "As the tongue speaketh to the ear so the gesture speaketh to the eye." And, therefore, a number of subtle persons, whose eyes do dwell upon the faces and fashions of men, do well know the advantage of this observation, as being most part of their ability; neither can it be denied, but that it is a great discovery of dissimulations, and a great direction in business.

(3) The latter branch, touching impression, hath not been collected into art, but hath been handled dispersedly; and it hath the same relation or antistrophe that the former hath. For the consideration is double—either how and how far the humours and affects of the body do alter or work upon the mind, or, again, how and how far the passions or apprehensions of the mind do alter or work upon the body. The former of these hath been inquired and considered as a part and appendix of medicine, but much more as a part of religion or superstition. For the physician prescribeth cures of the mind in frenzies and melancholy passions, and pretendeth also to exhibit medicines to exhilarate the mind, to control the courage, to clarify the wits, to corroborate the memory, and the like; but the scruples and superstitions of diet and other regiment of the body in the sect of the Pythagoreans, in the heresy of the Manichees, and in the law of Mahomet, do exceed. So likewise the ordinances in the ceremonial law, interdicting the eating of the blood and the fat, distinguishing between beasts clean and unclean for meat, are many and strict; nay, the faith itself being clear and serene from all clouds of ceremony, yet retaineth the use of fastlings, abstinences, and other macerations and humiliations of the body, as things real, and not figurative. The root and life of all which prescripts is (besides the ceremony) the consideration of that dependency which the affections of the mind are submitted unto upon the state and disposition of the body. And if any man of weak judgment do conceive that this suffering of the mind from the body doth either question the immortality, or derogate from the sovereignty of the soul, he may be taught, in easy

instances, that the infant in the mother's womb is compatible with the mother, and yet separable; and the most absolute monarch is sometimes led by his servants, and yet without subjection. As for the reciprocal knowledge, which is the operation of the conceits and passions of the mind upon the body, we see all wise physicians, in the prescriptions of their regiments to their patients, do ever consider accidentia animi, as of great force to further or hinder remedies or recoveries: and more specially it is an inquiry of great depth and worth concerning imagination, how and how far it altereth the body proper of the imaginant; for although it hath a manifest power to hurt, it followeth not it hath the same degree of power to help. No more than a man can conclude, that because there be pestilent airs, able suddenly to kill a man in health, therefore there should be sovereign airs, able suddenly to cure a man in sickness. But the inquisition of this part is of great use, though it needeth, as Socrates said, "a Delian diver," being difficult and profound. But unto all this knowledge de communi vinculo, of the concordances between the mind and the body, that part of inquiry is most necessary which considereth of the seats and domiciles which the several faculties of the mind do take and occupate in the organs of the body; which knowledge hath been attempted, and is controverted, and deserveth to be much better inquired. For the opinion of Plato, who placed the understanding in the brain, animosity (which he did unfitly call anger, having a greater mixture with pride) in the heart, and concupiscence or sensuality in the liver, deserveth not to be despised, but much less to be allowed. So, then, we have constituted (as in our own wish and advice) the inquiry touching human nature entire, as a just portion of knowledge to be handled apart.

X

(1) The knowledge that concerneth man's body is divided as the good of man's body is divided, unto which it referreth. The good of man's body is of four kinds—health, beauty, strength, and pleasure: so the knowledges are medicine, or art of cure; art of decoration, which is called cosmetic; art of activity, which is called athletic; and art voluptuary, which Tacitus truly calleth eruditus luxus. This subject of man's body is, of all other things in nature, most susceptible of remedy; but then that remedy is most susceptible of error; for the same subtlety of the subject doth cause large possibility and easy failing, and therefore the inquiry ought to be the more exact.

(2) To speak, therefore, of medicine, and to resume that we have said, ascending a little higher: the ancient opinion that man was microcosmus—an abstract or model of the world—hath been fantastically strained by Paracelsus and the alchemists, as if there were to be found in man's body certain correspondences and parallels, which should have respect to all varieties of things, as stars, planets, minerals, which are extant in the great world. But thus much is evidently true, that of all substances which nature hath produced, man's body is the most extremely compounded. For we see herbs and plants are nourished by earth and water; beasts for the most part by herbs and fruits; man by the flesh of beasts, birds, fishes, herbs, grains, fruits, water, and the manifold alterations, dressings, and preparations of these several bodies before they come to be his food and aliment. Add hereunto that beasts have a more simple order of life, and less change of affections to work upon their bodies, whereas man in his mansion, sleep, exercise, passions, hath infinite variations: and it cannot be denied but that the body of man of all other things is of the most compounded mass. The soul, on the other side, is the simplest of substances, as is well expressed:

> *"Purumque reliquit*
> *Æthereum sensum àtque auraï simplicis ignem."*

So that it is no marvel though the soul so placed enjoy no rest, if that principle be true, that Motus rerum est rapidus extra locum, placidus in loco. But to the purpose. This variable composition of man's body hath made it as an instrument easy to distemper; and, therefore, the poets did well to conjoin music and medicine in Apollo, because the office of medicine is but to tune this curious harp of man's body and to reduce it to harmony. So, then, the subject being so variable hath made the art by consequent more conjectural; and the art being conjectural hath made so much the more place to be left for imposture. For almost all other arts and sciences are judged by acts or masterpieces, as I may term them, and not by the successes and events. The lawyer is judged by the virtue of his pleading, and not by the issue of the cause; this master in this ship is judged by the directing his course aright, and not by the fortune of the voyage; but the physician, and perhaps this politique, hath no particular acts demonstrative of his ability, but is judged most by the event, which is ever but as it is taken: for who can tell, if a patient die or recover, or if a state be preserved or ruined, whether it be art or accident? And therefore many times the impostor is prized, and the man of virtue taxed. Nay, we see [the] weakness and credulity of men is such, as they will often refer a mountebank or witch before a learned physician. And therefore the poets were clear-sighted in discerning this extreme folly when they made Æsculapius and Circe, brother and sister, both children of the sun, as in the verses—

"Ipse repertorem medicinæ talis et artis
Fulmine Phœbigenam Stygias detrusit ad undas."
And again—

"Dives inaccessos ubi Solis filia lucos," &c.

For in all times, in the opinion of the multitude, witches and old women and impostors, have had a competition with physicians. And what followeth? Even this, that physicians say to themselves, as Solomon expresseth it.upon a higher occasion, "If it befall to me as befalleth to the fools, why should I labour to be more wise?" And therefore I cannot much blame physicians that they use commonly to intend some other art or practice, which they fancy more than their profession; for you shall have of them antiquaries, poets, humanists, statesmen, merchants, divines, and in every of these better seen than in their profession; and no doubt upon this ground that they find that mediocrity and excellency in their art maketh no difference in profit or reputation towards their fortune: for the weakness of patients, and sweetness of life, and nature of hope, maketh men depend upon physicians with all their defects. But, nevertheless, these things which we have spoken of are courses begotten between a little occasion and a great deal of sloth and default; for if we will excite and awake our observation, we shall see in familiar instances what a predominant faculty the subtlety of spirit hath over the variety of matter or form. Nothing more variable than faces and countenances, yet men can bear in memory the infinite distinctions of them; nay, a painter, with a few shells of colours, and the benefit of his eye, and habit of his imagination, can imitate them all that ever have been, are, or may be, if they were brought before him. Nothing more variable than voices, yet men can likewise discern them personally: nay, you shall have a buffon or pantomimus will express as many as he pleaseth. Nothing more variable than the differing sounds of words; yet men have found the way to reduce them to a few simple letters. So that it is not the insufficiency or incapacity of man's mind, but it is the remote standing or placing thereof that breedeth these mazes and incomprehensions; for as the sense afar off is full of mistaking, but is exact at hand, so is it of the understanding, the remedy whereof is, not to quicken or strengthen the organ, but to go nearer to the object; and therefore there is no doubt but if the physicians will learn and use the true approaches and avenues of nature, they may assume as much as the poet saith:

"Et quoniam variant morbi, variabimus artes;
Mille mali species, mille salutis erunt."

Which that they should do, the nobleness of their art doth deserve: well shadowed by the poets, in that they made Æsculapius to be the son of [the] sun, the one being the fountain of life, the other as the second-stream; but infinitely more honoured by the example of our Saviour, who made the body of man the object of His miracles, as the soul was the object of His doctrine. For we read not that ever He vouchsafed to do any miracle about honour or money (except that one for giving tribute to Cæsar), but only about the preserving, sustaining, and healing the body of man.

(3) Medicine is a science which hath been (as we have said) more professed than laboured, and yet more laboured than advanced; the labour having been, in my judgment, rather in circle than in progression. For I find much iteration, but small addition. It considereth causes of diseases, with the occasions or impulsions; the diseases themselves, with the accidents; and the cures, with the preservations. The deficiences which I think good to note, being a few of many, and those such as are of a more open and manifest nature, I will enumerate and not place.

(4) The first is the discontinuance of the ancient and serious diligence of Hippocrates, which used to set down a narrative of the special cases of his patients, and how they proceeded, and how they were judged by recovery or death. Therefore having an example proper in the father of the art, I shall not need to allege an example foreign, of the wisdom of the lawyers, who are careful to report new cases and decisions, for the direction of future judgments. This continuance of medicinal history I find deficient; which I understand neither to be so infinite as to extend to every common case, nor so reserved as to admit none but wonders: for many things are new in this manner, which are not new in the kind; and if men will intend to observe, they shall find much worthy to observe.

(5) In the inquiry which is made by anatomy, I find much deficience: for they inquire of the parts, and their substances, figures, and collocations; but they inquire not of the diversities of the parts, the secrecies of the passages, and the seats or nestling of the humours, nor much of the footsteps and impressions of diseases. The reason of which omission I suppose to be, because the first inquiry may be satisfied in the view of one or a few anatomies; but the latter, being comparative and casual, must arise from the view of many. And as to the diversity of parts, there is no doubt but the facture or framing of the inward parts is as full of difference as the outward, and in that is the cause continent of many diseases; which not being observed, they quarrel many times with the humours, which are not in fault;

the fault being in the very frame and mechanic of the part, which cannot be removed by medicine alterative, but must be accommodated and palliated by diets and medicines familiar. And for the passages and pores, it is true which was anciently noted, that the more subtle of them appear not in anatomies, because they are shut and latent in dead bodies, though they be open and manifest in life: which being supposed, though the inhumanity of anatomia vivorum was by Celsus justly reproved; yet in regard of the great use of this observation, the inquiry needed not by him so slightly to have been relinquished altogether, or referred to the casual practices of surgery; but might have been well diverted upon the dissection of beasts alive, which notwithstanding the dissimilitude of their parts may sufficiently satisfy this inquiry. And for the humours, they are commonly passed over in anatomies as purgaments; whereas it is most necessary to observe, what cavities, nests, and receptacles the humours do find in the parts, with the differing kind of the humour so lodged and received. And as for the footsteps of diseases, and their devastations of the inward parts, impostumations, exulcerations, discontinuations, putrefactions, consumptions, contractions, extensions, convulsions, dislocations, obstructions, repletions, together with all preternatural substances, as stones, carnosities, excrescences, worms, and the like; they ought to have been exactly observed by multitude of anatomies, and the contribution of men's several experiences, and carefully set down both historically according to the appearances, and artificially with a reference to the diseases and symptoms which resulted from them, in case where the anatomy is of a defunct patient; whereas now upon opening of bodies they are passed over slightly and in silence.

(6) In the inquiry of diseases, they do abandon the cures of many, some as in their nature incurable, and others as past the period of cure; so that Sylla and the Triumvirs never proscribed so many men to die, as they do by their ignorant edicts: whereof numbers do escape with less difficulty than they did in the Roman prescriptions. Therefore I will not doubt to note as a deficience, that they inquire not the perfect cures of many diseases, or extremities of diseases; but pronouncing them incurable do enact a law of neglect, and exempt ignorance from discredit.

(7) Nay further, I esteem it the office of a physician not only to restore health, but to mitigate pain and dolors; and not only when such mitigation may conduce to recovery, but when it may serve to make a fair and easy passage. For it is no small felicity which Augustus Cæsar was wont to wish to himself, that same Euthanasia; and which was specially noted in the death of Antoninus Pius, whose death was after the fashion, and semblance of a kindly and pleasant sleep. So it is written of Epicurus, that after his disease was judged desperate, he drowned his stomach and senses with a large draught and ingurgitation of wine; whereupon the epigram was made, Hinc Stygias ebrius hausit aquas; he was not sober enough to taste any bitterness of the Stygian water. But the physicians contrariwise do make a kind of scruple and religion to stay with the patient after the disease is deplored; whereas in my judgment they ought both to inquire the skill, and to give the attendances, for the facilitating and assuaging of the pains and agonies of death.

(5) In the consideration of the cures of diseases, I find a deficience in the receipts of propriety, respecting the particular cures of diseases: for the physicians have frustrated the fruit of tradition and experience by their magistralities, in adding and taking out and changing quid pro qua in their receipts, at their pleasures; commanding so over the medicine, as the medicine cannot command over the disease. For except it be treacle and mithridatum, and of late diascordium, and a few more, they tie themselves to no receipts severely and religiously. For as to the confections of sale which are in the shops, they are for readiness and not for propriety. For they are upon general intentions of purging, opening, comforting, altering, and not much appropriate to particular diseases. And this is the cause why empirics and old women àre more happy many times in their cures than learned physicians, because they are more religious in holding their medicines. Therefore here is the deficience which I find, that physicians have not, partly out of their own practice, partly out of the constant probations reported in books, and partly out of the traditions of empirics, set down and delivered over certain experimental medicines for the cure of particular diseases, besides their own conjectural and magistral descriptions. For as they were the men of the best composition in the state of Rome, which either being consuls inclined to the people, or being tribunes inclined to the senate; so in the matter we now handle, they be the best physicians, which being learned incline to the traditions of experience, or being empirics incline to the methods of learning.

(9) In preparation of medicines I do find strange, specially considering how mineral medicines have been extolled, and that they are safer for the outward than inward parts, that no man hath sought to make an imitation by art of natural baths and medicinable fountains: which nevertheless are confessed to receive their virtues from minerals; and not so only, but discerned and distinguished from what particular mineral they receive tincture, as sulphur, vitriol, steel, or the like; which nature, if it may be reduced to compositions of art, both the variety of them will be increased, and the temper of them will be more commanded.

(10) But lest I grow to be more particular than is agreeable either to my intention or to proportion, I will conclude this part with the note of one deficience more, which seemeth to me of greatest

Francis Bacon

consequence: which is, that the prescripts in use are too compendious to attain their end; for, to my understanding, it is a vain and flattering opinion to think any medicine can be so sovereign or so happy, as that the receipt or miss of it can work any great effect upon the body of man. It were a strange speech which spoken, or spoken oft, should reclaim a man from a vice to which he were by nature subject. It is order, pursuit, sequence, and interchange of application, which is mighty in nature; which although it require more exact knowledge in prescribing, and more precise obedience in observing, yet is recompensed with the magnitude of effects. And although a man would think, by the daily visitations of the physicians, that there were a pursuance in the cure, yet let a man look into their prescripts and ministrations, and he shall find them but inconstancies and every day's devices, without any settled providence or project. Not that every scrupulous or superstitious prescript is effectual, no more than every straight way is the way to heaven; but the truth of the direction must precede severity of observance.

(11) For cosmetic, it hath parts civil, and parts effeminate: for cleanness of body was ever esteemed to proceed from a due reverence to God, to society, and to ourselves. As for artificial decoration, it is well worthy of the deficiences which it hath; being neither fine enough to deceive, nor handsome to use, nor wholesome to please.

(12) For athletic, I take the subject of it largely, that is to say, for any point of ability whereunto the body of man may be brought, whether it be of activity, or of patience; whereof activity hath two parts, strength and swiftness; and patience likewise hath two parts, hardness against wants and extremities, and endurance of pain or torment; whereof we see the practices in tumblers, in savages, and in those that suffer punishment. Nay, if there be any other faculty which falls not within any of the former divisions, as in those that dive, that obtain a strange power of containing respiration, and the like, I refer it to this part. Of these things the practices are known, but the philosophy that concerneth them is not much inquired; the rather, I think, because they are supposed to be obtained, either by an aptness of nature, which cannot be taught, or only by continual custom, which is soon prescribed which though it be not true, yet I forbear to note any deficiences; for the Olympian games are down long since, and the mediocrity of these things is for use; as for the excellency of them it serveth for the most part but for mercenary ostentation.

(13) For arts of pleasure sensual, the chief deficience in them is of laws to repress them. For as it hath been well observed, that the arts which flourish in times while virtue is in growth, are military; and while virtue is in state, are liberal; and while virtue is in declination, are voluptuary: so I doubt that this age of the world is somewhat upon the descent of the wheel. With arts voluptuary I couple practices joculary; for the deceiving of the senses is one of the pleasures of the senses. As for games of recreation, I hold them to belong to civil life and education. And thus much of that particular human philosophy which concerns the body, which is but the tabernacle of the mind.

XI

(1) For human knowledge which concerns the mind, it hath two parts; the one that inquireth of the substance or nature of the soul or mind, the other that inquireth of the faculties or functions thereof. Unto the first of these, the considerations of the original of the soul, whether it be native or adventive, and how far it is exempted from laws of matter, and of the immortality thereof, and many other points, do appertain: which have been not more laboriously inquired than variously reported; so as the travail therein taken seemeth to have been rather in a maze than in a way. But although I am of opinion that this knowledge may be more really and soundly inquired, even in nature, than it hath been, yet I hold that in the end it must be bounded by religion, or else it will be subject to deceit and delusion. For as the substance of the soul in the creation was not extracted out of the mass of heaven and earth by the benediction of a producat, but was immediately inspired from God, so it is not possible that it should be (otherwise than by accident) subject to the laws of heaven and earth, which are the subject of philosophy; and therefore the true knowledge of the nature and state of the soul must come by the same inspiration that gave the substance. Unto this part of knowledge touching the soul there be two appendices; which, as they have been handled, have rather vapoured forth fables than kindled truth: divination and fascination.

(2) Divination hath been anciently and fitly divided into artificial and natural: whereof artificial is, when the mind maketh a prediction by argument, concluding upon signs and tokens; natural is, when the mind hath a presention by an internal power, without the inducement of a sign. Artificial is of two sorts: either when the argument is coupled with a derivation of causes, which is rational; or when it is only grounded upon a coincidence of the effect, which is experimental: whereof the latter for the most part is superstitious, such as were the heathen observations upon the inspection of sacrifices, the flights of birds, the swarming of bees; and such as was the Chaldean astrology, and the like. For artificial divination, the several kinds thereof are distributed amongst particular knowledges. The astronomer

hath his predictions, as of conjunctions, aspects, eclipses, and the like. The physician hath his predictions, of death, of recovery, of the accidents and issues of diseases. The politique hath his predictions; O urbem venalem, et cito perituram, si emptorem invenerit! which stayed not long to be performed, in Sylla first, and after in Cæsar: so as these predictions are now impertinent, and to be referred over. But the divination which springeth from the internal nature of the soul is that which we now speak of; which hath been made to be of two sorts, primitive and by influxion. Primitive is grounded upon the supposition that the mind, when it is withdrawn and collected into itself, and not diffused into the organs of the body, hath some extent and latitude of prenotion; which therefore appeareth most in sleep, in ecstasies, and near death, and more rarely in waking apprehensions; and is induced and furthered by those abstinences and observances which make the mind most to consist in itself. By influxion, is grounded upon the conceit that the mind, as a mirror or glass, should take illumination from the foreknowledge of God and spirits: unto which the same regiment doth likewise conduce. For the retiring of the mind within itself is the state which is most susceptible of divine influxions; save that it is accompanied in this case with a fervency and elevation (which the ancients noted by fury), and not with a repose and quiet, as it is in the other.

(3) Fascination is the power and act of imagination intensive upon other bodies than the body of the imaginant, for of that we spake in the proper place. Wherein the school of Paracelsus, and the disciples of pretended natural magic, have been so intemperate, as they have exalted the power of the imagination to be much one with the power of miracle-working faith. Others, that draw nearer to probability, calling to their view the secret passages of things, and specially of the contagion that passeth from body to body, do conceive it should likewise be agreeable to nature that there should be some transmissions and operations from spirit to spirit without the mediation of the senses; whence the conceits have grown (now almost made civil) of the mastering spirit, and the force of confidence, and the like. Incident unto this is the inquiry how to raise and fortify the imagination; for if the imagination fortified have power, then it is material to know how to fortify and exalt it. And herein comes in crookedly and dangerously a palliation of a great part of ceremonial magic. For it may be pretended that ceremonies, characters, and charms do work, not by any tacit or sacramental contract with evil spirits, but serve only to strengthen the imagination of him that useth it; as images are said by the Roman Church to fix the cogitations and raise the devotions of them that pray before them. But for mine own judgment, if it be admitted that imagination hath power, and that ceremonies fortify imagination, and that they be used sincerely and intentionally for that purpose; yet I should hold them unlawful, as opposing to that first edict which God gave unto man, In sudore vultus comedes panem tuum. For they propound those noble effects, which God hath set forth unto man to be bought at the price of labour, to be attained by a few easy and slothful observances. Deficiences in these knowledges I will report none, other than the general deficience, that it is not known how much of them is verity, and how much vanity.

XII

(1) The knowledge which respecteth the faculties of the mind of man is of two kinds—the one respecting his understanding and reason, and the other his will, appetite, and affection; whereof the former produceth position or decree, the latter action or execution. It is true that the imagination is an agent or nuncius in both provinces, both the judicial and the ministerial. For sense sendeth over to imagination before reason have judged, and reason sendeth over to imagination before the decree can be acted. For imagination ever precedeth voluntary motion. Saving that this Janus of imagination hath differing faces: for the face towards reason hath the print of truth, but the face towards action hath the print of good; which nevertheless are faces,

"Quales decet esse sororum."

Neither is the imagination simply and only a messenger; but is invested with, or at least wise usurpeth no small authority in itself, besides the duty of the message. For it was well said by Aristotle, "That the mind hath over the body that commandment, which the lord hath over a bondman; but that reason hath over the imagination that commandment which a magistrate hath over a free citizen," who may come also to rule in his turn. For we see that, in matters of faith and religion, we raise our imagination above our reason, which is the cause why religion sought ever access to the mind by similitudes, types, parables, visions, dreams. And again, in all persuasions that are wrought by eloquence, and other impressions of like nature, which do paint and disguise the true appearance of things, the chief recommendation unto reason is from the imagination. Nevertheless, because I find not any science that doth properly or fitly pertain to the imagination, I see no cause to alter the former division. For as for poesy, it is rather a pleasure or play of imagination than a work or duty thereof.

And if it be a work, we speak not now of such parts of learning as the imagination produceth, but of such sciences as handle and consider of the imagination. No more than we shall speak now of such knowledges as reason produceth (for that extendeth to all philosophy), but of such knowledges as do handle and inquire of the faculty of reason: so as poesy had his true place. As for the power of the imagination in nature, and the manner of fortifying the same, we have mentioned it in the doctrine De Anima, whereunto most fitly it belongeth. And lastly, for imaginative or insinuative reason, which is the subject of rhetoric, we think it best to refer it to the arts of reason. So therefore we content ourselves with the former division, that human philosophy, which respecteth the faculties of the mind of man, hath two parts, rational and moral.

(2) The part of human philosophy which is rational is of all knowledges, to the most wits, the least delightful, and seemeth but a net of subtlety and spinosity. For as it was truly said, that knowledge is pabulum animi; so in the nature of men's appetite to this food most men are of the taste and stomach of the Israelites in the desert, that would fain have returned ad ollas carnium, and were weary of manna; which, though it were celestial, yet seemed less nutritive and comfortable. So generally men taste well knowledges that are drenched in flesh and blood, civil history, morality, policy, about the which men's affections, praises, fortunes do turn and are conversant. But this same lumen siccum doth parch and offend most men's watery and soft natures. But to speak truly of things as they are in worth, rational knowledges are the keys of all other arts, for as Aristotle saith aptly and elegantly, "That the hand is the instrument of instruments, and the mind is the form of forms;" so these be truly said to be the art of arts. Neither do they only direct, but likewise confirm and strengthen; even as the habit of shooting doth not only enable to shoot a nearer shoot, but also to draw a stronger bow.

(3) The arts intellectual are four in number, divided according to the ends whereunto they are referred—for man's labour is to invent that which is sought or propounded; or to judge that which is invented; or to retain that which is judged; or to deliver over that which is retained. So as the arts must be four—art of inquiry or invention; art of examination or judgment; art of custody or memory; and art of elocution or tradition.

XIII

(1) Invention is of two kinds much differing—the one of arts and sciences, and the other of speech and arguments. The former of these I do report deficient; which seemeth to me to be such a deficience as if, in the making of an inventory touching the state of a defunct, it should be set down that there is no ready money. For as money will fetch all other commodities, so this knowledge is that which should purchase all the rest. And like as the West Indies had never been discovered if the use of the mariner's needle had not been first discovered, though the one be vast regions, and the other a small motion; so it cannot be found strange if sciences be no further discovered, if the art itself of invention and discovery hath been passed over.

(2) That this part of knowledge is wanting, to my judgment standeth plainly confessed; for first, logic doth not pretend to invent sciences, or the axioms of sciences, but passeth it over with a cuique in sua arte credendum. And Celsus acknowledgeth it gravely, speaking of the empirical and dogmatical sects of physicians, "That medicines and cures were first found out, and then after the reasons and causes were discoursed; and not the causes first found out, and by light from them the medicines and cures discovered." And Plato in his "Theætetus" noteth well, "That particulars are infinite, and the higher generalities give no sufficient direction; and that the pith of all sciences, which maketh the artsman differ from the inexpert, is in the middle propositions, which in every particular knowledge are taken from tradition and experience." And therefore we see, that they which discourse of the inventions and originals of things refer them rather to chance than to art, and rather to beasts, birds, fishes, serpents, than to men.

> "Dictamnum genetrix Cretæa carpit ab Ida,
> Puberibus caulem foliis et flore camantem
> Purpureo; non illa feris incognita capris
> Gramina, cum tergo volucres hæsere sagittæ."

So that it was no marvel (the manner of antiquity being to consecrate inventors) that the Egyptians had so few human idols in their temples, but almost all brute:

> "Omnigenumque Deum monstra, et latrator Anubis,
> Contra Neptunum, et Venerem, contraque Minervam, &c."

And if you like better the tradition of the Grecians, and ascribe the first inventions to men, yet you will rather believe that Prometheus first stroke the flints, and marvelled at the spark, than that when he first stroke the flints he expected the spark; and therefore we see the West Indian Prometheus had no intelligence with the European, because of the rareness with them of flint, that gave the first occasion. So as it should seem, that hitherto men are rather beholden to a wild goat for surgery, or to a nightingale for music, or to the ibis for some part of physic, or to the pot-lid that flew open for artillery, or generally to chance or anything else than to logic for the invention of arts and sciences. Neither is the form of invention which Virgil describeth much other:

> *"Ut varias usus meditande extunderet artes*
> *Paulatim."*

For if you observe the words well, it is no other method than that which brute beasts are capable of, and do put in ure; which is a perpetual intending or practising some one thing, urged and imposed by an absolute necessity of conservation of being. For so Cicero saith very truly, Usus uni rei deditus et naturam et artem sæpe vincit. And therefore if it be said of men,

> *"Labor omnia vincit*
> *Improbus, et duris urgens in rebus egestas,"*

it is likewise said of beasts, Quis psittaco docuit suum χαιρε? Who taught the raven in a drought to throw pebbles into a hollow tree, where she spied water, that the water might rise so as she might come to it? Who taught the bee to sail through such a vast sea or air, and to find the way from a field in a flower a great way off to her hive? Who taught the ant to bite every grain of corn that she burieth in her hill, lest it should take root and grow? Add then the word extundere, which importeth the extreme difficulty, and the word paulatim, which importeth the extreme slowness, and we are where we were, even amongst the Egyptians' gods; there being little left to the faculty of reason, and nothing to the duty or art, for matter of invention.

(3) Secondly, the induction which the logicians speak of, and which seemeth familiar with Plato, whereby the principles of sciences may be pretended to be invented, and so the middle propositions by derivation from the principles; their form of induction, I say, is utterly vicious and incompetent; wherein their error is the fouler, because it is the duty of art to perfect and exalt nature; but they contrariwise have wronged, abused, and traduced nature. For he that shall attentively observe how the mind doth gather this excellent dew of knowledge, like unto that which the poet speaketh of, Aërei mellis cælestia dona, distilling and contriving it out of particulars natural and artificial, as the flowers of the field and garden, shall find that the mind of herself by nature doth manage and act an induction much better than they describe it. For to conclude upon an enumeration of particulars, without instance contradictory, is no conclusion, but a conjecture; for who can assure (in many subjects) upon those particulars which appear of a side, that there are not other on the contrary side which appear not? As if Samuel should have rested upon those sons of Jesse which were brought before him, and failed of David which was in the field. And this form (to say truth), is so gross, as it had not been possible for wits so subtle as have managed these things to have offered it to the world, but that they hasted to their theories and dogmaticals, and were imperious and scornful toward particulars; which their manner was to use but as lictores and viatores, for sergeants and whifflers, ad summovendam turbam, to make way and make room for their opinions, rather than in their true use and service. Certainly it is a thing may touch a man with a religious wonder, to see how the footsteps of seducement are the very same in divine and human truth; for, as in divine truth man cannot endure to become as a child, so in human, they reputed the attending the inductions (whereof we speak), as if it were a second infancy or childhood.

(4) Thirdly, allow some principles or axioms were rightly induced, yet, nevertheless, certain it is that middle propositions cannot be deduced from them in subject of nature by syllogism—that is, by touch and reduction of them to principles in a middle term. It is true that in sciences popular, as moralities, laws, and the like, yea, and divinity (because it pleaseth God to apply Himself to the capacity of the simplest), that form may have use; and in natural philosophy likewise, by way of argument or satisfactory reason, Quæ assensum parit operis effæta est; but the subtlety of nature and operations will not be enchained in those bonds. For arguments consist of propositions, and propositions of words, and words are but the current tokens or marks of popular notions of things; which notions, if they be grossly and variably collected out of particulars, it is not the laborious examination either of consequences of arguments, or of the truth of propositions, that can ever correct that error, being (as the physicians speak) in the first digestion. And, therefore, it was not without cause, that so many excellent philosophers became sceptics and academics, and denied any certainty of knowledge or comprehension; and held opinion that the knowledge of man extended only to appearances and probabilities. It is true that in Socrates it was supposed to be but a form of irony, Scientiam dissimulando simulavit; for he used to disable his knowledge, to the end to enhance his knowledge; like the humour of Tiberius in his

beginnings, that would reign, but would not acknowledge so much. And in the later academy, which Cicero embraced, this opinion also of acatalepsia (I doubt) was not held sincerely; for that all those which excelled in copy of speech seem to have chosen that sect, as that which was fittest to give glory to their eloquence and variable discourses; being rather like progresses of pleasure than journeys to an end. But assuredly many scattered in both academies did hold it in subtlety and integrity. But here was their chief error: they charged the deceit upon the senses; which in my judgment (notwithstanding all their cavillations) are very sufficient to certify and report truth, though not always immediately, yet by comparison, by help of instrument, and by producing and urging such things as are too subtle for the sense to some effect comprehensible by the sense, and other like assistance. But they ought to have charged the deceit upon the weakness of the intellectual powers, and upon the manner of collecting and concluding upon the reports of the senses. This I speak, not to disable the mind of man, but to stir it up to seek help; for no man, be he never so cunning or practised, can make a straight line or perfect circle by steadiness of hand, which may be easily done by help of a ruler or compass.

(5) This part of invention, concerning the invention of sciences, I purpose (if God give me leave) hereafter to propound, having digested it into two parts: whereof the one I term experientia literata, and the other interpretatio naturæ; the former being but a degree and rudiment of the latter. But I will not dwell too long, nor speak too great upon a promise.

(6) The invention of speech or argument is not properly an invention; for to invent is to discover that we know not, and not to recover or resummon that which we already know; and the use of this invention is no other but, out of the knowledge whereof our mind is already possessed to draw forth or call before us that which may be pertinent to the purpose which we take into our consideration. So as to speak truly, it is no invention, but a remembrance or suggestion, with an application; which is the cause why the schools do place it after judgment, as subsequent and not precedent. Nevertheless, because we do account it a chase as well of deer in an enclosed park as in a forest at large, and that it hath already obtained the name, let it be called invention; so as it be perceived and discerned, that the scope and end of this invention is readiness and present use of our knowledge, and not addition or amplification thereof.

(7) To procure this ready use of knowledge there are two courses, preparation and suggestion. The former of these seemeth scarcely a part of knowledge, consisting rather of diligence than of any artificial erudition. And herein Aristotle wittily, but hurtfully, doth deride the sophists near his time, saying, "They did as if one that professed the art of shoemaking should not teach how to make up a shoe, but only exhibit in a readiness a number of shoes of all fashions and sizes." But yet a man might reply, that if a shoemaker should have no shoes in his shop, but only work as he is bespoken, he should be weakly customed. But our Saviour, speaking of divine knowledge, saith, "That the kingdom of heaven is like a good householder, that bringeth forth both new and old store;" and we see the ancient writers of rhetoric do give it in precept, that pleaders should have the places, whereof they have most continual use, ready handled in all the variety that may be; as that, to speak for the literal interpretation of the law against equity, and contrary; and to speak for presumptions and inferences against testimony, and contrary. And Cicero himself, being broken unto it by great experience, delivereth it plainly, that whatsoever a man shall have occasion to speak of (if he will take the pains), he may have it in effect premeditate and handled in thesi. So that when he cometh to a particular he shall have nothing to do, but to put to names, and times, and places, and such other circumstances of individuals. We see likewise the exact diligence of Demosthenes; who, in regard of the great force that the entrance and access into causes hath to make a good impression, had ready framed a number of prefaces for orations and speeches. All which authorities and precedents may overweigh Aristotle's opinion, that would have us change a rich wardrobe for a pair of shears.

(8) But the nature of the collection of this provision or preparatory store, though it be common both to logic and rhetoric, yet having made an entry of it here, where it came first to be spoken of, I think fit to refer over the further handling of it to rhetoric.

(9) The other part of invention, which I term suggestion, doth assign and direct us to certain marks, or places, which may excite our mind to return and produce such knowledge as it hath formerly collected, to the end we may make use thereof. Neither is this use (truly taken) only to furnish argument to dispute, probably with others, but likewise to minister unto our judgment to conclude aright within ourselves. Neither may these places serve only to apprompt our invention, but also to direct our inquiry. For a faculty of wise interrogating is half a knowledge. For as Plato saith, "Whosoever seeketh, knoweth that which he seeketh for in a general notion; else how shall he know it when he hath found it?" And, therefore, the larger your anticipation is, the more direct and compendious is your search. But the same places which will help us what to produce of that which we know already, will also help us, if a man of experience were before us, what questions to ask; or, if we have books and authors to instruct us, what points to search and revolve; so as I cannot report that this part of invention, which is that which the schools call topics, is deficient.

(10) Nevertheless, topics are of two sorts, general and special. The general we have spoken to; but

the particular hath been touched by some, but rejected generally as inartificial and variable. But leaving the humour which hath reigned too much in the schools (which is, to be vainly subtle in a few things which are within their command, and to reject the rest), I do receive particular topics; that is, places or directions of invention and inquiry in every particular knowledge, as things of great use, being mixtures of logic with the matter of sciences. For in these it holdeth ars inveniendi adolescit cum inventis; for as in going of a way, we do not only gain that part of the way which is passed, but we gain the better sight of that part of the way which remaineth, so every degree of proceeding in a science giveth a light to that which followeth; which light, if we strengthen by drawing it forth into questions or places of inquiry, we do greatly advance our pursuit.

XIV

(1) Now we pass unto the arts of judgment, which handle the natures of proofs and demonstrations, which as to induction hath a coincidence with invention; for all inductions, whether in good or vicious form, the same action of the mind which inventeth, judgeth—all one as in the sense. But otherwise it is in proof by syllogism, for the proof being not immediate, but by mean, the invention of the mean is one thing, and the judgment of the consequence is another; the one exciting only, the other examining. Therefore, for the real and exact form of judgment, we refer ourselves to that which we have spoken of interpretation of Nature.

(2) For the other judgment by syllogism, as it is a thing most agreeable to the mind of man, so it hath been vehemently end excellently laboured. For the nature of man doth extremely covet to have somewhat in his understanding fixed and unmovable, and as a rest and support of the mind. And, therefore, as Aristotle endeavoureth to prove, that in all motion there is some point quiescent; and as he elegantly expoundeth the ancient fable of Atlas (that stood fixed, and bare up the heaven from falling) to be meant of the poles or axle-tree of heaven, whereupon the conversion is accomplished, so assuredly men have a desire to have an Atlas or axle-tree within to keep them from fluctuation, which is like to a perpetual peril of falling. Therefore men did hasten to set down some principles about which the variety of their disputatious might turn.

(3) So, then, this art of judgment is but the reduction of propositions to principles in a middle term. The principles to be agreed by all and exempted from argument; the middle term to be elected at the liberty of every man's invention; the reduction to be of two kinds, direct and inverted: the one when the proposition is reduced to the principle, which they term a probation ostensive; the other, when the contradictory of the proposition is reduced to the contradictory of the principle, which is that which they call per incommodum, or pressing an absurdity; the number of middle terms to be as the proposition standeth degrees more or less removed from the principle.

(4) But this art hath two several methods of doctrine, the one by way of direction, the other by way of caution: the former frameth and setteth down a true form of consequence, by the variations and deflections from which errors and inconsequences may be exactly judged. Toward the composition and structure of which form it is incident to handle the parts thereof, which are propositions, and the parts of propositions, which are simple words. And this is that part of logic which is comprehended in the Analytics.

(5) The second method of doctrine was introduced for expedite use and assurance sake, discovering the more subtle forms of sophisms and illaqueations with their redargutions, which is that which is termed elenches. For although in the more gross sorts of fallacies it happeneth (as Seneca maketh the comparison well) as in juggling feats, which, though we know not how they are done, yet we know well it is not as it seemeth to be; yet the more subtle sort of them doth not only put a man besides his answer, but doth many times abuse his judgment.

(6) This part concerning elenches is excellently handled by Aristotle in precept, but more excellently by Plato in example; not only in the persons of the sophists, but even in Socrates himself, who, professing to affirm nothing, but to infirm that which was affirmed by another, hath exactly expressed all the forms of objection, fallace, and redargution. And although we have said that the use of this doctrine is for redargution, yet it is manifest the degenerate and corrupt use is for caption and contradiction, which passeth for a great faculty, and no doubt is of very great advantage, though the difference be good which was made between orators and sophisters, that the one is as the greyhound, which hath his advantage in the race, and the other as the hare, which hath her advantage in the turn, so as it is the advantage of the weaker creature.

(7) But yet further, this doctrine of elenches hath a more ample latitude and extent than is perceived; namely, unto divers parts of knowledge, whereof some are laboured and other omitted. For first, I conceive (though it may seem at first somewhat strange) that that part which is variably referred, sometimes to logic, sometimes to metaphysic, touching the common adjuncts of essences, is but an elenche; for the great sophism of all sophisms being equivocation or ambiguity of words and phrase,

specially of such words as are most general and intervene in every inquiry, it seemeth to me that the true and fruitful use (leaving vain subtleties and speculations) of the inquiry of majority, minority, priority, posteriority, identity, diversity, possibility, act, totality, parts, existence, privation, and the like, are but wise cautions against ambiguities of speech. So, again, the distribution of things into certain tribes, which we call categories or predicaments, are but cautions against the confusion of definitions and divisions.

(8) Secondly, there is a seducement that worketh by the strength of the impression, and not by the subtlety of the illaqueation—not so much perplexing the reason, as overruling it by power of the imagination. But this part I think more proper to handle when I shall speak of rhetoric.

(9) But lastly, there is yet a much more important and profound kind of fallacies in the mind of man, which I find not observed or inquired at all, and think good to place here, as that which of all others appertaineth most to rectify judgment, the force whereof is such as it doth not dazzle or snare the understanding in some particulars, but doth more generally and inwardly infect and corrupt the state thereof. For the mind of man is far from the nature of a clear and equal glass, wherein the beams of things should reflect according to their true incidence; nay, it is rather like an enchanted glass, full of superstition and imposture, if it be not delivered and reduced. For this purpose, let us consider the false appearances that are imposed upon us by the general nature of the mind, beholding them in an example or two; as first, in that instance which is the root of all superstition, namely, that to the nature of the mind of all men it is consonant for the affirmative or active to affect more than the negative or privative. So that a few times hitting or presence countervails ofttimes failing or absence, as was well answered by Diagoras to him that showed him in Neptune's temple the great number of pictures of such as had escaped shipwreck, and had paid their vows to Neptune, saying, "Advise now, you that think it folly to invocate Neptune in tempest." "Yea, but," saith Diagoras, "where are they painted that are drowned?" Let us behold it in another instance, namely, that the spirit of man, being of an equal and uniform substance, doth usually suppose and feign in nature a greater equality and uniformity than is in truth. Hence it cometh that the mathematicians cannot satisfy themselves except they reduce the motions of the celestial bodies to perfect circles, rejecting spiral lines, and labouring to be discharged of eccentrics. Hence it cometh that whereas there are many things in Nature as it were monodica, sui juris, yet the cogitations of man do feign unto them relatives, parallels, and conjugates, whereas no such thing is; as they have feigned an element of fire to keep square with earth, water, and air, and the like. Nay, it is not credible, till it be opened, what a number of fictions and fantasies the similitude of human actions and arts, together with the making of man communis mensura, have brought into natural philosophy; not much better than the heresy of the Anthropomorphites, bred in the cells of gross and solitary monks, and the opinion of Epicurus, answerable to the same in heathenism, who supposed the gods to be of human shape. And, therefore, Velleius the Epicurean needed not to have asked why God should have adorned the heavens with stars, as if He had been an ædilis, one that should have set forth some magnificent shows or plays. For if that great Work-master had been of a human disposition, He would have cast the stars into some pleasant and beautiful works and orders like the frets in the roofs of houses; whereas one can scarce find a posture in square, or triangle, or straight line, amongst such an infinite number, so differing a harmony there is between the spirit of man and the spirit of Nature.

(10) Let us consider again the false appearances imposed upon us by every man's own individual nature and custom in that feigned supposition that Plato maketh of the cave; for certainly if a child were continued in a grot or cave under the earth until maturity of age, and came suddenly abroad, he would have strange and absurd imaginations. So, in like manner, although our persons live in the view of heaven, yet our spirits are included in the caves of our own complexions and customs, which minister unto us infinite errors and vain opinions if they be not recalled to examination. But hereof we have given many examples in one of the errors, or peccant humours, which we ran briefly over in our first book.

(11) And lastly, let us consider the false appearances that are imposed upon us by words, which are framed and applied according to the conceit and capacities of the vulgar sort; and although we think we govern our words, and prescribe it well loquendum ut vulgus sentiendum ut sapientes, yet certain it is that words, as a Tartar's bow, do shoot back upon the understanding of the wisest, and mightily entangle and pervert the judgment. So as it is almost necessary in all controversies and disputations to imitate the wisdom of the mathematicians, in setting down in the very beginning the definitions of our words and terms, that others may know how we accept and understand them, and whether they concur with us or no. For it cometh to pass, for want of this, that we are sure to end there where we ought to have begun, which is, in questions and differences about words. To conclude, therefore, it must be confessed that it is not possible to divorce ourselves from these fallacies and false appearances because they are inseparable from our nature and condition of life; so yet, nevertheless, the caution of them (for all elenches, as was said, are but cautions) doth extremely import the true conduct of human judgment. The particular elenches or cautions against these three false appearances I find altogether deficient.

(12) There remaineth one part of judgment of great excellency which to mine understanding is so

slightly touched, as I may report that also deficient; which is the application of the differing kinds of proofs to the differing kinds of subjects. For there being but four kinds of demonstrations, that is, by the immediate consent of the mind or sense, by induction, by syllogism, and by congruity, which is that which Aristotle calleth demonstration in orb or circle, and not a notioribus, every of these hath certain subjects in the matter of sciences, in which respectively they have chiefest use; and certain others, from which respectively they ought to be excluded; and the rigour and curiosity in requiring the more severe proofs in some things, and chiefly the facility in contenting ourselves with the more remiss proofs in others, hath been amongst the greatest causes of detriment and hindrance to knowledge. The distributions and assignations of demonstrations according to the analogy of sciences I note as deficient.

XV

(1) The custody or retaining of knowledge is either in writing or memory; whereof writing hath two parts, the nature of the character and the order of the entry. For the art of characters, or other visible notes of words or things, it hath nearest conjugation with grammar, and, therefore, I refer it to the due place; for the disposition and collocation of that knowledge which we preserve in writing, it consisteth in a good digest of common-places, wherein I am not ignorant of the prejudice imputed to the use of common-place books, as causing a retardation of reading, and some sloth or relaxation of memory. But because it is but a counterfeit thing in knowledges to be forward and pregnant, except a man be deep and full, I hold the entry of common-places to be a matter of great use and essence in studying, as that which assureth copy of invention, and contracteth judgment to a strength. But this is true, that of the methods of common-places that I have seen, there is none of any sufficient worth, all of them carrying merely the face of a school and not of a world; and referring to vulgar matters and pedantical divisions, without all life or respect to action.

(2) For the other principal part of the custody of knowledge, which is memory, I find that faculty in my judgment weakly inquired of. An art there is extant of it; but it seemeth to me that there are better precepts than that art, and better practices of that art than those received. It is certain the art (as it is) may be raised to points of ostentation prodigious; but in use (as is now managed) it is barren, not burdensome, nor dangerous to natural memory, as is imagined, but barren, that is, not dexterous to be applied to the serious use of business and occasions. And, therefore, I make no more estimation of repeating a great number of names or words upon once hearing, or the pouring forth of a number of verses or rhymes extempore, or the making of a satirical simile of everything, or the turning of everything to a jest, or the falsifying or contradicting of everything by cavil, or the like (whereof in the faculties of the mind there is great copy, and such as by device and practice may be exalted to an extreme degree of wonder), than I do of the tricks of tumblers, funambuloes, baladines; the one being the same in the mind that the other is in the body, matters of strangeness without worthiness.

(3) This art of memory is but built upon two intentions; the one prenotion, the other emblem. Prenotion dischargeth the indefinite seeking of that we would remember, and directeth us to seek in a narrow compass, that is, somewhat that hath congruity with our place of memory. Emblem reduceth conceits intellectual to images sensible, which strike the memory more; out of which axioms may be drawn much better practice than that in use; and besides which axioms, there are divers more touching help of memory not inferior to them. But I did in the beginning distinguish, not to report those things deficient, which are but only ill managed.

XVI

(1) There remaineth the fourth kind of rational knowledge, which is transitive, concerning the expressing or transferring our knowledge to others, which I will term by the general name of tradition or delivery. Tradition hath three parts: the first concerning the organ of tradition; the second concerning the method of tradition; and the third concerning the illustration of tradition.

(2) For the organ of tradition, it is either speech or writing; for Aristotle saith well, "Words are the images of cogitations, and letters are the images of words." But yet it is not of necessity that cogitations be expressed by the medium of words. For whatsoever is capable of sufficient differences, and those perceptible by the sense, is in nature competent to express cogitations. And, therefore, we see in the commerce of barbarous people that understand not one another's language, and in the practice of divers that are dumb and deaf, that men's minds are expressed in gestures, though not exactly, yet to serve the turn. And we understand further, that it is the use of China and the kingdoms of the High Levant to write in characters real, which express neither letters nor words in gross, but things or notions; insomuch as countries and provinces which understand not one another's language can nevertheless read one another's writings, because the characters are accepted more generally than the languages do

extend; and, therefore, they have a vast multitude of characters, as many, I suppose, as radical words.

(3) These notes of cogitations are of two sorts: the one when the note hath some similitude or congruity with the notion; the other ad placitum, having force only by contract or acceptation. Of the former sort are hieroglyphics and gestures. For as to hieroglyphics (things of ancient use and embraced chiefly by the Egyptians, one of the most ancient nations), they are but as continued impresses and emblems. And as for gestures, they are as transitory hieroglyphics, and are to hieroglyphics as words spoken are to words written, in that they abide not; but they have evermore, as well as the other, an affinity with the things signified. As Periander, being consulted with how to preserve a tyranny newly usurped, bid the messenger attend and report what he saw him do; and went into his garden and topped all the highest flowers, signifying that it consisted in the cutting off and keeping low of the nobility and grandees. Ad placitum, are the characters real before mentioned, and words: although some have been willing by curious inquiry, or rather by apt feigning, to have derived imposition of names from reason and intendment; a speculation elegant, and, by reason it searcheth into antiquity, reverent, but sparingly mixed with truth, and of small fruit. This portion of knowledge touching the notes of things and cogitations in general, I find not inquired, but deficient. And although it may seem of no great use, considering that words and writings by letters do far excel all the other ways; yet because this part concerneth, as it were, the mint of knowledge (for words are the tokens current and accepted for conceits, as moneys are for values, and that it is fit men be not ignorant that moneys may be of another kind than gold and silver), I thought good to propound it to better inquiry.

(4) Concerning speech and words, the consideration of them hath produced the science of grammar. For man still striveth to reintegrate himself in those benedictions, from which by his fault he hath been deprived; and as he hath striven against the first general curse by the invention of all other arts, so hath he sought to come forth of the second general curse (which was the confusion of tongues) by the art of grammar; whereof the use in a mother tongue is small, in a foreign tongue more; but most in such foreign tongues as have ceased to be vulgar tongues, and are turned only to learned tongues. The duty of it is of two natures: the one popular, which is for the speedy and perfect attaining languages, as well for intercourse of speech as for understanding of authors; the other philosophical, examining the power and nature of words, as they are the footsteps and prints of reason: which kind of analogy between words and reason is handled sparsim, brokenly though not entirely; and, therefore, I cannot report it deficient, though I think it very worthy to be reduced into a science by itself.

(5) Unto grammar also belongeth, as an appendix, the consideration of the accidents of words; which are measure, sound, and elevation or accent, and the sweetness and harshness of them: whence hath issued some curious observations in rhetoric, but chiefly poesy, as we consider it, in respect of the verse and not of the argument. Wherein though men in learned tongues do tie themselves to the ancient measures, yet in modern languages it seemeth to me as free to make new measures of verses as of dances; for a dance is a measured pace, as a verse is a measured speech. In these things this sense is better judge than the art:

> *"Cœnæ fercula nostræ*
> *Mallem convivis quam placuisse cocis."*

And of the servile expressing antiquity in an unlike and an unfit subject, it is well said, "Quod tempore antiquum videtur, id incongruitate est maxime novum."

(6) For ciphers, they are commonly in letters or alphabets, but may be in words. The kinds of ciphers (besides the simple ciphers, with changes, and intermixtures of nulls and non-significants) are many, according to the nature or rule of the infolding, wheel-ciphers, key-ciphers, doubles, &c. But the virtues of them, whereby they are to be preferred, are three; that they be not laborious to write and read; that they be impossible to decipher; and, in some cases, that they be without suspicion. The highest degree whereof is to write omnia per omnia; which is undoubtedly possible, with a proportion quintuple at most of the writing infolding to the writing infolded, and no other restraint whatsoever. This art of ciphering hath for relative an art of deciphering, by supposition unprofitable, but, as things are, of great use. For suppose that ciphers were well managed, there be multitudes of them which exclude the decipherer. But in regard of the rawness and unskilfulness of the hands through which they pass, the greatest matters are many times carried in the weakest ciphers.

(7) In the enumeration of these private and retired arts it may be thought I seek to make a great muster-roll of sciences, naming them for show and ostentation, and to little other purpose. But let those, which are skilful in them, judge whether I bring them in only for appearance, or whether in that which I speak of them (though in few words) there be not some seed of proficience. And this must be remembered, that as there be many of great account in their countries and provinces, which, when they come up to the seat of the estate, are but of mean rank and scarcely regarded; so these arts, being here placed with the principal and supreme sciences, seem petty things: yet to such as have chosen them to spend their labours and studies in them, they seem great matters.

XVII

(1) For the method of tradition, I see it hath moved a controversy in our time. But as in civil business, if there be a meeting, and men fall at words, there is commonly an end of the matter for that time, and no proceeding at all; so in learning, where there is much controversy, there is many times little inquiry. For this part of knowledge of method seemeth to me so weakly inquired as I shall report it deficient.

(2) Method hath been placed, and that not amiss, in logic, as a part of judgment. For as the doctrine of syllogisms comprehendeth the rules of judgment upon that which is invented, so the doctrine of method containeth the rules of judgment upon that which is to be delivered; for judgment precedeth delivery, as it followeth invention. Neither is the method or the nature of the tradition material only to the use of knowledge, but likewise to the progression of knowledge: for since the labour and life of one man cannot attain to perfection of knowledge, the wisdom of the tradition is that which inspireth the felicity of continuance and proceeding. And therefore the most real diversity of method is of method referred to use, and method referred to progression: whereof the one may be termed magistral, and the other of probation.

(3) The latter whereof seemeth to be via deserta et interclusa. For as knowledges are now delivered, there is a kind of contract of error between the deliverer and the receiver. For he that delivereth knowledge desireth to deliver it in such form as may be best believed, and not as may be best examined; and he that receiveth knowledge desireth rather present satisfaction than expectant inquiry; and so rather not to doubt, than not to err: glory making the author not to lay open his weakness, and sloth making the disciple not to know his strength.

(4) But knowledge that is delivered as a thread to be spun on ought to be delivered and intimated, if it were possible, in the same method wherein it was invented: and so is it possible of knowledge induced. But in this same anticipated and prevented knowledge, no man knoweth how he came to the knowledge which he hath obtained. But yet, nevertheless, secundum majus et minus, a man may revisit and descend unto the foundations of his knowledge and consent; and so transplant it into another, as it grew in his own mind. For it is in knowledges as it is in plants: if you mean to use the plant, it is no matter for the roots—but if you mean to remove it to grow, then it is more assured to rest upon roots than slips: so the delivery of knowledges (as it is now used) is as of fair bodies of trees without the roots; good for the carpenter, but not for the planter. But if you will have sciences grow, it is less matter for the shaft or body of the tree, so you look well to the taking up of the roots. Of which kind of delivery the method of the mathematics, in that subject, hath some shadow: but generally I see it neither put in use nor put in inquisition, and therefore note it for deficient.

(5) Another diversity of method there is, which hath some affinity with the former, used in some cases by the discretion of the ancients, but disgraced since by the impostures of many vain persons, who have made it as a false light for their counterfeit merchandises; and that is enigmatical and disclosed. The pretence whereof is, to remove the vulgar capacities from being admitted to the secrets of knowledges, and to reserve them to selected auditors, or wits of such sharpness as can pierce the veil.

(6) Another diversity of method, whereof the consequence is great, is the delivery of knowledge in aphorisms, or in methods; wherein we may observe that it hath been too much taken into custom, out of a few axioms or observations upon any subject, to make a solemn and formal art, filling it with some discourses, and illustrating it with examples, and digesting it into a sensible method. But the writing in aphorisms hath many excellent virtues, whereto the writing in method doth not approach.

(7) For first, it trieth the writer, whether he be superficial or solid: for aphorisms, except they should be ridiculous, cannot be made but of the pith and heart of sciences; for discourse of illustration is cut off; recitals of examples are cut off; discourse of connection and order is cut off; descriptions of practice are cut off. So there remaineth nothing to fill the aphorisms but some good quantity of observation; and therefore no man can suffice, nor in reason will attempt, to write aphorisms, but he that is sound and grounded. But in methods,

> "Tantum series juncturaque pollet,
> Tantum de medio sumptis accedit honoris,"

as a man shall make a great show of an art, which, if it were disjointed, would come to little. Secondly, methods are more fit to win consent or belief, but less fit to point to action; for they carry a kind of demonstration in orb or circle, one part illuminating another, and therefore satisfy. But particulars being dispersed do best agree with dispersed directions. And lastly, aphorisms, representing a knowledge broken, do invite men to inquire further; whereas methods, carrying the show of a total, do secure men, as if they were at furthest.

(8) Another diversity of method, which is likewise of great weight, is the handling of knowledge by assertions and their proofs, or by questions and their determinations. The latter kind whereof, if it be immoderately followed, is as prejudicial to the proceeding of learning as it is to the proceeding of an army to go about to besiege every little fort or hold. For if the field be kept, and the sum of the enterprise pursued, those smaller things will come in of themselves: indeed a man would not leave some important piece enemy at his back. In like manner, the use of confutation in the delivery of sciences ought to be very sparing; and to serve to remove strong preoccupations and prejudgments, and not to minister and excite disputatious and doubts.

(9) Another diversity of method is, according to the subject or matter which is handled. For there is a great difference in.delivery of the mathematics, which are the most abstracted of knowledges, and policy, which is the most immersed. And howsoever contention hath been moved, touching a uniformity of method in multiformity of matter, yet we see how that opinion, besides the weakness of it, hath been of ill desert towards learning, as that which taketh the way to reduce learning to certain empty and barren generalities; being but the very husks and shells of sciences, all the kernel being forced out and expulsed with the torture and press of the method. And, therefore, as I did allow well of particular topics for invention, so I do allow likewise of particular methods of tradition.

(10) Another diversity of judgment in the delivery and teaching of knowledge is, according unto the light and presuppositions of that which is delivered. For that knowledge which is new, and foreign from opinions received, is to be delivered in another form than that that is agreeable and familiar; and therefore Aristotle, when he thinks to tax Democritus, doth in truth commend him, where he saith "If we shall indeed dispute, and not follow after similitudes," &c. For those whose conceits are seated in popular opinions need only but to prove or dispute; but those whose conceits are beyond popular opinions, have a double labour; the one to make themselves conceived, and the other to prove and demonstrate. So that it is of necessity with them to have recourse to similitudes and translations to express themselves. And therefore in the infancy of learning, and in rude times when those conceits which are now trivial were then new, the world was full of parables and similitudes; for else would men either have passed over without mark, or else rejected for paradoxes that which was offered, before they had understood or judged. So in divine learning, we see how frequent parables and tropes are, for it is a rule, that whatsoever science is not consonant to presuppositions must pray in aid of similitudes.

(11) There be also other diversities of methods vulgar and received: as that of resolution or analysis, of constitution or systasis, of concealment or cryptic, &c., which I do allow well of, though I have stood upon those which are least handled and observed. All which I have remembered to this purpose, because I would erect and constitute one general inquiry (which seems to me deficient) touching the wisdom of tradition.

(12) But unto this part of knowledge, concerning method, doth further belong not only the architecture of the whole frame of a work, but also the several beams and columns thereof; not as to their stuff, but as to their quantity and figure. And therefore method considereth not only the disposition of the argument or subject, but likewise the propositions: not as to their truth or matter, but as to their limitation and manner. For herein Ramus merited better a great deal in reviving the good rules of propositions—Καθολον πρωτον, κυτα παντος &c.—than he did in introducing the canker of epitomes; and yet (as it is the condition of human things that, according to the ancient fables, "the most precious things have the most pernicious keepers") it was so, that the attempt of the one made him fall upon the other. For he had need be well conducted that should design to make axioms convertible, if he make them not withal circular, and non-promovent, or incurring into themselves; but yet the intention was excellent.

(13) The other considerations of method, concerning propositions, are chiefly touching the utmost propositions, which limit the dimensions of sciences: for every knowledge may be fitly said, besides the profundity (which is the truth and substance of it, that makes it solid), to have a longitude and a latitude; accounting the latitude towards other sciences, and the longitude towards action; that is, from the greatest generality to the most particular precept. The one giveth rule how far one knowledge ought to intermeddle within the province of another, which is the rule they call Καθαυτο; the other giveth rule unto what degree of particularity a knowledge should descend: which latter I find passed over in silence, being in my judgment the more material. For certainty there must be somewhat left to practice; but how much is worthy the inquiry? We see remote and superficial generalities do but offer knowledge to scorn of practical men; and are no more aiding to practice than an Ortelius' universal map is to direct the way between London and York. The better sort of rules have been not unfitly compared to glasses of steel unpolished, where you may see the images of things, but first they must be filed: so the rules will help if they be laboured and polished by practice. But how crystalline they may be made at the first, and how far forth they may be polished aforehand, is the question, the inquiry whereof seemeth to me deficient.

(14) There hath been also laboured and put in practice a method, which is not a lawful method, but a method of imposture: which is, to deliver knowledges in such manner as men may speedily come to make a show of learning, who have it not. Such was the travail of Raymundus Lullius in making that

art which bears his name; not unlike to some books of typocosmy, which have been made since; being nothing but a mass of words of all arts, to give men countenance, that those which use the terms might be thought to understand the art; which collections are much like a fripper's or broker's shop, that hath ends of everything, but nothing of worth.

XVIII

(1) Now we descend to that part which concerneth the illustration of tradition, comprehended in that science which we call rhetoric, or art of eloquence, a science excellent, and excellently well laboured. For although in true value it is inferior to wisdom (as it is said by God to Moses, when he disabled himself for want of this faculty, "Aaron shall be thy speaker, and thou shalt be to him as God"), yet with people it is the more mighty; for so Solomon saith, Sapiens corde appellabitur prudens, sed dulcis eloquio majora reperiet, signifying that profoundness of wisdom will help a man to a name or admiration, but that it is eloquence that prevaileth in an active life. And as to the labouring of it, the emulation of Aristotle with the rhetoricians of his time, and the experience of Cicero, hath made them in their works of rhetoric exceed themselves. Again, the excellency of examples of eloquence in the orations of Demosthenes and Cicero, added to the perfection of the precepts of eloquence, hath doubled the progression in this art; and therefore the deficiences which I shall note will rather be in some collections, which may as handmaids attend the art, than in the rules or use of the art itself.

(2) Notwithstanding, to stir the earth a little about the roots of this science, as we have done of the rest, the duty and office of rhetoric is to apply reason to imagination for the better moving of the will. For we see reason is disturbed in the administration thereof by three means—by illaqueation or sophism, which pertains to logic; by imagination or impression, which pertains to rhetoric; and by passion or affection, which pertains to morality. And as in negotiation with others, men are wrought by cunning, by importunity, and by vehemency; so in this negotiation within ourselves, men are undermined by inconsequences, solicited and importuned by impressions or observations, and transported by passions. Neither is the nature of man so unfortunately built, as that those powers and arts should have force to disturb reason, and not to establish and advance it. For the end of logic is to teach a form of argument to secure reason, and not to entrap it; the end of morality is to procure the affections to obey reason, and not to invade it; the end of rhetoric is to fill the imagination to second reason, and not to oppress it; for these abuses of arts come in but ex oblique, for caution.

(3) And therefore it was great injustice in Plato, though springing out of a just hatred to the rhetoricians of his time, to esteem of rhetoric but as a voluptuary art, resembling it to cookery, that did mar wholesome meats, and help unwholesome by variety of sauces to the pleasure of the taste. For we see that speech is much more conversant in adorning that which is good than in colouring that which is evil; for there is no man but speaketh more honestly than he can do or think; and it was excellently noted by Thucydides, in Cleon, that because he used to hold on the bad side in causes of estate, therefore he was ever inveighing against eloquence and good speech, knowing that no man can speak fair of courses sordid and base. And therefore, as Plato said elegantly, "That virtue, if she could be seen, would move great love and affection;" so seeing that she cannot be showed to the sense by corporal shape, the next degree is to show her to the imagination in lively representation; for to show her to reason only in subtlety of argument was a thing ever derided in Chrysippus and many of the Stoics, who thought to thrust virtue upon men by sharp disputations and conclusions, which have no sympathy with the will of man.

(4) Again, if the affections in themselves were pliant and obedient to reason, it were true there should be no great use of persuasions and insinuations to the will, more than of naked proposition and proofs; but in regard of the continual mutinies and seditious of the affections—

"Video meliora, proboque,
Deteriora sequor,"

reason would become captive and servile, if eloquence of persuasions did not practise and win the imagination from the affections' part, and contract a confederacy between the reason and imagination against the affections; for the affections themselves carry ever an appetite to good, as reason doth. The difference is, that the affection beholdeth merely the present; reason beholdeth the future and sum of time. And, therefore, the present filling the imagination more, reason is commonly vanquished; but after that force of eloquence and persuasion hath made things future and remote appear as present, then upon the revolt of the imagination reason prevaileth.

(5) We conclude, therefore, that rhetoric can be no more charged with the colouring of the worst part, than logic with sophistry, or morality with vice; for we know the doctrines of contraries are the same, though the use be opposite. It appeareth also that logic differeth from rhetoric, not only as the fist

242

from the palm—the one close, the other at large—but much more in this, that logic handleth reason exact and in truth, and rhetoric handleth it as it is planted in popular opinions and manners. And therefore Aristotle doth wisely place rhetoric as between logic on the one side, and moral or civil knowledge on the other, as participating of both; for the proofs and demonstrations of logic are toward all men indifferent and the same, but the proofs and persuasions of rhetoric ought to differ according to the auditors:

"Orpheus in sylvis, inter delphinas Arion."

Which application in perfection of idea ought to extend so far that if a man should speak of the same thing to several persons, he should speak to them all respectively and several ways; though this politic part of eloquence in private speech it is easy for the greatest orators to want: whilst, by the observing their well-graced forms of speech, they leese the volubility of application; and therefore it shall not be amiss to recommend this to better inquiry, not being curious whether we place it here or in that part which concerneth policy.

(6) Now therefore will I descend to the deficiences, which, as I said, are but attendances; and first, I do not find the wisdom and diligence of Aristotle well pursued, who began to make a collection of the popular signs and colours of good and evil, both simple and comparative, which are as the sophisms of rhetoric (as I touched before). For example—

"Sophisma.

Quod laudatur, bonum: quod vituperatur, malum.

Redargutio.

Laudat venales qui vult extrudere merces."

Malum est, malum est (inquit emptor): sed cum recesserit, tum gloriabitur! The defects in the labour of Aristotle are three—one, that there be but a few of many; another, that there elenches are not annexed; and the third, that he conceived but a part of the use of them: for their use is not only in probation, but much more in impression. For many forms are equal in signification which are differing in impression, as the difference is great in the piercing of that which is sharp and that which is flat, though the strength of the percussion be the same. For there is no man but will be a little more raised by hearing it said, "Your enemies will be glad of this"—

"Hoc Ithacus velit, et magno mercentur Atridæ."
than by hearing it said only, *"This is evil for you."*

(7) Secondly, I do resume also that which I mentioned before, touching provision or preparatory store for the furniture of speech and readiness of invention, which appeareth to be of two sorts: the one in resemblance to a shop of pieces unmade up, the other to a shop of things ready made up; both to be applied to that which is frequent and most in request. The former of these I will call antitheta, and the latter formulæ.

(8) Antitheta are théses argued pro et contra, wherein men may be more large and laborious; but (in such as are able to do it) to avoid prolixity of entry, I wish the seeds of the several arguments to be cast up into some brief and acute sentences, not to be cited, but to be as skeins or bottoms of thread, to be unwinded at large when they come to be used; supplying authorities and examples by reference.

"Pro verbis legis.
Non est interpretatio, sed divinatio, quæ recedit a litera:
Cum receditur a litera, judex transit in legislatorem.

Pro sententia legis.
Ex omnibus verbis est eliciendus sensus qui interpretatur singula."

(9) Formulæ are but decent and apt passages or conveyances of speech, which may serve indifferently for differing subjects; as of preface, conclusion, digression, transition, excusation, &c. For as in buildings there is great pleasure and use in the well casting of the staircases, entries, doors, windows, and the like; so in speech, the conveyances and passages are of special ornament and effect.

"A conclusion in a deliberative.

>   *So may we redeem the faults passed, and prevent the inconveniences
>   future."*

XIX

(1) There remain two appendices touching the tradition of knowledge, the one critical, the other pedantical. For all knowledge is either delivered by teachers, or attained by men's proper endeavours: and therefore as the principal part of tradition of knowledge concerneth chiefly writing of books, so the relative part thereof concerneth reading of books; whereunto appertain incidently these considerations. The first is concerning the true correction and edition of authors; wherein nevertheless rash diligence hath done great prejudice. For these critics have often presumed that that which they understand not is false set down: as the priest that, where he found it written of St. Paul Demissus est per sportam, mended his book, and made it Demissus est per portam; because sporta was a hard word, and out of his reading: and surely their errors, though they be not so palpable and ridiculous, yet are of the same kind. And therefore, as it hath been wisely noted, the most corrected copies are commonly the least correct.

The second is concerning the exposition and explication of authors, which resteth in annotations and commentaries: wherein it is over usual to blanch the obscure places and discourse upon the plain.

The third is concerning the times, which in many cases give great light to true interpretations.

The fourth is concerning some brief censure and judgment of the authors; that men thereby may make some election unto themselves what books to read.

And the fifth is concerning the syntax and disposition of studies; that men may know in what order or pursuit to read.

(2) For pedantical knowledge, it containeth that difference of tradition which is proper for youth; whereunto appertain divers considerations of great fruit.

As first, the timing and seasoning of knowledges; as with what to initiate them, and from what for a time to refrain them.

Secondly, the consideration where to begin with the easiest, and so proceed to the more difficult; and in what courses to press the more difficult, and then to turn them to the more easy; for it is one method to practise swimming with bladders, and another to practise dancing with heavy shoes.

A third is the application of learning according unto the propriety of the wits; for there is no defect in the faculties intellectual, but seemeth to have a proper cure contained in some studies: as, for example, if a child be bird-witted, that is, hath not the faculty of attention, the mathematics giveth a remedy thereunto; for in them, if the wit be caught away but a moment, one is new to begin. And as sciences have a propriety towards faculties for cure and help, so faculties or powers have a sympathy towards sciences for excellency or speedy profiting: and therefore it is an inquiry of great wisdom, what kinds of wits and natures are most apt and proper for what sciences.

Fourthly, the ordering of exercises is matter of great consequence to hurt or help: for, as is well observed by Cicero, men in exercising their faculties, if they be not well advised, do exercise their faults and get ill habits as well as good; so as there is a great judgment to be had in the continuance and intermission of exercises. It were too long to particularise a number of other considerations of this nature, things but of mean appearance, but of singular efficacy. For as the wronging or cherishing of seeds or young plants is that that is most important to their thriving, and as it was noted that the first six kings being in truth as tutors of the state of Rome in the infancy thereof was the principal cause of the immense greatness of that state which followed, so the culture and manurance of minds in youth hath such a forcible (though unseen) operation, as hardly any length of time or contention of labour can countervail it afterwards. And it is not amiss to observe also how small and mean faculties gotten by education, yet when they fall into great men or great matters, do work great and important effects: whereof we see a notable example in Tacitus of two stage players, Percennius and Vibulenus, who by their faculty of playing put the Pannonian armies into an extreme tumult and combustion. For there arising a mutiny amongst them upon the death of Augustus Cæsar, Blæsus the lieutenant had committed some of the mutineers, which were suddenly rescued; whereupon Vibulenus got to be heard speak, which he did in this manner:—"These poor innocent wretches appointed to cruel death, you have restored to behold the light; but who shall restore my brother to me, or life unto my brother, that was sent hither in message from the legions of Germany, to treat of the common cause? and he hath murdered him this last night by some of his fencers and ruffians, that he hath about him for his executioners upon soldiers. Answer, Blæsus, what is done with his body? The mortalest enemies do not deny burial. When I have performed my last duties to the corpse with kisses, with tears, command me to be slain besides him; so that these my fellows, for our good meaning and our true hearts to the legions, may have leave to bury us." With which speech he put the army into an infinite fury and uproar: whereas truth was he had no brother, neither was there any such matter; but he played it merely as if he had been upon the stage.

(3) But to return: we are now come to a period of rational knowledges; wherein if I have made the divisions other than those that are received, yet would I not be thought to disallow all those divisions which I do not use. For there is a double necessity imposed upon me of altering the divisions. The one, because it differeth in end and purpose, to sort together those things which are next in nature, and those things which are next in use. For if a secretary of estate should sort his papers, it is like in his study or general cabinet he would sort together things of a nature, as treaties, instructions, &c. But in his boxes or particular cabinet he would sort together those that he were like to use together, though of several natures. So in this general cabinet of knowledge it was necessary for me to follow the divisions of the nature of things; whereas if myself had been to handle any particular knowledge, I would have respected the divisions fittest for use. The other, because the bringing in of the deficiences did by consequence alter the partitions of the rest. For let the knowledge extant (for demonstration sake) be fifteen. Let the knowledge with the deficiences be twenty; the parts of fifteen are not the parts of twenty; for the parts of fifteen are three and five; the parts of twenty are two, four, five, and ten. So as these things are without contradiction, and could not otherwise be.

XX

(1) We proceed now to that knowledge which considereth of the appetite and will of man: whereof Solomon saith, Ante omnia, fili, custodi cor tuum: nam inde procedunt actiones vitæ. In the handling of this science, those which have written seem to me to have done as if a man, that professed to teach to write, did only exhibit fair copies of alphabets and letters joined, without giving any precepts or directions for the carriage of the hand and framing of the letters. So have they made good and fair exemplars and copies, carrying the draughts and portraitures of good, virtue, duty, felicity; propounding them well described as the true objects and scopes of man's will and desires. But how to attain these excellent marks, and how to frame and subdue the will of man to become true and conformable to these pursuits, they pass it over altogether, or slightly and unprofitably. For it is not the disputing that moral virtues are in the mind of man by habit and not by nature, or the distinguishing that generous spirits are won by doctrines and persuasions, and the vulgar sort by reward and punishment, and the like scattered glances and touches, that can excuse the absence of this part.

(2) The reason of this omission I suppose to be that hidden rock whereupon both this and many other barks of knowledge have been cast away; which is, that men have despised to be conversant in ordinary and common matters, the judicious direction whereof nevertheless is the wisest doctrine (for life consisteth not in novelties nor subtleties), but contrariwise they have compounded sciences chiefly of a certain resplendent or lustrous mass of matter, chosen to give glory either to the subtlety of disputations, or to the eloquence of discourses. But Seneca giveth an excellent check to eloquence, Nocet illis eloquentia, quibus non rerum cupiditatem facit, sed sui. Doctrine should be such as should make men in love with the lesson, and not with the teacher; being directed to the auditor's benefit, and not to the author's commendation. And therefore those are of the right kind which may be concluded as Demosthenes concludes his counsel, Quæ si feceritis, non oratorem dumtaxat in præsentia laudabitis, sed vosmetipsos etiam non ita multo post statu rerum vestraram meliore.

(3) Neither needed men of so excellent parts to have despaired of a fortune, which the poet Virgil promised himself, and indeed obtained, who got as much glory of eloquence, wit, and learning in the expressing of the observations of husbandry, as of the heroical acts of Æneas:

> *"Nec sum animi dubius, verbis ea vincere magnum*
> *Quam sit, et angustis his addere rebus honorem."*

And surely, if the purpose be in good earnest, not to write at leisure that which men may read at leisure, but really to instruct and suborn action and active life, these Georgics of the mind, concerning the husbandry and tillage thereof, are no less worthy than the heroical descriptions of virtue, duty, and felicity. Wherefore the main and primitive division of moral knowledge seemeth to be into the exemplar or platform of good, and the regiment or culture of the mind: the one describing the nature of good, the other prescribing rules how to subdue, apply, and accommodate the will of man thereunto.

(4) The doctrine touching the platform or nature of good considereth it either simple or compared; either the kinds of good, or the degrees of good; in the latter whereof those infinite disputatious which were touching the supreme degree thereof, which they term felicity, beatitude, or the highest good, the doctrines concerning which were as the heathen divinity, are by the Christian faith discharged. And as Aristotle saith, "That young men may be happy, but not otherwise but by hope;" so we must all acknowledge our minority, and embrace the felicity which is by hope of the future world.

(5) Freed therefore and delivered from this doctrine of the philosopher's heaven, whereby they feigned a higher elevation of man's nature than was (for we see in what height of style Seneca writeth,

Vere magnum, habere fragilitatem hominis, securitatem Dei), we may with more sobriety and truth receive the rest of their inquiries and labours. Wherein for the nature of good positive or simple, they have set it down excellently in describing the forms of virtue and duty, with their situations and postures; in distributing them into their kinds, parts, provinces, actions, and administrations, and the like: nay further, they have commended them to man's nature and spirit with great quickness of argument and beauty of persuasions; yea, and fortified and entrenched them (as much as discourse can do) against corrupt and popular opinions. Again, for the degrees and comparative nature of good, they have also excellently handled it in their triplicity of good, in the comparisons between a contemplative and an active life, in the distinction between virtue with reluctation and virtue secured, in their encounters between honesty and profit, in their balancing of virtue with virtue, and the like; so as this part deserveth to be reported for excellently laboured.

(6) Notwithstanding, if before they had come to the popular and received notions of virtue and vice, pleasure and pain, and the rest, they had stayed a little longer upon the inquiry concerning the roots of good and evil, and the strings of those roots, they had given, in my opinion, a great light to that which followed; and specially if they had consulted with nature, they had made their doctrines less prolix and more profound: which being by them in part omitted and in part handled with much confusion, we will endeavour to resume and open in a more clear manner.

(7) There is formed in everything a double nature of good—the one, as everything is a total or substantive in itself; the other, as it is a part or member of a greater body; whereof the latter is in degree the greater and the worthier, because it tendeth to the conservation of a more general form. Therefore we see the iron in particular sympathy moveth to the loadstone; but yet if it exceed a certain quantity, it forsaketh the affection to the loadstone, and like a good patriot moveth to the earth, which is the region and country of massy bodies; so may we go forward, and see that water and massy bodies move to the centre of the earth; but rather than to suffer a divulsion in the continuance of nature, they will move upwards from the centre of the earth, forsaking their duty to the earth in regard of their duty to the world. This double nature of good, and the comparative thereof, is much more engraven upon man, if he degenerate not, unto whom the conservation of duty to the public ought to be much more precious than the conservation of life and being; according to that memorable speech of Pompeius Magnus, when being in commission of purveyance for a famine at Rome, and being dissuaded with great vehemency and instance by his friends about him, that he should not hazard himself to sea in an extremity of weather, he said only to them, Necesse est ut eam, non ut vivam. But it may be truly affirmed that there was never any philosophy, religion, or other discipline, which did so plainly and highly exalt the good which is communicative, and depress the good which is private and particular, as the Holy Faith; well declaring that it was the same God that gave the Christian law to men, who gave those laws of nature to inanimate creatures that we spake of before; for we read that the elected saints of God have wished themselves anathematised and razed out of the book of life, in an ecstasy of charity and infinite feeling of communion.

(8) This being set down and strongly planted, doth judge and determine most of the controversies wherein moral philosophy is conversant. For first, it decideth the question touching the preferment of the contemplative or active life, and decideth it against Aristotle. For all the reasons which he bringeth for the contemplative are private, and respecting the pleasure and dignity of a man's self (in which respects no question the contemplative life hath the pre-eminence), not much unlike to that comparison which Pythagoras made for the gracing and magnifying of philosophy and contemplation, who being asked what he was, answered, "That if Hiero were ever at the Olympian games, he knew the manner, that some came to try their fortune for the prizes, and some came as merchants to utter their commodities, and some came to make good cheer and meet their friends, and some came to look on; and that he was one of them that came to look on." But men must know, that in this theatre of man's life it is reserved only for God and angels to be lookers on. Neither could the like question ever have been received in the Church, notwithstanding their Pretiosa in oculis Domini mors sanctorum ejus, by which place they would exalt their civil death and regular professions, but upon this defence, that the monastical life is not simple contemplative, but performeth the duty either of incessant prayers and supplications, which hath been truly esteemed as an office in the Church, or else of writing or taking instructions for writing concerning the law of God, as Moses did when he abode so long in the mount. And so we see Enoch, the seventh from Adam, who was the first contemplative and walked with God, yet did also endow the Church with prophecy, which Saint Jude citeth. But for contemplation which should be finished in itself, without casting beams upon society, assuredly divinity knoweth it not.

(9) It decideth also the controversies between Zeno and Socrates, and their schools and successions, on the one side, who placed felicity in virtue simply or attended, the actions and exercises whereof do chiefly embrace and concern society; and on the other side, the Cyrenaics and Epicureans, who placed it in pleasure, and made virtue (as it is used in some comedies of errors, wherein the mistress and the maid change habits) to be but as a servant, without which pleasure cannot be served and attended; and the reformed school of the Epicureans, which placed it in serenity of mind and

freedom from perturbation; as if they would have deposed Jupiter again, and restored Saturn and the first age, when there was no summer nor winter, spring nor autumn, but all after one air and season; and Herillus, which placed felicity in extinguishment of the disputes of the mind, making no fixed nature of good and evil, esteeming things according to the clearness of the desires, or the reluctation; which opinion was revived in the heresy of the Anabaptists, measuring things according to the motions of the spirit, and the constancy or wavering of belief; all which are manifest to tend to private repose and contentment, and not to point of society.

(10) It censureth also the philosophy of Epictetus, which presupposeth that felicity must be placed in those things which are in our power, lest we be liable to fortune and disturbance; as if it were not a thing much more happy to fail in good and virtuous ends for the public, than to obtain all that we can wish to ourselves in our proper fortune: as Consalvo said to his soldiers, showing them Naples, and protesting he had rather die one foot forwards, than to have his life secured for long by one foot of retreat. Whereunto the wisdom of that heavenly leader hath signed, who hath affirmed that "a good conscience is a continual feast;" showing plainly that the conscience of good intentions, howsoever succeeding, is a more continual joy to nature than all the provision which can be made for security and repose.

(11) It censureth likewise that abuse of philosophy which grew general about the time of Epictetus, in converting it into an occupation or profession; as if the purpose had been, not to resist and extinguish perturbations, but to fly and avoid the causes of them, and to shape a particular kind and course of life to that end; introducing such a health of mind, as was that health of body of which Aristotle speaketh of Herodicus, who did nothing all his life long but intend his health; whereas if men refer themselves to duties of society, as that health of body is best which is ablest to endure all alterations and extremities, so likewise that health of mind is most proper which can go through the greatest temptations and perturbations. So as Diogenes' opinion is to be accepted, who commended not them which abstained, but them which sustained, and could refrain their mind in præcipitio, and could give unto the mind (as is used in horsemanship) the shortest stop or turn.

(12) Lastly, it censureth the tenderness and want of application in some of the most ancient and reverend philosophers and philosophical men, that did retire too easily from civil business, for avoiding of indignities and perturbations; whereas the resolution of men truly moral ought to be such as the same Consalvo said the honour of a soldier should be, e telâ crassiore, and not so fine as that everything should catch in it and endanger it.

XXI

(1) To resume private or particular good, it falleth into the division of good active and passive; for this difference of good (not unlike to that which amongst the Romans was expressed in the familiar or household terms of promus and condus) is formed also in all things, and is best disclosed in the two several appetites in creatures; the one to preserve or continue themselves, and the other to dilate or multiply themselves, whereof the latter seemeth to be the worthier; for in nature the heavens, which are the more worthy, are the agent, and the earth, which is the less worthy, is the patient. In the pleasures of living creatures, that of generation is greater than that of food. In divine doctrine, beatius est dare quam accipere. And in life, there is no man's spirit so soft, but esteemeth the effecting of somewhat that he hath fixed in his desire, more than sensuality, which priority of the active good is much upheld by the consideration of our estatè to be mortal and exposed to fortune. For if we might have a perpetuity and certainty in our pleasures, the state of them would advance their price. But when we see it is but magni æstimamus mori tardius, and ne glorieris de crastino, nescis partum diei, it maketh us to desire to have somewhat secured and exempted from time, which are only our deeds and works; as it is said, Opera eorum sequuntur eos. The pre-eminence likewise of this active good is upheld by the affection which is natural in man towards variety and proceeding, which in the pleasures of the sense, which is the principal part of passive good, can have no great latitude. Cogita quamdiu eadem feceris; cibus, somnus, ludus per hunc circulum curritur; mori velle non tantum fortis, aut miser, aut prudens, sed etiam fastidiosus potest. But in enterprises, pursuits, and purposes of life, there is much variety; whereof men are sensible with pleasure in their inceptions, progressions, recoils, reintegrations, approaches and attainings to their ends. So as it was well said, Vita sine proposito languida et vaga est. Neither hath this active good an identity with the good of society, though in some cases it hath an incidence into it. For although it do many times bring forth acts of beneficence, yet it is with a respect private to a man's own power, glory, amplification, continuance; as appeareth plainly, when it findeth a contrary subject. For that gigantine state of mind which possesseth the troublers of the world, such as was Lucius Sylla and infinite other in smaller model, who would have all men happy or unhappy as they were their friends or enemies, and would give form to the world, according to their own humours (which is the true theomachy), pretendeth and aspireth to active good, though it recedeth furthest from

good of society, which we have determined to be the greater.

(2) To resume passive good, it receiveth a subdivision of conservative and effective. For let us take a brief review of that which we have said: we have spoken first of the good of society, the intention whereof embraceth the form of human nature, whereof we are members and portions, and not our own proper and individual form; we have spoken of active good, and supposed it as a part of private and particular good. And rightly, for there is impressed upon all things a triple desire or appetite proceeding from love to themselves: one of preserving and continuing their form; another of advancing and perfecting their form; and a third of multiplying and extending their form upon other things: whereof the multiplying, or signature of it upon other things, is that which we handled by the name of active good. So as there remaineth the conserving of it, and perfecting or raising of it, which latter is the highest degree of passive good. For to preserve in state is the less, to preserve with advancement is the greater. So in man,

"*Igneus est ollis vigor, et cœlestis origo.*"

His approach or assumption to divine or angelical nature is the perfection of his form; the error or false imitation of which good is that which is the tempest of human life; while man, upon the instinct of an advancement, formal and essential, is carried to seek an advancement local. For as those which are sick, and find no remedy, do tumble up and down and change place, as if by a remove local they could obtain a remove internal, so is it with men in ambition, when failing of the mean to exalt their nature, they are in a perpetual estuation to exalt their place. So then passive good is, as was said, either conservative or perfective.

(3) To resume the good of conservation or comfort, which consisteth in the fruition of that which is agreeable to our natures; it seemeth to be most pure and natural of pleasures, but yet the softest and lowest. And this also receiveth a difference, which hath neither been well judged of, nor well inquired; for the good of fruition or contentment is placed either in the sincereness of the fruition, or in the quickness and vigour of it; the one superinduced by equality, the other by vicissitude; the one having less mixture of evil, the other more impression of good. Whether of these is the greater good is a question controverted; but whether man's nature may not be capable of both is a question not inquired.

(4) The former question being debated between Socrates and a sophist, Socrates placing felicity in an equal and constant peace of mind, and the sophist in much desiring and much enjoying, they fell from argument to ill words: the sophist saying that Socrates' felicity was the felicity of a block or stone; and Socrates saying that the sophist's felicity was the felicity of one that had the itch, who did nothing but itch and scratch. And both these opinions do not want their supports. For the opinion of Socrates is much upheld by the general consent even of the epicures themselves, that virtue beareth a great part in felicity; and if so, certain it is, that virtue hath more use in clearing perturbations then in compassing desires. The sophist's opinion is much favoured by the assertion we last spake of, that good of advancement is greater than good of simple preservation; because every obtaining a desire hath a show of advancement, as motion though in a circle hath a show of progression.

(5) But the second question, decided the true way, maketh the former superfluous. For can it be doubted, but that there are some who take more pleasure in enjoying pleasures than some other, and yet, nevertheless, are less troubled with the loss or leaving of them? So as this same, Non uti ut non appetas, non appetere ut non metuas, sunt animi pusilli et diffidentis. And it seemeth to me that most of the doctrines of the philosophers are more fearful and cautious than the nature of things requireth. So have they increased the fear of death in offering to cure it. For when they would have a man's whole life to be but a discipline or preparation to die, they must needs make men think that it is a terrible enemy, against whom there is no end of preparing. Better saith the poet:—

"*Qui finem vitæ extremum inter munera ponat*
Naturæ."

So have they sought to make men's minds too uniform and harmonical, by not breaking them sufficiently to contrary motions; the reasons whereof I suppose to be, because they themselves were men dedicated to a private, free, and unapplied course of life. For as we see, upon the lute or like instrument, a ground, though it be sweet and have show of many changes, yet breaketh not the hand to such strange and hard stops and passages, as a set song or voluntary; much after the same manner was the diversity between a philosophical and civil life. And, therefore, men are to imitate the wisdom of jewellers: who, if there be a grain, or a cloud, or an ice which may be ground forth without taking too much of the stone, they help it; but if it should lessen and abate the stone too much, they will not meddle with it: so ought men so to procure serenity as they destroy not magnanimity.

(6) Having therefore deduced the good of man which is private and particular, as far as seemeth fit, we will now return to that good of man which respecteth and beholdeth society, which we may term

duty; because the term of duty is more proper to a mind well framed and disposed towards others, as the term of virtue is applied to a mind well formed and composed in itself; though neither can a man understand virtue without some relation to society, nor duty without an inward disposition. This part may seem at first to pertain to science civil and politic; but not if it be well observed. For it concerneth the regiment and government of every man over himself, and not over others. And as in architecture the direction of framing the posts, beams, and other parts of building, is not the same with the manner of joining them and erecting the building; and in mechanicals, the direction how to frame an instrument or engine is not the same with the manner of setting it on work and employing it; and yet, nevertheless, in expressing of the one you incidently express the aptness towards the other; so the doctrine of conjugation of men in society differeth from that of their conformity thereunto.

(7) This part of duty is subdivided into two parts: the common duty of every man, as a man or member of a state; the other, the respective or special duty of every man in his profession, vocation, and place. The first of these is extant and well laboured, as hath been said. The second likewise I may report rather dispersed than deficient; which manner of dispersed writing in this kind of argument I acknowledge to be best. For who can take upon him to write of the proper duty, virtue, challenge, and right of every several vocation, profession, and place? For although sometimes a looker on may see more than a gamester, and there be a proverb more arrogant than sound, "That the vale best discovereth the hill;" yet there is small doubt but that men can write best and most really and materially in their own professions; and that the writing of speculative men of active matter for the most part doth seem to men of experience, as Phormio's argument of the wars seemed to Hannibal, to be but dreams and dotage. Only there is one vice which accompanieth them that write in their own professions, that they magnify them in excess. But generally it were to be wished (as that which would make learning indeed solid and fruitful) that active men would or could become writers.

(8) In which kind I cannot but mention, honoris causa, your Majesty's excellent book touching the duty of a king; a work richly compounded of divinity, morality, and policy, with great aspersion of all other arts; and being in some opinion one of the most sound and healthful writings that I have read: not distempered in the heat of invention, nor in the coldness of negligence; not sick of dizziness, as those are who leese themselves in their order, nor of convulsions, as those which cramp in matters impertinent; not savouring of perfumes and paintings, as those do who seek to please the reader more than nature beareth; and chiefly well disposed in the spirits thereof, being agreeable to truth and apt for action; and far removed from that natural infirmity, whereunto I noted those that write in their own professions to be subject—which is, that they exalt it above measure. For your Majesty hath truly described, not a king of Assyria or Persia in their extern glory, but a Moses or a David, pastors of their people. Neither can I ever leese out of my remembrance what I heard your Majesty in the same sacred spirit of government deliver in a great cause of judicature, which was, "That kings ruled by their laws, as God did by the laws of nature; and ought as rarely to put in use their supreme prerogative as God doth His power of working miracles." And yet notwithstanding in your book of a free monarchy, you do well give men to understand, that you know the plenitude of the power and right of a king, as well as the circle of his office and duty. Thus have I presumed to allege this excellent writing of your Majesty, as a prime or eminent example of tractates concerning special and respective duties; wherein I should have said as much, if it had been written a thousand years since. Neither am I moved with certain courtly decencies, which esteem it flattery to praise in presence. No, it is flattery to praise in absence— that is, when either the virtue is absent, or the occasion is absent; and so the praise is not natural, but forced, either in truth or in time. But let Cicero be read in his oration pro Marcello, which is nothing but an excellent table of Cæsàr's virtue, and made to his face; besides the example of many other excellent persons, wiser a great deal than such observers; and we will never doubt, upon a full occasion, to give just praises to present or absent.

(9) But to return; there belongeth further to the handling of this part, touching the duties of professions and vocations, a relative or opposite, touching the frauds, cautels, impostures, and vices of every profession, which hath been likewise handled; but how? rather in a satire and cynically, than seriously and wisely; for men have rather sought by wit to deride and traduce much of that which is good in professions, than with judgment to discover and sever that which is corrupt. For, as Solomon saith, he that cometh to seek after knowledge with a mind to scorn and censure shall be sure to find matter for his humour, but no matter for his instruction: Quærenti derisori scientiam ipsa se abscondit; sed studioso fit obviam. But the managing of this argument with integrity and truth, which I note as deficient, seemeth to me to be one of the best fortifications for honesty and virtue that can be planted. For, as the fable goeth of the basilisk—that if he see you first, you die for it; but if you see him first, he dieth—so is it with deceits and evil arts, which, if they be first espied they leese their life; but if they prevent, they endanger. So that we are much beholden to Machiavel and others, that write what men do, and not what they ought to do. For it is not possible to join serpentine wisdom with the columbine innocency, except men know exactly all the conditions of the serpent; his baseness and going upon his belly, his volubility and lubricity, his envy and sting, and the rest—that is, all forms and natures of evil.

For without this, virtue lieth open and unfenced. Nay, an honest man can do no good upon those that are wicked, to reclaim them, without the help of the knowledge of evil. For men of corrupted minds presuppose that honesty groweth out of simplicity of manners, and believing of preachers, schoolmasters, and men's exterior language. So as, except you can make them perceive that you know the utmost reaches of their own corrupt opinions, they despise all morality. Non recipit stultus verba prudentiæ, nisi ea dixeris quæ, versantur in corde ejus.

(10) Unto this part, touching respective duty, doth also appertain the duties between husband and wife, parent and child, master and servant. So likewise the laws of friendship and gratitude, the civil bond of companies, colleges, and politic bodies, of neighbourhood, and all other proportionate duties; not as they are parts of government and society, but as to the framing of the mind of particular persons.

(11) The knowledge concerning good respecting society doth handle it also, not simply alone, but comparatively; whereunto belongeth the weighing of duties between person and person, case and case, particular and public. As we see in the proceeding of Lucius Brutus against his own sons, which was so much extolled, yet what was said?

"Infelix, utcunque ferent ea fata minores."

So the case was doubtful, and had opinion on both sides. Again, we see when M. Brutus and Cassius invited to a supper certain whose opinions they meant to feel, whether they were fit to be made their associates, and cast forth the question touching the killing of a tyrant being a usurper, they were divided in opinion; some holding that servitude was the extreme of evils, and others that tyranny was better than a civil war: and a number of the like cases there are of comparative duty. Amongst which that of all others is the most frequent, where the question is of a great deal of good to ensue of a small injustice. Which Jason of Thessalia determined against the truth: Aliqua sunt injuste facienda, ut multa juste fieri possint. But the reply is good: Auctorem præsentis justitiæ habes, sponsorem futuræ non habes. Men must pursue things which are just in present, and leave the future to the Divine Providence. So then we pass on from this general part touching the exemplar and description of good.

XXII

(1) Now, therefore, that we have spoken of this fruit of life, it remaineth to speak of the husbandry that belongeth thereunto, without which part the former seemeth to be no better than a fair image or statue, which is beautiful to contemplate, but is without life and motion; whereunto Aristotle himself subscribeth in these words: Necesse est scilicet de virtute dicere, et quid sit, et ex quibus gignatur. Inutile enum fere fuerit virtutem quidem nosse, acquirendæ autem ejus modos et vias ignorare. Non enum de virtute tantum, qua specie sit, quærendum est, sed et quomodo sui copiam faciat: utrumque enum volumeus, et rem ipsam nosse, et ejus compotes fieri: hoc autem ex voto non succedet, nisi sciamus et ex quibus et quomodo. In such full words and with such iteration doth he inculcate this part. So saith Cicero in great commendation of Cato the second, that he had applied himself to philosophy, Non ita disputandi causa, sed ita vivendi. And although the neglect of our times, wherein few men do hold any consultations touching the reformation of their life (as Seneca excellently saith, De partibus vitæ quisque deliberat, de summa nemo), may make this part seem superfluous; yet I must conclude with that aphorism of Hippocrates, Qui gravi morbo correpti dolores non sentiunt, iis mens ægrotat. They need medicine, not only to assuage the disease, but to awake the sense. And if it be said that the cure of men's minds belongeth to sacred divinity, it is most true; but yet moral philosophy may be preferred unto her as a wise servant and humble handmaid. For as the Psalm saith, "That the eyes of the handmaid look perpetually towards the mistress," and yet no doubt many things are left to the discretion of the handmaid to discern of the mistress' will; so ought moral philosophy to give a constant attention to the doctrines of divinity, and yet so as it may yield of herself (within due limits) many sound and profitable directions.

(2) This part, therefore, because of the excellency thereof, I cannot but find exceeding strange that it is not reduced to written inquiry; the rather, because it consisteth of much matter, wherein both speech and action is often conversant; and such wherein the common talk of men (which is rare, but yet cometh sometimes to pass) is wiser than their books. It is reasonable, therefore, that we propound it in the more particularity, both for the worthiness, and because we may acquit ourselves for reporting it deficient, which seemeth almost incredible, and is otherwise conceived and presupposed by those themselves that have written. We will, therefore, enumerate some heads or points thereof, that it may appear the better what it is, and whether it be extant.

(3) First, therefore, in this, as in all things which are practical we ought to cast up our account, what is in our power, and what not; for the one may be dealt with by way of alteration, but the other by way of application only. The husbandman cannot command neither the nature of the earth nor the

seasons of the weather; no more can the physician the constitution of the patient nor the variety of accidents. So in the culture and cure of the mind of man, two things are without our command: points of Nature, and points of fortune. For to the basis of the one, and the conditions of the other, our work is limited and tied. In these things, therefore, it is left unto us to proceed by application:—

"Vincenda est omnis fertuna ferendo:"
and so likewise,

"Vincenda est omnis Natura ferendo."

But when that we speak of suffering, we do not speak of a dull and neglected suffering, but of a wise and industrious suffering, which draweth and contriveth use and advantage out of that which seemeth adverse and contrary; which is that properly which we call accommodating or applying. Now the wisdom of application resteth principally in the exact and distinct knowledge of the precedent state or disposition, unto which we do apply; for we cannot fit a garment except we first take measure of the body.

(4) So, then, the first article of this knowledge is to set down sound and true distributions and descriptions of the several characters and tempers of men's natures and dispositions, specially having regard to those differences which are most radical in being the fountains and causes of the rest, or most frequent in concurrence or commixture; wherein it is not the handling of a few of them in passage, the better to describe the mediocrities of virtues, that can satisfy this intention. For if it deserve to be considered, that there are minds which are proportioned to great matters, and others to small (which Aristotle handleth, or ought to have bandied, by the name of magnanimity), doth it not deserve as well to be considered that there are minds proportioned to intend many matters, and others to few? So that some can divide themselves: others can perchance do exactly well, but it must be but in few things at once; and so there cometh to be a narrowness of mind, as well as a pusillanimity. And again, that some minds are proportioned to that which may be dispatched at once, or within a short return of time; others to that which begins afar off, and is to be won with length of pursuit:—

"Jam tum tenditqus fovetque."

So that there may be fitly said to be a longanimity, which is commonly also ascribed to God as a magnanimity. So further deserved it to be considered by Aristotle, "That there is a disposition in conversation (supposing it in things which do in no sort touch or concern a man's self) to soothe and please, and a disposition contrary to contradict and cross;" and deserveth it not much better to be considered. "That there is a disposition, not in conversation or talk, but in matter of more serious nature (and supposing it still in things merely indifferent), to take pleasure in the good of another; and a disposition contrariwise, to take distaste at the good of another?" which is that properly which we call good nature or ill nature, benignity or malignity; and, therefore, I cannot sufficiently marvel that this part of knowledge, touching the several characters of natures and dispositions, should be omitted both in morality and policy, considering it is of so great ministry and suppeditation to them both. A man shall find in the traditions of astrology some pretty and apt divisions of men's natures, according to the predominances of the planets: lovers of quiet, lovers of action, lovers of victory, lovers of honour, lovers of pleasure, lovers of arts, lovers of change, and so forth. A man shall find in the wisest sort of these relations which the Italians make touching conclaves, the natures of the several cardinals handsomely and lively painted forth. A man shall meet with in every day's conference the denominations of sensitive, dry, formal, real, humorous, certain, huomo di prima impressione, huomo di ultima impressione, and the like; and yet, nevertheless, this kind of observations wandereth in words, but is not fixed in inquiry. For the distinctions are found (many of them), but we conclude no precepts upon them: wherein our fault is the greater, because both history, poesy, and daily experience are as goodly fields where these observations grow; whereof we make a few posies to hold in our hands, but no man bringeth them to the confectionary that receipts might be made of them for use of life.

(5) Of much like kind are those impressions of Nature, which are imposed upon the mind by the sex, by the age, by the region, by health and sickness, by beauty and deformity, and the like, which are inherent and not extern; and again, those which are caused by extern fortune, as sovereignty, nobility, obscure birth, riches, want, magistracy, privateness, prosperity, adversity, constant fortune, variable fortune, rising per saltum, per gradus, and the like. And, therefore, we see that Plautus maketh it a wonder to see an old man beneficent, benignitas hujis ut adolescentuli est. Saint Paul concludeth that severity of discipline was to be used to the Cretans, increpa eos dure, upon the disposition of their country, Cretensus semper mendaces, malæ bestiæ, ventres. Sallust noteth that it is usual with kings to desire contradictories: Sed plerumque regiæ voluntates, ut vehementes sunt, sic mobiles, sæpeque ipsæ sibi advers. Tacitus observeth how rarely raising of the fortune mendeth the disposition: solus

Vespasianus mutatus in melius. Pindarus maketh an observation, that great and sudden fortune for the most part defeateth men qui magnam felicitatem concoquere non possunt. So the Psalm showeth it is more easy to keep a measure in the enjoying of fortune, than in the increase of fortune; Divitiæ si affluant, nolite cor apponere. These observations and the like I deny not but are touched a little by Aristotle as in passage in his Rhetorics, and are handled in some scattered discourses; but they were never incorporate into moral philosophy, to which they do essentially appertain; as the knowledge of this diversity of grounds and moulds doth to agriculture, and the knowledge of the diversity of complexions and constitutions doth to the physician, except we mean to follow the indiscretion of empirics, which minister the same medicines to all patients.

(6) Another article of this knowledge is the inquiry touching the affections; for as in medicining of the body, it is in order first to know the divers complexions and constitutions; secondly, the diseases; and lastly, the cures: so in medicining of the mind, after knowledge of the divers characters of men's natures, it followeth in order to know the diseases and infirmities of the mind, which are no other than the perturbations and distempars of the affections. For as the ancient politiques in popular estates were wont to compare the people to the sea, and the orators to the winds; because as the sea would of itself be calm and quiet, if the winds did not move and trouble it; so the people would be peaceable and tractable if the seditious orators did not set them in working and agitation: so it may be fitly said, that the mind in the nature thereof would be temperate and stayed, if the affections, as winds, did not put it into tumult and perturbation. And here again I find strange, as before, that Aristotle should have written divers volumes of Ethics, and never handled the affections which is the principal subject thereof; and yet in his Rhetorics, where they are considered but collaterally and in a second degree (as they may be moved by speech), he findeth place for them, and handleth them well for the quantity; but where their true place is he pretermitteth them. For it is not his disputations about pleasure and pain that can satisfy this inquiry, no more than he that should generally handle the nature of light can be said to handle the nature of colours; for pleasure and pain are to the particular affections as light is to particular colours. Better travails, I suppose, had the Stoics taken in this argument, as far as I can gather by that which we have at second hand. But yet it is like it was after their manner, rather in subtlety of definitions (which in a subject of this nature are but curiosities), than in active and ample descriptions and observations. So likewise I find some particular writings of an elegant nature, touching some of the affections: as of anger, of comfort upon adverse accidents, of tenderness of countenance, and other. But the poets and writers of histories are the best doctors of this knowledge; where we may find painted forth, with great life, how affections are kindled and incited; and how pacified and refrained; and how again contained from act and further degree; how they disclose themselves; how they work; how they vary; how they gather and fortify: how they are enwrapped one within another; and how they do fight and encounter one with another; and other the like particularities. Amongst the which this last is of special use in moral and civil matters; how, I say, to set affection against affection, and to master one by another; even as we used to hunt beast with beast, and fly bird with bird, which otherwise percase we could not so easily recover: upon which foundation is erected that excellent use of præmium and pæna, whereby civil states consist: employing the predominant affections of fear and hope, for the suppressing and bridling the rest. For as in the government of states it is sometimes necessary to bridle one faction with another, so it is in the government within.

(7) Now come we to those points which are within our own command, and have force and operation upon the mind, to affect the will and appetite, and to alter manners: wherein they ought to have handled custom, exercise, habit, education, example, imitation, emulation, company, friends, praise, reproof, exhortation, fame, laws, books, studies: these as they have determinate use in moralities, from these the mind suffereth, and of these are such receipts and regiments compounded and described, as may serve to recover or preserve the health and good estate of the mind, as far as pertaineth to human medicine: of which number we will insist upon some one or two, as an example of the rest, because it were too long to prosecute all; and therefore we do resume custom and habit to speak of.

(8) The opinion of Aristotle seemeth to me a negligent opinion, that of those things which consist by Nature, nothing can be changed by custom; using for example, that if a stone be thrown ten thousand times up it will not learn to ascend; and that by often seeing or hearing we do not learn to see or hear the better. For though this principle be true in things wherein Nature is peremptory (the reason whereof we cannot now stand to discuss), yet it is otherwise in things wherein Nature admitteth a latitude. For he might see that a strait glove will come more easily on with use; and that a wand will by use bend otherwise than it grew; and that by use of the voice we speak louder and stronger; and that by use of enduring heat or cold we endure it the better, and the like: which latter sort have a nearer resemblance unto that subject of manners he handleth, than those instances which he allegeth. But allowing his conclusion, that virtues and vices consist in habit, he ought so much the more to have taught the manner of superinducing that habit: for there be many precepts of the wise ordering the exercises of the mind, as there is of ordering the exercises of the body, whereof we will recite a few.

(9) The first shall be, that we beware we take not at the first either too high a strain or too weak:

for if too high, in a diffident nature you discourage, in a confident nature you breed an opinion of facility, and so a sloth; and in all natures you breed a further expectation than can hold out, and so an insatisfaction in the end: if too weak, of the other side, you may not look to perform and overcome any great task.

(10) Another precept is to practise all things chiefly at two several times, the one when the mind is best disposed, the other when it is worst disposed; that by the one you may gain a great step, by the other you may work out the knots and stonds of the mind, and make the middle times the more easy and pleasant.

(11) Another precept is that which Aristotle mentioneth by the way, which is to bear ever towards the contrary extreme of that whereunto we are by nature inclined; like unto the rowing against the stream, or making a wand straight by bending him contrary to his natural crookedness.

(12) Another precept is that the mind is brought to anything better, and with more sweetness and happiness, if that whereunto you pretend be not first in the intention, but tanquam aliud agendo, because of the natural hatred of the mind against necessity and constraint. Many other axioms there are touching the managing of exercise and custom, which being so conducted doth prove indeed another nature; but, being governed by chance, doth commonly prove but an ape of Nature, and bringeth forth that which is lame and counterfeit.

(13) So if we should handle books and studies, and what influence and operation they have upon manners, are there not divers precepts of great caution and direction appertaining thereunto? Did not one of the fathers in great indignation call poesy vinum dæmonum, because it increaseth temptations, perturbations, and vain opinions? Is not the opinion of Aristotle worthy to be regarded, wherein he saith, "That young men are no fit auditors of moral philosophy, because they are not settled from the boiling heat of their affections, nor attempered with time and experience"? And doth it not hereof come, that those excellent books and discourses of the ancient writers (whereby they have persuaded unto virtue most effectually, by representing her in state and majesty, and popular opinions against virtue in their parasites' coats fit to be scorned and derided), are of so little effect towards honesty of life, because they are not read and revolved by men in their mature and settled years, but confined almost to boys and beginners? But is it not true also, that much less young men are fit auditors of matters of policy, till they have been thoroughly seasoned in religion and morality; lest their judgments be corrupted, and made apt to think that there are no true differences of things, but according to utility and fortune, as the verse describes it, Prosperum et felix scelus virtus vocatur; and again, Ille crucem pretium sceleris tulit, hic diadema: which the poets do speak satirically and in indignation on virtue's behalf; but books of policy do speak it seriously and positively; for so it pleaseth Machiavel to say, "That if Cæsar had been overthrown, he would have been more odious than ever was Catiline;" as if there had been no difference, but in fortune, between a very fury of lust and blood, and the most excellent spirit (his ambition reserved) of the world? Again, is there not a caution likewise to be given of the doctrines of moralities themselves (some kinds of them), lest they make men too precise, arrogant, incompatible; as Cicero saith of Cato, In Marco Catone hæc bona quæ videmus divina et egregia, ipsius scitote esse propria; quæ nonunquam requirimus ea sunt omnia non a natura, sed a magistro? Many other axioms and advices there are touching those proprieties and effects, which studies do infuse and instil into manners. And so, likewise, is there touching the use of all those other points, of company, fame, laws, and the rest, which we recited in the beginning in the doctrine of morality.

(14) But there is a kind of culture of the mind that seemeth yet more accurate and elaborate than the rest, and is built upon this ground; that the minds of all men are at some times in a state more perfect, and at other times in a state more depraved. The purpose, therefore, of this practice is to fix and cherish the good hours of the mind, and to obliterate and take forth the evil. The fixing of the good hath been practised by two means, vows or constant resolutions, and observances or exercises; which are not to be regarded so much in themselves, as because they keep the mind in continual obedience. The obliteration of the evil hath been practised by two means, some kind of redemption or expiation of that which is past, and an inception or account de novo for the time to come. But this part seemeth sacred and religious, and justly; for all good moral philosophy (as was said) is but a handmaid to religion.

(15) Wherefore we will conclude with that last point, which is of all other means the most compendious and summary, and again, the most noble and effectual to the reducing of the mind unto virtue and good estate; which is, the electing and propounding unto a man's self good and virtuous ends of his life, such as may be in a reasonable sort within his compass to attain. For if these two things be supposed, that a man set before him honest and good ends, and again, that he be resolute, constant, and true unto them; it will follow that he shall mould himself into all virtue at once. And this indeed is like the work of nature; whereas the other course is like the work of the hand. For as when a carver makes an image, he shapes only that part whereupon he worketh; as if he be upon the face, that part which shall be the body is but a rude stone still, till such times as he comes to it. But contrariwise when nature makes a flower or living creature, she formeth rudiments of all the parts at one time. So in obtaining

virtue by habit, while a man practiseth temperance, he doth not profit much to fortitude, nor the like but when he dedicateth and applieth himself to good ends, look, what virtue soever the pursuit and passage towards those ends doth commend unto him, he is invested of a precedent disposition to conform himself thereunto. Which state of mind Aristotle doth excellently express himself, that it ought not to be called virtuous, but divine. His words are these: Immanitati autem consentaneum est opponere eam, quæ supra humanitatem est, heroicam sive divinam virtutem; and a little after, Nam ut feræ neque vitium neque virtus est, swic neque Dei: sed hic quidem status altius quiddam virtute est, ille aluid quiddam a vitio. And therefore we may see what celsitude of honour Plinius Secundus attributeth to Trajan in his funeral oration, where he said, "That men needed to make no other prayers to the gods, but that they would continue as good lords to them as Trajan had been;" as if he had not been only an imitation of divine nature, but a pattern of it. But these be heathen and profane passages, having but a shadow of that divine state of mind, which religion and the holy faith doth conduct men unto, by imprinting upon their souls charity, which is excellently called the bond of perfection, because it comprehendeth and fasteneth all virtues together. And as it is elegantly said by Menander of vain love, which is but a false imitation of divine love, Amor melior Sophista lœvo ad humanam vitam—that love teacheth a man to carry himself better than the sophist or preceptor; which he calleth left-handed, because, with all his rules and preceptions, he cannot form a man so dexterously, nor with that facility to prize himself and govern himself, as love can do: so certainly, if a man's mind be truly inflamed with charity, it doth work him suddenly into greater perfection than all the doctrine of morality can do, which is but a sophist in comparison of the other. Nay, further, as Xenophon observed truly, that all other affections, though they raise the mind, yet they do it by distorting and uncomeliness of ecstasies or excesses; but only love doth exalt the mind, and nevertheless at the same instant doth settle and compose it: so in all other excellences, though they advance nature, yet they are subject to excess. Only charity admitteth no excess. For so we see, aspiring to be like God in power, the angels transgressed and fell; Ascendam, et ero similis altissimo: by aspiring to be like God in knowledge, man transgressed and fell; Eritis sicut Dii, scientes bonum et malum: but by aspiring to a similitude of God in goodness or love, neither man nor angel ever transgressed, or shall transgress. For unto that imitation we are called: Diligite inimicos vestros, benefacite eis qui oderunt vos, et orate pro persequentibus et calumniantibus vos, ut sitis filii Patris vestri qui in cœlis est, qui solem suum oriri facit super bonos et malos, et pluit super justos et injustos. So in the first platform of the divine nature itself, the heathen religion speaketh thus, Optimus Maximus: and the sacred Scriptures thus, Miscericordia ejus super omnia opera ejus.

(16) Wherefore I do conclude this part of moral knowledge, concerning the culture and regiment of the mind; wherein if any man, considering the arts thereof which I have enumerated, do judge that my labour is but to collect into an art or science that which hath been pretermitted by others, as matter of common sense and experience, he judgeth well. But as Philocrates sported with Demosthenes, "You may not marvel (Athenians) that Demosthenes and I do differ; for he drinketh water, and I drink wine;" and like as we read of an ancient parable of the two gates of sleep—

> *"Sunt geminæ somni portæ: quarum altera fertur*
> *Cornea, qua veris facilis datur exitus umbris:*
> *Altera candenti perfecta nitens elephanto,*
> *Sed falsa ad cœlum mittunt insomnia manes:"*

so if we put on sobriety and attention, we shall find it a sure maxim in knowledge, that the more pleasant liquor ("of wine") is the more vaporous, and the braver gate ("of ivory") sendeth forth the falser dreams.

(17) But we have now concluded that general part of human philosophy, which contemplateth man segregate, and as he consisteth of body and spirit. Wherein we may further note, that there seemeth to be a relation or conformity between the good of the mind and the good of the body. For as we divided the good of the body into health, beauty, strength, and pleasure, so the good of the mind, inquired in rational and moral knowledges, tendeth to this, to make the mind sound, and without perturbation; beautiful, and graced with decency; and strong and agile for all duties of life. These three, as in the body, so in the mind, seldom meet, and commonly sever. For it is easy to observe, that many have strength of wit and courage, but have neither health from perturbations, nor any beauty or decency in their doings; some again have an elegancy and fineness of carriage which have neither soundness of honesty nor substance of sufficiency; and some again have honest and reformed minds, that can neither become themselves nor manage business; and sometimes two of them meet, and rarely all three. As for pleasure, we have likewise determined that the mind ought not to be reduced to stupid, but to retain pleasure; confined rather in the subject of it, than in the strength and vigour of it.

XXIII

(1) Civil knowledge is conversant about a subject which of all others is most immersed in matter, and hardliest reduced to axiom. Nevertheless, as Cato the Censor said, "That the Romans were like sheep, for that a man were better drive a flock of them, than one of them; for in a flock, if you could get but some few go right, the rest would follow:" so in that respect moral philosophy is more difficile than policy. Again, moral philosophy propoundeth to itself the framing of internal goodness; but civil knowledge requireth only an external goodness; for that as to society sufficeth. And therefore it cometh oft to pass that there be evil times in good governments: for so we find in the Holy story, when the kings were good, yet it is added, Sed adhuc poulus non direxerat cor suum ad Dominum Deum patrum suorum. Again, states, as great engines, move slowly, and are not so soon put out of frame: for as in Egypt the seven good years sustained the seven bad, so governments for a time well grounded do bear out errors following; but the resolution of particular persons is more suddenly subverted. These respects do somewhat qualify the extreme difficulty of civil knowledge.

(2) This knowledge hath three parts, according to the three summary actions of society; which are conversation, negotiation, and government. For man seeketh in society comfort, use, and protection; and they be three wisdoms of divers natures which do often sever—wisdom of the behaviour, wisdom of business, and wisdom of state.

(3) The wisdom of conversation ought not to be over much affected, but much less despised; for it hath not only an honour in itself, but an influence also into business and government. The poet saith, Nec vultu destrue verba tuo: a man may destroy the force of his words with his countenance; so may he of his deeds, saith Cicero, recommending to his brother affability and easy access; Nil interest habere ostium apertum, vultum clausum: it is nothing won to admit men with an open door, and to receive them with a shut and reserved countenance. So we see Atticus, before the first interview between Cæsar and Cicero, the war depending, did seriously advise Cicero touching the composing and ordering of his countenance and gesture. And if the government of the countenance be of such effect, much more is that of the speech, and other carriage appertaining to conversation; the true model whereof seemeth to me well expressed by Livy, though not meant for this purpose: Ne aut arrogans videar, aut obnoxius; quorum alterum est àlienæ libertatis obliti, alterum suæ: the sum of behaviour is to retain a man's own dignity, without intruding upon the liberty of others. On the other side, if behaviour and outward carriage be intended too much, first it may pass into affectation, and then Quid deformius quam scenam in vitam transferre—to act a man's life? But although it proceed not to that extreme, yet it consumeth time, and employeth the mind too much. And therefore as we use to advise young students from company keeping, by saying, Amici fures temporis: so certainly the intending of the discretion of behaviour is a great thief of meditation. Again, such as are accomplished in that form of urbanity please themselves in it, and seldom aspire to higher virtue; whereas those that have defect in it do seek comeliness by reputation; for where reputation is, almost everything becometh; but where that is not, it must be supplied by puntos and compliments. Again, there is no greater impediment of action than an over-curious observance of decency, and the guide of decency, which is time and season. For as Solomon saith, Qui respicit ad ventos, non seminat; et qui respicit ad nubes, non metet: a man must make his opportunity, as oft as find it. To conclude, behaviour seemeth to me as a garment of the mind, and to have the conditions of a garment. For it ought to be made in fashion; it ought not to be too curious; it ought to be shaped so as to set forth any good making of the mind and hide any deformity; and above all, it ought not to be too strait or restrained for exercise or motion. But this part of civil knowledge hath been elegantly handled, and therefore I cannot report it for deficient.

(4) The wisdom touching negotiation or business hath not been hitherto collected into writing, to the great derogation of learning and the professors of learning. For from this root springeth chiefly that note or opinion, which by us is expressed in adage to this effect, that there is no great concurrence between learning and wisdom. For of the three wisdoms which we have set down to pertain to civil life, for wisdom of behaviour, it is by learned men for the most part despised, as an inferior to virtue and an enemy to meditation; for wisdom of government, they acquit themselves well when they are called to it, but that happeneth to few; but for the wisdom of business, wherein man's life is most conversant, there be no books of it, except some few scattered advertisements, that have no proportion to the magnitude of this subject. For if books were written of this as the other, I doubt not but learned men with mean experience would far excel men of long experience without learning, and outshoot them in their own bow.

(5) Neither needeth it at all to be doubted, that this knowledge should be so variable as it falleth not under precept; for it is much less infinite than science of government, which we see is laboured and in some part reduced. Of this wisdom it seemeth some of the ancient Romans in the saddest and wisest times were professors; for Cicero reporteth, that it was then in use for senators that had name and opinion for general wise men, as Coruncanius, Curius, Lælius, and many others, to walk at certain hours

in the Place, and to give audience to those that would use their advice; and that the particular citizens would resort unto them, and consult with them of the marriage of a daughter, or of the employing of a son, or of a purchase or bargain, or of an accusation, and every other occasion incident to man's life. So as there is a wisdom of counsel and advice even in private causes, arising out of a universal insight into the affairs of the world; which is used indeed upon particular causes propounded, but is gathered by general observation of causes of like nature. For so we see in the book which Q. Cicero writeth to his brother, De petitione consulatus (being the only book of business that I know written by the ancients), although it concerned a particular action then on foot, yet the substance thereof consisteth of many wise and politic axioms, which contain not a temporary, but a perpetual direction in the case of popular elections. But chiefly we may see in those aphorisms which have place amongst divine writings, composed by Solomon the king, of whom the Scriptures testify that his heart was as the sands of the sea, encompassing the world and all worldly matters, we see, I say, not a few profound and excellent cautions, precepts, positions, extending to much variety of occasions; whereupon we will stay a while, offering to consideration some number of examples.

(6) Sed et cunctis sermonibus qui dicuntur ne accommodes aurem tuam, ne forte audias servum tuum maledicentem tibi. Here is commended the provident stay of inquiry of that which we would be loth to find: as it was judged great wisdom in Pompeius Magnus that he burned Sertorius' papers unperused.

Vir sapiens, si cum stulto contenderit, sive irascatur, sive rideat, non inveniet requiem. Here is described the great disadvantage which a wise man hath in undertaking a lighter person than himself; which is such an engagement as, whether a man turn the matter to jest, or turn it to heat, or howsoever he change copy, he can no ways quit himself well of it.

Qui delicate a pueritia nutrit servum suum, postea sentiet eum contumacem. Here is signified, that if a man begin too high a pitch in his favours, it doth commonly end in unkindness and unthankfulness.

Vidisti virum velocem in opere suo? coram regibus stabit, nec erit inter ignobiles. Here is observed, that of all virtues for rising to honour, quickness of despatch is the best; for superiors many times love not to have those they employ too deep or too sufficient, but ready and diligent.

Vidi cunctos viventes qui ambulant sub sole, cum adolescente secundo qui consurgit pro eo. Here is expressed that which was noted by Sylla first, and after him by Tiberius. Plures adorant solem orientem quam occidentem vel meridianum.

Si spiritus potestatem habentis ascenderit super te, locum tuum ne demiseris; quia curatio faciet cessare peccata maxima. Here caution is given, that upon displeasure, retiring is of all courses the unfittest; for a man leaveth things at worst, and depriveth himself of means to make them better.

Erat civitas parva, et pauci in ea viri: venit contra eam rex magnus, et vallavit eam, instruxitque munitones per gyrum, et perfecta est obsidio; inventusque est in ea vir pauper et sapiens, et liberavit eam per sapientiam suam; et nullus deinceps recordatus est huminis illius pauperis. Here the corruption of states is set forth, that esteem not virtue or merit longer than they have use of it.

Millis responsio frangit iram. Here is noted that silence or rough answer exasperateth; but an answer present and temperate pacifieth.

Iter pigrorum quasi sepes spinarum. Here is lively represented how laborious sloth proveth in the end; for when things are deferred till the last instant, and nothing prepared beforehand, every step findeth a briar or impediment, which catcheth or stoppeth.

Melior est finis orationis quam principium. Here is taxed the vanity of formal speakers, that study more about prefaces and inducements, than upon the conclusions and issues of speech.

Qui cognoscit in judicio faciem, non bene facit; iste et pro buccella panis deseret veritatem. Here is noted, that a judge were better be a briber than a respecter of persons; for a corrupt judge offendeth not so lightly as a facile.

Vir pauper calumnians pauperes simils est imbri vehementi, in quo paratur fames. Here is expressed the extremity of necessitous extortions, figured in the ancient fable of the full and the hungry horseleech.

Fons turbatus pede, et vena corrupta, est justus cadens coram impio. Here is noted, that one judicial and exemplar iniquity in the face of the world doth trouble the fountains of justice more than many particular injuries passed over by connivance.

Qui subtrahit aliquid a patre et a matre, et dicit hoc non esse peccatum, particeps est homicidii. Here is noted that, whereas men in wronging their best friends use to extenuate their fault, as if they might presume or be bold upon them, it doth contrariwise indeed aggravate their fault, and turneth it from injury to impiety.

Noli esse amicus homini iracundo, nec ambulato cum homine furioso. Here caution is given, that in the election of our friends we do principally avoid those which are impatient, as those that will espouse us to many factions and quarrels.

Qui conturbat domum suam, possidebit ventum. Here is noted, that in domestical separations and breaches men do promise to themselves quieting of their mind and contentment; but still they are deceived of their expectation, and it turneth to wind.

Filius sapiens lætificat patrem: filius vero stultus mæstitia est matri suæ. Here is distinguished, that fathers have most comfort of the good proof of their sons; but mothers have most discomfort of their ill proof, because women have little discerning of virtue, but of fortune.

Qui celat delictum, quærit amicitiam; sed qui altero sermone repetit, separat fæderatos. Here caution is given, that reconcilement is better managed by an amnesty, and passing over that which is past, than by apologies and excuses.

In omni opere bono erit abundantia; ubi autem verba sunt plurima, ibi frequenter egestas. Here is noted, that words and discourse aboundeth most where there is idleness and want.

Primus in sua causa justus: sed venit altera pars, et inquiret in eum. Here is observed, that in all causes the first tale possesseth much; in sort, that the prejudice thereby wrought will be hardly removed, except some abuse or falsity in the information be detected.

Verba bilinguis quasi simplicia, et ipsa perveniunt ad interiora ventris. Here is distinguished, that flattery and insinuation, which seemeth set and artificial, sinketh not far; but that entereth deep which hath show of nature, liberty, and simplicity.

Qui erudit derisorem, ipse sibi injuriam facit; et qui arguit impium, sibi maculam generat. Here caution is given how we tender reprehension to arrogant and scornful natures, whose manner is to esteem it for contumely, and accordingly to return it.

Da sapienti occasionem, et addetur ei sapientia. Here is distinguished the wisdom brought into habit, and that which is but verbal and swimming only in conceit; for the one upon the occasion presented is quickened and redoubled, the other is amazed and confused.

Quomodo in aquis resplendent vultus prospicientium, sic corda hominum manifesta sunt prudentibus. Here the mind of a wise man is compared to a glass, wherein the images of all diversity of natures and customs are represented; from which representation proceedeth that application,

"Qui sapit, innumeris moribus aptus erit."

(7) Thus have I stayed somewhat longer upon these sentences politic of Solomon than is agreeable to the proportion of an example; led with a desire to give authority to this part of knowledge, which I noted as deficient, by so excellent a precedent; and have also attended them with brief observations, such as to my understanding offer no violence to the sense, though I know they may be applied to a more divine use: but it is allowed, even in divinity, that some interpretations, yea, and some writings, have more of the eagle than others; but taking them as instructions for life, they might have received large discourse, if I would have broken them and illustrated them by deducements and examples.

(8) Neither was this in use only with the Hebrews, but it is generally to be found in the wisdom of the more ancient times; that as men found out any observation that they thought was good for life, they would gather it and express it in parable or aphorism or fable. But for fables, they were vicegerents and supplies where examples failed: now that the times abound with history, the aim is better when the mark is alive. And therefore the form of writing which of all others is fittest for this variable argument of negotiation and occasions is that which Machiavel chose wisely and aptly for government; namely, discourse upon histories or examples. For knowledge drawn freshly and in our view out of particulars, knoweth the way best to particulars again. And it hath much greater life for practice when the discourse attendeth upon the example, than when the example attendeth upon the discourse. For this is no point of order, as it seemeth at first, but of substance. For when the example is the ground, being set down in a history at large, it is set down with all circumstances, which may sometimes control the discourse thereupon made, and sometimes supply it, as a very pattern for action; whereas the examples alleged for the discourse's sake are cited succinctly, and without particularity, and carry a servile aspect towards the discourse which they are brought in to make good.

(9) But this difference is not amiss to be remembered, that as history of times is the best ground for discourse of government, such as Machiavel handleth, so histories of lives is the most popular for discourse of business, because it is more conversant in private actions. Nay, there is a ground of discourse for this purpose fitter than them both, which is discourse upon letters, such as are wise and weighty, as many are of Cicero ad Atticum, and others. For letters have a great and more particular representation of business than either chronicles or lives. Thus have we spoken both of the matter and form of this part of civil knowledge, touching negotiation, which we note to be deficient.

(10) But yet there is another part of this part, which differeth as much from that whereof we have spoken as sapere and sibi sapere, the one moving as it were to the circumference, the other to the centre. For there is a wisdom of counsel, and again there is a wisdom of pressing a man's own fortune; and they do sometimes meet, and often sever. For many are wise in their own ways that are weak for government or counsel; like ants, which is a wise creature for itself, but very hurtful for the garden. This wisdom the Romans did take much knowledge of: Nam pol sapiens (saith the comical poet) fingit fortunam sibi; and it grew to an adage, Faber quisque fortunæ propriæ; and Livy attributed it to Cato the

first, In hoc viro tanta vis animi et ingenii inerat, ut quocunque loco natus esset sibi ipse fortunam facturus videretur.

(11) This conceit or position, if it be too much declared and professed, hath been thought a thing impolitic and unlucky, as was observed in Timotheus the Athenian, who, having done many great services to the state in his government, and giving an account thereof to the people as the manner was, did conclude every particular with this clause, "And in this fortune had no part." And it came so to pass, that he never prospered in anything he took in hand afterwards. For this is too high and too arrogant, savouring of that which Ezekiel saith of Pharaoh, Dicis, Fluvius est neus et ego feci memet ipsum; or of that which another prophet speaketh, that men offer sacrifices to their nets and snares; and that which the poet expresseth,

> *"Dextra mihi Deus, et telum quod missile libro,*
> *Nunc adsint!"*

For these confidences were ever unhallowed, and unblessed; and, therefore, those that were great politiques indeed ever ascribed their successes to their felicity and not to their skill or virtue. For so Sylla surnamed himself Felix, not Magnus. So Cæsar said to the master of the ship, Cæsarem portas et fortunam ejus.

(12) But yet, nevertheless, these positions, Faber quisque fortunæ suæ: Sapiens dominabitur astris: Invia virtuti null est via, and the like, being taken and used as spurs to industry, and not as stirrups to insolency, rather for resolution than for the presumption or outward declaration, have been ever thought sound and good; and are no question imprinted in the greatest minds, who are so sensible of this opinion as they can scarce contain it within. As we see in Augustus Cæsar (who was rather diverse from his uncle than inferior in virtue), how when he died he desired his friends about him to give him a plaudite, as if he were conscious to himself that he had played his part well upon the stage. This part of knowledge we do report also as deficient; not but that it is practised too much, but it hath not been reduced to writing. And, therefore, lest it should seem to any that it is not comprehensible by axiom, it is requisite, as we did in the former, that we set down some heads or passages of it.

(13) Wherein it may appear at the first a new and unwonted argument to teach men how to raise and make their fortune; a doctrine wherein every man perchance will be ready to yield himself a disciple, till he see the difficulty: for fortune layeth as heavy impositions as virtue; and it is as hard and severe a thing to be a true politique, as to be truly moral. But the handling hereof concerneth learning greatly, both in honour and in substance. In honour, because pragmatical men may not go away with an opinion that learning is like a lark, that can mount and sing, and please herself, and nothing else; but may know that she holdeth as well of the hawk, that can soar aloft, and can also descend and strike upon the prey. In substance, because it is the perfect law of inquiry of truth, that nothing be in the globe of matter, which should not be likewise in the globe of crystal or form; that is, that there be not anything in being and action which should not be drawn and collected into contemplation and doctrine. Neither doth learning admire or esteem of this architecture of fortune otherwise than as of an inferior work, for no man's fortune can be an end worthy of his being, and many times the worthiest men do abandon their fortune willingly for better respects: but nevertheless fortune as an organ of virtue and merit deserveth the consideration.

(14) First, therefore, the precept which I conceive to be most summary towards the prevailing in fortune, is to obtain that window which Momus did require; who seeing in the frame of man's heart such angles and recesses, found fault there was not a window to look into them; that is, to procure good informations of particulars touching persons, their natures, their desires and ends, their customs and fashions, their helps and advantages, and whereby they chiefly stand, so again their weaknesses and disadvantages, and where they lie most open and obnoxious, their friends, factions, dependences; and again their opposites, enviers, competitors, their moods and times, Sola viri molles aditus et tempora noras; their principles, rules, and observations, and the like: and this not only of persons but of actions; what are on foot from time to time, and how they are conducted, favoured, opposed, and how they import, and the like. For the knowledge of present actions is not only material in itself, but without it also the knowledge of persons is very erroneous: for men change with the actions; and whilst they are in pursuit they are one, and when they return to their nature they are another. These informations of particulars, touching persons and actions, are as the minor propositions in every active syllogism; for no excellency of observations (which are as the major propositions) can suffice to ground a conclusion, if there be error and mistaking in the minors.

(15) That this knowledge is possible, Solomon is our surety, who saith, Consilium in corde viri tanquam aqua profunda; sed vir prudens exhauriet illud. And although the knowledge itself falleth not under precept because it is of individuals, yet the instructions for the obtaining of it may.

(16) We will begin, therefore, with this precept, according to the ancient opinion, that the sinews of wisdom are slowness of belief and distrust; that more trust be given to countenances and deeds than

to words; and in words rather to sudden passages and surprised words than to set and purposed words. Neither let that be feared which is said, Fronti nulla fides, which is meant of a general outward behaviour, and not of the private and subtle motions and labours of the countenance and gesture; which, as Q. Cicero elegantly saith, is Animi janua, "the gate of the mind." None more close than Tiberius, and yet Tacitus saith of Gallus, Etenim vultu offensionem conjectaverat. So again, noting the differing character and manner of his commending Germanicus and Drusus in the Senate, he saith, touching his fashion wherein he carried his speech of Germanicus, thus: Magis in speciem adornatis verbis, quam ut penitus sentire crederetur; but of Drusus thus: Paucioribus sed intentior, et fida oratione; and in another place, speaking of his character of speech when he did anything that was gracious and popular, he saith, "That in other things he was velut eluctantium verborum;" but then again, solutius loquebatur quando subveniret. So that there is no such artificer of dissimulation, nor no such commanded countenance (vultus jussus), that can sever from a feigned tale some of these fashions, either a more slight and careless fashion, or more set and formal, or more tedious and wandering, or coming from a man more drily and hardly.

(17) Neither are deeds such assured pledges as that they may be trusted without a judicious consideration of their magnitude and nature: Fraus sibi in parvis fidem præstruit ut majore emolumento fallat; and the Italian thinketh himself upon the point to be bought and sold, when he is better used than he was wont to be without manifest cause. For small favours, they do but lull men to sleep, both as to caution and as to industry; and are, as Demosthenes calleth them, Alimenta socordiæ. So again we see how false the nature of some deeds are, in that particular which Mutianus practised upon Antonius Primus, upon that hollow and unfaithful reconcilement which was made between them; whereupon Mutianus advanced many of the friends of Antonius, Simul amicis ejus præfecturas et tribunatus largitur: wherein, under pretence to strengthen him, he did desolate him, and won from him his dependents.

(18) As for words, though they be like waters to physicians, full of flattery and uncertainty, yet they are not to be despised specially with the advantage of passion and affection. For so we see Tiberius, upon a stinging and incensing speech of Agrippina, came a step forth of his dissimulation when he said, "You are hurt because you do not reign;" of which Tacitus saith, Audita hæc raram occulti pectoris vocem elicuere: correptamque Græco versu admonuit, ideo lædi quia non regnaret. And, therefore, the poet doth elegantly call passions tortures that urge men to confess their secrets:—

"Vino torus et ira."

And experience showeth there are few men so true to themselves and so settled but that, sometimes upon heat, sometimes upon bravery, sometimes upon kindness, sometimes upon trouble of mind and weakness, they open themselves; specially if they be put to it with a counter-dissimulation, according to the proverb of Spain, Di mentira, y sacar as verdad: "Tell a lie and find a truth."

(19) As for the knowing of men which is at second hand from reports: men's weaknesses and faults are best known from their enemies, their virtues and abilities from their friends, their customs and times from their servants, their conceits and opinions from their familiar friends, with whom they discourse most. General fame is light, and the opinions conceived by superiors or equals are deceitful; for to such men are more masked: Verior fama e domesticis emanat.

(20) But the soundest disclosing and expounding of men is by their natures and ends, wherein the weakest sort of men are best interpreted by their natures, and the wisest by their ends. For it was both pleasantly and wisely said (though I think very untruly) by a nuncio of the Pope, returning from a certain nation where he served as lidger; whose opinion being asked touching the appointment of one to go in his place, he wished that in any case they did not send one that was too wise; because no very wise man would ever imagine what they in that country were like to do. And certainly it is an error frequent for men to shoot over, and to suppose deeper ends and more compass reaches than are: the Italian proverb being elegant, and for the most part true:—

"Di danari, di senno, e di fede,
C'è ne manco che non credi."
"There is commonly less money, less wisdom, and less good faith than men do account upon."

(21) But princes, upon a far other reason, are best interpreted by their natures, and private persons by their ends. For princes being at the top of human desires, they have for the most part no particular ends whereto they aspire, by distance from which a man might take measure and scale of the rest of their actions and desires; which is one of the causes that maketh their hearts more inscrutable. Neither is it sufficient to inform ourselves in men's ends and natures of the variety of them only, but also of the predominancy, what humour reigneth most, and what end is principally sought. For so we see, when Tigellinus saw himself outstripped by Petronius Turpilianus in Nero's humours of pleasures, metus ejus

rimatur, he wrought upon Nero's fears, whereby he broke the other's neck.

(22) But to all this part of inquiry the most compendious way resteth in three things; the first, to have general acquaintance and inwardness with those which have general acquaintance and look most into the world; and specially according to the diversity of business, and the diversity of persons, to have privacy and conversation with some one friend at least which is perfect and well-intelligenced in every several kind. The second is to keep a good mediocrity in liberty of speech and secrecy; in most things liberty; secrecy where it importeth; for liberty of speech inviteth and provoketh liberty to be used again, and so bringeth much to a man's knowledge; and secrecy on the other side induceth trust and inwardness. The last is the reducing of a man's self to this watchful and serene habit, as to make account and purpose, in every conference and action, as well to observe as to act. For as Epictetus would have a philosopher in every particular action to say to himself, Et hoc volo, et etiam institutum servare; so a politic man in everything should say to himself, Et hoc volo, ac etiam aliquid addiscere. I have stayed the longer upon this precept of obtaining good information because it is a main part by itself, which answereth to all the rest. But, above all things, caution must be taken that men have a good stay and hold of themselves, and that this much knowing do not draw on much meddling; for nothing is more unfortunate than light and rash intermeddling in many matters. So that this variety of knowledge tendeth in conclusion but only to this, to make a better and freer choice of those actions which may concern us, and to conduct them with the less error and the more dexterity.

(23) The second precept concerning this knowledge is, for men to take good information touching their own person, and well to understand themselves; knowing that, as St. James saith, though men look oft in a glass, yet they do suddenly forget themselves; wherein as the divine glass is the Word of God, so the politic glass is the state of the world, or times wherein we live, in the which we are to behold ourselves.

(24) For men ought to take an impartial view of their own abilities and virtues; and again of their wants and impediments; accounting these with the most, and those other with the least; and from this view and examination to frame the considerations following.

(25) First, to consider how the constitution of their nature sorteth with the general state of the times; which if they find agreeable and fit, then in all things to give themselves more scope and liberty; but if differing and dissonant, then in the whole course of their life to be more close retired, and reserved; as we see in Tiberius, who was never seen at a play, and came not into the senate in twelve of his last years; whereas Augustus Cæsar lived ever in men's eyes, which Tacitus observeth, alia Tiberio morum via.

(26) Secondly, to consider how their nature sorteth with professions and courses of life, and accordingly to make election, if they be free; and, if engaged, to make the departure at the first opportunity; as we see was done by Duke Valentine, that was designed by his father to a sacerdotal profession, but quitted it soon after in regard of his parts and inclination; being such, nevertheless, as a man cannot tell well whether they were worse for a prince or for a priest.

(27) Thirdly, to consider how they sort with those whom they are like to have competitors and concurrents; and to take that course wherein there is most solitude, and themselves like to be most eminent; as Cæsar Julius did, who at first was an orator or pleader; but when he saw the excellency of Cicero, Hortensius, Catulus, and others for eloquence, and saw there was no man of reputation for the wars but Pompeius, upon whom the state was forced to rely, he forsook his course begun towards a civil and popular greatness, and transferred his designs to a martial greatness.

(28) Fourthly, in the choice of their friends and dependents, to proceed according to the composition of their own nature; as we may see in Cæsar, all whose friends and followers were men active and effectual, but not solemn, or of reputation.

(29) Fifthly, to take special heed how they guide themselves by examples, in thinking they can do as they see others do; whereas perhaps their natures and carriages are far differing. In which error it seemeth Pompey was, of whom Cicero saith that he was wont often to say, Sylla potuit, ego non potero? Wherein he was much abused, the natures and proceedings of himself and his example being the unlikest in the world; the one being fierce, violent, and pressing the fact; the other solemn, and full of majesty and circumstance, and therefore the less effectual.

But this precept touching the politic knowledge of ourselves hath many other branches, whereupon we cannot insist.

(30) Next to the well understanding and discerning of a man's self, there followeth the well opening and revealing a man's self; wherein we see nothing more usual than for the more able man to make the less show. For there is a great advantage in the well setting forth of a man's virtues, fortunes, merits; and again, in the artificial covering of a man's weaknesses, defects, disgraces; staying upon the one, sliding from the other; cherishing the one by circumstances, gracing the other by exposition, and the like. Wherein we see what Tacitus saith of Mutianus, who was the greatest politique of his time, Omnium quæ dixerat feceratque arte quadam ostentator, which requireth indeed some art, lest it turn tedious and arrogant; but yet so, as ostentation (though it be to the first degree of vanity) seemeth to me

rather a vice in manners than in policy; for as it is said, Audacter calumniare, semper aliquid hæret; so, except it be in a ridiculous degree of deformity, Audacter te vendita, semper aluquid hæret. For it will stick with the more ignorant and inferior sort of men, though men of wisdom and rank do smile at it and despise it; and yet the authority won with many doth countervail the disdain of a few. But if it be carried with decency and government, as with a natural, pleasant, and ingenious fashion; or at times when it is mixed with some peril and unsafety (as in military persons); or at times when others are most envied; or with easy and careless passage to it and from it, without dwelling too long, or being too serious; or with an equal freedom of taxing a man's self, as well as gracing himself; or by occasion of repelling or putting down others' injury or insolency; it doth greatly add to reputation: and surely not a few solid natures, that want this ventosity and cannot sail in the height of the winds, are not without some prejudice and disadvantage by their moderation.

(31) But for these flourishes and enhancements of virtue, as they are not perchance unnecessary, so it is at least necessary that virtue be not disvalued and embased under the just price, which is done in three manners—by offering and obtruding a man's self, wherein men think he is rewarded when he is accepted; by doing too much, which will not give that which is well done leave to settle, and in the end induceth satiety; and by finding too soon the fruit of a man's virtue, in commendation, applause, honour, favour; wherein if a man be pleased with a little, let him hear what is truly said: Cave ne insuetus rebus majoribus videaris, si hæc te res parva sicuti magna delectat.

(32) But the covering of defects is of no less importance than the valuing of good parts; which may be done likewise in three manners—by caution, by colour, and by confidence. Caution is when men do ingeniously and discreetly avoid to be put into those things for which they are not proper; whereas contrariwise bold and unquiet spirits will thrust themselves into matters without difference, and so publish and proclaim all their wants. Colour is when men make a way for themselves to have a construction made of their faults or wants, as proceeding from a better cause or intended for some other purpose. For of the one it is well said,

"*Sæpe latet vitium proximitate boni,*"

and therefore whatsoever want a man hath, he must see that he pretend the virtue that shadoweth it; as if he be dull, he must affect gravity; if a coward, mildness; and so the rest. For the second, a man must frame some probable cause why he should not do his best, and why he should dissemble his abilities; and for that purpose must use to dissemble those abilities which are notorious in him, to give colour that his true wants are but industries and dissimulations. For confidence, it is the last but the surest remedy—namely, to depress and seem to despise whatsoever a man cannot attain; observing the good principle of the merchants, who endeavour to raise the price of their own commodities, and to beat down the price of others. But there is a confidence that passeth this other, which is to face out a man's own defects, in seeming to conceive that he is best in those things wherein he is failing; and, to help that again, to seem on the other side that he hath least opinion of himself in those things wherein he is best: like as we shall see it commonly in poets, that if they show their verses, and you except to any, they will say, "That that line cost them more labour than any of the rest;" and presently will seem to disable and suspect rather some other line, which they know well enough to be the best in the number. But above all, in this righting and helping of a man's self in his own carriage, he must take heed he show not himself dismantled and exposed to scorn and injury, by too much dulceness, goodness, and facility of nature; but show some sparkles of liberty, spirit, and edge. Which kind of fortified carriage, with a ready rescussing of a man's self from scorns, is sometimes of necessity imposed upon men by somewhat in their person or fortune; but it ever succeedeth with good felicity.

(33) Another precept of this knowledge is by all possible endeavour to frame the mind to be pliant and obedient to occasion; for nothing hindereth men's fortunes so much as this: Idem manebat, neque idem decebat—men are where they were, when occasions turn: and therefore to Cato, whom Livy maketh such an architect of fortune, he addeth that he had versatile ingenium. And thereof it cometh that these grave solemn wits, which must be like themselves and cannot make departures, have more dignity than felicity. But in some it is nature to be somewhat vicious and enwrapped, and not easy to turn. In some it is a conceit that is almost a nature, which is, that men can hardly make themselves believe that they ought to change their course, when they have found good by it in former experience. For Machiavel noted wisely how Fabius Maximus would have been temporising still, according to his old bias, when the nature of the war was altered and required hot pursuit. In some other it is want of point and penetration in their judgment, that they do not discern when things have a period, but come in too late after the occasion; as Demosthenes compareth the people of Athens to country fellows, when they play in a fence school, that if they have a blow, then they remove their weapon to that ward, and not before. In some other it is a lothness to lose labours passed, and a conceit that they can bring about occasions to their ply; and yet in the end, when they see no other remedy, then they come to it with disadvantage; as Tarquinius, that gave for the third part of Sibylla's books the treble price, when he

might at first have had all three for the simple. But from whatsoever root or cause this restiveness of mind proceedeth, it is a thing most prejudicial; and nothing is more politic than to make the wheels of our mind concentric and voluble with the wheels of fortune.

(34) Another precept of this knowledge, which hath some affinity with that we last spoke of, but with difference, is that which is well expressed, Fatis accede deisque, that men do not only turn with the occasions, but also run with the occasions, and not strain their credit or strength to over-hard or extreme points; but choose in their actions that which is most passable: for this will preserve men from foil, not occupy them too much about one matter, win opinion of moderation, please the most, and make a show of a perpetual felicity in all they undertake: which cannot but mightily increase reputation.

(35) Another part of this knowledge seemeth to have some repugnancy with the former two, but not as I understand it; and it is that which Demosthenes uttereth in high terms: Et quemadmodum receptum est, ut exercitum ducat imperator, sic et a cordatis viris res ipsæ ducendæ; ut quæipsis videntur, ea gerantur, et non ipsi eventus persequi cogantur. For if we observe we shall find two differing kinds of sufficiency in managing of business: some can make use of occasions aptly and dexterously, but plot little; some can urge and pursue their own plots well, but cannot accommodate nor take in; either of which is very imperfect without the other.

(36) Another part of this knowledge is the observing a good mediocrity in the declaring or not declaring a man's self: for although depth of secrecy, and making way (qualis est via navis in mari, which the French calleth sourdes menées, when men set things in work without opening themselves at all), be sometimes both prosperous and admirable; yet many times dissimulatio errores parit, qui dissimulatorem ipsum illaqueant. And therefore we see the greatest politiques have in a natural and free manner professed their desires, rather than been reserved and disguised in them. For so we see that Lucius Sylla made a kind of profession, "that he wished all men happy or unhappy, as they stood his friends or enemies." So Cæsar, when he went first into Gaul, made no scruple to profess "that he had rather be first in a village than second at Rome." So again, as soon as he had begun the war, we see what Cicero saith of him, Alter (meaning of Cæsar) non recusat, sed quodammodo postulat, ut (ut est) sic appelletur tyrannus. So we may see in a letter of Cicero to Atticus, that Augustus Cæsar, in his very entrance into affairs, when he was a darling of the senate, yet in his harangues to the people would swear, Ita parentis honores consequi liceat (which was no less than the tyranny), save that, to help it, he would stretch forth his hand towards a statue of Cæsar's that was erected in the place: and men laughed and wondered, and said, "Is it possible?" or, "Did you ever hear the like?" and yet thought he meant no hurt; he did it so handsomely and ingenuously. And all these were prosperous: whereas Pompey, who tended to the same ends, but in a more dark and dissembling manner as Tacitus saith of him, Occultior non melior, wherein Sallust concurreth, Ore probo, animo inverecundo, made it his design, by infinite secret engines, to cast the state into an absolute anarchy and confusion, that the state might cast itself into his arms for necessity and protection, and so the sovereign power be put upon him, and he never seen in it: and when he had brought it (as he thought) to that point when he was chosen consul alone, as never any was, yet he could make no great matter of it, because men understood him not; but was fain in the end to go the beaten track of getting arms into his hands, by colour of the doubt of Cæsar's designs: so tedious, casual, and unfortunate are these deep dissimulations: whereof it seemeth Tacitus made this judgment, that they were a cunning of an inferior form in regard of true policy; attributing the one to Augustus, the other to Tiberius; where, speaking of Livia, he saith, Et cum artibus mariti simulatione filii bene compostia: for surely the continual habit of dissimulation is but a weak and sluggish cunning, and not greatly politic.

(37) Another precept of this architecture of fortune is to accustom our minds to judge of the proportion or value of things, as they conduce and are material to our particular ends; and that to do substantially and not superficially. For we shall find the logical part (as I may term it) of some men's minds good, but the mathematical part erroneous; that is, they can well judge of consequences, but not of proportions and comparison, preferring things of show and sense before things of substance and effect. So some fall in love with access to princes, others with popular fame and applause, supposing they are things of great purchase, when in many cases they are but matters of envy, peril, and impediment. So some measure things according to the labour and difficulty or assiduity which are spent about them; and think, if they be ever moving, that they must needs advance and proceed; as Cæsar saith in a despising manner of Cato the second, when he describeth how laborious and indefatigable he was to no great purpose, Hæc omnia magno studio agebat. So in most things men are ready to abuse themselves in thinking the greatest means to be best, when it should be the fittest.

(38) As for the true marshalling of men's pursuits towards their fortune, as they are more or less material, I hold them to stand thus. First the amendment of their own minds. For the removal of the impediments of the mind will sooner clear the passages of fortune than the obtaining fortune will remove the impediments of the mind. In the second place I set down wealth and means; which I know most men would have placed first, because of the general use which it beareth towards all variety of occasions. But that opinion I may condemn with like reason as Machiavel doth that other, that moneys

were the sinews of the wars; whereas (saith he) the true sinews of the wars are the sinews of men's arms, that is, a valiant, populous, and military nation: and he voucheth aptly the authority of Solon, who, when Crœsus showed him his treasury of gold, said to him, that if another came that had better iron, he would be master of his gold. In like manner it may be truly affirmed that it is not moneys that are the sinews of fortune, but it is the sinews and steel of men's minds, wit, courage, audacity, resolution, temper, industry, and the like. In the third place I set down reputation, because of the peremptory tides and currents it hath; which, if they be not taken in their due time, are seldom recovered, it being extreme hard to play an after-game of reputation. And lastly I place honour, which is more easily won by any of the other three, much more by all, than any of them can be purchased by honour. To conclude this precept, as there is order and priority in matter, so is there in time, the preposterous placing whereof is one of the commonest errors: while men fly to their ends when they should intend their beginnings, and do not take things in order of time as they come on, but marshal them according to greatness and not according to instance; not observing the good precept, Quod nunc instat agamus.

(39) Another precept of this knowledge is not to embrace any matters which do occupy too great a quantity of time, but to have that sounding in a man's ears, Sed fugit interea fugit irreparabile tempus: and that is the cause why those which take their course of rising by professions of burden, as lawyers, orators, painful divines, and the like, are not commonly so politic for their own fortune, otherwise than in their ordinary way, because they want time to learn particulars, to wait occasions, and to devise plots.

(40) Another precept of this knowledge is to imitate nature, which doth nothing in vain; which surely a man may do if he do well interlace his business, and bend not his mind too much upon that which he principally intendeth. For a man ought in every particular action so to carry the motions of his mind, and so to have one thing under another, as if he cannot have that he seeketh in the best degree, yet to have it in a second, or so in a third; and if he can have no part of that which he purposed, yet to turn the use of it to somewhat else; and if he cannot make anything of it for the present, yet to make it as a seed of somewhat in time to come; and if he can contrive no effect or substance from it, yet to win some good opinion by it, or the like. So that he should exact an account of himself of every action, to reap somewhat, and not to stand amazed and confused if he fail of that he chiefly meant: for nothing is more impolitic than to mind actions wholly one by one. For he that doth so loseth infinite occasions which intervene, and are many times more proper and propitious for somewhat that he shall need afterwards, than for that which he urgeth for the present; and therefore men must be perfect in that rule, Hæc oportet facere, et illa non imittere.

(41) Another precept of this knowledge is, not to engage a man's self peremptorily in anything, though it seem not liable to accident; but ever to have a window to fly out at, or a way to retire: following the wisdom in the ancient fable of the two frogs, which consulted when their plash was dry whither they should go; and the one moved to go down into a pit, because it was not likely the water would dry there; but the other answered, "True, but if it do, how shall we get out again?"

(42) Another precept of this knowledge is that ancient precept of Bias, construed not to any point of perfidiousness, but to caution and moderation, Et ama tanquam inimicus futurus et odi tanquam amaturus. For it utterly betrayeth all utility for men to embark themselves too far into unfortunate friendships, troublesome spleens, and childish and humorous envies or emulations.

(43) But I continue this beyond the measure of an example; led, because I would not have such knowledges, which I note as deficient, to be thought things imaginative or in the air, or an observation or two much made of, but things of bulk and mass, whereof an end is more hardly made than a beginning. It must be likewise conceived, that in these points which I mention and set down, they are far from complete tractates of them, but only as small pieces for patterns. And lastly, no man I suppose will think that I mean fortunes are not obtained without all this ado; for I know they come tumbling into some men's laps; and a number obtain good fortunes by diligence in a plain way, little intermeddling, and keeping themselves from gross errors.

(44) But as Cicero, when he setteth down an idea of a perfect orator, doth not mean that every pleader should be such; and so likewise, when a prince or a courtier hath been described by such as have handled those subjects, the mould hath used to be made according to the perfection of the art, and not according to common practice: so I understand it, that it ought to be done in the description of a politic man, I mean politic for his own fortune.

(45) But it must be remembered all this while, that the precepts which we have set down are of that kind which may be counted and called Bonæ Artes. As for evil arts, if a man would set down for himself that principle of Machiavel, "That a man seek not to attain virtue itself, but the appearance only thereof; because the credit of virtue is a help, but the use of it is cumber:" or that other of his principles, "That he presuppose that men are not fitly to be wrought otherwise but by fear; and therefore that he seek to have every man obnoxious, low, and in straits," which the Italians call seminar spine, to sow thorns: or that other principle, contained in the verse which Cicero citeth, Cadant amici, dummodo inimici intercidant, as the triumvirs, which sold every one to other the lives of their friends for the

deaths of their enemies: or that other protestation of L. Catilina, to set on fire and trouble states, to the end to fish in droumy waters, and to unwrap their fortunes, *Ego si quid in fortunis meis excitatum sit incendium, id non aqua sed ruina restinguam:* or that other principle of Lysander, "That children are to be deceived with comfits, and men with oaths:" and the like evil and corrupt positions, whereof (as in all things) there are more in number than of the good: certainly with these dispensations from the laws of charity and integrity, the pressing of a man's fortune may be more hasty and compendious. But it is in life as it is in ways, the shortest way is commonly the foulest, and surely the fairer way is not much about.

(46) But men, if they be in their own power, and do bear and sustain themselves, and be not carried away with a whirlwind or tempest of ambition, ought in the pursuit of their own fortune to set before their eyes not only that general map of the world, "That all things are vanity and vexation of spirit," but many other more particular cards and directions: chiefly that, that being without well-being is a curse, and the greater being the greater curse; and that all virtue is most rewarded and all wickedness most punished in itself: according as the poet saith excellently:

> *"Quæ vobis, quæ digna, viri pro laudibus istis*
> *Præmia posse rear solvi? pulcherrima primum*
> *Dii moresque dabunt vestri."*

And so of the contrary. And secondly they ought to look up to the Eternal Providence and Divine Judgment, which often subverteth the wisdom of evil plots and imaginations, according to that scripture, "He hath conceived mischief, and shall bring forth a vain thing." And although men should refrain themselves from injury and evil arts, yet this incessant and Sabbathless pursuit of a man's fortune leaveth not tribute which we owe to God of our time; who (we see) demandeth a tenth of our substance, and a seventh, which is more strict, of our time: and it is to small purpose to have an erected face towards heaven, and a perpetual grovelling spirit upon earth, eating dust as doth the serpent, *Atque affigit humo divinæ particulam auræ.* And if any man flatter himself that he will employ his fortune well, though he should obtain it ill, as was said concerning Augustus Cæsar, and after of Septimius Severus, "That either they should never have been born, or else they should never have died," they did so much mischief in the pursuit and ascent of their greatness, and so much good when they were established; yet these compensations and satisfactions are good to be used, but never good to be purposed. And lastly, it is not amiss for men, in their race towards their fortune, to cool themselves a little with that conceit which is elegantly expressed by the Emperor Charles V., in his instructions to the king his son, "That fortune hath somewhat of the nature of a woman, that if she he too much wooed she is the farther off." But this last is but a remedy for those whose tastes are corrupted: let men rather build upon that foundation which is as a corner-stone of divinity and philosophy, wherein they join close, namely that same Primum quærite. For divinity saith, Primum quærite regnum Dei, et ista omnia adjicientur vobis: and philosophy saith, Primum quærite bona animi; cætera aut aderunt, aut non oberunt. And although the human foundation hath somewhat of the sands, as we see in M. Brutus, when he broke forth into that speech,

> *"Te colui (Virtus) ut rem; ast tu nomen inane es;"*

yet the divine foundation is upon the rock. But this may serve for a taste of that knowledge which I noted as deficient.

(47) Concerning government, it is a part of knowledge secret and retired in both these respects in which things are deemed secret; for some things are secret because they are hard to know, and some because they are not fit to utter. We see all governments are obscure and invisible:

> *"Totamque infusa per artus*
> *Mens agitat molem, et magno se corpore miscet."*

Such is the description of governments. We see the government of God over the world is hidden, insomuch as it seemeth to participate of much irregularity and confusion. The government of the soul in moving the body is inward and profound, and the passages thereof hardly to be reduced to demonstration. Again, the wisdom of antiquity (the shadows whereof are in the poets) in the description of torments and pains, next unto the crime of rebellion, which was the giants' offence, doth detest the offence of futility, as in Sisyphus and Tantalus. But this was meant of particulars: nevertheless even unto the general rules and discourses of policy and government there is due a reverent and reserved handling.

(48) But contrariwise in the governors towards the governed, all things ought as far as the frailty of man permitteth to be manifest and revealed. For so it is expressed in the Scriptures touching the

government of God, that this globe, which seemeth to us a dark and shady body, is in the view of God as crystal: Et in conspectu sedis tanquam mare vitreum simile crystallo. So unto princes and states, and specially towards wise senates and councils, the natures and dispositions of the people, their conditions and necessities, their factions and combinations, their animosities and discontents, ought to be, in regard of the variety of their intelligences, the wisdom of their observations, and the height of their station where they keep sentinel, in great part clear and transparent. Wherefore, considering that I write to a king that is a master of this science, and is so well assisted, I think it decent to pass over this part in silence, as willing to obtain the certificate which one of the ancient philosophers aspired unto; who being silent, when others contended to make demonstration of their abilities by speech, desired it might be certified for his part, "That there was one that knew how to hold his peace."

(49) Notwithstanding, for the more public part of government, which is laws, I think good to note only one deficiency; which is, that all those which have written of laws have written either as philosophers or as lawyers, and none as statesmen. As for the philosophers, they make imaginary laws for imaginary commonwealths, and their discourses are as the stars, which give little light because they are so high. For the lawyers, they write according to the states where they live what is received law, and not what ought to be law; for the wisdom of a law-maker is one, and of a lawyer is another. For there are in nature certain fountains of justice whence all civil laws are derived but as streams; and like as waters do take tinctures and tastes from the soils through which they run, so do civil laws vary according to the regions and governments where they are planted, though they proceed from the same fountains. Again, the wisdom of a law-maker consisteth not only in a platform of justice, but in the application thereof; taking into consideration by what means laws may be made certain, and what are the causes and remedies of the doubtfulness and uncertainty of law; by what means laws may be made apt and easy to be executed, and what are the impediments and remedies in the execution of laws; what influence laws touching private right of meum and tuum have into the public state, and how they may be made apt and agreeable; how laws are to be penned and delivered, whether in texts or in Acts, brief or large, with preambles or without; how they are to be pruned and reformed from time to time, and what is the best means to keep them from being too vast in volume, or too full of multiplicity and crossness; how they are to be expounded, when upon causes emergent and judicially discussed, and when upon responses and conferences touching general points or questions; how they are to be pressed, rigorously or tenderly; how they are to be mitigated by equity and good conscience, and whether discretion and strict law are to be mingled in the same courts, or kept apart in several courts; again, how the practice, profession, and erudition of law is to be censured and governed; and many other points touching the administration and (as I may term it) animation of laws. Upon which I insist the less, because I purpose (if God give me leave), having begun a work of this nature in aphorisms, to propound it hereafter, noting it in the meantime for deficient.

(50) And for your Majesty's laws of England, I could say much of their dignity, and somewhat of their defect; but they cannot but excel the civil laws in fitness for the government, for the civil law was nonhos quæsitum munus in usus; it was not made for the countries which it governeth. Hereof I cease to speak because I will not intermingle matter of action with matter of general learning.

XXIV

Thus have I concluded this portion of learning touching civil knowledge; and with civil knowledge have concludéd human philosophy; and with human philosophy, philosophy in general. And being now at some pause, looking back into that I have passed through, this writing seemeth to me (si nunquam fallit imago), as far as a man can judge of his own work, not much better than that noise or sound which musicians make while they are in tuning their instruments, which is nothing pleasant to hear, but yet is a cause why the music is sweeter afterwards. So have I been content to tune the instruments of the Muses, that they may play that have better hands. And surely, when I set before me the condition of these times, in which learning hath made her third visitation or circuit in all the qualities thereof; as the excellency and vivacity of the wits of this age; the noble helps and lights which we have by the travails of ancient writers; the art of printing, which communicateth books to men of all fortunes; the openness of the world by navigation, which hath disclosed multitudes of experiments, and a mass of natural history; the leisure wherewith these times abound, not employing men so generally in civil business, as the states of Græcia did, in respect of their popularity, and the state of Rome, in respect of the greatness of their monarchy; the present disposition of these times at this instant to peace; the consumption of all that ever can be said in controversies of religion, which have so much diverted men from other sciences; the perfection of your Majesty's learning, which as a phœnix may call whole volleys of wits to follow you; and the inseparable propriety of time, which is ever more and more to disclose truth; I cannot but be raised to this persuasion, that this third period of time will far surpass that of the Grecian and Roman learning; only if men will know their own strength and their own weakness both; and take, one from the other, light of invention, and not fire of contradiction; and esteem of the inquisition of truth as of an enterprise, and not as of a quality or ornament; and employ wit and magnificence to things of worth and excellency, and not to things vulgar and of popular estimation. As for my labours, if any man shall please himself or others in the reprehension of them, they shall make that ancient and patient request, Verbera, sed audi: let men reprehend them, so they observe and weigh them. For the appeal is lawful (though it may be it shall not be needful) from the first cogitations of men to their second, and from the nearer times to the times further off. Now let us come to that learning, which both the former times were not so blessed as to know, sacred and inspired divinity, the Sabbath and port of all men's labours and peregrinations.

XXV

(1) The prerogative of God extendeth as well to the reason as to the will of man: so that as we are to obey His law, though we find a reluctation in our will, so we are to believe His word, though we find a reluctation in our reason. For if we believe only that which is agreeable to our sense we give consent to the matter, and not to the author; which is no more than we would do towards a suspected and discredited witness; but that faith which was accounted to Abraham for righteousness was of such a point as whereat Sarah laughed, who therein was an image of natural reason.

(2) Howbeit (if we will truly consider of it) more worthy it is to believe than to know as we now know. For in knowledge man's mind suffereth from sense: but in belief it suffereth from spirit, such one as it holdeth for more authorised than itself and so suffereth from the worthier agent. Otherwise it is of the state of man glorified; for then faith shall cease, and we shall know as we are known.

(3) Wherefore we conclude that sacred theology (which in our idiom we call divinity) is grounded only upon the word and oracle of God, and not upon the light of nature: for it is written, Cæli enarrant gloriam Dei; but it is not written, Cæli enarrant voluntatem Dei: but of that it is said, Ad legem et testimonium: si non fecerint secundum verbum istud, &c. This holdeth not only in those points of faith which concern the great mysteries of the Deity, of the creation, of the redemption, but likewise those which concern the law moral, truly interpreted: "Love your enemies: do good to them that hate you; be like to your heavenly Father, that suffereth His rain to fall upon the just and unjust." To this it ought to be applauded, Nec vox hominem sonat: it is a voice beyond the light of nature. So we see the heathen poets, when they fall upon a libertine passion, do still expostulate with laws and moralities, as if they were opposite and malignant to nature: Et quod natura remittit, invida jura negant. So said Dendamis the Indian unto Alexander's messengers, that he had heard somewhat of Pythagoras, and some other of the wise men of Græcia, and that he held them for excellent men: but that they had a fault, which was that they had in too great reverence and veneration a thing they called law and manners. So it must be confessed that a great part of the law moral is of that perfection whereunto the light of nature cannot aspire: how then is it that man is said to have, by the light and law of nature, some notions and conceits of virtue and vice, justice and wrong, good and evil? Thus, because the light of nature is used in two several senses: the one, that which springeth from reason, sense, induction, argument, according to the laws of heaven and earth; the other, that which is imprinted upon the spirit of man by an inward instinct, according to the law of conscience, which is a sparkle of the purity of his first estate: in which latter sense only he is participant of some light and discerning touching the perfection of the moral law; but how? sufficient to check the vice but not to inform the duty. So then the doctrine of religion, as well moral as mystical, is not to be attained but by inspiration and revelation from God.

(4) The use notwithstanding of reason in spiritual things, and the latitude thereof, is very great and general: for it is not for nothing that the apostle calleth religion "our reasonable service of God;" insomuch as the very ceremonies and figures of the old law were full of reason and signification, much more than the ceremonies of idolatry and magic, that are full of non-significants and surd characters. But most specially the Christian faith, as in all things so in this, deserveth to be highly magnified; holding and preserving the golden mediocrity in this point between the law of the heathen and the law of Mahomet, which have embraced the two extremes. For the religion of the heathen had no constant belief or confession, but left all to the liberty of agent; and the religion of Mahomet on the other side interdicteth argument altogether: the one having the very face of error, and the other of imposture; whereas the Faith doth both admit and reject disputation with difference.

(5) The use of human reason in religion is of two sorts: the former, in the conception and apprehension of the mysteries of God to us revealed; the other, in the inferring and deriving of doctrine and direction thereupon. The former extendeth to the mysteries themselves; but how? by way of illustration, and not by way of argument. The latter consisteth indeed of probation and argument. In the former we see God vouchsafeth to descend to our capacity, in the expressing of His mysteries in sort as may be sensible unto us; and doth graft His revelations and holy doctrine upon the notions of our reason, and applieth His inspirations to open our understanding, as the form of the key to the ward of

the lock. For the latter there is allowed us a use of reason and argument, secondary and respective, although not original and absolute. For after the articles and principles of religion are placed and exempted from examination of reason, it is then permitted unto us to make derivations and inferences from and according to the analogy of them, for our better direction. In nature this holdeth not; for both the principles are examinable by induction, though not by a medium or syllogism; and besides, those principles or first positions have no discordance with that reason which draweth down and deduceth the inferior positions. But yet it holdeth not in religion alone, but in many knowledges, both of greater and smaller nature, namely, wherein there are not only posita but placita; for in such there can be no use of absolute reason. We see it familiarly in games of wit, as chess, or the like. The draughts and first laws of the game are positive, but how? merely ad placitum, and not examinable by reason; but then how to direct our play thereupon with best advantage to win the game is artificial and rational. So in human laws there be many grounds and maxims which are placita juris, positive upon authority, and not upon reason, and therefore not to be disputed: but what is most just, not absolutely but relatively, and according to those maxims, that affordeth a long field of disputation. Such therefore is that secondary reason, which hath place in divinity, which is grounded upon the placets of God.

(6) Here therefore I note this deficiency, that there hath not been, to my understanding, sufficiently inquired and handled the true limits and use of reason in spiritual things, as a kind of divine dialectic: which for that it is not done, it seemeth to me a thing usual, by pretext of true conceiving that which is revealed, to search and mine into that which is not revealed; and by pretext of enucleating inferences and contradictories, to examine that which is positive. The one sort falling into the error of Nicodemus, demanding to have things made more sensible than it pleaseth God to reveal them, Quomodo possit homo nasci cum sit senex? The other sort into the error of the disciples, which were scandalised at a show of contradiction, Quid est hoc quod dicit nobis? Modicum et non videbitis me; et iterum, modicum, et videbitis me, &c.

(7) Upon this I have insisted the more, in regard of the great and blessed use thereof; for this point well laboured and defined of would in my judgment be an opiate to stay and bridle not only the vanity of curious speculations, wherewith the schools labour, but the fury of controversies, wherewith the Church laboureth. For it cannot but open men's eyes to see that many controversies do merely pertain to that which is either not revealed or positive; and that many others do grow upon weak and obscure inferences or derivations: which latter sort, if men would revive the blessed style of that great doctor of the Gentiles, would be carried thus, ego, non dominus; and again, secundum consilium meum, in opinions and counsels, and not in positions and oppositions. But men are now over-ready to usurp the style, non ego, sed dominus; and not so only, but to bind it with the thunder and denunciation of curses and anathemas, to the terror of those which have not sufficiently learned out of Solomon that "The causeless curse shall not come."

(8) Divinity hath two principal parts: the matter informed or revealed, and the nature of the information or revelation; and with the latter we will begin, because it hath most coherence with that which we have now last handled. The nature of the information consisteth of three branches: the limits of the information, the sufficiency of the information, and the acquiring or obtaining the information. Unto the limits of the information belong these considerations: how far forth particular persons continue to be inspired; how far forth the Church is inspired; and how far forth reason may be used; the last point whereof I have noted as deficient. Unto the sufficiency of the information belong two considerations: what points of religion are fundamental, and what perfective, being matter of further building and perfection upon one and the same foundation; and again, how the gradations of light according to the dispensation of times are material to the sufficiency of belief.

(9) Here again I may rather give it in advice than note it as deficient, that the points fundamental, and the points of further perfection only, ought to be with piety and wisdom distinguished; a subject tending to much like end as that I noted before; for as that other were likely to abate the number of controversies, so this is likely to abate the heat of many of them. We see Moses when he saw the Israelite and the Egyptian fight, he did not say, "Why strive you?" but drew his sword and slew the Egyptian; but when he saw the two Israelites fight, he said, "You are brethren, why strive you?" If the point of doctrine be an Egyptian, it must be slain by the sword of the Spirit, and not reconciled; but if it be an Israelite, though in the wrong, then, "Why strive you?" We see of the fundamental points, our Saviour penneth the league thus, "He that is not with us is against us;" but of points not fundamental, thus, "He that is not against us is with us." So we see the coat of our Saviour was entire without seam, and so is the doctrine of the Scriptures in itself; but the garment of the Church was of divers colours and yet not divided. We see the chaff may and ought to be severed from the corn in the ear, but the tares may not be pulled up from the corn in the field. So as it is a thing of great use well to define what, and of what latitude, those points are which do make men mere aliens and disincorporate from the Church of God.

(10) For the obtaining of the information, it resteth upon the true and sound interpretation of the Scriptures, which are the fountains of the water of life. The interpretations of the Scriptures are of two

sorts: methodical, and solute or at large. For this divine water, which excelleth so much that of Jacob's well, is drawn forth much in the same kind as natural water useth to be out of wells and fountains; either it is first forced up into a cistern, and from thence fetched and derived for use; or else it is drawn and received in buckets and vessels immediately where it springeth. The former sort whereof, though it seem to be the more ready, yet in my judgment is more subject to corrupt. This is that method which hath exhibited unto us the scholastical divinity; whereby divinity hath been reduced into an art, as into a cistern, and the streams of doctrine or positions fetched and derived from thence.

(11) In this men have sought three things, a summary brevity, a compacted strength, and a complete perfection; whereof the two first they fail to find, and the last they ought not to seek. For as to brevity, we see in all summary methods, while men purpose to abridge, they give cause to dilate. For the sum or abridgment by contraction becometh obscure; the obscurity requireth exposition, and the exposition is deduced into large commentaries, or into commonplaces and titles, which grow to be more vast than the original writings, whence the sum was at first extracted. So we see the volumes of the schoolmen are greater much than the first writings of the fathers, whence the master of the sentences made his sum or collection. So in like manner the volumes of the modern doctors of the civil law exceed those of the ancient jurisconsults, of which Tribonian compiled the digest. So as this course of sums and commentaries is that which doth infallibly make the body of sciences more immense in quantity, and more base in substance.

(12) And for strength, it is true that knowledges reduced into exact methods have a show of strength, in that each part seemeth to support and sustain the other; but this is more satisfactory than substantial, like unto buildings which stand by architecture and compaction, which are more subject to ruin than those that are built more strong in their several parts, though less compacted. But it is plain that the more you recede from your grounds, the weaker do you conclude; and as in nature, the more you remove yourself from particulars, the greater peril of error you do incur; so much more in divinity, the more you recede from the Scriptures by inferences and consequences, the more weak and dilute are your positions.

(13) And as for perfection or completeness in divinity, it is not to be sought, which makes this course of artificial divinity the more suspect. For he that will reduce a knowledge into an art will make it round and uniform; but in divinity many things must be left abrupt, and concluded with this: O altitudo sapientiæ et scientiæ Dei! quam incomprehensibilia sunt juducua ejus, et non investigabiles viæ ejus. So again the apostle saith, Ex parte scimus: and to have the form of a total, where there is but matter for a part, cannot be without supplies by supposition and presumption. And therefore I conclude that the true use of these sums and methods hath place in institutions or introductions preparatory unto knowledge; but in them, or by deducement from them, to handle the main body and substance of a knowledge is in all sciences prejudicial, and in divinity dangerous.

(14) As to the interpretation of the Scriptures solute and at large, there have been divers kinds introduced and devised; some of them rather curious and unsafe than sober and warranted. Notwithstanding, thus much must be confessed, that the Scriptures, being given by inspiration and not by human reason, do differ from all other books in the Author, which by consequence doth draw on some difference to be used by the expositor. For the Inditer of them did know four things which no man attains to know; which are—the mysteries of the kingdom of glory, the perfection of the laws of nature, the secrets of the heart of man, and the future succession of all ages. For as to the first it is said, "He that presseth into the light shall be oppressed of the glory." And again, "No man shall see My face and live." To the second, "When He prepared the heavens I was present, when by law and compass He enclosed the deep." To the third, "Neither was it needful that any should bear witness to Him of man, for He knew well what was in man." And to the last, "From the beginning are known to the Lord all His works."

(15) From the former two of these have been drawn certain senses and expositions of Scriptures, which had need be contained within the bounds of sobriety—the one anagogical, and the other philosophical. But as to the former, man is not to prevent his time: Videmus nunc per speculum in ænigmate, tunc autem facie ad faciem; wherein nevertheless there seemeth to be a liberty granted, as far forth as the polishing of this glass, or some moderate explication of this enigma. But to press too far into it cannot but cause a dissolution and overthrow of the spirit of man. For in the body there are three degrees of that we receive into it—aliment, medicine, and poison; whereof aliment is that which the nature of man can perfectly alter and overcome; medicine is that which is partly converted by nature, and partly converteth nature; and poison is that which worketh wholly upon nature, without that nature can in any part work upon it. So in the mind, whatsoever knowledge reason cannot at all work upon and convert is a mere intoxication, and endangereth a dissolution of the mind and understanding.

(16) But for the latter, it hath been extremely set on foot of late time by the school of Paracelsus, and some others, that have pretended to find the truth of all natural philosophy in the Scriptures; scandalising and traducing all other philosophy as heathenish and profane. But there is no such enmity between God's Word and His works; neither do they give honour to the Scriptures, as they suppose, but

much embase them. For to seek heaven and earth in the Word of God, whereof it is said, "Heaven and earth shall pass, but My word shall not pass," is to seek temporary things amongst eternal: and as to seek divinity in philosophy is to seek the living amongst the dead, so to seek philosophy in divinity is to seek the dead amongst the living: neither are the pots or lavers, whose place was in the outward part of the temple, to be sought in the holiest place of all where the ark of the testimony was seated. And again, the scope or purpose of the Spirit of God is not to express matters of nature in the Scriptures, otherwise than in passage, and for application to man's capacity and to matters moral or divine. And it is a true rule, Auctoris aliud agentis parva auctoritas. For it were a strange conclusion, if a man should use a similitude for ornament or illustration sake, borrowed from nature or history according to vulgar conceit, as of a basilisk, a unicorn, a centaur, a Briareus, a hydra, or the like, that therefore he must needs be thought to affirm the matter thereof positively to be true. To conclude therefore these two interpretations, the one by reduction or enigmatical, the other philosophical or physical, which have been received and pursued in imitation of the rabbins and cabalists, are to be confined with a a noli akryn sapere, sed time.

(17) But the two latter points, known to God and unknown to man, touching the secrets of the heart and the successions of time, doth make a just and sound difference between the manner of the exposition of the Scriptures and all other books. For it is an excellent observation which hath been made upon the answers of our Saviour Christ to many of the questions which were propounded to Him, how that they are impertinent to the state of the question demanded: the reason whereof is, because not being like man, which knows man's thoughts by his words, but knowing man's thoughts immediately, He never answered their words, but their thoughts. Much in the like manner it is with the Scriptures, which being written to the thoughts of men, and to the succession of all ages, with a foresight of all heresies, contradictions, differing estates of the Church, yea, and particularly of the elect, are not to be interpreted only according to the latitude of the proper sense of the place, and respectively towards that present occasion whereupon the words were uttered, or in precise congruity or contexture with the words before or after, or in contemplation of the principal scope of the place; but have in themselves, not only totally or collectively, but distributively in clauses and words, infinite springs and streams of doctrine to water the Church in every part. And therefore as the literal sense is, as it were, the main stream or river, so the moral sense chiefly, and sometimes the allegorical or typical, are they whereof the Church hath most use; not that I wish men to be bold in allegories, or indulgent or light in allusions: but that I do much condemn that interpretation of the Scripture which is only after the manner as men use to interpret a profane book.

(18) In this part touching the exposition of the Scriptures, I can report no deficiency; but by way of remembrance this I will add. In perusing books of divinity I find many books of controversies, and many of commonplaces and treatises, a mass of positive divinity, as it is made an art: a number of sermons and lectures, and many prolix commentaries upon the Scriptures, with harmonies and concordances. But that form of writing in divinity which in my judgment is of all others most rich and precious is positive divinity, collected upon particular texts of Scriptures in brief observations; not dilated into commonplaces, not chasing after controversies, not reduced into method of art; a thing abounding in sermons, which will vanish, but defective in books which will remain, and a thing wherein this age excelleth. For I am persuaded, and I may speak it with an absit invidia verbo, and nowise in derogation of antiquity, but as in a good emulation between the vine and the olive, that if the choice and best of those observations upon texts of Scriptures which have been made dispersedly in sermons within this your Majesty's Island of Brittany by the space of these forty years and more (leaving out the largeness of exhortations and applications thereupon) had been set down in a continuance, it had been the best work in divinity which had been written since the Apostles' times.

(19) The matter informed by divinity is of two kinds: matter of belief and truth of opinion, and matter of service and adoration; which is also judged and directed by the former—the one being as the internal soul of religion, and the other as the external body thereof. And, therefore, the heathen religion was not only a worship of idols, but the whole religion was an idol in itself; for it had no soul; that is, no certainty of belief or confession: as a man may well think, considering the chief doctors of their church were the poets; and the reason was because the heathen gods were no jealous gods, but were glad to be admitted into part, as they had reason. Neither did they respect the pureness of heart, so they might have external honour and rites.

(20) But out of these two do result and issue four main branches of divinity: faith, manners, liturgy, and government. Faith containeth the doctrine of the nature of God, of the attributes of God, and of the works of God. The nature of God consisteth of three persons in unity of Godhead. The attributes of God are either common to the Deity, or respective to the persons. The works of God summary are two, that of the creation and that of the redemption; and both these works, as in total they appertain to the unity of the Godhead, so in their parts they refer to the three persons: that of the creation, in the mass of the matter, to the Father; in the disposition of the form, to the Son; and in the continuance and conservation of the being, to the Holy Spirit. So that of the redemption, in the election

and counsel, to the Father; in the whole act and consummation, to the Son; and in the application, to the Holy Spirit; for by the Holy Ghost was Christ conceived in flesh, and by the Holy Ghost are the elect regenerate in spirit. This work likewise we consider either effectually, in the elect; or privately, in the reprobate; or according to appearance, in the visible Church.

(21) For manners, the doctrine thereof is contained in the law, which discloseth sin. The law itself is divided, according to the edition thereof, into the law of nature, the law moral, and the law positive; and according to the style, into negative and affirmative, prohibitions and commandments. Sin, in the matter and subject thereof, is divided according to the commandments; in the form thereof it referreth to the three persons in Deity: sins of infirmity against the Father, whose more special attribute is power; sins of ignorance against the Son, whose attribute is wisdom; and sins of malice against the Holy Ghost, whose attribute is grace or love. In the motions of it, it either moveth to the right hand or to the left; either to blind devotion or to profane and libertine transgression; either in imposing restraint where God granteth liberty, or in taking liberty where God imposeth restraint. In the degrees and progress of it, it divideth itself into thought, word, or act. And in this part I commend much the deducing of the law of God to cases of conscience; for that I take indeed to be a breaking, and not exhibiting whole of the bread of life. But that which quickeneth both these doctrines of faith and manners is the elevation and consent of the heart; whereunto appertain books of exhortation, holy meditation, Christian resolution, and the like.

(22) For the liturgy or service, it consisteth of the reciprocal acts between God and man; which, on the part of God, are the preaching of the word, and the sacraments, which are seals to the covenant, or as the visible word; and on the part of man, invocation of the name of God; and under the law, sacrifices; which were as visible prayers or confessions: but now the adoration being in spiritu et veritate, there remaineth only vituli labiorum; although the use of holy vows of thankfulness and retribution may be accounted also as sealed petitions.

(23) And for the government of the Church, it consisteth of the patrimony of the Church, the franchises of the Church, and the offices and jurisdictions of the Church, and the laws of the Church directing the whole; all which have two considerations, the one in themselves, the other how they stand compatible and agreeable to the civil estate.

(24) This matter of divinity is handled either in form of instruction of truth, or in form of confutation of falsehood. The declinations from religion, besides the privative, which is atheism and the branches thereof, are three—heresies, idolatry, and witchcraft: heresies, when we serve the true God with a false worship; idolatry, when we worship false gods, supposing them to be true; and witchcraft, when we adore false gods, knowing them to be wicked and false. For so your Majesty doth excellently well observe, that witchcraft is the height of idolatry. And yet we see though these be true degrees, Samuel teacheth us that they are all of a nature, when there is once a receding from the Word of God; for so he saith, Quasi peccatum ariolandi est repugnare, et quasi scelus idololatriæ nolle acquiescere.

(25) These things I have passed over so briefly because I can report no deficiency concerning them: for I can find no space or ground that lieth vacant and unsown in the matter of divinity, so diligent have men been either in sowing of good seed, or in sowing of tares.

***

Thus have I made as it were a small globe of the intellectual world, as truly and faithfully as I could discover; with a note and description of those parts which seem to me not constantly occupate, or not well converted by the labour of man. In which, if I have in any point receded from that which is commonly received, it hath been with a purpose of proceeding in melius, and not in aliud; a mind of amendment and proficiency, and not of change and difference. For I could not be true and constant to the argument I handle if I were not willing to go beyond others; but yet not more willing than to have others go beyond me again: which may the better appear by this, that I have propounded my opinions naked and unarmed, not seeking to preoccupate the liberty of men's judgments by confutations. For in anything which is well set down, I am in good hope that if the first reading move an objection, the second reading will make an answer. And in those things wherein I have erred, I am sure I have not prejudiced the right by litigious arguments; which certainly have this contrary effect and operation, that they add authority to error, and destroy the authority of that which is well invented. For question is an honour and preferment to falsehood, as on the other side it is a repulse to truth. But the errors I claim and challenge to myself as mine own. The good, it any be, is due tanquam adeps sacrificii, to be incensed to the honour, first of the Divine Majesty, and next of your Majesty, to whom on earth I am most bounden.

Footnotes:

1. Stoops in the rice and takes the speeding gold. Ovid. Metam, x. 667.

The New Atlantis

INTRODUCTORY NOTE

Bacon's literary executor, Dr. Rowley, published "The New Atlantis" in 1627, the year after the author's death. It seems to have been written about 1623, during that period of literary activity which followed Bacon's political fall. None of Bacon's writings gives in short apace so vivid a picture of his tastes and aspirations as this fragment of the plan of an ideal commonwealth. The generosity and enlightenment, the dignity and splendor, the piety and public spirit, of the inhabitants of Bensalem represent the ideal qualities which Bacon the statesman desired rather than hoped to see characteristic of his own country; and in Solomon's House we have Bacon the scientist indulging without restriction his prophetic vision of the future of human knowledge. No reader acquainted in any degree with the processes and results of modern scientific inquiry can fail to be struck by the numerous approximations made by Bacon's imagination to the actual achievements of modern times. The plan and organization of his great college lay down the main lines of the modern research university; and both in pure and applied science he anticipates a strikingly large number of recent inventions and discoveries. In still another way is "The New Atlantis" typical of Bacon's attitude. In spite of the enthusiastic and broad-minded schemes he laid down for the pursuit of truth, Bacon always had an eye to utility. The advancement of science which he sought was conceived by him as a means to a practical end the increase of man's control over nature, and the comfort and convenience of humanity. For pure metaphysics, or any form of abstract thinking that yielded no "fruit," he had little interest; and this leaning to the useful is shown in the practical applications of the discoveries made by the scholars of Solomon's House. Nor does the interest of the work stop here. It contains much, both in its political and in its scientific ideals, that we have as yet by no means achieved, but which contain valuable elements of suggestion and stimulus for the future.

THE NEW ATLANTIS

We sailed from Peru, (where we had continued for the space of one whole year) for China and Japan, by the South Sea; taking with us victuals for twelve months; and had good winds from the east, though soft and weak, for five months space, and more. But the wind came about, and settled in the west for many days, so as we could make little or no way, and were sometime in purpose to turn back. But then again there arose strong and great winds from the south, with a point east, which carried us up (for all that we could do) towards the north; by which time our victuals failed us, though we had made good spare of them. So that finding ourselves, in the midst of the greatest wilderness of waters in the world, without victuals, we gave ourselves for lost men and prepared for death. Yet we did lift up our hearts and voices to God above, who showeth his wonders in the deep, beseeching him of his mercy, that as in the beginning he discovered the face of the deep, and brought forth dry land, so he would now discover land to us, that we might not perish.

And it came to pass that the next day about evening we saw within a kenning before us, towards the north, as it were thick clouds, which did put us in some hope of land; knowing how that part of the South Sea was utterly unknown; and might have islands, or continents, that hitherto were not come to light. Wherefore we bent our course thither, where we saw the appearance of land, all that night; and in the dawning of the next day, we might plainly discern that it was a land; flat to our sight, and full of boscage; which made it show the more dark. And after an hour and a half's sailing, we entered into a good haven, being the port of a fair city; not great indeed, but well built, and that gave a pleasant view from the sea: and we thinking every minute long, till we were on land, came close to the shore, and offered to land. But straightways we saw divers of the people, with bastons in their hands (as it were) forbidding us to land; yet without any cries of fierceness, but only as warning us off, by signs that they made. Whereupon being not a little discomforted, we were advising with ourselves, what we should do.

During which time, there made forth to us a small boat, with about eight persons in it; whereof one of them had in his hand a tipstaff of a yellow cane, tipped at both ends with blue, who came aboard our ship, without any show of distrust at all. And when he saw one of our number, present himself somewhat before the rest, he drew forth a little scroll of parchment (somewhat yellower than our parchment, and shining like the leaves of writing tables, but otherwise soft and flexible,) and delivered it to our foremost man. In which scroll were written in ancient Hebrew, and in ancient Greek, and in good Latin of the school, and in Spanish, these words: Land ye not, none of you; and provide to be gone from this coast, within sixteen days, except you have further time given you. Meanwhile, if you want fresh water or victuals, or help for your sick, or that your ship needeth repairs, write down your wants, and you shall have that, which belongeth to mercy. This scroll was signed with a stamp of cherubim: wings, not spread, but hanging downwards; and by them a cross. This being delivered, the officer returned, and left only a servant with us to receive our answer.

Consulting hereupon amongst ourselves, we were much perplexed. The denial of landing and hasty warning us away troubled us much; on the other side, to find that the people had languages, and were so full of humanity, did comfort us not a little. And above all, the sign of the cross to that instrument was to us a great rejoicing, and as it were a certain presage of good. Our answer was in the Spanish tongue; that for our ship, it was well; for we had rather met with calms and contrary winds than any tempests. For our sick, they were many, and in very ill case; so that if they were not permitted to land, they ran danger of their lives. Our other wants we set down in particular; adding, That we had some little store of merchandise, which if it pleased them to deal for, it might supply our wants, without being chargeable unto them. We offered some reward in pistolets unto the servant, and a piece of crimson velvet to be presented to the officer; but the servant took them not, nor would scarce look upon them; and so left us, and went back in another little boat, which was sent for him.

About three hours after we had dispatched our answer, there came towards us a person (as it seemed) of place. He had on him a gown with wide sleeves, of a kind of water chamolet, of an excellent azure colour, fair more glossy than ours; his under apparel was green; and so was his hat, being in the form of a turban, daintily made, and not so huge as the Turkish turbans; and the locks of his hair came down below the brims of it. A reverend man was he to behold. He came in a boat, gilt in some part of it, with four persons more only in that boat; and was followed by another boat, wherein were some twenty. When he was come within a flightshot of our ship, signs were made to us, that we should send forth some to meet him upon the water; which we presently did in our ship-boat, sending the principal man amongst us save one, and four of our number with him.

When we were come within six yards of their boat, they called to us to stay, and not to approach farther; which we did. And thereupon the man, whom I before described, stood up, and with a loud voice, in Spanish, asked, "Are ye Christians?" We answered, "We were;" fearing the less, because of the cross we had seen in the subscription. At which answer the said person lifted up his right hand towards Heaven, and drew it softly to his mouth (which is the gesture they use, when they thank God;) and then said: "If ye will swear (all of you) by the merits of the Saviour, that ye are no pirates, nor have shed blood, lawfully, nor unlawfully within forty days past, you may have licence to come on land." We said, "We were all ready to take that oath." Whereupon one of those that were with him, being (as it seemed) a notary, made an entry of this act. Which done, another of the attendants of the great person which was with him in the same boat, after his Lord had spoken a little to him, said aloud: "My Lord would have you know, that it is not of pride, or greatness, that he cometh not aboard your ship; but for that in your answer you declare that you have many sick amongst you, he was warned by the Conservator of Health of the city that he should keep a distance." We bowed ourselves towards him, and answered, "We were his humble servants; and accounted for great honour, and singular humanity towards us, that which was already done; but hoped well, that the nature of the sickness of our men was not infectious." So he returned; and a while after came the Notary to us aboard our ship; holding in his hand a fruit of that country, like an orange, but of color between orange-tawney and scarlet; which cast a most excellent odour. He used it (as it seemeth) for a preservative against infection. He gave us our oath; "By the name of Jesus, and his merits:" and after told us, that the next day, by six of the Clock, in the Morning, we should be sent to, and brought to the Strangers' House, (so he called it,) where we should be accommodated of things, both for our whole, and for our sick. So he left us; and when we offered him some pistolets, he smiling said, "He must not be twice paid for one labour:" meaning (as I take it) that he had salary sufficient of the State for his service. For (as I after learned) they call an officer that taketh rewards, "twice paid."

The next morning early, there came to us the same officer that came to us at first with his cane, and told us, He came to conduct us to the Strangers' House; and that he had prevented the hour, because we might have the whole day before us, for our business. "For," said he, "if you will follow my advice, there shall first go with me some few of you, and see the place, and how it may be made convenient for you; and then you may send for your sick, and the rest of your number, which ye will bring on land." We thanked him, and said, "That this care, which he took of desolate strangers, God would reward." And so six of us went on land with him: and when we were on land, he went before us, and turned to us, and said, "He was but our servant, and our guide." He led us through three fair streets; and all the way we went, there were gathered some people on both sides, standing in a row; but in so civil a fashion, as if it had been, not to wonder at us, but to welcome us: and divers of them, as we passed by them, put their arms a little abroad; which is their gesture, when they did bid any welcome.

The Strangers' House is a fair and spacious house, built of brick, of somewhat a bluer colour than our brick; and with handsome windows, some of glass, some of a kind of cambric oiled. He brought us first into a fair parlour above stairs, and then asked us, "What number of persons we were? And how many sick?" We answered, "We were in all, (sick and whole,) one and fifty persons, whereof our sick were seventeen." He desired us to have patience a little, and to stay till he came back to us; which was about an hour after; and then he led us to see the chambers which were provided for us, being in number nineteen: they having cast it (as it seemeth) that four of those chambers, which were better than the rest, might receive four of the principal men of our company; and lodge them alone by themselves; and the other fifteen chambers were to lodge us two and two together. The chambers were handsome and cheerful chambers, and furnished civilly. Then he led us to a long gallery, like a dorture, where he showed us all along the one side (for the other side was but wall and window), seventeen cells, very neat ones, having partitions of cedar wood. Which gallery and cells, being in all forty, many more than we needed, were instituted as an infirmary for sick persons. And he told us withal, that as any of our sick waxed well, he might be removed from his cell, to a chamber; for which purpose there were set forth ten spare chambers, besides the number we spake of before. This done, he brought us back to the parlour, and lifting up his cane a little, (as they do when they give any charge or command) said to us, "Ye are to know, that the custom of the land requireth, that after this day and to-morrow, (which we give you for removing of your people from your ship,) you are to keep within doors for three days. But let it not trouble you, nor do not think yourselves restrained, but rather left to your rest and ease. You shall want nothing, and there are six of our people appointed to attend you, for any business you may have abroad." We gave him thanks, with all affection and respect, and said, "God surely is manifested in this land." We offered him also twenty pistolets; but he smiled, and only said; "What? twice paid!" And so he left us.

Soon after our dinner was served in; which was right good viands, both for bread and treat: better than any collegiate diet, that I have known in Europe. We had also drink of three sorts, all wholesome and good; wine of the grape; a drink of grain, such as is with us our ale, but more clear: And a kind of cider made of a fruit of that country; a wonderful pleasing and refreshing drink. Besides, there were brought in to us, great store of those scarlet oranges, for our sick; which (they said) were an assured remedy for sickness taken at sea. There was given us also, a box of small gray, or whitish pills, which they wished our sick should take, one of the pills, every night before sleep; which (they said) would hasten their recovery.

The next day, after that our trouble of carriage and removing of our men and goods out of our ship, was somewhat settled and quiet, I thought good to call our company together; and when they were assembled, said unto them; "My dear friends, let us know ourselves, and how it standeth with us. We are men cast on land, as Jonas was, out of the whale's belly, when we were as buried in the deep: and now we are on land, we are but between death and life; for we are beyond, both the old world, and the new; and whether ever we shall see Europe, God only knoweth. It is a kind of miracle hath brought us hither: and it must be little less, that shall bring us hence. Therefore in regard of our deliverance past, and our danger present, and to come, let us look up to God, and every man reform his own ways. Besides we are come here amongst a Christian people, full of piety and humanity: let us not bring that confusion of face upon ourselves, as to show our vices, or unworthiness before them. Yet there is more. For they have by commandment, (though in form of courtesy) cloistered us within these wall, for three days: who knoweth, whether it be not, to take some taste of our manners and conditions? And if they find them bad, to banish us straightways; if good, to give us further time. For these men that they have given us for attendance, may withal have an eye upon us. Therefore for God's love, and as we love the weal of our souls and bodies, let us so behave ourselves, as we may be at peace with God, and may find grace in the eyes of this people." Our company with one voice thanked me for my good admonition, and promised me to live soberly and civilly, and without giving any the least occasion of offence. So we spent our three days joyfully, and without care, in expectation what would be done with us, when they were expired. During which time, we had every hour joy of the amendment of our sick; who thought themselves cast into some divine pool of healing; they mended so kindly, and so fast.

The morrow after our three days were past, there came to us a new man, that we had not seen before, clothed in blue as the former was, save that his turban was white, with a small red cross on the top. He had also a tippet of fine linen. At his coming in, he did bend to us a little, and put his arms abroad. We of our parts saluted him in a very lowly and submissive manner; as looking that from him, we should receive sentence of life, or death: he desired to speak with some few of us: whereupon six of us only staid, and the rest avoided the room. He said, "I am by office governor of this House of Strangers, and by vocation I am a Christian priest: and therefore am come to you to offer you my service, both as strangers and chiefly as Christians. Some things I may tell you, which I think you will not be unwilling to hear. The State hath given you license to stay on land, for the space of six weeks; and let it not trouble you, if your occasions ask further time, for the law in this point is not precise; and I do not doubt, but my self shall be able, to obtain for you such further time, as may be convenient. Ye shall also understand, that the Strangers' House is at this time rich, and much aforehand; for it hath laid up revenue these thirty-seven years; for so long it is since any stranger arrived in this part: and therefore take ye no care; the State will defray you all the time you stay; neither shall you stay one day the less for that. As for any merchandise ye have brought, ye shall be well used, and have your return, either in merchandise, or in gold and silver: for to us it is all one. And if you have any other request to make, hide it not. For ye shall find we will not make your countenance to fall by the answer ye shall receive. Only this I must tell you, that none of you must go above a karan," (that is with them a mile and an half) "from the walls of the city, without especial leave."

We answered, after we had looked awhile one upon another, admiring this gracious and parent-like usage; "That we could not tell what to say: for we wanted words to express our thanks; and his noble free offers left us nothing to ask. It seemed to us, that we had before us a picture of our salvation in Heaven; for we that were a while since in the jaws of death, were now brought into a place, where we found nothing but consolations. For the commandment laid upon us, we would not fail to obey it, though it was impossible but our hearts should be enflamed to tread further upon this happy and holy ground." We added, "That our tongues should first cleave to the roofs of our mouths, ere we should forget, either his reverend person, or this whole nation, in our prayers." We also most humbly besought him, to accept of us as his true servants; by as just a right as ever men on earth were bounden; laying and presenting, both our persons, and all we had, at his feet. He said; "He was a priest, and looked for a priest's reward; which was our brotherly love, and the good of our souls and bodies." So he went from us, not without tears of tenderness in his eyes; and left us also confused with joy and kindness, saying amongst ourselves; "That we were come into a land of angels, which did appear to us daily, and prevent us with comforts, which we thought not of, much less expected."

The next day about ten of the clock, the Governor came to us again, and after salutations, said familiarly; "That he was come to visit us;" and called for a chair, and sat him down: and we, being some ten of us, (the rest were of the meaner sort, or else gone abroad,) sat down with him, And when we were set, he began thus: "We of this island of Bensalem," (for so they call it in their language,) "have this; that by means of our solitary situation; and of the laws of secrecy, which we have for our travellers, and our rare admission of strangers; we know well most part of the habitable world, and are ourselves unknown. Therefore because he that knoweth least is fittest to ask questions, it is more reason, for the entertainment of the time, that ye ask me questions, than that I ask you."

We answered; "That we humbly thanked him that he would give us leave so to do: and that we conceived by the taste we had already, that there was no worldly thing on earth, more worthy to be known than the state of that happy land. But above all," (we said,) "since that we were met from the several ends of the world, and hoped assuredly that we should meet one day in the kingdom of Heaven, (for that we were both parts Christians,) we desired to know, (in respect that land was so remote, and so divided by vast and unknown seas, from the land where our Saviour walked on earth,) who was the apostle of that nation, and how it was converted to the faith?" It appeared in his face that he took great contentment in this our question: he said; "Ye knit my heart to you, by asking this question in the first place; for it sheweth that you first seek the kingdom of heaven; and I shall gladly, and briefly, satisfy your demand.

"About twenty years after the ascension of our Saviour, it came to pass, that there was seen by the people of Renfusa, (a city upon the eastern coast of our island,) within night, (the night was cloudy, and calm,) as it might be some mile into the sea, a great pillar of light; not sharp, but in form of a column, or cylinder, rising from the sea a great way up towards heaven; and on the top of it was seen a large cross of light, more bright and resplendent than the body of the pillar. Upon which so strange a spectacle, the people of the city gathered apace together upon the sands, to wonder; and so after put themselves into a number of small boats, to go nearer to this marvellous sight. But when the boats were come within (about) sixty yards of the pillar, they found themselves all bound, and could go no further; yet so as they might move to go about, but might not approach nearer: so as the boats stood all as in a theatre, beholding this light as an heavenly sign. It so fell out, that there was in one of the boats one of the wise men, of the society of Salomon's House; which house, or college (my good brethren) is the very eye of this kingdom; who having awhile attentively and devoutly viewed and contemplated this pillar and cross, fell down upon his face; and then raised himself upon his knees, and lifting up his hands to heaven, made his prayers in this manner.

"'LORD God of heaven and earth, thou hast vouchsafed of thy grace to those of our order, to know thy works of Creation, and the secrets of them: and to discern (as far as appertaineth to the generations of men) between divine miracles, works of nature, works of art, and impostures and illusions of all sorts. I do here acknowledge and testify before this people, that the thing which we now see before our eyes is thy Finger and a true Miracle. And forasmuch as we learn in our books that thou never workest miracles, but to divine and excellent end, (for the laws of nature are thine own laws, and thou exceedest them not but upon great cause,) we most humbly beseech thee to prosper this great sign, and to give us the interpretation and use of it in mercy; which thou dost in some part secretly promise by sending it unto us.'

"When he had made his prayer, he presently found the boat he was in, moveable and unbound; whereas all the rest remained still fast; and taking that for an assurance of leave to approach, he caused the boat to be softly and with silence rowed towards the pillar. But ere he came near it, the pillar and cross of light brake up, and cast itself abroad, as it were, into a firmament of many stars; which also vanished soon after, and there was nothing left to be seen, but a small ark, or chest of cedar, dry, and not wet at all with water, though it swam. And in the fore-end of it, which was towards him, grew a small green branch of palm; and when the wise man had taken it, with all reverence, into his boat, it opened of itself, and there were found in it a Book and a Letter; both written in fine parchment, and wrapped in sindons of linen. The Book contained all the canonical books of the Old and New Testament, according as you have them; (for we know well what the churches with you receive); and the Apocalypse itself, and some other books of the New Testament, which were not at that time written, were nevertheless in the Book. And for the Letter, it was in these words:

"'I, Bartholomew, a servant of the Highest, and Apostle of Jesus Christ, was warned by an angel that appeareth to me, in a vision of glory, that I should commit this ark to the floods of the sea. Therefore I do testify and declare unto that people where God shall ordain this ark to come to land, that in the same day is come unto them salvation and peace and good-will, from the Father, and from the Lord Jesus.'

"There was also in both these writings, as well the Book, as the Letter, wrought a great miracle, conform to that of the Apostles, in the original Gift of Tongues. For there being at that time in this land Hebrews, Persians, and Indians, besides the natives, every one read upon the Book, and Letter, as if they had been written in his own language. And thus was this land saved from infidelity (as the remainder of the old world was from water) by an ark, through the apostolical and miraculous evangelism of Saint Bartholomew." And here he paused, and a messenger came, and called him from us. So this was all that passed in that conference.

The next day, the same governor came again to us, immediately after dinner, and excused himself, saying; "That the day before he was called from us, somewhat abruptly, but now he would make us amends, and spend time with us if we held his company and conference agreeable." We answered, "That we held it so agreeable and pleasing to us, as we forgot both dangers past and fears to come, for the time we hear him speak; and that we thought an hour spent with him, was worth years of our former life." He bowed himself a little to us, and after we were set again, he said; "Well, the questions are on your part."

One of our number said, after a little pause; that there was a matter, we were no less desirous to know, than fearful to ask, lest we might presume too far. But encouraged by his rare humanity towards us, (that could scarce think ourselves strangers, being his vowed and professed servants,) we would take the hardiness to propound it: humbly beseeching him, if he thought it not fit to be answered, that he would pardon it, though he rejected it. We said; "We well observed those his words, which he formerly spake, that this happy island, where we now stood, was known to few, and yet knew most of the nations of the world; which we found to be true, considering they had the languages of Europe, and knew much of our state and business; and yet we in Europe, (notwithstanding all the remote discoveries and navigations of this last age), never heard of the least inkling or glimpse of this island. This we found wonderful strange; for that all nations have inter-knowledge one of another, either by voyage into foreign parts, or by strangers that come to them: and though the traveller into a foreign country, doth commonly know more by the eye, than he that stayeth at home can by relation of the traveller; yet both ways suffice to make a mutual knowledge, in some degree, on both parts. But for this island, we never heard tell of any ship of theirs that had been seen to arrive upon any shore of Europe; nor of either the East or West Indies; nor yet of any ship of any other part of the world, that had made return from them. And yet the marvel rested not in this. For the situation of it (as his lordship said) in the secret conclave' of such a vast sea might cause it. But then, that they should have knowledge of the languages, books, affairs, of those that lie such a distance from them, it was a thing we could not tell what to make of; for that it seemed to us a conditioner and propriety of divine powers and beings, to be hidden and unseen to others, and yet to have others open and as in a light to them."

At this speech the Governor gave a gracious smile, and said; "That we did well to ask pardon for this question we now asked: for that it imported, as if we thought this land, a land of magicians, that sent forth spirits of the air into all parts, to bring them news and intelligence of other countries." It was answered by us all, in all possible humbleness, but yet with a countenance taking knowledge, that we knew that he spake it but merrily, "That we were apt enough to think there was somewhat supernatural in this island; but yet rather as angelical than magical. But to let his lordship know truly what it was that made us tender and doubtful to ask this question, it was not any such conceit, but because we remembered, he had given a touch in his former speech, that this land had laws of secrecy touching strangers." To this he said; "You remember it aright and therefore in that I shall say to you, I must reserve some particulars, which it is not lawful for me to reveal; but there will be enough left, to give you satisfaction."

"You shall understand (that which perhaps you will scarce think credible) that about three thousand years ago, or somewhat more, the navigation of the world, (especially for remote voyages,) was greater than at this day. Do not think with yourselves, that I know not how much it is increased with you, within these six-score years: I know it well: and yet I say greater then than now; whether it was, that the example of the ark, that saved the remnant of men from the universal deluge, gave men confidence to adventure upon the waters; or what it was; but such is the truth. The Phoenicians, and especially the Tyrians, had great fleets. So had the Carthaginians their colony, which is yet further west. Toward the east the shipping of Egypt and of Palestine was likewise great. China also, and the great Atlantis, (that you call America,) which have now but junks and canoes, abounded then in tall ships. This island, (as appeareth by faithful registers of those times,) had then fifteen hundred strong ships, of great content. Of all this, there is with you sparing memory, or none; but we have large knowledge thereof.

"At that time, this land was known and frequented by the ships and vessels of all the nations before named. And (as it cometh to pass) they had many times men of other countries, that were no sailors, that came with them; as Persians, Chaldeans, Arabians; so as almost all nations of might and fame resorted hither; of whom we have some stirps, and little tribes with us at this day. And for our own ships, they went sundry voyages, as well to your straits, which you call the Pillars of Hercules, as to other parts in the Atlantic and Mediterrane Seas; as to Paguin, (which is the same with Cambaline,) and Quinzy, upon the Oriental Seas, as far as to the borders of the East Tartary.

"At the same time, and an age after, or more, the inhabitants of the great Atlantis did flourish. For though the narration and description, which is made by a great man with you; that the descendants of Neptune planted there; and of the magnificent temple, palace, city, and hill; and the manifold streams of goodly navigable rivers, (which as so many chains environed the same site and temple); and the several degrees of ascent, whereby men did climb up to the same, as if it had been a scala coeli, be all poetical and fabulous: yet so much is true, that the said country of Atlantis, as well that of Peru, then called Coya, as that of Mexico, then named Tyrambel, were mighty and proud kingdoms in arms, shipping and riches: so mighty, as at one time (or at least within the space of ten years) they both made two great expeditions; they of Tyrambel through the Atlantic to the Mediterrane Sea; and they of Coya through the South Sea upon this our island: and for the former of these, which was into Europe, the same author amongst you (as it seemeth) had some relation from the Egyptian priest whom he cited. For assuredly such a thing there was. But whether it were the ancient Athenians that had the glory of the repulse and resistance of those forces, I can say nothing: but certain it is, there never came back either ship or man from that voyage. Neither had the other voyage of those of Coya upon us had better fortune, if they had not met with enemies of greater clemency. For the king of this island, (by name Altabin,) a wise man and a great warrior, knowing well both his own strength and that of his enemies, handled the matter so, as he cut off their land-forces from their ships; and entoiled both their navy and their tamp with a greater power than theirs, both by sea and land: arid compelled them to render themselves without striking stroke and after they were at his mercy, contenting himself only with their oath that they should no more bear arms against him, dismissed them all in safety.

"But the divine revenge overtook not long after those proud enterprises. For within less than the space of one hundred years, the great Atlantis was utterly lost and destroyed: not by a great earthquake, as your man saith; (for that whole tract is little subject to earthquakes;) but by a particular' deluge or inundation; those countries having, at this day, far greater rivers and far higher mountains to pour down waters, than any part of the old world. But it is true that the same inundation was not deep; not past forty foot, in most places, from the ground; so that although it destroyed man and beast generally, yet some few wild inhabitants of the wood escaped. Birds also were saved by flying to the high trees and woods. For as for men, although they had buildings in many places, higher than the depth of the water, yet that inundation, though it were shallow, had a long continuance; whereby they of the vale that were not drowned, perished for want of food and other things necessary.

"So as marvel you not at the thin population of America, nor at the rudeness and ignorance of the people; for you must account your inhabitants of America as a young people; younger a thousand years, at the least, than the rest of the world: for that there was so much time between the universal flood and their particular inundation. For the poor remnant of human seed, which remained in their mountains, peopled the country again slowly, by little and little; and being simple and savage people, (not like Noah and his sons, which was the chief family of the earth;) they were not able to leave letters, arts, and civility to their posterity; and having likewise in their mountainous habitations been used (in respect of the extreme cold of those regions) to clothe themselves with the skins of tigers, bears, and great hairy goats, that they have in those parts; when after they came down into the valley, and found the intolerable heats which are there, and knew no means of lighter apparel, they were forced to begin the custom of going naked, which continueth at this day. Only they take great pride and delight in the feathers of birds; and this also they took from those their ancestors of the mountains, who were invited unto it by the infinite flights of birds that came up to the high grounds, while the waters stood below. So you see, by this main accident of time, we lost our traffic with the Americans, with whom of, all others, in regard they lay nearest to us, we had most commerce.

"As for the other parts of the world, it is most manifest that in the ages following (whether it were in respect of wars, or by a natural revolution of time,) navigation did every where greatly decay; and specially far voyages (the rather by the use of galleys, and such vessels as could hardly brook the ocean,) were altogether left and omitted. So then, that part of intercourse which could be from other nations to sail to us, you See how it hath long since ceased; except it were by some rare accident, as this of yours. But now of the cessation of that other part of intercourse, which might be by our sailing to other nations, I must yield you some other cause. For I cannot say (if I shall say truly,) but our shipping, for number, strength, mariners, pilots, and all things that appertain to navigation, is as great as ever; and therefore why we should sit at home, I shall now give you an account by itself: and it will draw nearer to give you satisfaction to your principal question.

"There reigned in this land, about nineteen hundred years ago, a king, whose memory of all others we most adore; not superstitiously, but as a divine instrument, though a mortal man; his name was Solamona: and we esteem him as the lawgiver of our nation. This king had a large heart, inscrutable for good; and was wholly bent to make his kingdom and people happy. He therefore, taking into consideration how sufficient and substantive this land was to maintain itself without any aid (at all) of the foreigner; being five thousand six hundred miles in circuit, and of rare fertility of soil in the greatest part thereof; and finding also the shipping of this country might be plentifully set on work, both by fishing and by transportations from port to port, and likewise by sailing unto some small islands that are not far from us, and are under the crown and laws of this state; and, recalling into his memory the happy and flourishing estate wherein this land then was; so as it might be a thousand ways altered to the worse, but scarce any one way to the better; thought nothing wanted to his noble and heroical intentions, but only (as far as human foresight might reach) to give perpetuity to that which was in his time so happily established. Therefore amongst his other fundamental laws of this kingdom, he did ordain the interdicts and prohibitions which we have touching entrance of strangers; which at that time (though it was after the calamity of America) was frequent; doubting novelties, and commixture of manners. It is true, the like law against the admission of strangers without licence is an ancient law in the kingdom of China, and yet continued in use. But there it is a poor thing; and hath made them a curious, ignorant, fearful, foolish nation. But our lawgiver made his law of another temper. For first, he hath preserved all points of humanity, in taking order and making provision for the relief of strangers distressed; whereof you have tasted."

At which speech (as reason was) we all rose up and bowed ourselves. He went on.

"That king also, still desiring to join humanity and policy together; and thinking it against humanity, to detain strangers here against their wills, and against policy that they should return and discover their knowledge of this estate, he took this course: he did ordain that of the strangers that should be permitted to land, as many (at all times) might depart as would; but as many as would stay should have very good conditions and means to live from the state. Wherein he saw so far, that now in so many ages since the prohibition, we have memory not of one ship that ever returned, and but of thirteen persons only, at several times, that chose to return in our bottoms. What those few that returned may have reported abroad I know not. But you must think, whatsoever they have said could be taken where they came but for a dream. Now for our travelling from henna into parts abroad, our Lawgiver thought fit altogether to restrain it. So is it not in China. For the Chinese sail where they will or can; which sheweth that their law of keeping out strangers is a law of pusillanimity and fear. But this restraint of ours hath one only exception, which is admirable; preserving the good which cometh by communicating with strangers, and avoiding the hurt; and I will now open it to you. And here I shall seem a little to digress, but you will by and by find it pertinent.

"Ye shall understand (my dear friends) that amongst the excellent acts of that king, one above all hath the pre-eminence. It was the erection and institution of an Order or Society, which we call Salomon's House; the noblest foundation (as we think) that ever was upon the earth; and the lanthorn of this kingdom. It is dedicated to the study of the works and creatures of God. Some think it beareth the founder's name a little corrupted, as if it should be Solamona's House. But the records write it as it is spoken. So as I take it to be denominate of the king of the Hebrews, which is famous with you, and no stranger to us. For we have some parts of his works, which with you are lost; namely, that natural history, which he wrote, of all plants, from the cedar of Libanus to the moss that groweth out of the wall, and of all things that have life and motion. This maketh me think that our king, finding himself to symbolize in many things with that king of the Hebrews (which lived many years before him), honored him with the title of this foundation. And I am rather induced to be of this opinion, for that I find in ancient records this Order or Society is sometimes called Salomon's House, and sometimes the College of the Six Days Works; whereby I am satisfied that our excellent king had learned from the Hebrews that God had created the world and all that therein is within six days: and therefore he instituting that House for the finding out of the true nature of all things, (whereby God might have the more glory in the workmanship of them, and insert the more fruit in the use of them), did give it also that second name.

"But now to come to our present purpose. When the king had forbidden to all his people navigation into any part that was not under his crown, he made nevertheless this ordinance; that every twelve years there should be set forth, out of this kingdom two ships, appointed to several voyages; That in either of these ships there should be a mission of three of the Fellows or Brethren of Salomon's House; whose errand was only to give us knowledge of the affairs and state of those countries to which they were designed, and especially of the sciences, arts, manufactures, and inventions of all the world; and withal to bring unto us books, instruments, and patterns in every kind: That the ships, after they had landed the brethren, should return; and that the brethren should stay abroad till the new mission. These ships are not otherwise fraught, than with store of victuals, and good quantity of treasure to remain with the brethren, for the buying of such things and rewarding of such persons as they should think fit. Now for me to tell you how the vulgar sort of mariners are contained from being discovered at land; and how they that must be put on shore for any time, color themselves under the names of other nations; and to what places these voyages have been designed; and what places of rendezvous are appointed for the new missions; and the like circumstances of the practique; I may not do it: neither is it much to your desire. But thus you see we maintain a trade not for gold, silver, or jewels; nor for silks; nor for spices; nor any other commodity of matter; but only for God's first creature, which was Light: to have light (I say) of the growth of all parts of the world."

And when he had said this, he was silent; and so were we all. For indeed we were all astonished to hear so strange things so probably told. And he, perceiving that we were willing to say somewhat but had it not ready in great courtesy took us off, and descended to ask us questions of our voyage and fortunes and in the end concluded, that we might do well to think with ourselves what time of stay we would demand of the state; and bade us not to scant ourselves; for he would procure such time as we desired: Whereupon we all rose up, and presented ourselves to kiss the skirt of his tippet; but he would not suffer us; and so took his leave. But when it came once amongst our people that the state used to offer conditions to strangers that would stay, we had work enough to get any of our men to look to our ship; and to keep them from going presently to the governor to crave conditions. But with much ado we refrained them, till we might agree what course to take.

We took ourselves now for free men, seeing there was no danger of our utter perdition; and lived most joyfully, going abroad and seeing what was to be seen in the city and places adjacent within our tedder; and obtaining acquaintance with many of the city, not of the meanest quality; at whose hands we found such humanity, and such a freedom and desire to take strangers as it were into their bosom, as was enough to make us forget all that was dear to us in our own countries: and continually we met with many things right worthy of observation and relation: as indeed, if there be a mirror in the world worthy to hold men's eyes, it is that country.

One day there were two of our company bidden to a Feast of the Family, as they call it. A most natural, pious, and reverend custom it is, shewing that nation to be compounded of all goodness. This is the manner of it. It is granted to any man that shall live to see thirty persons descended of his body alive together, and all above three years old, to make this feast which is done at the cost of the state. The Father of the Family, whom they call the Tirsan, two days before the feast, taketh to him three of such friends as he liketh to choose; and is assisted also by the governor of the city or place where the feast is celebrated; and all the persons of the family, of both sexes, are summoned to attend him. These two days the Tirsan sitteth in consultation concerning the good estate of the family. There, if there be any discord or suits between any of the family, they are compounded and appeased. There, if any of the family be distressed or decayed, order is taken for their relief and competent means to live. There, if any be subject to vice, or take ill courses, they are reproved and censured. So likewise direction is given touching marriages, and the courses of life, which any of them should take, with divers other the like orders and advices. The governor assisteth, to the end to put in execution by his public authority the decrees and orders of the Tirsan, if they should be disobeyed; though that seldom needeth; such reverence and obedience they give to the order of nature. The Tirsan doth also then ever choose one man from among his sons, to live in house with him; who is called ever after the Son of the Vine. The reason will hereafter appear.

On the feast day, the father or Tirsan cometh forth after divine service into a large room where the feast is celebrated; which room hath an half-pace at the upper end. Against the wall, in the middle of the half-pace, is a chair placed for him, with a table and carpet before it. Over the chair is a state, made round or oval, and it is of ivy; an ivy somewhat whiter than ours, like the leaf of a silver asp; but more shining; for it is green all winter. And the state is curiously wrought with silver and silk of divers colors, broiding or binding in the ivy; and is ever of the work of some of the daughters of the family; and veiled over at the top with a fine net of silk and silver. But the substance of it is true ivy; whereof, after it is taken down, the friends of the family are desirous to have some leaf or sprig to keep.

The Tirsan cometh forth with all his generation or linage, the males before him, and the females following him; and if there be a mother from whose body the whole linage is descended, there is a traverse placed in a loft above on the right hand of the chair, with a privy door, and a carved window of glass, leaded with gold and blue; where she sitteth, but is not seen. When the Tirsan is come forth, he sitteth down in the chair; and all the linage place themselves against the wall, both at his back and upon the return of the half-pace, in order of their years without difference of sex; and stand upon their feet. When he is set; the room being always full of company, but well kept and without disorder; after some pause, there cometh in from the lower end of the room, a taratan (which is as much as an herald) and on either side of him two young lads; whereof one carrieth a scroll of their shining yellow parchment; and the other a cluster of grapes of gold, with a long foot or stalk. The herald and children are clothed with mantles of sea-water green satin; but the herald's mantle is streamed with gold, and hath a train.

Then the herald with three curtesies, or rather inclinations, cometh up as far as the half-pace; and there first taketh into his hand the scroll. This scroll is the king's charter, containing gifts of revenew, and many privileges, exemptions, and points of honour, granted to the Father of the Family; and is ever styled and directed, To such do one our well beloved friend and creditor: which is a title proper only to this case. For they say the king is debtor to no man, but for propagation of his subjects. The seal set to the king's charter is the king's image, imbossed or moulded in gold; and though such charters be expedited of course, and as of right, yet they are varied by discretion, according to the number and dignity of the family. This charter the herald readeth aloud; and while it is read, the father or Tirsan standeth up supported by two of his sons, such as he chooseth. Then the herald mounteth the half-pace and delivereth the charter into his hand: and with that there is an acclamation by all that are present in their language, which is thus much: Happy are the people of Bensalem.

Then the herald taketh into his hand from the other child the cluster of grapes, which is of gold, both the stalk and the grapes. But the grapes are daintily enamelled; and if the males of the family be the greater number, the grapes are enamelled purple, with a little sun set on the top; if the females, then they are enamelled into a greenish yellow, with a crescent on the top. The grapes are in number as many as there are descendants of the family. This golden cluster the herald delivereth also to the Tirsan; who presently delivereth it over to that son that he had formerly chosen to be in house with him: who beareth it before his father as an ensign of honour when he goeth in public, ever after; and is thereupon called the Son of the Vine.

After the ceremony endeth the father or Tirsan retireth; and after some time cometh forth again to dinner, where he sitteth alone under the state, as before; and none of his descendants sit with him, of what degree or dignity soever, except he hap to be of Salomon's House. He is served only by his own children, such as are male; who perform unto him all service of the table upon the knee; and the women only stand about him, leaning against the wall. The room below the half-pace hath tables on the sides for the guests that are bidden; who are served with great and comely order; and towards the end of dinner (which in the greatest feasts with them lasteth never above an hour and an half) there is an hymn sung, varied according to the invention of him that composeth it (for they have excellent posy) but the subject of it is (always) the praises of Adam and Noah and Abraham; whereof the former two peopled the world, and the last was the Father of the Faithful: concluding ever with a thanksgiving for the nativity of our Saviour, in whose birth the births of all are only blessed.

Dinner being done, the Tirsan retireth again; and having withdrawn himself alone into a place, where he makes some private prayers, he cometh forth the third time, to give the blessing with all his descendants, who stand about him as at the first. Then he calleth them forth by one and by one, by name, as he pleaseth, though seldom the order of age be inverted. The person that is called (the table being before removed) kneeleth down before the chair, and the father layeth his hand upon his head, or her head, and giveth the blessing in these words: Son of Bensalem, (or daughter of Bensalem,) thy father with it: the man by whom thou hast breath and life speaketh the word: the blessing of the everlasting Father, the Prince of Peace, and the Holy Dove, be upon thee, and make the days of thy pilgrimage good and many. This he saith to every of them; and that done, if there be any of his sons of eminent merit and virtue, (so they be not above two,) he calleth for them again; and saith, laying his arm over their shoulders, they standing; Sons, it is well ye are born, give God the praise, and persevere to the end. And withall delivereth to either of them a jewel, made in the figure of an ear of wheat, which they ever after wear in the front of their turban or hat. This done, they fall to music and dances, and other recreations, after their manner, for the rest of the day. This is the full order of that feast.

By that time six or seven days were spent, I was fallen into straight acquaintance with a merchant of that city, whose name was Joabin. He was a Jew and circumcised: for they have some few stirps of Jews yet remaining among them, whom they leave to their own religion. Which they may the better do, because they are of a far differing disposition from the Jews in other parts. For whereas they hate the name of Christ; and have a secret inbred rancour against the people among whom they live: these (contrariwise) give unto our Saviour many high attributes, and love the nation of Bensalem extremely. Surely this man of whom I speak would ever acknowledge that Christ was born of a virgin and that he was more than a man; and he would tell how God made him ruler of the seraphims which guard his throne; and they call him also the Milken Way, and the Eliah of the Messiah; and many other high names; which though they be inferior to his divine majesty, yet they are far from the language of other Jews.

And for the country of Bensalem, this man would make no end of commending it; being desirous, by tradition among the Jews there, to have it believed that the people thereof were of the generations of Abraham, by another son, whom they call Nachoran; and that Moses by a secret Cabala ordained the Laws of Bensalem which they now use; and that when the Messiah should come, and sit in his throne at Hierusalem, the king of Bensalem should sit at his feet, whereas other kings should keep a great distance. But yet setting aside these Jewish dreams, the man was a wise man, and learned, and of great policy, and excellently seen in the laws and customs of that nation.

Amongst other discourses, one day I told him I was much affected with the relation I had, from some of the company, of their custom, in holding the Feast of the Family; for that (methought) I had never heard of a solemnity wherein nature did so much preside. And because propagation of families proceedeth from the nuptial copulation, I desired to know of him what laws and customs they had concerning marriage; and whether they kept marriage well and whether they were tied to one wife; for that where population is so much affected,' and such as with them it seemed to be, there is commonly permission of plurality of wives.

To this he said, "You have reason for to commend that excellent institution of the Feast of the Family. And indeed we have experience that those families that are partakers of the blessing of that feast do flourish and prosper ever after in an extraordinary manner. But hear me now, and I will tell you what I know. You shall understand that there is not under the heavens so chaste a nation as this of Bensalem; nor so free from all pollution or foulness. It is the virgin of the world. I remember I have read in one of your European books, of an holy hermit amongst you that desired to see the Spirit of Fornication; and there appeared to him a little foul ugly Aethiop. But if he had desired to see the Spirit of Chastity of Bensalem, it would have appeared to him in the likeness of a fair beautiful Cherubim. For there is nothing amongst mortal men more fair and admirable, than the chaste minds of this people. Know therefore, that with them there are no stews, no dissolute houses, no courtesans, nor anything of that kind. Nay they wonder (with detestation) at you in Europe, which permit such things. They say ye have put marriage out of office: for marriage is ordained a remedy for unlawful concupiscence; and natural concupiscence seemeth as a spar to marriage. But when men have at hand a remedy more agreeable to their corrupt will, marriage is almost expulsed. And therefore there are with you seen infinite men that marry not, but chose rather a libertine and impure single life, than to be yoked in marriage; and many that do marry, marry late, when the prime and strength of their years is past. And when they do marry, what is marriage to them but a very bargain; wherein is sought alliance, or portion, or reputation, with some desire (almost indifferent) of issue; and not the faithful nuptial union of man and wife, that was first instituted. Neither is it possible that those that have cast away so basely so much of their strength, should greatly esteem children, (being of the same matter,) as chaste men do. So likewise during marriage, is the case much amended, as it ought to be if those things were tolerated only for necessity? No, but they remain still as a very affront to marriage. The haunting of those dissolute places, or resort to courtesans, are no more punished in married men than in bachelors. And the depraved custom of change, and the delight in meretricious embracements, (where sin is turned into art,) maketh marriage a dull thing, and a kind of imposition or tax. They hear you defend these things, as done to avoid greater evils; as advoutries, deflowering of virgins, unnatural lust, and the like. But they say this is a preposterous wisdom; and they call it Lot's offer, who to save his guests from abusing, offered his daughters: nay they say farther that there is little gained in this; for that the same vices and appetites do still remain and abound; unlawful lust being like a furnace, that if you stop the flames altogether, it will quench; but if you give it any vent, it will rage. As for masculine love, they have no touch of it; and yet there are not so faithful and inviolate friendships in the world again as are there; and to speak generally, (as I said before,) I have not read of any such chastity, in any people as theirs. And their usual saying is, That whosoever is unchaste cannot reverence himself; and they say, That the reverence of a man's self, is, next to religion, the chiefest bridle of all vices."

And when he had said this, the good Jew paused a little; whereupon I, far more willing to hear him speak on than to speak myself, yet thinking it decent that upon his pause of speech I should not be altogether silent, said only this; "That I would say to him, as the widow of Sarepta said to Elias; that he was come to bring to memory our sins; and that I confess the righteousness of Bensalem was greater than the righteousness of Europe." At which speech he bowed his head, and went on in this manner:

"They have also many wise and excellent laws touching marriage. They allow no polygamy. They have ordained that none do intermarry or contract, until a month be past from their first interview. Marriage without consent of parents they do not make void, but they mulct it in the inheritors: for the children of such marriages are not admitted to inherit above a third part of their parents' inheritance. I have read in a book of one of your men, of a Feigned Commonwealth, where the married couple are permitted, before they contract, to see one another naked. This they dislike; for they think it a scorn to give a refusal after so familiar knowledge: but because of many hidden defects in men and women's bodies, they have a more civil way; for they have near every town a couple of pools, (which they call Adam and Eve's pools,) where it is permitted to one of the friends of the men, and another of the friends of the woman, to see them severally bathe naked."

And as we were thus in conference, there came one that seemed to be a messenger, in a rich huke, that spake with the Jew: whereupon he turned to me and said; "You will pardon me, for I am commanded away in haste." The next morning he came to me again, joyful as it seemed, and said; "There is word come to the Governor of the city, that one of the Fathers of Salomon's House will be here this day seven-night: we have seen none of them this dozen years. His coming is in state; but the cause of his coming is secret. I will provide you and your fellows of a good standing to see his entry." I thanked him, and told him, I was most glad of the news.

The day being come, he made his entry. He was a man of middle stature and age, comely of person, and had an aspect as if he pitied men. He was clothed in a robe of fine black cloth, with wide sleeves and a cape. His under garment was of excellent white linen down to the foot, girt with a girdle of the same; and a sindon or tippet of the same about his neck. He had gloves, that were curious," and set with stone; and shoes of peach-coloured velvet. His neck was bare to the shoulders. His hat was like a helmet, or Spanish montera; and his locks curled below it decently: they were of colour brown. His beard was cut round, and of the same colour with his hair, somewhat lighter. He was carried in a rich chariot without wheels, litter-wise; with two horses at either end, richly trapped in blue velvet embroidered; and two footmen on each side in the like attire. The chariot was all of cedar, gilt, and adorned with crystal; save that the fore-end had panels of sapphires, set in borders of gold; and the hinder-end the like of emeralds of the Peru colour. There was also a sun of gold, radiant, upon the top, in the midst; and on the top before, a small cherub of gold, with wings displayed. The chariot was covered with cloth of gold tissued upon blue. He had before him fifty attendants, young men all, in white satin loose coats to the mid leg; and stockings of white silk; and shoes of blue velvet; and hats of blue velvet; with fine plumes of diverse colours, set round like hat-bands. Next before the chariot, went two men, bare-headed, in linen garments down the foot, girt, and shoes of blue velvet; who carried, the one a crosier, the other a pastoral staff like a sheep-hook; neither of them of metal, but the crosier of balm-wood, the pastoral staff of cedar. Horsemen he had none, neither before nor behind his chariot: as it seemeth, to avoid all tumult and trouble. Behind his chariot went all the officers and principals of the companies of the city. He sat alone, upon cushions of a kind of excellent plush, blue; and under his foot curious carpets of silk of diverse colours, like the Persian, but far finer. He held up his bare hand as he went, as blessing the people, but in silence. The street was wonderfully well kept: so that there was never any army had their men stand in better battle-array than the people stood. The windows likewise were not crowded, but every one stood in them as if they had been placed.

When the shew was past, the Jew said to me; "I shall not be able to attend you as I would, in regard of some charge the city hath laid upon me, for the entertaining of this great person." Three days after the Jew came to me again, and said; "Ye are happy men; for the Father of Salomon's House taketh knowledge of your being here, and commanded me to tell you that he will admit all your company to his presence, and have private conference with one of you, that ye shall choose: and for this hath appointed the next day after to-morrow. And because he meaneth to give you his blessing, he hath appointed it in the forenoon."

We came at our day and hour, and I was chosen by my fellows for the private access. We found him in a fair chamber, richly hanged, and carpeted under foot without any degrees to the state. He was set upon a low Throne richly adorned, and a rich cloth of state over his head, of blue satin embroidered. He was alone, save that he had two pages of honour, on either hand one, finely attired in white. His under garments were the like that we saw him wear in the chariot; but instead of his gown, he had on him a mantle with a cape, of the same fine black, fastened about him. When we came in, as we were taught, we bowed low at our first entrance; and when we were come near his chair, he stood up, holding forth his hand ungloved, and in posture of blessing; and we every one of us stooped down, and kissed the hem of his tippet. That done, the rest departed, and I remained. Then he warned the pages forth of the room, and caused me to sit down beside him, and spake to me thus in the Spanish tongue.

"God bless thee, my son; I will give thee the greatest jewel I have. For I will impart unto thee, for the love of God and men, a relation of the true state of Salomon's House. Son, to make you know the true state of Salomon's House, I will keep this order. First, I will set forth unto you the end of our foundation. Secondly, the preparations and instruments we have for our works. Thirdly, the several employments and functions whereto our fellows are assigned. And fourthly, the ordinances and rites which we observe.

"The end of our foundation is the knowledge of causes, and secret motions of things; and the enlarging of the bounds of human empire, to the effecting of all things possible.

"The Preparations and Instruments are these. We have large and deep caves of several depths: the deepest are sunk six hundred fathom: and some of them are digged and made under great hills and mountains: so that if you reckon together the depth of the hill and the depth of the cave, they are (some of them) above three miles deep. For we find, that the depth of a hill, and the depth of a cave from the flat, is the same thing; both remote alike, from the sun and heaven's beams, and from the open air. These caves we call the Lower Region; and we use them for all coagulations, indurations, refrigerations, and conservations of bodies. We use them likewise for the imitation of natural mines; and the producing also of new artificial metals, by compositions and materials which we use, and lay there for many years. We use them also sometimes, (which may seem strange,) for curing of some diseases, and for prolongation of life in some hermits that choose to live there, well accommodated of all things necessary, and indeed live very long; by whom also we learn many things.

"We have burials in several earths, where we put diverse cements, as the Chineses do their porcellain. But we have them in greater variety, and some of them more fine. We have also great variety of composts and soils, for the making of the earth fruitful.

"We have high towers; the highest about half a mile in height; and some of them likewise set upon high mountains; so that the vantage of the hill with the tower is in the highest of them three miles at least. And these places we call the Upper Region; accounting the air between the high places and the low, as a Middle Region. We use these towers, according to their several heights, and situations, for insolation, refrigeration, conservation; and for the view of divers meteors; as winds, rain, snow, hail; and some of the fiery meteors also. And upon them, in some places, are dwellings of hermits, whom we visit sometimes, and instruct what to observe.

"We have great lakes, both salt, and fresh; whereof we have use for the fish and fowl. We use them also for burials of some natural bodies: for we find a difference in things buried in earth or in air below the earth, and things buried in water. We have also pools, of which some do strain fresh water out of salt; and others by art do turn fresh water into salt. We have also some rocks in the midst of the sea, and some bays upon the shore for some works, wherein is required the air and vapor of the sea. We have likewise violent streams and cataracts, which serve us for many motions: and likewise engines for multiplying and enforcing of winds, to set also on going diverse motions.

"We have also a number of artificial wells and fountains, made in imitation of the natural sources and baths; as tincted upon vitriol, sulphur, steel, brass, lead, nitre, and other minerals. And again we have little wells for infusions of many things, where the waters take the virtue quicker and better, than in vessels or basins. And amongst them we have a water which we call Water of Paradise, being, by that we do to it made very sovereign for health, and prolongation of life.

"We have also great and spacious houses where we imitate and demonstrate meteors; as snow, hail, rain, some artificial rains of bodies and not of water, thunders, lightnings; also generations of bodies in air; as frogs, flies, and divers others.

"We have also certain chambers, which we call Chambers of Health, where we qualify the air as we think good and proper for the cure of divers diseases, and preservation of health.

"We have also fair and large baths, of several mixtures, for the cure of diseases, and the restoring of man's body from arefaction: and others for the confirming of it in strength of sinewes, vital parts, and the very juice and substance of the body.

"We have also large and various orchards and gardens; wherein we do not so much respect beauty, as variety of ground and soil, proper for divers trees and herbs: and some very spacious, where trees and berries are set whereof we make divers kinds of drinks, besides the vineyards. In these we practise likewise all conclusions of grafting, and inoculating as well of wild-trees as fruit-trees, which produceth many effects. And we make (by art) in the same orchards and gardens, trees and flowers to come earlier or later than their seasons; and to come up and bear more speedily than by their natural course they do. We make them also by art greater much than their nature; and their fruit greater and sweeter and of differing taste, smell, colour, and figure, from their nature. And many of them we so order, as they become of medicinal use.

"We have also means to make divers plants rise by mixtures of earths without seeds; and likewise to make divers new plants, differing from the vulgar; and to make one tree or plant turn into another.

"We have also parks and enclosures of all sorts of beasts and birds which we use not only for view or rareness, but likewise for dissections and trials; that thereby we may take light what may be wrought upon the body of man. Wherein we find many strange effects; as continuing life in them, though divers parts, which you account vital, be perished and taken forth; resuscitating of some that seem dead in appearance; and the like. We try also all poisons and other medicines upon them, as well of chirurgery, as physic. By art likewise, we make them greater or taller than their kind is; and contrariwise dwarf them, and stay their growth: we make them more fruitful and bearing than their kind is; and contrariwise barren and not generative. Also we make them differ in colour, shape, activity, many ways. We find means to make commixtures and copulations of different kinds; which have produced many new kinds, and them not barren, as the general opinion is. We make a number of kinds of serpents, worms, flies, fishes, of putrefaction; whereof some are advanced (in effect) to be perfect creatures, like bests or birds; and have sexes, and do propagate. Neither do we this by chance, but we know beforehand, of what matter and commixture what kind of those creatures will arise.

"We have also particular pools, where we make trials upon fishes, as we have said before of beasts and birds.

"We have also places for breed and generation of those kinds of worms and flies which are of special use; such as are with you your silk-worms and bees.

"I will not hold you long with recounting of our brewhouses, bake-houses, and kitchens, where are made divers drinks, breads, and meats, rare and of special effects. Wines we have of grapes; and drinks of other juice of fruits, of grains, and of roots; and of mixtures with honey, sugar, manna, and fruits dried, and decocted; Also of the tears or woundings of trees; and of the pulp of canes. And these drinks are of several ages, some to the age or last of forty years. We have drinks also brewed with several herbs, and roots, and spices; yea with several fleshes, and white-meats; whereof some of the drinks are such, as they are in effect meat and drink both: so that divers, especially in age, do desire to live with them, with little or no meat or bread. And above all, we strive to have drink of extreme thin parts, to insinuate into the body, and yet without all biting, sharpness, or fretting; insomuch as some of them put upon the back of your hand will, with a little stay, pass through to the palm, and yet taste mild to the mouth. We have also waters which we ripen in that fashion, as they become nourishing; so that they are indeed excellent drink; and many will use no other. Breads we have of several grains, roots, and kernels; yea and some of flesh and fish dried; with divers kinds of leavenings and seasonings: so that some do extremely move appetites; some do nourish so, as divers do live of them, without any other meat; who live very long. So for meats, we have some of them so beaten and made tender and mortified,' yet without all corrupting, as a weak heat of the stomach will turn them into good chylus; as well as a strong heat would meat otherwise prepared. We have some meats also and breads and drinks, which taken by men enable them to fast long after; and some other, that used make the very flesh of men's bodies sensibly' more hard and tough and their strength far greater than otherwise it would be.

"We have dispensatories, or shops of medicines. Wherein you may easily think, if we have such variety of plants and living creatures more than you have in Europe, (for we know what you have,) the simples, drugs, and ingredients of medicines, must likewise be in so much the greater variety. We have them likewise of divers ages, and long fermentations. And for their preparations, we have not only all manner of exquisite distillations and separations, and especially by gentle heats and percolations through divers strainers, yea and substances; but also exact forms of composition, whereby they incorporate almost, as they were natural simples.

"We have also divers mechanical arts, which you have not; and stuffs made by them; as papers, linen, silks, tissues; dainty works of feathers of wonderful lustre; excellent dies, and, many others; and shops likewise, as well for such as are not brought into vulgar use amongst us as for those that are. For you must know that of the things before recited, many of them are grown into use throughout the kingdom; but yet, if they did flow from our invention, we have of them also for patterns and principals.

"We have also furnaces of great diversities, and that keep great diversity of heats; fierce and quick; strong and constant; soft and mild; blown, quiet; dry, moist; and the like. But above all, we have heats, in imitation of the Sun's and heavenly bodies' heats, that pass divers inequalities, and (as it were) orbs, progresses, and returns, whereby we produce admirable effects. Besides, we have heats of dungs; and of bellies and maws of living creatures, and of their bloods and bodies; and of hays and herbs laid up moist; of lime unquenched; and such like. Instruments also which generate heat only by motion. And farther, places for strong insulations; and again, places under the earth, which by nature, or art, yield heat. These divers heats we use, as the nature of the operation, which we intend, requireth.

"We have also perspective-houses, where we make demonstrations of all lights and radiations; and of all colours: and out of things uncoloured and transparent, we can represent unto you all several colours; not in rain-bows, (as it is in gems, and prisms,) but of themselves single. We represent also all multiplications of light, which we carry to great distance, and make so sharp as to discern small points and lines. Also all colourations of light; all delusions and deceits of the sight, in figures, magnitudes, motions, colours all demonstrations of shadows. We find also divers means, yet unknown to you, of producing of light originally from divers bodies. We procure means of seeing objects afar off; as in the heaven and remote places; and represent things near as afar off; and things afar off as near; making feigned distances. We have also helps for the sight, far above spectacles and glasses in use. We have also glasses and means to see small and minute bodies perfectly and distinctly; as the shapes and colours of small flies and worms, grains and flaws in gems, which cannot otherwise be seen, observations in urine and blood not otherwise to be seen. We make artificial rain-bows, halo's, and circles about light. We represent also all manner of reflexions, refractions, and multiplications of visual beams of objects.

"We have also precious stones of all kinds, many of them of great beauty, and to you unknown; crystals likewise; and glasses of divers kinds; and amongst them some of metals vitrificated, and other materials besides those of which you make glass. Also a number of fossils, and imperfect minerals, which you have not. Likewise loadstones of prodigious virtue; and other rare stones, both natural and artificial.

"We have also sound-houses, where we practise and demonstrate all sounds, and their generation. We have harmonies which you have not, of quarter-sounds, and lesser slides of sounds. Divers instruments of music likewise to you unknown, some sweeter than any you have, together with bells and rings that are dainty and sweet. We represent small sounds as great and deep; likewise great sounds extenuate and sharp; we make divers tremblings and warblings of sounds, which in their original are entire. We represent and imitate all articulate sounds and letters, and the voices and notes of beasts and birds. We have certain helps which set to the ear do further the hearing greatly. We have also divers strange and artificial echoes, reflecting the voice many times, and as it were tossing it: and some that give back the voice louder than it came, some shriller, and some deeper; yea, some rendering the voice differing in the letters or articulate sound from that they receive. We have also means to convey sounds in trunks and pipes, in strange lines and distances.

"We have also perfume-houses; wherewith we join also practices of taste. We multiply smells, which may seem strange. We imitate smells, making all smells to breathe out of other mixtures than those that give them. We make divers imitations of taste likewise, so that they will deceive any man's taste. And in this house we contain also a confiture-house; where we make all sweet-meats, dry and moist; and divers pleasant wines, milks, broths, and sallets; in far greater variety than you have.

"We have also engine-houses, where are prepared engines and instruments for all sorts of motions. There we imitate and practise to make swifter motions than any you have, either out of your muskets or any engine that you have: and to make them and multiply them more easily, and with small force, by wheels and other means: and to make them stronger and more violent than yours are; exceeding your greatest cannons and basilisks. We represent also ordnance and instruments of war, and engines of all kinds: and likewise new mixtures and compositions of gun-powder, wild-fires burning in water, and unquenchable. Also fireworks of all variety both for pleasure and use. We imitate also flights of birds;

we have some degrees of flying in the air. We have ships and boats for going under water, and brooking of seas; also swimming-girdles and supporters. We have divers curious clocks, and other like motions of return: and some perpetual motions. We imitate also motions of living creatures, by images, of men, beasts, birds, fishes, and serpents. We have also a great number of other various motions, strange for equality, fineness, and subtilty.

"We have also a mathematical house, where are represented all instruments, as well of geometry as astronomy, exquisitely made.

"We have also houses of deceits of the senses; where we represent all manner of feats of juggling, false apparitions, impostures, and illusions; and their fallacies. And surely you will easily believe that we that have so many things truly natural which induce admiration, could in a world of particulars deceive the senses, if we would disguise those things and labour to make them seem more miraculous. But we do hate all impostures, and lies; insomuch as we have severely forbidden it to all our fellows, under pain of ignominy and fines, that they do not show any natural work or thing, adorned or swelling; but only pure as it is, and without all affectation of strangeness.

"These are (my son) the riches of Salomon's House.

"For the several employments and offices of our fellows; we have twelve that sail into foreign countries, under the names of other nations, (for our own we conceal); who bring us the books, and abstracts, and patterns of experiments of all other parts. These we call Merchants of Light.

"We have three that collect the experiments which are in all books. These we call Depredators.

"We have three that collect the experiments of all mechanical arts; and also of liberal sciences; and also of practices which are not brought into arts. These we call Mystery-men.

"We have three that try new experiments, such as themselves think good. These we call Pioneers or Miners.

"We have three that draw the experiments of the former four into titles and tables, to give the better light for the drawing of observations and axioms out of them. These we call Compilers.

"We have three that bend themselves, looking into the experiments of their fellows, and cast about how to draw out of them things of use and practise for man's life, and knowledge, as well for works as for plain demonstration of causes, means of natural divinations, and the easy and clear discovery of the virtues and parts of bodies. These we call Dowry-men or Benefactors.

"Then after divers meetings and consults of our whole number, to consider of the former labours and collections, we have three that take care, out of them, to direct new experiments, of a higher light, more penetrating into nature than the former. These we call Lamps.

"We have three others that do execute the experiments so directed, and report them. These we call Inoculators.

"Lastly, we have three that raise the former discoveries by experiments into greater observations, axioms, and aphorisms. These we call Interpreters of Nature.

"We have also, as you must think, novices and apprentices, that the succession of the former employed men do not fail; besides, a great number of servants and attendants, men and women. And this we do also: we have consultations, which of the inventions and experiences which we have discovered shall be published, and which not: and take all an oath of secrecy, for the concealing of those which we think fit to keep secret: though some of those we do reveal sometimes to the state and some not.

"For our ordinances and rites: we have two very long and fair galleries: in one of these we place patterns and samples of all manner of the more rare and excellent inventions in the other we place the statues of all principal inventors. There we have the statue of your Columbus, that discovered the West Indies: also the inventor of ships: your monk that was the inventor of ordnance and of gunpowder: the inventor of music: the inventor of letters: the inventor of printing: the inventor of observations of astronomy: the inventor of works in metal: the inventor of glass: the inventor of silk of the worm: the inventor of wine: the inventor of corn and bread: the inventor of sugars: and all these, by more certain tradition than you have. Then have we divers inventors of our own, of excellent works; which since you have not seen, it were too long to make descriptions of them; and besides, in the right understanding of those descriptions you might easily err. For upon every invention of value, we erect a statue to the inventor, and give him a liberal and honourable reward. These statues are some of brass; some of marble and touch-stone; some of cedar and other special woods gilt and adorned; some of iron; some of silver; some of gold.

"We have certain hymns and services, which we say daily, of Lord and thanks to God for his marvellous works: and forms of prayers, imploring his aid and blessing for the illumination of our labours, and the turning of them into good and holy uses.

"Lastly, we have circuits or visits of divers principal cities of the kingdom; where, as it cometh to pass, we do publish such new profitable inventions as we think good. And we do also declare natural divinations of diseases, plagues, swarms-of hurtful creatures, scarcity, tempests, earthquakes, great inundations, comets, temperature of the year, and divers other things; and we give counsel thereupon, what the people shall do for the prevention and remedy of them."

And when he had said this, he stood up; and I, as I had been taught, kneeled down, and he laid his right hand upon my head, and said; "God bless thee, my son; and God bless this relation, which I have made. I give thee leave to publish it for the good of other nations; for we here are in God's bosom, a land unknown." And so he left me; having assigned a value of about two thousand ducats, for a bounty to me and my fellows. For they give great largesses where they come upon all occasions.

The rest was not perfected.

Of Gardens
An Essay

Originally Published By:
 LONDON; HACON AND RICKETTS; CRAVEN ST., STRAND. MCMII.

God almighty first planted a garden: and indeed it is the purest of human pleasures. It is the greatest refreshment of the spirits of man; without which, buildings or palaces are but gross handy-works: and a man shall ever see, that when ages grow to civility or elegancy, men come to build stately, sooner than to garden finely, as if gardening were the greater perfection. I do hold it, in the royal ordering of gardens, there ought to be gardens for all the months in the year: in which, severally, things of beauty may be then in season. For December and January, or the latter part of November, you must take such things as are green all winter; holly, ivy, bays, juniper, cypress-trees, yew, pine-apple trees, fir trees, rosemary, lavender, periwinkle (the white, the purple, and the blue), germander, flags, orange trees, lemon trees, and myrtles, if they be stoved, and sweet marjoram, warm set. There followeth, for the latter part of January and February, the mezereon tree, which then blossoms; crocus vernus, both the yellow and the gray; primroses, anemonies, the early tulip, hyacinthus orientalis, chamaïris, fritellaria. For March there come violets, especially the single blue, which are the earliest; the yellow daffodil, the daisy, the almond tree in blossom, the peach tree in blossom, the cornelian tree in blossom, sweet briar. In April follow the double white violet, the wallflower the stock-gilliflower, the cowslip, flower-de-luces, and lilies of all natures, rosemary-flowers, the tulip, the double piony, the pale daffodil, the French honeysuckle, the cherry tree in blossom, the damascene and plum trees in blossom, the white-thorn in leaf, the lilach-tree. In May and June come pinks of all sorts, especially the blush pink; roses of all kinds, except the musk, which comes later; honeysuckles, strawberries, bugloss, columbine, the French marygold, flos Africanus, cherry-tree in fruit, ribes, figs in fruit, rasps, vine-flowers, lavender in flowers, the sweet satyrian, with the white flower; herba muscaria, lilium convalium, the apple tree in blossom. In July come gilliflowers of all varieties, musk roses, the lime tree in blossom, early pears and plums in fruit, gennitings, codlins. In August come plums of all sorts in fruit, pears, apricots, berberries, filberds, musk melons, monks-hoods, of all colours. In September comes grapes, apples, poppies of all colours, peaches, melo-cotones, nectarines, cornelians, wardens, quinces. In October, and the beginning of November, come services, medlars, bullaces, roses cut or removed to come late, holly oaks, and such-like. These particulars are for the climate of London: but my meaning is perceived, that you may have 'ver perpetuum,' as the place affords.

And because the breath of flowers is far sweeter in the air, where it comes and goes, like the warbling of music, than in the hand, therefore nothing is more fit for that delight, than to know what be the flowers and plants that do best perfume the air. Roses, damask and red, are fast flowers of their smells; so that you may walk by a whole row of them, and find nothing of their sweetness: yea, though it be in a morning's dew. Bays likewise yield no smell as they grow; rosemary little; nor sweet marjoram. That which above all others yields the sweetest smell in the air, is the violet; especially the white double violet, which comes twice a year, about the middle of April, and about Bartholomew-tide. Next to that is the muskrose; then the strawberry leaves dying, with a most excellent cordial smell; then the flower of the vines—it is a little dust, like the dust of a bent, which grows upon the cluster, in the first coming forth; then sweet-brier; then wallflowers, which are very delightful, to be set under a parlour, or lower chamber window; then pinks and gilliflowers, especially the matted pink and clove-gilli-flower; then the flowers of the lime tree; then the honeysuckles, so they be somewhat afar off. Of bean flowers I speak not, because they are field flowers; but those which perfume the air most delightfully, not passed by as the rest, but being trodden upon and crushed, are three; that is, burnet, wild thyme, and water mints. Therefore you are to set whole alleys of them, to have the pleasure when you walk or tread.

For gardens, speaking of those which are indeed princelike, as we have done of buildings, the contents ought not well to be under thirty acres of ground, and to be divided into three parts: a green in the entrance; a heath or desert in the going forth; and the main garden in the midst; besides alleys on both sides. And I like well, that four acres of ground be assigned to the green, six to the heath, four and four to either side, and twelve to the main garden. The green hath two pleasures: the one, because nothing is more pleasant to the eye than green grass kept finely shorn; the other, because it will give you a fair alley in the midst; by which you may go in front upon a stately hedge, which is to inclose the garden. But because the alley will be long, and in great heat of the year or day, you ought not to buy the shade in the garden by going in the sun through the green; therefore, you are, of either side the green, to plant a covert alley, upon carpenter's work, about twelve foot in height, by which you may go in shade into the garden. As for the making of knots or figures, with divers coloured earths, that they may lie under the windows of the house, on that side which the garden stands, they be but toys; you may see as good sights, many times, in tarts. The garden is best to be square, encompassed on all the four sides with a stately arched hedge: the arches to be upon pillars of carpenter's work, of some ten foot high, and six foot broad; and the spaces between of the same dimension with the breadth of the arch. Over the arches let there be an entire hedge, of some four foot high, framed also upon carpenter's work; and upon the upper hedge, over every arch, a little turret, with a belly enough to receive a cage of birds; and over every space between the arches, some other little figure, with broad plates of round coloured glass, gilt, for the sun to play upon. But this hedge I intend to be raised upon a bank, not steep, but gently slope, of some six foot, set all with flowers. Also I understand, that this square of the garden should not be the whole breadth of the ground, but to leave on either side ground enough for diversity of side alleys; into which the two covert alleys of the green may deliver you; but there must be no alleys with hedges at either end of this great enclosure; not at the hither end, for letting your prospect upon the fair hedge from the green; nor at the further end, for letting your prospect from the hedge, through the arches, upon the heath.

For the ordering of the ground within the great hedge, I leave it to variety of device; advising nevertheless, that whatsoever form you cast it into, first it be not too busy, or full of work; wherein I, for my part, do not like images cut out in juniper or other garden stuff; they be for children. Little low hedges round, like welts, with some pretty pyramids, I like well; and in some places, fair columns upon frames of carpenter's work. I would also have the alleys spacious and fair. You may have closer alleys upon the side grounds, but none in the main garden. I wish also, in the very middle, a fair mount, with three ascents and alleys, enough for four to walk a-breast; which I would have to be perfect circles, without any bulwarks or embossments; and the whole amount to be thirty foot high; and some fine banqueting house, with some chimneys neatly cast, and without too much glass.

For fountains, they are a great beauty and refreshment; but pools mar all, and make the garden unwholesome, and full of flies and frogs. Fountains I intend to be of two natures: the one that sprinkleth or spouteth water; the other a fair receipt of water, of some thirty or forty foot square, but without fish, or slime, or mud. For the first, the ornaments of images gilt, or of marble, which are in use, do well: but the main matter is so to convey the water, as it never stay either in the bowls, or in the cistern; that the water be never by rest discoloured, green or red, or the like; or gather any mossiness or putrefaction. Besides that, it is to be cleansed every day by the hand. Also some steps up to it, and some fine pavement about it doth well. As for the other kind of fountain, which we may call a bathing pool, it may admit much curiosity and beauty, wherewith we will not trouble ourselves; as, that the bottom be finely paved, and with images; the sides likewise: and withal embellished with coloured glass, and such things of lustre; encompassed also with fine rails of low statues. But the main point is the same which we mentioned in the former kind of fountain; which is, that the water be in perpetual motion, fed by a water higher than the pool, and delivered into it by fair spouts, and then discharged away under ground by some equality of bores, that it stay little. And for fine devices of arching water without spilling, and making it rise in several forms, of feathers, drinking glasses, canopies, and the like, they be pretty things to look on, but nothing to health and sweetness.

For the health, which was the third part of our plot, I wish it to be framed as much as may be to a natural wildness. Trees I would have none in it, but some thickets made only of sweet-brier and honeysuckle, and some wild vine amongst; and the ground set with violets, strawberries, and primroses. For these are sweet, and prosper in the shade. And these to be in the heath here & there, not in any order. I like also little heaps, in the nature of mole-hills, such as are in wild heaths, to be set, some with wild thyme, some with pinks, some with germander, that gives a good flower to the eye, some with periwinkle, some with violets, some with strawberries, some with cowslips, some with daisies, some with red roses, some with lilium convallium, some with sweet-williams red, some with bears-foot, and the like low flowers, being withal sweet and sightly. Part of which heaps to be with standards of little bushes, pricked upon their top, and part without. The standards to be roses, juniper, holly, berberries, but here and there, because of the smell of their blossom, red currants, gooseberries, rosemary, bays, sweet-brier, and such-like. But these standards to be kept with cutting, that they may not grow out of course.

For the side grounds, you are to fill them with variety of alleys, private, to give a full shade, some of them, wheresoever the sun be. You are to frame some of them likewise for shelter, that when the wind blows sharp, you may walk as in a gallery. And those alleys must be likewise hedged at both ends, to keep out the wind; and these closer alleys must be ever finely gravelled, and no grass, because of going wet. In many of these alleys likewise, you are to set fruit trees of all sorts; as well upon the walls as in ranges. And this would be generally observed, that the borders wherein you plant your fruit trees be fair and large, and low, and not steep; and set with fine flowers, but thin and sparingly, lest they deceive the trees. At the end of both the side grounds, I would have a mount of some pretty height, leaving the wall of the inclosure breast high, to look abroad into the fields.

For the main garden, I do not deny but there should be some fair alleys, ranged on both sides, with fruit trees, and some pretty tufts of fruit trees, and arbours with seats, set in some decent order; but these to be by no means set too thick, but to leave the main garden so as it be not close, but the air open and free. For as for shade, I would have you rest upon the alleys of the side grounds, there to walk, if you be disposed, in the heat of the year or day; but to make account, that the main garden is for the more temperate parts of the year; and in the heat of summer, for the morning and the evening, or overcast days.

For aviaries, I like them not, except they be of that largeness, as they may be turfed, and have living plants and bushes set in them; that the birds may have more scope, and natural nestling, and that no foulness appear in the floor of the aviary.

So I have made a platform of a princely garden, partly by precept, partly by drawing; not a model, but some general lines of it; and in this I have spared for no cost. But it is nothing for great princes, that for the most part, taking advice with workmen, with no less cost set their things together; and sometimes add statues, and such things, for state and magnificence, but nothing to the true pleasure of a garden.

THE END.

www.ingramcontent.com/pod-product-compliance
Lightning Source LLC
Chambersburg PA
CBHW050455190326
41458CB00005B/1295